A

Aquatic Chemistry

ADVANCES IN CHEMISTRY SERIES **244**

Aquatic Chemistry
Interfacial and Interspecies Processes

Chin Pao Huang, EDITOR
University of Delaware

Charles R. O'Melia, EDITOR
The Johns Hopkins University

James J. Morgan, EDITOR
California Institute of Technology

Developed from a symposium sponsored
by the Division of Environmental Chemistry, Inc.
at the 203rd National Meeting
of the American Chemical Society,
San Francisco, California
April 5–10, 1992

American Chemical Society, Washington, DC 1995

Library of Congress Cataloging-in-Publication Data

Aquatic chemistry : interfacial and interspecies processes / Chin Pao Huang, Charles R. O'Melia, James J. Morgan, [editors].
    p.   cm.—(Advances in chemistry series, ISSN 0065-2393; 244)

"Developed from a symposium sponsored by the Division of Environmental Chemistry, Inc., at the 203rd National Meeting of the American Chemical Society, San Francisco, California, April 5-10, 1992."

Includes bibliographical references and index.

ISBN 0-8412-2921-X

1. Water chemistry—Congresses.
I. Huang, C. P. (Chin P.)   II. O'Melia, Charles R.   III. Morgan, James J. 1932–   .   IV. American Chemical Society. Division of Environmental Chemistry, Inc. V. American Chemical Society. Meeting (203rd : 1992 : San Francisco, Calif.)   VI. Series.

QD1.A355 no. 244
[GB855]
540 s—dc20
[551.48'01'54]
                                                          94-41862
                                                             CIP

The paper used in this publication meets the minimum requirements of American National Standard for Information Sciences—Permanence of Paper for Printed Library Materials, ANSI Z39.48-1984. ∞

Copyright © 1995

American Chemical Society

All Rights Reserved. The appearance of the code at the bottom of the first page of each chapter in this volume indicates the copyright owner's consent that reprographic copies of the chapter may be made for personal or internal use or for the personal or internal use of specific clients. This consent is given on the condition, however, that the copier pay the stated per-copy fee through the Copyright Clearance Center, Inc., 27 Congress Street, Salem, MA 01970, for copying beyond that permitted by Sections 107 or 108 of the U.S. Copyright Law. This consent does not extend to copying or transmission by any means—graphic or electronic—for any other purpose, such as for general distribution, for advertising or promotional purposes, for creating a new collective work, for resale, or for information storage and retrieval systems. The copying fee for each chapter is indicated in the code at the bottom of the first page of the chapter.

The citation of trade names and/or names of manufacturers in this publication is not to be construed as an endorsement or as approval by ACS of the commercial products or services referenced herein; nor should the mere reference herein to any drawing, specification, chemical process, or other data be regarded as a license or as a conveyance of any right or permission to the holder, reader, or any other person or corporation, to manufacture, reproduce, use, or sell any patented invention or copyrighted work that may in any way be related thereto. Registered names, trademarks, etc., used in this publication, even without specific indication thereof, are not to be considered unprotected by law.

PRINTED IN THE UNITED STATES OF AMERICA

1994 Advisory Board
Advances in Chemistry Series
M. Joan Comstock, *Series Editor*

Robert J. Alaimo
Procter & Gamble Pharmaceuticals

Mark Arnold
University of Iowa

David Baker
University of Tennessee

Arindam Bose
Pfizer Central Research

Robert F. Brady, Jr.
Naval Research Laboratory

Margaret A. Cavanaugh
National Science Foundation

Arthur B. Ellis
University of Wisconsin at Madison

Dennis W. Hess
Lehigh University

Hiroshi Ito
IBM Almaden Research Center

Madeleine M. Joullie
University of Pennsylvania

Lawrence P. Klemann
Nabisco Foods Group

Gretchen S. Kohl
Dow-Corning Corporation

Bonnie Lawlor
Institute for Scientific Information

Douglas R. Lloyd
The University of Texas at Austin

Cynithia A. Maryanoff
R. W. Johnson Pharmaceutical Research Institute

Julius J. Menn
Western Cotton Research Laboratory, U.S. Department of Agriculture

Roger A. Minear
University of Illinois at Urbana-Champaign

Vincent Pecoraro
University of Michigan

Marshall Phillips
Delmont Laboratories

George W. Roberts
North Carolina State University

A. Truman Schwartz
Macalaster College

John R. Shapley
University of Illinois at Urbana-Champaign

L. Somasundaram
DuPont

Michael D. Taylor
Parke-Davis Pharmaceutical Research

Peter Willett
University of Sheffield (England)

FOREWORD

The ADVANCES IN CHEMISTRY SERIES was founded in 1949 by the American Chemical Society as an outlet for symposia and collections of data in special areas of topical interest that could not be accommodated in the Society's journals. It provides a medium for symposia that would otherwise be fragmented because their papers would be distributed among several journals or not published at all.

Papers are reviewed critically according to ACS editorial standards and receive the careful attention and processing characteristic of ACS publications. Volumes in the ADVANCES IN CHEMISTRY SERIES maintain the integrity of the symposia on which they are based; however, verbatim reproductions of previously published papers are not accepted. Papers may include reports of research as well as reviews, because symposia may embrace both types of presentation.

DEDICATION

This book collects papers presented at the 1992 ACS National Meeting in San Francisco honoring Werner Stumm—a pioneer of aquatic chemistry. Born in Switzerland in 1924, Professor Stumm received his Ph.D. in Chemistry from the University of Zurich in 1952. His U.S. academic career began in 1954 as a research fellow in sanitary engineering at Harvard University. He was appointed assistant professor in 1956 and later was promoted to Gordon McKay Professor of Applied Chemistry, a position that he held until 1970 when he returned to his native Switzerland to direct the Institute of Water Resources and Water Pollution Control, Swiss Federal Institute of Technology, and to teach as Professor of Aquatic Chemistry. He served in this position until retirement in 1992. Currently, he is Professor Emeritus, Swiss Federal Institute of Technology in Zurich, and Adjunct Professor, Department of Geography and Environmental Engineering, The Johns Hopkins University.

Stumm is the recipient of the American Chemical Society's Monsanto Prize for Pollution Control in 1976, the Association of Environmental Engineering Professors' Outstanding Publication Award in 1983 and 1984, the World Cultural Council's Albert Einstein World Award of Science in 1985, the Tyler Prize for Environmental Achievement in 1986, the American Society of Civil Engineer's S. W. Freese Award in 1991, the Swiss Confederation's Marcel-Benoist Prize in 1991, and the

European Science Prize in 1992. Stumm has also received numerous honorary doctoral degrees from various institutes throughout the world, including University of Geneva in Switzerland, Royal Institute of Technology in Sweden, University of Crete in Greece, Northwestern University in the United States, and TECHNION in Israel. He is a member of the U.S. National Academy of Engineering and Academia Europaea.

He is the senior author of the popular *Aquatic Chemistry* (coauthored with J. J. Morgan), which has been translated into Japanese and Chinese. He is also the coauthor of the following books: *Gewässer als Ökosystem* (with R. Kummert), *Aquatische Chemie* (with L. Sigg), and *Chimie des Milieux Aquatiques* (with L. Sigg and Ph. Behra). His recent book, *Chemistry of Solid–Water Interface*, was published in 1992. He has also edited numerous books, including *Equilibrium Concepts in Natural Waters* in 1967, *Global Chemical Cycles and Their Alteration by Man* in 1977, *Chemical Processes in Lakes* in 1985, *Aquatic Surface Chemistry* in 1987, *Aquatic Chemical Kinetics* in 1990, and *Chemistry of Aquatic Systems: Local and Global Perspectives* in 1994.

ABOUT THE EDITORS

C. P. HUANG is the Distinguished Professor of Environmental Engineering at the University of Delaware. He received his Ph.D. and M.S. in environmental engineering from Harvard University and his B.S. in civil engineering from the National Taiwan University, Taipei, Taiwan. Huang has authored or coauthored over 150 research papers, book chapters, technical reports, and conference proceedings and is a citation classics author. His research expertise is in environmental physical–chemical processes, including the removal of heavy metals from dilute aqueous solutions by adsorption process, photooxidative dissolution of metal sulfide minerals, and surface acidity of hydrous solids. His recent research interests are advanced chemical oxidation for the treatment of hazardous organic wastes and in-situ treatment of contaminated soils and aquifers by electrochemical processes. Currently, he is on the editorial board of the *Journal of Environmental Engineering*, the Chinese Society of Environmental Engineering, and is Editorial Advisor of the Taiwanese Industrial Park Communication.

CHARLES R. O'MELIA is Professor of Environmental Engineering and Chairman of the Department of Geography and Environmental Engineering at The Johns Hopkins University in Baltimore, Maryland. He received his B.C.E. (1955) from Manhattan College and his M.S.E. (1956) and Ph.D. (1963) in Sanitary Engineering from the University of Michigan in Ann Arbor. He was employed by Hazen and Sawyer, Engineers in 1956–1957. From 1961 to 1964 he served as Assistant Professor of Sanitary Engineering at the Georgia Institute of Technology. From 1964 to 1966 he was a postdoctoral fellow and lecturer in water chemistry at Harvard University. He joined the University of North Carolina at Chapel Hill in 1966 as

Associate Professor and became Professor in 1970. From 1977 to 1980 he served as Deputy Chairman of the Department of Environmental Sciences and Engineering at UNC. In 1973–1974 he was Visiting Professor of Environmental Engineering Science at the California Institute of Technology while on sabbatical leave. He assumed his present position as Professor at Johns Hopkins in 1980 and was appointed Department Chairman in 1990. While on sabbatical leave from 1988 to 1990 he was a Guest Professor at ETH-Zürich with the Swiss Federal Institute for Water Resources and Water Pollution Control.

O'Melia was elected to the National Academy of Engineering in 1989. He has received many awards, including the 1982 Distinguished Lecturer of the Association of Environmental Engineering Professors. He is a member of many societies and organizations and has served as Director, Vice President, and President of the Association of Environmental Engineering Professors. O'Melia's research interests are in aquatic colloid chemistry, water and wastewater treatment, and modeling of natural surface and subsurface waters.

JAMES MORGAN is the Marvin L. Goldberger Professor of Environmental Engineering Science at the California Institute of Technology. He received his Ph.D. from Harvard University in 1964, his M.S. from the University of Michigan in 1956, and his B.C.E. from Manhattan College in 1954. He was the founding editor of the ACS Publication *Environmental Science & Technology* from 1966 through 1974. Among other awards, he received the Association of Environmental Engineering Professors' Research Publication Award with Werner Stumm in 1983. He was elected to the National Academy of Engineering in 1978, and received the ACS Award for Creative Advances in Environmental Science and Technology in 1980. Morgan has authored more than 80 articles and chapters dealing with the chemistry of natural water systems, oxidation processes in water, adsorption and surface chemistry, and other topics. He is coauthor of the book *Aquatic Chemistry*.

CONTENTS

Preface xiii

1. **The Inner-Sphere Surface Complex: A Key to Understanding Surface Reactivity** 1
   Werner Stumm

2. **Adsorption as a Problem in Coordination Chemistry: The Concept of the Surface Complex** 33
   Garrison Sposito

3. **Ion Exchange: The Contributions of Diffuse Layer Sorption and Surface Complexation** 59
   David A. Dzombak and Robert J. M. Hudson

4. **Interaction of Organic Matter with Mineral Surfaces: Effects on Geochemical Processes at the Mineral–Water Interface** 95
   Janet G. Hering

5. **Reaction Rates and Products of Manganese Oxidation at the Sediment–Water Interface** 111
   Bernhard Wehrli, Gabriela Friedl, and Alain Manceau

6. **Redox Chemistry of Iodine in Seawater: Frontier Molecular Orbital Theory Considerations** 135
   George W. Luther, III, Jingfeng Wu, and John B. Cullen

7. **Oxidation–Reduction Environments: The Suboxic Zone in the Black Sea** 157
   James W. Murray, Louis A. Codispoti, and Gernot E. Friederich

8. **Cycles of Trace Elements (Copper and Zinc) in a Eutrophic Lake: Role of Speciation and Sedimentation** 177
   Laura Sigg, Annette Kuhn, Hanbin Xue, Elke Kiefer, and David Kistler

9. **Metals and Microbiology: The Influence of Copper on Methane Oxidation** 195
   Mary E. Lidstrom and Jeremy D. Semrau

10. **Coagulation of Marine Algae** 203
    George A. Jackson

11. Diversity of Anaerobes and Their Biodegradative Capacities 219
    L. Y. Young and M. M. Häggblom

12. The Chemical Effects of Collapsing Cavitation Bubbles:
    Mathematical Modeling 233
    Anatassia Kotronarou and Michael R. Hoffmann

13. Photoreactions Providing Sinks and Sources of Halocarbons
    in Aquatic Environments 253
    Richard G. Zepp and Leroy F. Ritmiller

14. Photochemical Reductive Dissolution of Lepidocrocite:
    Effect of pH 279
    Barbara Sulzberger and Hansulrich Laubscher

15. Photocatalytic Degradation of 4-Chlorophenol in TiO_2 Aqueous
    Suspensions 291
    Chengdi Dong and Chin-Pao Huang

16. From Algae to Aquifers: Solid–Liquid Separation in Aquatic
    Systems 315
    Charles R. O'Melia

17. Surfactant Solubilization of Phenanthrene in Soil–Aqueous
    Systems and Its Effects on Biomineralization 339
    Shonali Laha, Zhongbao Liu, David A. Edwards, and Richard G. Luthy

18. Distributed Reactivity in the Sorption of Hydrophobic Organic
    Contaminants in Natural Aquatic Systems 363
    Walter J. Weber, Jr., Paul M. McGinley, and Lynn E. Katz

19. Interaction of Coagulation–Flocculation with Separation Processes ...383
    Hermann H. Hahn

Author Index 397
Affiliation Index 397
Subject Index 398

PREFACE

THE FIRST REFERENCE BOOK IN AQUATIC CHEMISTRY, *Equilibrium Concepts in Natural Water systems*, was published by the American Chemical Society in 1967. Since then, many advances, both theoretical and experimental, have been made. New concepts have been developed and verified because advanced instrumentation that was not available 25 years ago may now be found in a routine research facility. The field has flourished and diversified tremendously. Aquatic chemistry is no longer a subject dealing with the principles of dilute aqueous solution chemistry; it has evolved as the applied chemistry of multiphase and multicomponent environmental systems and as a highly multidisciplinary subject with a strong emphasis on interfacial phenomena. Interfaces are ubiquitous in natural waters as well as environmental engineering systems such as air, soil, water, and wastewater treatment facilities. The teaching of aquatic chemistry has spread across the North American continent and to many parts of the world. Although the field is thriving and progressing, many questions still await answers. How and why some chemical species are transformed and transported in aquatic systems are unknown; more efficient, safe, and ecologically sound ways to process our wastes are needed; and unknowns exist about how to better manage our total environment. Answers to these questions require a strong, multidisciplinary approach.

The symposium upon which this book is based was organized in honor of Werner Stumm, the founder of aquatic chemistry. A total of 21 invited papers and 30 posters were presented at this special symposium. A wide spectrum of scientists (surface chemists, soil chemists, geochemists, limnologists, and oceanographers) and engineers (environmental, civil, and chemical) attended. The symposium opened with a paper by Stumm titled "The Inner-Sphere Surface Complex: A Key to Understanding Surface Reactivity", which was followed by five key talks representing five major topics of the symposium: surface chemistry, earth sciences, biology, redox and photochemistry, and engineering. This book is structured to present these five key topics.

By no means is this book intended to provide a detailed account of all progress made in the past 25 years by aquatic chemists. Rather, chapters provide examples of recent developments in the field and contribute toward a better understanding of the mechanisms regulating the chemical composition of natural waters. Also, the transformation and transport of species (abiotic and biotic or soluble and insoluble) in aquatic systems (lakes, rivers, estuaries, aquifers, atmosphere, and oceans) through interphase and interspecies interactions are discussed. Moreover, principles discussed in the book can be useful to the design of air, soil, water, and wastewater treatment systems. For example, processes such as solute–solid interactions, solid–liquid separation, colloid stability, and redox and photochemical reactions that occur in the natural environment can also be applied to the design of air, soil, water and wastewater treatment processes. Finally, we hope that chapters in the book will provide readers with an opportunity to revisit concepts conceived 25 years ago, to witness some past achievements, and to contemplate future research needs.

Acknowledgments

Many have contributed to this book. Authors, speakers, and attendees at the symposium deserve our special gratitude. Their enthusiastic support has made our task in organizing the symposium a most pleasant one. To our reviewers, we are deeply in debt. Their timely review of the chapters was crucial to the completion of the project. A grant from the National Science Foundation (NSF) greatly eased the financial burden of symposium participants. For that, we thank Edward Bryan at NSF for his support and interest in this project. Throughout the preparation process, the staff of the ACS Books Department was most helpful. We wish to thank, in particular, Colleen Stamm and Rhonda Bitterli for their professional assistance.

C. P. HUANG
Department of Civil
  Engineering
University of Delaware
Newark, DE 19716

JAMES J. MORGAN
Department of Environmental
  Engineering Science
California Institute of Technology
Pasadena, CA 91125

CHARLES R. O'MELIA
Department of Geography and
  Environmental Engineering
The Johns Hopkins University
Baltimore, MD 21218

March 1994

The Inner-Sphere Surface Complex

A Key to Understanding Surface Reactivity

Werner Stumm

Swiss Federal Institute of Technology, Zürich; EAWAG (Institute for Environmental Science and Technology), CH–8600, Dübendorf, Switzerland

> *Functional groups on the interface of natural solids (minerals and particles) with water provide a diversity of interactions through the formation of coordinate bonds with H^+, metal ions, and ligands. The concept of active surface sites is essential in understanding the mechanism of many surface-controlled processes (nucleation and crystal growth, biomineralization, dissolution and weathering of minerals, soil formation, catalysis of redox processes, and photochemical reactions). The enhancement of the dissolution rate by a ligand implies that surface complex formation facilitates the release of ions from the surface to the adjacent solution. These ligands bring electron density within the coordinating sphere of the central ion. Surface species thus destabilize the bonds in the surface lattice; they are especially efficient in the dissolution of iron and aluminum oxides and of aluminum silicates. Ascorbate, phenols, and $S(-II)$ compounds, including H_2S, readily form surface complexes with $Fe(III)$ or $Mn(III,IV)$ (hydr)oxides that subsequently undergo electron transfer and the release of $Fe(II)$ or $Mn(II)$ into solution. Reductive and nonreductive dissolutions are markedly inhibited by competitive (ligand exchange) adsorption of inorganic oxoanions. These oxoanions can form bi- or multinuclear surface complexes. A better understanding of the electronic structure of the interface of solids and aquatic solutes would push the boundaries of aquatic surface chemistry.*

INTERACTION AT THE SOLID–WATER INTERFACE can be characterized in terms of the chemical and physical properties of water, the solute, and the sorbent. The two basic processes in the reaction of solutes with natural surfaces are the formation of coordinate bonds (surface complexation) and hydrophobic adsorption.

Hydrophobic adsorption is primarily driven by the incompatibility of nonpolar, hydrophobic substances with water. The formation of coordinate bonds is based on the generalization that the solids can be considered either inorganic or organic polymers; their surfaces can be seen as extending structures bearing surface functional groups. These functional groups contain the same donor atoms found in functional groups of solute ligands such as –OH, –SH, –SS, and –CO_2H. Such functional groups provide a diversity of interactions through the formation of coordinate bonds. Similarly, ligands can replace surface OH groups (ligand exchange) to form ligand surface complexes.

The concept of active sites has helped explain catalysis by enzymes and coenzymes. Although surface functional groups are less specific than enzymes, they form an array of surface complexes whose reactivities determine the mechanism of many surface-controlled processes. Many mechanisms can be described readily in terms of Brønsted acid sites or Lewis acid sites. Of course, the properties of the surfaces are influenced by the properties and conditions of the bulk structure, and the action of special surface structural entities will be influenced by the properties of both surface and bulk. List I gives an overview of the major concepts and important applications.

Surface chemistry of the oxide–water interface is emphasized here, not only because the oxides are of great importance at the mineral–water (including the clay–water) interface but also because its coordination chemistry is much better understood than that of other surfaces. Experimental studies on the surface interactions of carbonates, sulfides, disulfides, phosphates, and biological materials are only now emerging. The concepts of surface coordination chemistry can also be applied to these interfaces. This chapter is designed

- to briefly review surface complexation theory, reflecting on the nature of site-specific binding to H^+, metal ions, and ligands
- to discuss the need for assessing the bonding between solids and solutes to understand better the reactivity of the solid–water interface and to illustrate this reactivity in terms of surface-controlled dissolution of oxides and silicates
- to present exemplifying experimental evidence on various factors that enhance or inhibit dissolution to make the point that we need a better appreciation of the electronic structure and the geometry of the bonding at the solid–water interface to predict reactivity
- to exemplify some applications of the effects of surface complex formation, surface reactivity enhancement, and inhibition of dis-

List I. Coordination Chemistry of the Solid–Water Interface: Concepts and Applications in Natural and Technical Systems

Surface Complex Formation	Applications: Distribution of Solutes between Water and Solid Surface	Applications: Rate Dependence on Surface Speciation
Interaction with H⁺, OH⁻ Metal ions Ligands (ligand exchange)	**Binding of Reactive Elements to Aquatic Particles in Natural Systems** Regulation of metals in soil, sediment, and water systems Regulation of oxyanions of P, As, Se, and Si in water and soil systems Interaction with phenols, carboxylates, and humic acids Transport of reactive elements including radionuclides in soils and aquifers	**Natural Systems** **Dissolution of Oxides, Silicates, Carbonates, and Other Minerals** Weathering of minerals Proton- and ligand-promoted dissolution Reductive dissolution of Fe(III) and Mn(III,IV) oxides
Thermodynamics of Surface Complex Formation K (mass law constants, corrected for electrostatic effects ΔG, ΔH		**Formation of Solid Phases** Heterogeneous nucleation Surface precipitation, crystal growth Biomineralization
Kinetics of Surface Complex Formation Rates of sorption and desorption		
Structure of Surface Compounds (Surface Speciation) Inner-sphere versus outer-sphere Mononuclear versus binuclear Monodentate versus bidentate	**Binding of Cations, Anions, and Weak Acids to Particles in Technical Systems** Corrosion, passive films Processing of ores, flotation Coagulation, flocculation, filtration Ceramics, cements Photoelectrochemistry (electrodes, oxide electrodes, and semiconductors)	**Surface-Catalyzed Processes** (Photo)redox processes Hydrolysis of esters Transformations of organic matter by Fe and Mn (photo)redox cycles Oxygenation of Fe(II), Mn(II), Cu(I), and V(IV)
Establishment of Surface Charge (Structure of Lattice) Defect sites Adatoms, kinks, steps, ledges Lattice statistics		**Technical Systems** Passive films (corrosion) Photoredox processes with colloidal semiconductor particles as photocatalyst (e.g., degradation of refractory organic substances) Photoelectrochemistry (e.g., photoredox processes at semiconductor electrodes)
Microtopography	**Surface Charge Resulting from the Sorption of Solutes** Particle–particle interaction; coagulation, filtration	

SOURCE: Modified from reference 1.

solution in natural weathering processes, in heterogeneous photochemical processes, and in technical systems (corrosion and dissolution of passive iron oxide films)

Surface Coordination Chemistry

Inner- and Outer-Sphere Complexes. As illustrated in Figure 1, a cation can associate with a surface as an inner-sphere or an outer-sphere complex, depending on whether a chemical bond is formed (i.e., a largely covalent bond between the metal and the electron-donating oxygen ions, as in an inner-sphere complex) or whether a cation of opposite charge approaches the surface groups within a critical distance. As with solute ion pairs, the cation and the base are separated by one or more water molecules (1, 2). Furthermore, ions may exist in the diffuse swarm of the double layer.

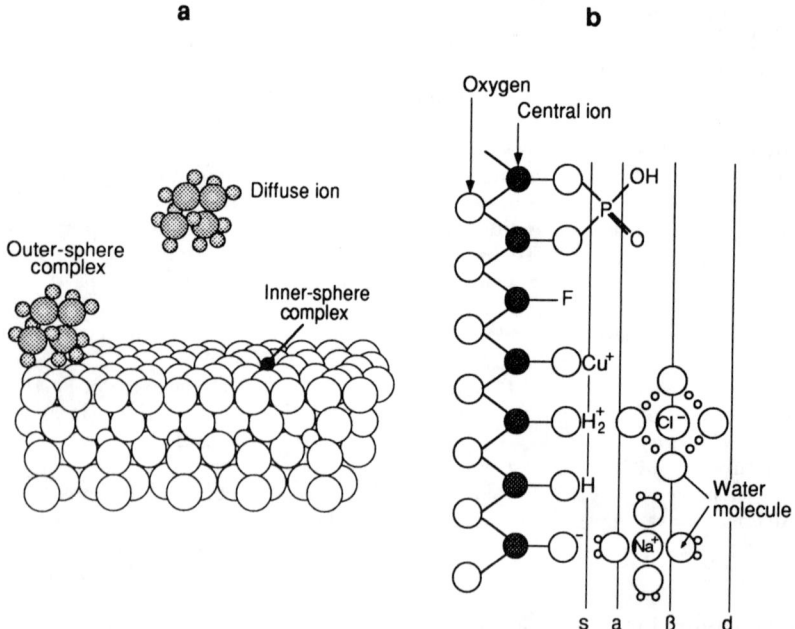

Figure 1. Part a: Surface complex formation of an ion (e.g., cation) on a hydrous oxide surface. The ion may form an inner-sphere complex (chemical bond), an outer-sphere complex (ion pair), or be in the diffuse swarm of the electric double layer. (Reproduced with permission from reference 2. Copyright 1984.) Part b: Schematic portrayal of the hydrous oxide surface, showing planes associated with surface hydroxyl groups (s), inner-sphere complexes (a), outer-sphere complexes (β), and the diffuse ion swarm (d). In the case of an inner-sphere complex with a ligand (e.g., F^- or HPO_4^{2-}), the surface hydroxyl groups are replaced by the ligand (ligand exchange). (Modified from reference 3.)

It is important to distinguish between outer-sphere and inner-sphere complexes. In inner-sphere complexes the surface hydroxyl groups act as σ-donor ligands, which increase the electron density of the coordinated metal ion. Cu(II) bound in an inner-sphere complex is a different chemical entity from Cu(II) bound in an outer-sphere complex or present in the diffuse part of the double layer. The inner-spheric Cu(II) has different chemical properties; for example, it has a different redox potential with respect to Cu(I), and its equatorial water is expected to exchange faster than that in Cu(II) bound in an outer-sphere complex. As we shall see, the reactivity of a surface is affected, above all, by inner-sphere complexes.

List II summarizes schematically the type of surface complex formation equilibria that characterize the adsorption of H^+, OH^-, cations, and ligands at a hydrous oxide surface. The various surface hydroxyls formed at a hydrous oxide surface may not be fully equivalent structurally and chemically. However, to facilitate the schematic representation of reactions and of equilibria, we will consider the chemical reaction of «a» surface hydroxyl group, S–OH. The following surface groups can be envisaged.

$$\mathrm{S{>}OH} \;,\; \mathrm{S{-}OH} \quad \mathrm{S{<}^{OH_2}_{OH}} \quad \mathrm{S{\lessdot}^{OH}_{OH}{}_{OH}}$$

These functional groups have donor properties similar to those of their counterparts in dissolved solutes such as hydroxides or carboxylates. Thus,

List II. Adsorption (Surface Complex Formation Equilibria)

Acid-base equilibria

$$\mathrm{S{-}OH} + H^+ \rightleftharpoons \mathrm{S{-}OH_2^+}$$
$$\mathrm{S{-}OH}\ (+\ OH^-) \rightleftharpoons \mathrm{S{-}O^-}\ (+\ H_2O)$$

Metal binding

$$\mathrm{S{-}OH} + M^{z+} \rightleftharpoons \mathrm{S{-}OM}^{(z-1)+} + H^+$$
$$2\,\mathrm{S{-}OH} + M^{z+} \rightleftharpoons \mathrm{(S{-}O)_2M}^{(z-2)+} + 2\,H^+$$
$$\mathrm{S{-}OH} + M^{z+} + H_2O \rightleftharpoons \mathrm{S{-}OMOH}^{(z-2)+} + 2\,H^+$$

Ligand exchange (L^- = ligand)

$$\mathrm{S{-}OH} + L^- \rightleftharpoons \mathrm{S{-}L} + OH^-$$
$$2\,\mathrm{S{-}OH} + L^- \rightleftharpoons \mathrm{S_2{-}L^+} + 2\,OH^-$$

Ternary surface complex formation

$$\mathrm{S{-}OH} + L^- + M^{z+} \rightleftharpoons \mathrm{S{-}L{-}M}^{z+} + OH^-$$
$$\mathrm{S{-}OH} + L^- + M^{z+} \rightleftharpoons \mathrm{S{-}OM{-}L}^{(z-2)+} + H^+$$

SOURCE: Modified from reference 4.

deprotonated surface groups (S–O$^-$) behave like Lewis bases and the sorption of metal ions (and protons) can be understood as competitive complex formation.

Adsorption of Ligands on Metal Oxides. The adsorption of ligands (anions and weak acids) on metal oxide and silicate surfaces can also be compared with complex formation reactions in solution.

$$Fe(OH)^{2+} + F^- \longrightarrow FeF^{2+} + OH^- \qquad (1a)$$

$$S\text{–}OH + F^- \longrightarrow S\text{–}F + OH^- \qquad (1b)$$

The central ion of a mineral surface acts as a Lewis acid and exchanges its structural OH with other ligands (ligand exchange). In this case consider the surface of Fe(III) oxide as an example. S–OH corresponds to ≡Fe–OH. A Lewis acid site is a surface site capable of receiving a pair of electrons from the adsorbate. (A Lewis base site has a free pair of electrons—like the oxygen donor atom in a surface OH$^-$ group—that can be transferred to the adsorbate.) The extent of surface complex formation (adsorption) for metal ions and anions is strongly dependent on pH and on the release of protons and OH$^-$ ions, respectively. In addition to monodentate surface complexes, bidentate (mononuclear or binuclear) surface complexes can be formed.

$$2S\text{–}OH + Cu^{2+} \longrightarrow (S\text{–}O)_2Cu + 2H^+ \qquad (2a)$$

$$\begin{array}{l} -S-OH \\ | \\ -S-OH \end{array} + Cu^{2+} \rightleftharpoons \begin{array}{l} -S-O \\ | \diagdown \\ -S-O \diagup \end{array} Cu + 2H^+ \qquad (2b)$$

$$\equiv FeOH + H C_2O_4^- \text{ (Oxalate)} \rightleftharpoons \equiv Fe\begin{array}{c} O-C \diagup O^- \\ | \\ O-C \diagdown O \end{array} + H_2O \qquad (3)$$

$$\begin{array}{l} \equiv FeOH \\ | \\ \equiv FeOH \end{array} + H_2PO_4^- \rightleftharpoons \begin{array}{l} \equiv Fe-O \diagdown \diagup O^- \\ | P \\ \equiv Fe-O \diagup \diagdown O \end{array} + 2H_2O \qquad (4)$$

The following criteria characterize all surface complexation models (5):

- Sorption takes place at specific surface coordination sites.
- Sorption reactions can be described by mass law equations.

- Surface charge results from the sorption (surface complex formation) reaction itself.
- The effect of surface charge on sorption (the extent of complex formation) can be taken into account by applying to the mass law constants for surface reactions a correction factor derived from the electric double-layer theory.

The extent of adsorption, or surface coordination, and its pH dependence can be accounted for by mass law equilibria (Figure 2). Their equilibrium constants reflect the affinity of the surface sites for H^+, metal ions, and ligands. The tendency to form surface complexes may be compared with the tendency to form corresponding (inner-sphere) solute complexes (4–6). Figure 3 shows the relation between the solute complex formation of $FeOH_2^+$ or $AlOH^{2+}$ with various ligands and the surface complexation of $\equiv$FeOH and $\equiv$AlOH surface groups with the same ligands. The reasonably good correlation obtained in this and similar linear free energy relation (LFER) plots (4–6) indicates that the same chemical mode of interaction occurs in solution and at the surface and that the available sorption data are consistent with one another. Therefore, such LFERs may be used to predict intrinsic sorption constants from solute complex formation constants and vice versa.

Surface Complex Formation on Carbonates. There are various possibilities for functional groups on the surface of carbonates, sulfides, phosphates, and similar compounds. By using a very simple approach similar to the one used for hydrous oxides (chemisorption of H_2O), one could postulate surface groups for carbonates (e.g., $FeCO_3$) as shown in List III.

As indicated in Scheme I, it is reasonable to assume that H^+, OH^-, HCO_3^-, CO_2(aq), and Fe^{2+} can interact with MCO_3(s) and affect its surface charge. Surface complex formation of the surface groups with ligands and metal ions can occur (9).

Surface Reactivity Dependence on Surface Structure

Many heterogeneous processes such as dissolution of minerals, formation of the solid phase (precipitation, nucleation, crystal growth, and biomineralization), redox processes at the solid–water interface (including light-induced reactions), and reductive and oxidative dissolutions are rate-controlled at the surface (and not by transport) (10). Because surfaces can adsorb oxidants and reductants and modify redox intensity, the solid–solution interface can catalyze many redox reactions. Surfaces can accelerate many organic reactions such as ester hydrolysis (11).

The mechanisms of most surface-controlled processes depend on the coordination environment at the solid–water interface. Above all, they depend

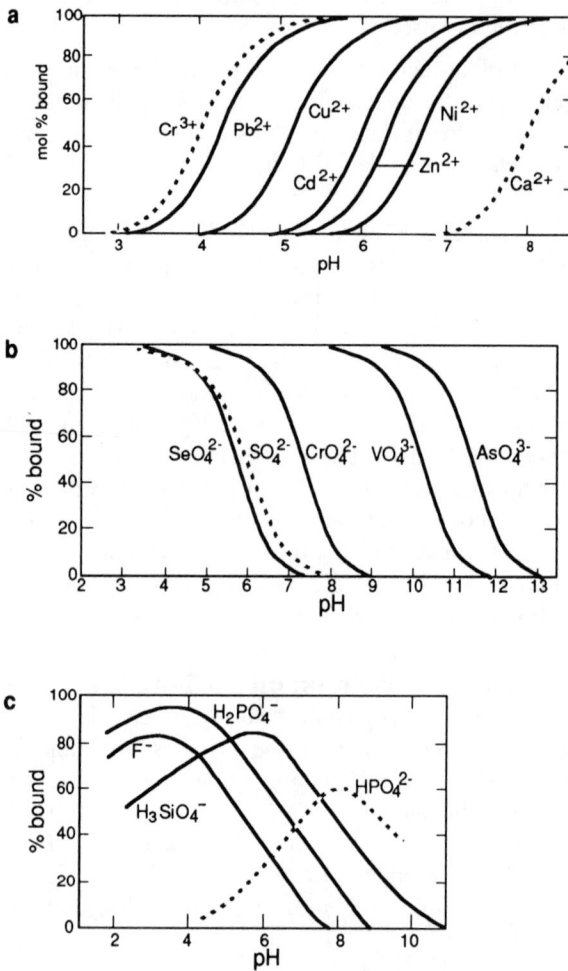

Figure 2. These curves were calculated with the help of experimentally determined equilibrium constants. Part a: Extent of surface complex formation as a function of pH (measured as mole percent of the metal ions in the system, adsorbed or surface-bound). Total ion concentration [TOTFe] = 10^{-3} M (2×10^{-4} mol/L of reactive sites; metal concentrations in solution = 5×10^{-7} M; I = 0.1 M $NaNO_3$. (The curves are based on data compiled by Dzombak and Morel in reference 5.) Part b: Surface complex formation with ligands (anions) as a function of pH. Binding of anions from dilute solutions (5×10^{-7} M) to hydrous ferric oxide; [TOTFe] = 10^{-3} M. I = 0.1. (Curves are based on data from Dzombak and Morel in reference 5.) Part c: Binding of phosphate, silicate, and fluoride on goethite (α-FeOOH); the species shown are surface species (6 g/L of FeOOH, $P_T = 10^{-3}$ M, $Si_T = 8 \times 10^{-4}$ M). (Reproduced with permission from reference 6. Copyright 1981.)

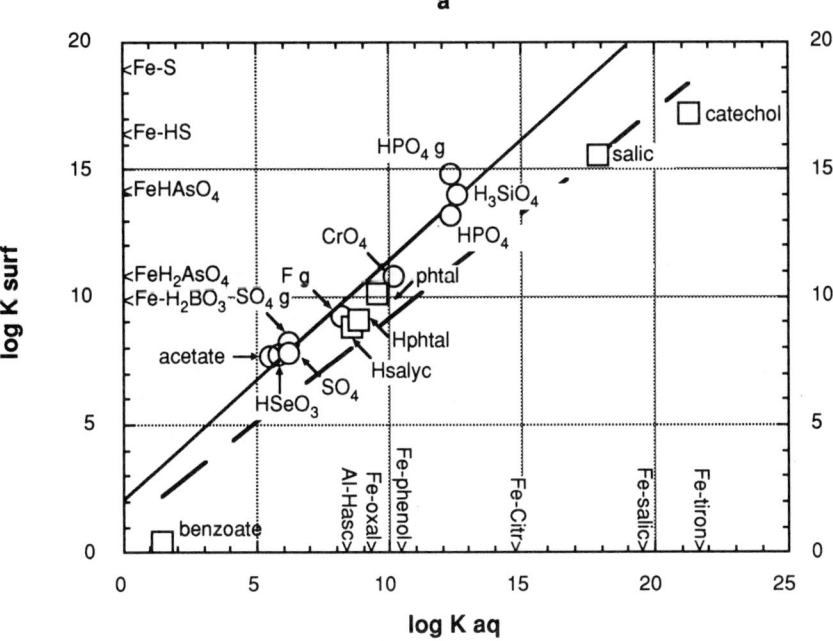

Figure 3a. Linear free relation between the tendency to form solute complexes of Fe(III)(aq) and Al(III)(aq)

$$MOH^{2+} + H^+ + A \longrightarrow MA + H_2O; \quad K_1(aq)$$

and the tendency to form surface complexes (intrinsic equilibrium constant) on γ-Al_2O_3 and hydrous ferric oxide or goethite surfaces

$$\equiv MOH + H^+ + A \longrightarrow \equiv MA + H_2O; \quad K^s(surf)$$

A is the actual species that forms the complex (e.g., $A = H_3SiO_4^-$ and $\equiv MA = \equiv FeH_3SiO_4$); charges are omitted for simplicity. Equilibrium constants in solution (I = 0) are from Smith and Martell (7). Constants given in Fe^{3+} were converted into constants valid for $FeOH^{2+}$ by log K = –2.2 for the reaction

$$Fe^{3+} + H_2O \longrightarrow FeOH^{2+} + H^+$$

Data for surface complex formation on hydrous ferric oxide (○) are from Dzombak and Morel (5), data for goethite (marked g) are from Sigg and Stumm (6), and data for γ-Al_2O_3 (□) are from Kummert and Stumm (8). These data are intrinsic equilibrium constants (i.e., extrapolated to zero surface charge). At the ordinate and abscissa a few relevant surface complex formation constants and solute equilibrium constants, respectively, are listed for which the constants in solution or at the surface are not known; they may be used to estimate the corresponding unknown constant.

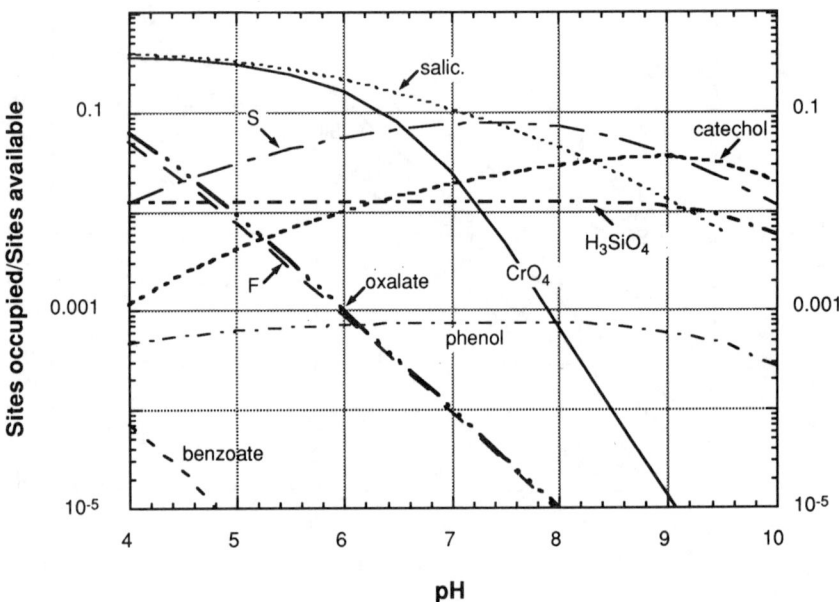

Figure 3b. Fractional surface coverage of ≡Fe(III) surface complexes as a function of pH. The calculation is based on the condition

$$[\equiv FeOH]_T = [X]_T = 10^{-6} \text{ M}; I = 10^{-2}$$

Electrostatic correction was made with the diffuse (Gouy–Chapman) double-layer model. The figure shows the effect of pH on the relative extent of surface complexation.

on the electronic structure of the bonding between solids and solutes and on the geometry of the coordinating shell of surface sites and reactants at the surface.

In aquatic chemistry, initial researchers realized that most important processes occur at interfaces. Also, the solid–water interface was discovered to play commanding roles in regulating the concentrations of most reactive elements

List III. Possible Surface Groups for FeCO$_3$

H	OH	H	OH	H	OH	water ↑
CO$_3$	Fe	CO$_3$	Fe	CO$_3$	Fe	solid ↓
Fe	CO$_3$	Fe	CO$_3$	Fe	CO$_3$	
CO$_3$	Fe	CO$_3$	Fe	CO$_3$	Fe	

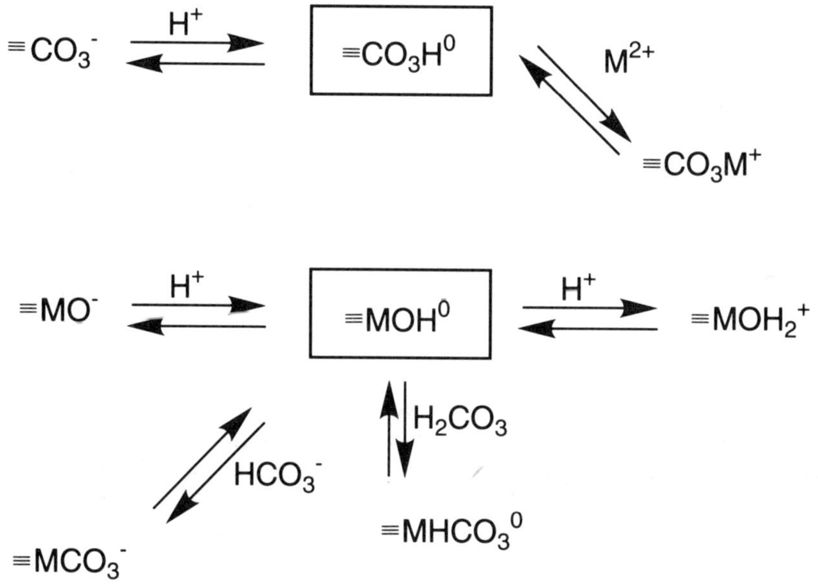

$$\equiv CO_3^- \xrightleftharpoons[]{H^+} \boxed{\equiv CO_3H^0} \xrightleftharpoons[]{M^{2+}} \equiv CO_3M^+$$

$$\equiv MO^- \xrightleftharpoons[]{H^+} \boxed{\equiv MOH^0} \xrightleftharpoons[]{H^+} \equiv MOH_2^+$$

$$\equiv MCO_3^- \xrightleftharpoons[]{HCO_3^-} \quad \equiv MHCO_3^0 \xrightleftharpoons[]{H_2CO_3}$$

Scheme I. *Interaction of functional groups to affect surface charge.*

in soil and natural water systems, in the coupling of various hydrogeochemical cycles, and in many processes in water technology. Surface complexation has successfully addressed many pragmatic questions about the distribution of solutes between the aqueous solution and the solid surface. Although these answers are useful and have predictive value, they often do not provide unique information about the ways in which molecules, atoms, and ions interact at the solid–water interface and about the electronic structure of the bonding.

In recent years new insights have come from spectroscopic methods. Motschi (12) used electron spin resonance spectroscopy to study Cu(II) surface complexes. Additional studies were carried out with electron nuclear double resonance (ENDOR) spectroscopy and electron spin echo envelope modulation (ESEEM) to elucidate structural aspects of surface-bound Cu(II), of ternary copper complexes in which coordinated water is replaced by ligands, and of vanadyl ions on δ-Al_2O_3. Application of ENDOR spectroscopy allows the resolution of weak interactions between the unpaired electron and nuclei within a distance of about 5 °. From these so-called hyperfine data, structural parameters can be derived (e.g., bond distances of the paramagnetic center to the coupling nuclei or ligands). In the ENDOR spectrum of adsorbed VO^{2+} on δ-Al_2O_3, signals caused by coupling with the surface Lewis center (^{27}Al) are more strongly split than is calculated from molecular modeling. The existence of an inner-sphere coordination between the hydrated oxide and the metal is confirmed experimentally (12). Attenuated Fourier transform infrared spectroscopy (FTIR) has also contributed significantly to elucidating the type of surface species present (e.g., see reference 13).

Direct in situ extended X-ray adsorption fine structure (EXAFS) measurements) from synchrotron radiation (*14–18*) permit the determination of species adsorbed to neighboring ions and to central ions on oxide surfaces in the presence of water. Such investigations showed, for example, that selenite is bound in an inner-sphere complex and selenate is bound in an outer-sphere complex to the central Fe(III) ions of a goethite surface. This technique also showed that Pb(II) is bound in an inner-sphere complex to δ-Al_2O_3 (*19*) and that Cr(III) is bound in an inner-sphere complex at the oxide–water interface of Mn(IV) oxides and ferric hydrous oxides (*17, 18*).

Dissolution of Oxides. The coordination environment of the metal changes in the dissolution reaction of an oxide mineral. For example, when an aluminum oxide layer dissolves, the Al^{3+} in the crystalline lattice exchanges its O^{2-} ligand for H_2O or another ligand L. As seen in Figure 4, the most important reactants participating in the dissolution of a solid mineral are H_2O, H^+, OH^-, ligands (surface complex building), and reductants and oxidants (for reducible or oxidizable minerals).

There is a considerable amount of empirical data available on surface reactivity in terms of rates of processes such as dissolution and reductive dissolution. One difficulty in relating this reaction rate information to surface

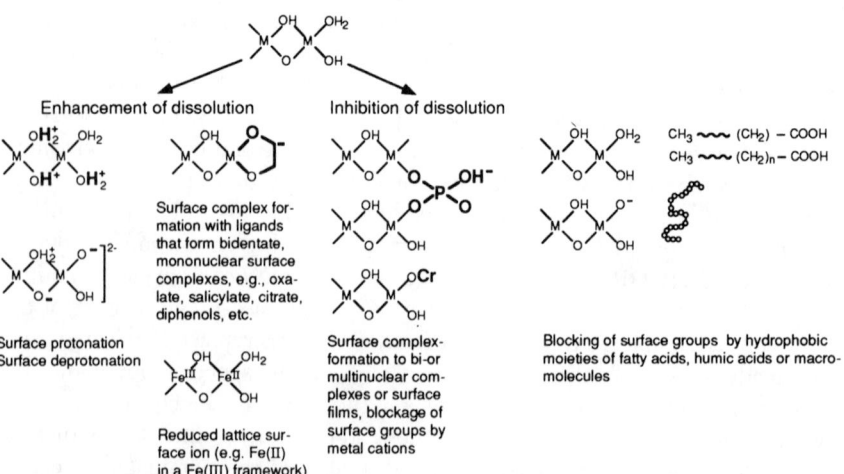

Figure 4. Effects of protonation, complex formation with ligands and metal ions, and reduction on dissolution rate. The structures given here are schematic shorthand notations to illustrate the principal features. They do not reveal either the structural properties or the coordination numbers of the oxides under consideration; charges given are relative.

structure is that we often do not have sufficiently detailed knowledge about the latter. Thus, much of the sought-after interdependence is still presumptive. We will first briefly review the rate laws and the corresponding mechanisms (1).

The reaction occurs schematically in the following sequence:

$$\text{surface sites} + \text{reactants (H}^+\text{, OH}^-\text{, or ligands)} \xrightarrow{\text{fast}} \text{surface species} \quad (5)$$

$$\text{surface species} \xrightarrow[\text{detachment of metal}]{\text{slow}} \text{metal(aq)} + \text{surface sites} \quad (6)$$

Although each sequence may consist of a series of smaller reaction steps, the rate law of surface-controlled dissolution is based on the idea

- that the attachment of reactants to the surface sites is fast
- that the subsequent detachment of the metal species from the surface of the crystalline lattice into the solution is slow and thus rate-limiting
- that the original surface sites are continuously reconstituted

In reaction 5 the dissolution reaction is initiated by the surface coordination with H^+, OH^-, and ligands, which polarizes, weakens, and tends to break the metal–oxygen bonds in the lattice of the surface. Because Reaction 6 is rate-limiting, the steady-state approach leads to a dependence of the rate law of the dissolution reaction on the concentration (activity) of the particular surface species, C_j (mol/m²):

$$\text{dissolution rate} \propto \langle\text{surface species}\rangle \quad (7)$$

We reach the same conclusion (eq 7) if we treat the reaction sequence according to the activated complex theory (ACT), often called the transition state theory. The particular surface species that has formed from the interaction of H^+, OH^-, or ligands with surface sites is the precursor of the activated complex:

$$\text{dissolution rate} \propto \langle\text{precursor of the activated complex}\rangle \quad (8a)$$

$$R = kC_j \quad (8b)$$

C_j is proportional to the density of surface sites (S) the mole fraction (χ_a) of these surface sites that are dissolution active, and the probability (P_j) of finding a site in which there is suitable coordinative arrangement of precursor complex:

$$C_j = \chi_a P_j S \tag{9}$$

Thus, the dissolution rate can be generalized (20) into

$$R = k\chi_a P_j S \tag{10}$$

where R is the dissolution rate (mol/m² per second), k is the appropriate rate constant per second, S is the surface concentration of sites (mol/m²); χ_a is the mole fraction of dissolution-active sites, and P_j is the probability of finding a specific site in the coordinate arrangement of the precursor complex.

In simple terms, the rate laws for the ligand-promoted (R_L) and proton-promoted (R_H) dissolution rate can be given (1, 21–23) by

$$R_L = k_L C_L^s = k_L \langle SL \rangle \tag{11a}$$

$$R_H = k_L (C_H^s)^j = k_H \langle SOH_2^+ \rangle^j \tag{11b}$$

The overall rate law for the dissolution (R) is given by the sum of the individual reaction rates

$$R = R_H + R_{L_1} + R_{L_2} + \cdots \tag{12a}$$

$$R = k_H (C_H^s)^j + k_{L_1} C_{L_1}^s + k_{L_2} C_{L_2}^s + \cdots \tag{12b}$$

assuming that the dissolution occurs in parallel at the various metal centers. C_L^s is the surface concentration of a ligand (L_1, L_2, etc.); C_H^s and C_L^s are the surface concentrations of protons (surface protonation, or concentrations of protons bound above pH_{PZC}) and of ligands, respectively; SL and SOH_2^+ are alternative shorthand notations; and j is an integer corresponding in ideal cases to the valency of the central ion.

Competitive occupation of the various metal centers

$$L_2 + SL_1 \longrightarrow SL_2 + L_1 \tag{13a}$$

$$OH^- + SL_2 \longrightarrow SOH + L_2 \tag{13b}$$

may be assumed to take place. Binding (absorption) of metal ions and of ligands affects surface protonation (6). It has been suggested (21) that cations and ligands occupy different types of surface sites, but more exacting data on this question are needed.

Reductive Dissolution. The reductive dissolution of an oxide such as Fe(III) (hydr)oxide can be accounted for by the following sequence involving the reductant R (24).

$$\equiv Fe^{III}OH + R \underset{}{\overset{fast}{\rightleftarrows}} \equiv Fe^{III}R + H_2O \quad (14a)$$

$$\equiv Fe^{III}R \underset{transfer}{\overset{electron}{\rightleftarrows}} \equiv Fe^{II}OH_2^+ + Ox^\cdot \quad (14b)$$

$$\equiv Fe^{II}OH_2^+ \xrightarrow{detachment} \text{new surface site} + Fe_{aq}^{2+} \quad (14c)$$

The foregoing equations suggest that either electron transfer (eq 14b) or detachment (eq 14c) is the rate-determining step. The oxidized reactant $Ox^\cdot$ is often a radical that may undergo further non-rate-determining reactions with the oxidant. Equations 14a and 14b may be coupled. The reaction sequence accounts for the observation (24–27) that the reaction rate, R, is proportional to the density of the surface concentration of the surface species, $\equiv Fe^{III}R$ (mol/m^2), provided that the concentration of the oxidized reactant $Ox^\cdot$ is at steady state or is negligible. The reaction rate is given by

$$R = \frac{d[Fe_{aq}^{2+}]}{A\,dt} = k\langle\equiv Fe^{III}R\rangle \quad (15)$$

where A is the surface area concentration in m^2/L, $[\equiv Fe^{III}R]$ is the surface concentration in mol/m^2, and k is the reaction rate coefficient per time.

Table I gives a survey of rate laws for various types of reactions. It illustrates that the rate laws for heterogeneous processes can be written in terms of the concentration (activity) of surface species.

Dissolution-Promoting and Dissolution-Inhibiting Ligands

If different reactants (ligands) compete for the available surface sites, the replacement of a dissolution-reactive ligand L_1 by a ligand that is less dissolution-reactive L_2 ($k_{L_1} \gg k_{L_2}$) diminishes the overall dissolution rate and constitutes an inhibition. The term inhibition is relative and depends on the reference conditions.

For a given mineral, for example an oxide, which factors influence k (eq 12)? General experience (20, 21, 32) has shown that ligands such as oxalate, salicylate, citrate, and ethylenediaminetetraacetate (EDTA) that form strong complexes with the Lewis acid metal centers of the hydrous oxide surface, such as Al(III) and Fe(III), enhance dissolution markedly. Figure 5 illustrates the effect of ligands on the dissolution reaction. These complex formers also are known to form complexes with these ions in solution. This reaction has

Table I. Surface Reactivity

Reaction Number	Dissolution Reactant	Surface Complex	Rate Law
1)	H^+	(M with OH_2^+, OH_2, OH^+, OH_2^+ groups)	$R \propto \langle \equiv MOH_2^+ \rangle^j$
2)	**Ligand** HO–/–O–	(M–OH, M–O with bridging ligand)	$R \propto \langle \equiv ML \rangle$
3)	HCO_3^-, CO_2	>Ca–OH, >CO_3H, Ca–HCO_3	$R \propto \langle \equiv CaHCO_3 \rangle$ at low pH: $R = \langle \equiv CO_3H \rangle$
4)	**Oxidation** O_2, Fe^{II}	$\equiv M-O$ $\rangle Fe^{II}$ $\equiv M-O$	$R \propto \langle (\equiv MO)_2 Fe^{II} \rangle p_{O_2}$
5)	**Reductive Dissolution of Fe^{III}(hydr)oxide** Ascorbate, HA — HO/–O	$\equiv Fe^{III}$–O/O	$R \propto \langle \equiv Fe^{III} HA \rangle$
6)	H_2S, HS^-	$\equiv Fe^{III}$–SH	$R \propto \langle \equiv Fe^{III}-SH \rangle$
7)	**Heterogeneous Nucleation** A^+, B^-	S–OB, S–B	$R \propto \theta_A \theta_B$

NOTES: Rate R depends on concentration of surface species. Reaction 1 is acid-promoted dissolution (21). Reaction 2 is a ligand-promoted dissolution. Dissolution rates are proportional to ligand surface complex (1, 21). In reaction 3, the dissolution of $CaCO_3$ in a given pH range is proportional to $\langle \equiv CaHCO_3 \rangle$ (9, 28). In reaction 4, the rate of oxidation of Fe(II) bonded to a hydrous oxide is proportional to the concentration of the adsorbed Fe(II) (29, 30). In reaction 5, the rate of reductive dissolution of Fe(III) (hydr)oxides and of Mn(III,IV) (hydr)oxides with organic reductants is proportional to the concentration of the adsorbed reductant (24, 31). In reaction 6, H_2S reduces Fe(III) (hydr)oxide in proportion to the concentration of the FeS and FeSH surface complex (27). In reaction 7, the rate of heterogeneous nucleation of the salt A^+B^- is proportional to $\langle SOA \rangle$ and $\langle SOB \rangle$ or $\theta_A \times \theta_B$.

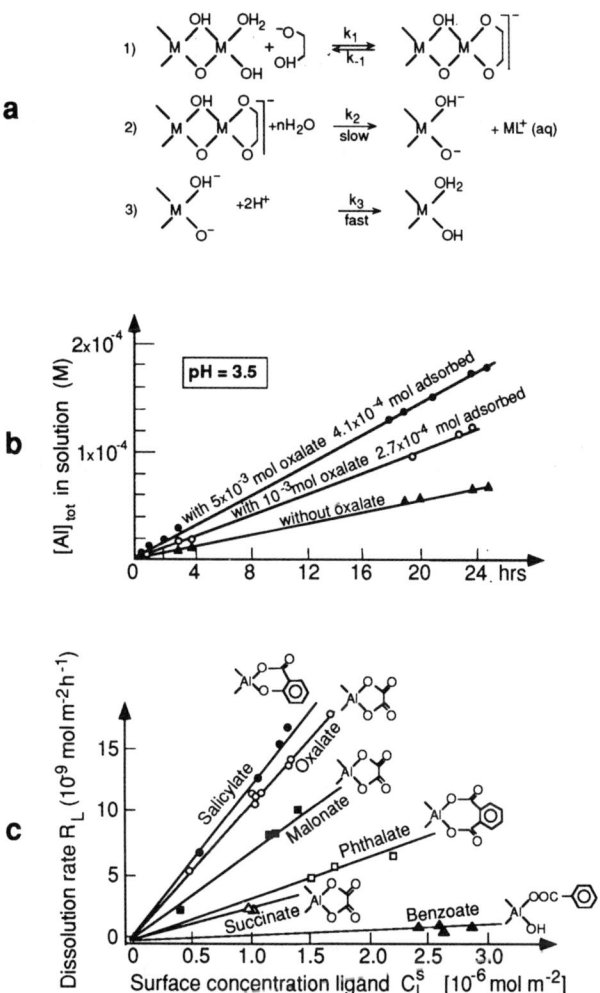

Figure 5. Promotion of the dissolution of an oxide by a ligand. The ligand illustrated here, in a shorthand notation, is a bidentate ligand with two oxygen donor atoms (such as in oxalate, salicylate, citrate, or diphenols). Part a: The ligand-catalyzed dissolution reaction of a M_2O_3 can be described by three elementary steps: A fast ligand adsorption step (equilibrium); a slow detachment process; and fast protonation subsequent to detachment, restoring the incipient surface configuration. Part b: The dissolution rate increases with increasing oxalate concentrations. Part c: In accord with the reaction scheme of (a) and of eq 11a, the rate of ligand-catalyzed dissolution of $\delta\text{-}Al_2O_3$ by the ligands, R_L (nmol/m^2 per hour) can be interpreted as linearly dependent on the surface concentrations of the complexes C_L^s. (Reproduced with permission from reference 21. Copyright 1986 Pergamon.)

no direct effect on the dissolution rate, however, because the dissolution is surface-controlled.

The enhancement of the dissolution rate by a ligand in a surface-controlled reaction implies that surface complex formation facilitates the release of ions from the surface to the adjacent solution. These ligands can bring electron density or negative charge into the coordination sphere of the surface central metal ions and thus lower their Lewis acidity. This charged species may labilize the critical metal–oxygen bonds and facilitate the detachment of the metal from the surface. Bidentate ligands (i.e., ligands with two donor atoms) such as dicarboxylates and hydroxycarboxylates can form relatively strong surface chelates (i.e., ring-type surface complexes).

trans Effect. The labilizing effect of a ligand on the bonds in the surface of the solid oxide phase of the central metal ions with oxygen or OH can also be interpreted in terms of the trans effect (i.e., the influence of the ligand on the strength of the bond that is trans to it). In our example it would be the effect of a ligand such as a dicarboxylate on the strength of the Al–oxygen bonds

The trans influence is a ground-state property. It is attributable to the fact that ligands trans to each other both participate in the orbital of the metal ion (in our example, Al). The more one ligand preempts this orbital, the weaker the bond to the other ligand will be.

Figure 6 gives the rate of the reductive dissolution of α-Fe_2O_3 (hematite) by H_2S. The reaction mechanism (27) implies that, in line with the scheme given in equations 14a–14c, surface complexes of $\equiv$FeS and of $\equiv$FeSH are formed and then undergo electron transfer. The dissolution rate, R (mol/m^2 per hour), is given by

$$R = k_e[\equiv\text{FeS}] + k'_e[\equiv\text{FeSH}] \qquad (16)$$

where [$\equiv$FeS] and [$\equiv$FeSH] are concentrations of the surface complexes in mol/m^2 and k_e and k'_e are the reaction rate constants.

Ligands such as phosphate, chromate, arsenate, selenate, selenite, sulfate, borate, and molybdate are less effective or inert in enhancing the dissolution rate. Although these anions are specifically adsorbed on oxide surfaces, their potential for donating electrons to the central ion must be less than that of organic bidentate oxygen-bearing ligands.

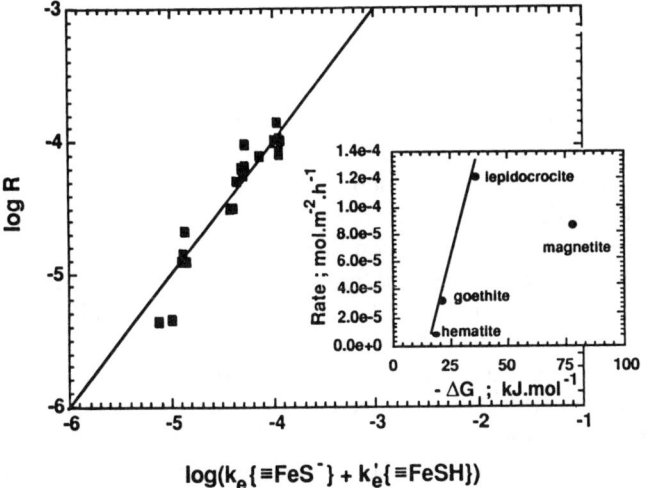

Figure 6. Experimental dissolution rate (mol/m² per hour) as a function of surface speciation (eq 17). Insert: dissolution rates (mol/m² per hour) for hematite, goethite, lepidocrocite, and magnetite as a function of the free energy (kJ/mol of electrons) of the reduction reactions

$$FeOOH + 3H^+ + e^- \rightleftharpoons Fe_{aq}^{2+} + 2H_2O$$

$$Fe_2O_3 + 6H^+ + 2e^- \rightleftharpoons 2Fe_{aq}^{2+} + 3H_2O$$

$$Fe_3O_4 + 8H^+ + 2e^- \rightleftharpoons 3Fe_{aq}^{2+} + 4H_2O$$

The rate was determined for pH_2S at 10^{-3} atm and pH 5.0. (Reproduced from reference 27. Copyright 1992 American Chemical Society.)

Their surface complexes provide less suitable leaving groups for detachment into water. As Figure 3 illustrates, inorganic oxoanions have complex-forming tendencies—as far as the first step in the complex formation to the monoligand complex (e.g., FeL) is concerned—very similar to those of bidentate organic oxygen-bearing ligands. Although these latter ligands tend to form multiligand complexes in solution (like $FeOx_3^{3-}$), oxoanions do not. They tend, at higher concentrations, to form dioxo- or polyoxoanions (e.g., $Cr_2O_7^{2-}$, $V_3O_9^{3-}$, $B_3O_6^{3-}$, $[(SiO_3)^{2-}]_n$, molybdate, and $P_2O_7^{4-}$). Correspondingly, these oxoanions are more likely to form binuclear or polynuclear surface complexes.

Hypothesis: Mononuclear Ligand Surface Complexes Enhance and Binuclear Surface Complexes Inhibit the Dissolution.

Binuclear surface complexes most likely are inert in promoting the dissolution reaction; much more energy is needed to detach two center metal ions from the surface lattice simultaneously. Because binuclear surface complexes occupy sites that otherwise might be occupied by dissolution-promoting (mono- or bidentate) mononuclear ligands, they act as relative inhibitors of dissolution.

Our present information on the effect of surface speciation on the reactivity of the surface (i.e., its tendency to dissolve) is summarized in Figure 4. Evidence for the formation of binuclear surface complexes is often circumstantial. Most researchers who modeled surface complex formation with oxyanions could fit the adsorption data only by assuming the formation of binuclear complexes, usually in addition to mononuclear ones.

Many of these oxoanions can form, depending on concentration and pH, various surface complexes. This ability may explain the different effects observed under different solution conditions. For example, Bondietti et al. (33) found that phosphate at low pH (where mononuclear complexes are probably formed) accelerated EDTA-promoted dissolution of lepidocrocite, whereas at near-neutral pH conditions (where binuclear complexes are presumably formed), phosphate was an efficient inhibitor. Furthermore, because of the several geometries involved, the extent of corner sharing or edge sharing by adsorbed oxoanions may differ with the type of oxide and with allotropic modifications of the same metal oxide.

We need spectroscopic evidence to suggest more explicitly under what conditions bi- or multinuclear surface complexes are formed. A few references providing spectroscopic and other evidence for the formation of bi- or multinuclear surface complexes are given in Table II.

Adsorption of Ligands and of Metal Ions: Change of Surface Protonation. Adsorbed mononuclear ligands enhance the dissolution rate directly (by bringing electron density into the coordination sphere of the surface central metal–oxygen bonds). These ligands facilitate the detachment of the central metal ion from the surface and increase the surface protonation (C_H^s; cf. eq 12). As illustrated by Figure 7, this latter effect results from the fact that specifically adsorbable anions increase the pH at the point of zero net proton charge, pH_{PZNPC}, but lower the pH of the point of zero charge,

Table II. Formation of Bi- or Multinuclear Fe(III) Surface Complexes with Oxyanions

Ligand	Adsorption Data Fitting	Spectroscopic Evidence
SO_4^{2-}	Goethite (6)	IR (34)
PO_4^{3-}	Goethite (6)	CIRFTIR (13)
SeO_3^{2-}	Goethite (35)	EXAFS (14)
AsO_3^{3-}		Lepidocrocite; EXAFS (36)
Benzoate		Goethite; CIRFTIR (13)
Cr^{3+}		Hydrous ferric oxide; EXAFS (18)
Co^{2+}		γ-Al_2O_3, TiO_2 (19)

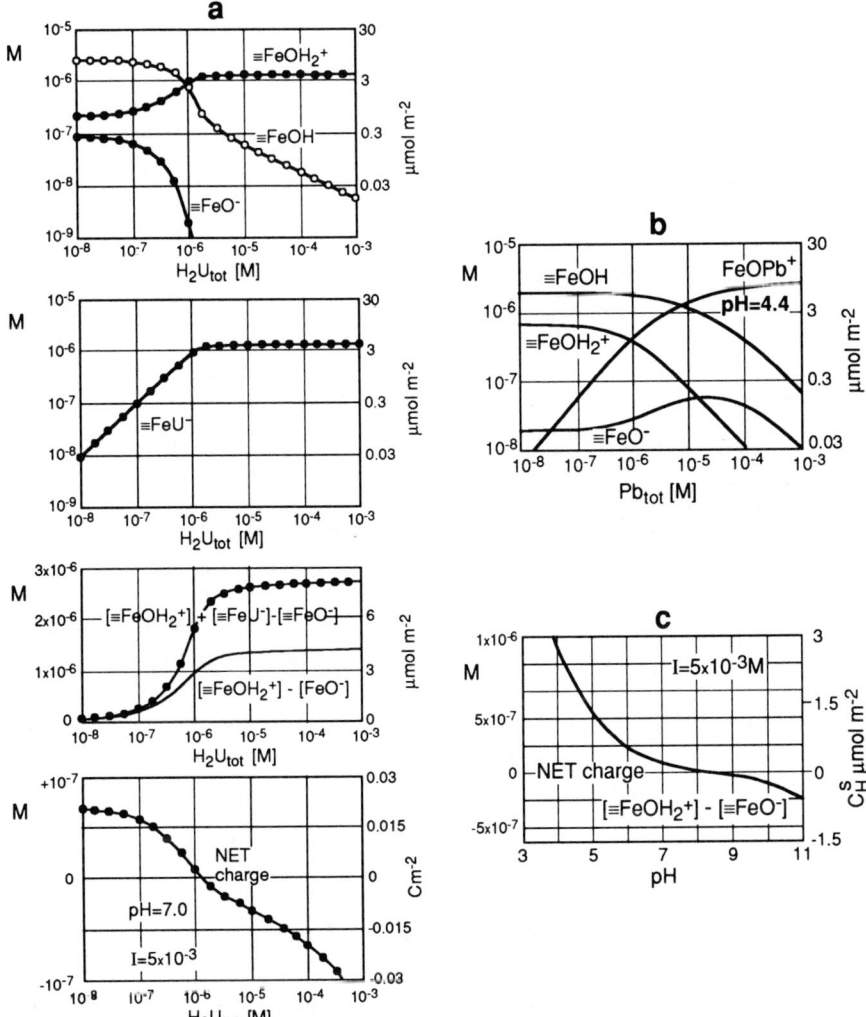

Figure 7. The effect of ligands and metal ions on surface protonation of a hydrous oxide is illustrated by two examples (1). Part a: Binding of a ligand (pH 7) to hematite, which increases surface protonation. Part b: Adsorption of Pb^{2+} to hematite (pH 4.4), which reduces surface protonation. Part c: Surface protonation of hematite alone as a function of pH (for comparison). All data were calculated with the following surface complex formation equilibria ($I = 5 \times 10^{-3}$ M). Electrostatic correction was made by diffuse double layer model.

$$\equiv Fe-OH + H_2U \longrightarrow \equiv FeU^- + H^+ + H_2O; \log K^s = 2$$
$$H_2U \longrightarrow H^+ + HU^-; \log K_1 = -5.0$$
$$HU^- \longrightarrow H^+ + U^{2-}; \log K_2 = -9.0$$
$$\equiv FeOH + Pb^{2+} \longrightarrow \equiv FeOPb^+ + H^+; \log K^s = 4.7$$

(Data courtesy of Stumm (1992).)

pH_{PZC} (this used to be called the isoelectric point). Correspondingly, adsorbed cations increase pH_{ZPC} but lower pH_{PZNPC}. The decrease in dissolution rate may be due to this reduction in C_H^s.

Case Examples. The effects of various oxoanions on EDTA-promoted dissolution of lepidocrocite (γ-FeOOH) have been studied by Bondietti et al. (33). EDTA was chosen as a reference system because it is dissolution-active over a relatively wide pH range. Phosphate, arsenate, and selenite markedly inhibit the dissolution at near-neutral pH values. At pH <5 phosphate, arsenate, and selenite accelerate the dissolution. It is presumed that the binuclear surface complexes formed at near-neutral pH values by these oxoanions (Table II) inhibit the dissolution. Figure 8a displays data on the effect of selenite on EDTA-promoted dissolution, and Figure 8b shows that calculations on surface speciation by Sposito et al. (35) support the preponderance of binuclear selenite surface complexes in the neutral-pH range. Mononuclear surface species prevail at lower pH values.

These oxoanions also inhibit reductive dissolution [e.g., the reductive dissolution of Fe(III) (hydr)oxides by H_2S]. The reaction rate (eq 15) is surface-controlled and is therefore retarded by solutes that compete with S(–II) for the surface sites to form surface complexes that are less dissolution-active. Figure 9 shows the effect of phosphate and sulfate in inhibiting dissolution by H_2S.

Figure 10 illustrates that Cr^{3+} effectively inhibits the proton-promoted dissolution of goethite. Cr(III) adsorbs even at low pH and, as bi- or polynuclear surface complexes, blocks surface sites from being protonated. Furthermore, isomorphically substituted Cr^{3+} ions, characterized by an extremely low water-exchange rate, impart inertness to the surface lattice bonds.

Some Applications in Nature and in Technology

Weathering and Natural Redox Cycling. Dissolution of minerals is significant in chemical weathering and in the cycling of iron and manganese. These processes control the global hydrogeochemical cycle of elements (38, 39). The roles of H^+ and of ligands in weathering (39–44) and in redox cycling (1, 44) have been extensively discussed. It is important to understand the factors that retard dissolution.

Naturally occurring oxoanions like SO_4^{2-} and $H_2PO_4^-$ at concentrations representative of those encountered in natural waters can inhibit dissolution and weathering reactions. A very low concentration of inhibitors can often be effective, because it may suffice to block the functional groups of solution-active sites (such as the kink sites). The effect of specifically adsorbable cations on the reduction of dissolution (weathering) rates of minerals is important. A case was documented by Grandstaff (32), who showed that thorium, Pb(II),

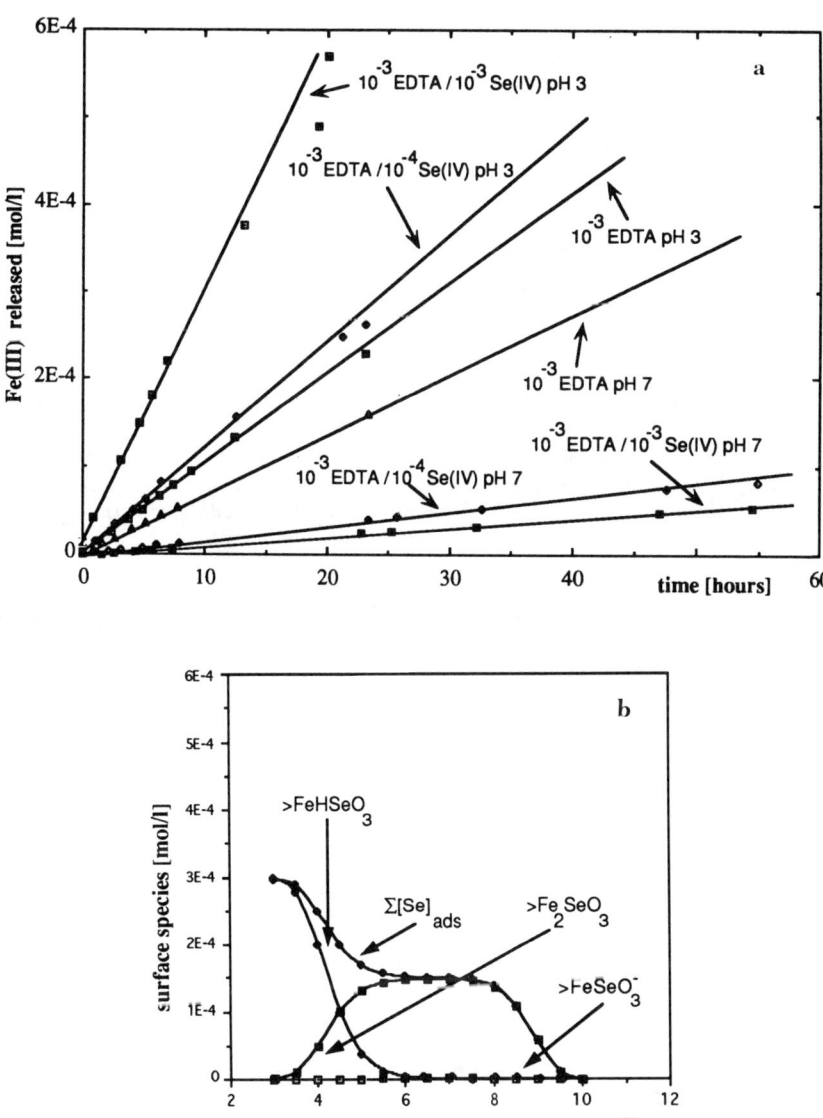

Figure 8. The effect of selenite on the EDTA-promoted dissolution of γ-FeOOH (0.5 g/L). Part a: At low pH the dissolution rate is increased by selenite; at pH 7 it is strongly inhibited. Concentration of the ligands is given in mol/L. Part b: Surface speciation on lepidocrocite as a function of pH according to Sposito et al. (35). These data suggest that binuclear selenite surface complexes are formed in the neutral pH range (from reference 33).

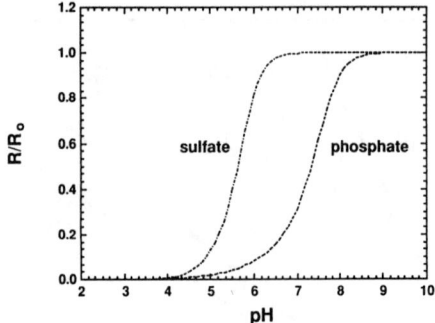

Figure 9. The relative dissolution rate, R/R_0, as a function of pH. Dashed lines were calculated by using the equilibrium and surface complex formation constants for pH_2S at 10^{-2} atm; -•-•- = $[SO_4^{2-}]$ = 10^{-3} M; and - - - - = $[H_2PO_4^-]$ = 10^{-4} M. R_0 is the dissolution rate observed in the absence of added SO_4^{2-} or $H_2PO_4^-$. Sulfate and phosphate, at these concentrations, are not specifically sorbed above pH 8.5 and 7, respectively. (Reproduced with permission from reference 37. Copyright 1994.)

and the rare earths associated with uraninite significantly retard the dissolution of UO_2.

Figure 11 shows the effect of some oxoanions and of complex formers on the reductive dissolution of goethite by H_2S (45). These data illustrate once more the competition among various dissolution-promoting and dissolution-

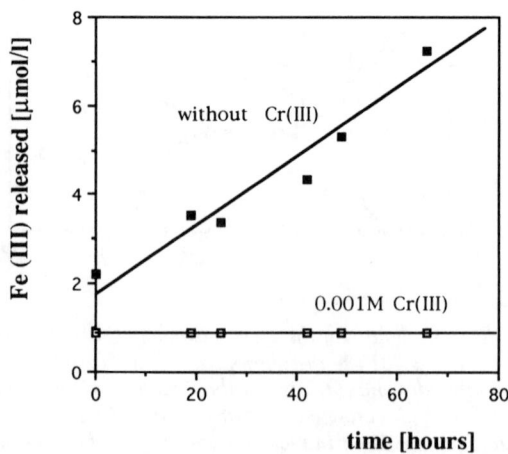

Figure 10. Effect of 10^{-3} M Cr(III) on the proton-promoted dissolution (pH 3) of α-FeOOH (0.5 g/L) in 0.1 M KNO_3. (Reproduced with permission from reference 33. Copyright 1993.)

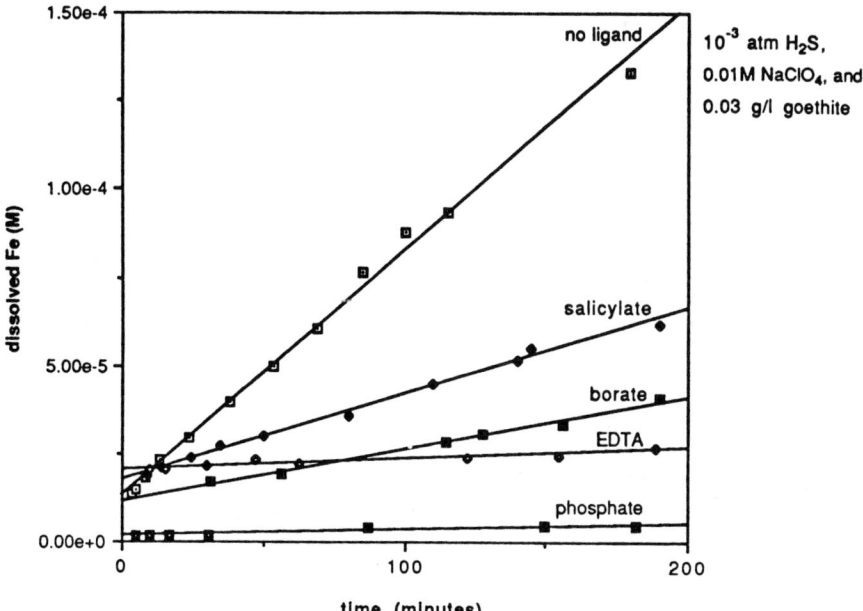

Figure 11. Effect of borate, phosphate, salicylate, and EDTA on the reductive dissolution of goethite by H_2S at pH 5, 10^{-3} atm H_2S, 0.01 M $NaClO_4$, and 0.03 g/L of goethite. (Reproduced from reference 45. Copyright 1994 American Chemical Society.)

inhibiting ligands. A strong naturally occurring complex former like oxalate [which promotes the nonreductive dissolution of Fe(III) (hydro)oxides] can relatively inhibit reductive dissolution by H_2S merely because the latter type of dissolution is much faster than nonreductive dissolution by oxalate alone. On the other hand, cases exist in which reductive dissolution is synergistically enhanced by complex formers (46).

Effect of Surface Complexes on Semiconductor-Mediated Photochemical Processes. In heterogeneous photoredox reactions, not only the solid phase (i.e., the semiconductor mineral) but also a surface species may act as the chromophore. Our discussion here is restricted to an exemplification of the role of inner-sphere ligand complexing on TiO_2. (For a review of the role of surface complexation in photochemistry and the cycling of iron in natural systems, see reference 44). As shown by Moser et al. (47), surface complexation of colloidal transparent TiO_2 (anatase) by salicylate shifts the UV–visible light absorption to longer wavelengths (Figure 12a); bright yellow is observed (48). This color indicates that the band in the visible range observed in the presence of salicylate or catechol corresponds to the charge-

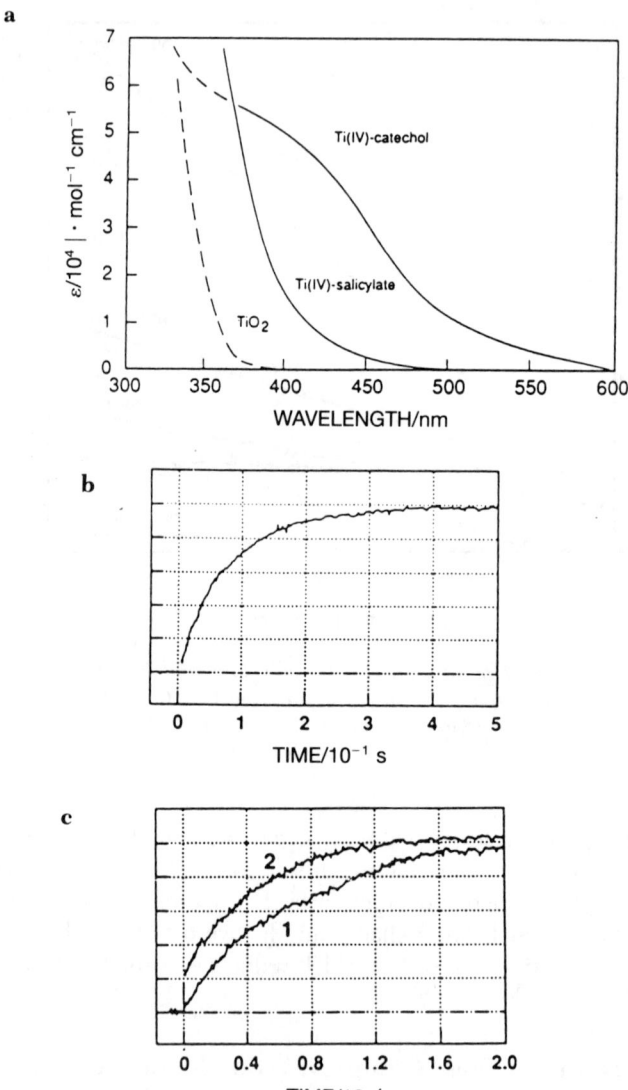

Figure 12. Effect of surface complexation on absorption spectra of TiO_2 transparent sols (0.5 g/L) and on the kinetics of electron transfer from the conduction band of TiO_2 to methyl viologen. Part a: Addition of salicylic acid and catechol (2×10^{-4} M) produces a red shift of the absorption onset to 500 and 600 nm, respectively. Part b: Oscillograms showing the temporal behavior of the 600-nm absorbance after laser excitation of water:methanol (90:10, v:v) degassed solutions containing colloidal TiO_2 (0.5 g/L), PVA (0.5 g/L), and 10^{-3} M MV^{2+} (bare TiO_2 particles) at pH 3.6. Part c: Same solution as in Figure 12b, but with 10^{-3} M isophthalic acid (1) and 10^{-3} M salicylic acid (2) added, respectively. (Reproduced from reference 47. Copyright 1991 American Chemical Society.)

transfer transition. Light promotes electron transfer from the surface complexant to the conduction band of TiO_2.

As shown by Moser et al. (47), surface complexation of colloidal TiO_2 accelerates electron transfer from the conduction band to methyl viologen. The enhancement of interfacial electron transfer is much more pronounced with the bidentate benzene derivates (Figures 12b and 12c) (1700 times faster with salicylate than in its absence). Similar results have been obtained (47) on the acceleration of electron transfer to oxygen by bidentate surface complexation.

In heterogeneous photochemistry, an increasing effort is going into the study of the photocatalytic degradation of organic pollutants with TiO_2 as a photocatalyst. In these oxidation processes either the organic compound is directly oxidized by the valence band holes (h_{vb}^+)

$$h_{vb}^+ + \text{organic compound} \longrightarrow \text{oxidized organic compound} \quad (17)$$

or the organic compound is oxidized by OH· radicals that are formed through the reaction of the functional OH groups of the TiO_2 with the photoholes.

$$h_{vb}^+ + {-}OH^-(\text{surf}) \longrightarrow OH^· \quad (18a)$$

$$OH^· + \text{organic compound} \longrightarrow \text{oxidized organic compound} \quad (18b)$$

As shown by Tunesi and Anderson (48), the efficiency of the photoredox process for organic compounds depends on their adsorption behavior. When direct charge transfer (inner-sphere complexes) occurs, this mechanism is more efficient than free radical attack. These authors interpret their results with salicylate at low pH as a direct electron transfer from the adsorbed organic molecule—assumed to be an orbital configuration of the chelate ring—to the semiconductor.

Corrosion and Passivity. The inhibition of dissolution is important in the corrosion of metals and building materials. Passivity is imparted to many metals by overlying oxides, the so-called passive films; the inhibition of the dissolution of these passive layers protects the underlying material. Figure 13 gives a schematic model of the hydrated passive film on iron.

Passive Films. The passive layer seems to vary in composition from Fe_3O_4 (magnetite) in oxygen-free solutions to $Fe_{2.67}O_4$ in the presence of oxygen. It may also be a duplex layer consisting of an inner layer of Fe_3O_4 and an outer layer of γ-Fe_2O_3. The coulometric reduction of the passive layer gives two waves, which are interpreted either by the reduction of two different layers, Fe_2O_3 and Fe_3O_4, or by successive reduction of Fe_3O_4 to lower valence oxides and its further reduction to metallic iron. Despite the disadvantage of

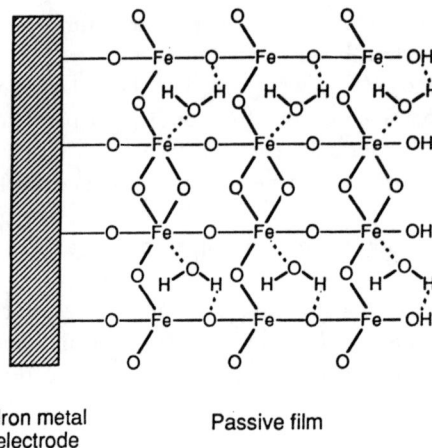

Figure 13. Schematic representation of the hydrated passive film on iron. (Reproduced with permission from reference 49. Copyright 1984.)

not knowing the exact composition and structure of the passive iron oxide film, some generalizations seem possible from a surface chemical point of view, because the hydrated passive film on iron displays the coordinative properties of the Fe(III) surface hydroxyl groups.

To what extent can our concepts of the coordination chemistry of the oxide–water interface and our knowledge of the factors that enhance and retard dissolution of Fe(III) oxides contribute toward an understanding of the properties of passive iron oxides? A review of the corrosion literature yields much phenomenological information that could be accounted for by surface-chemical theory. However, present passivity theories appear, with few exceptions, to be rather oblivious to the concepts of chemical surface reactivity. Thus, some perhaps speculative chemical ideas on the factors that enhance or reduce iron oxide passivity may be exposed to examination and discussion.

Protons and Ligands. The effects of acids on dissolution are obvious. For the passivation of iron, a critical current density, which increases with decreasing pH, must be exceeded. Acids that contain complex-forming ligands (HCl, oxalic acid, and H_3PO_4) enhance the dissolution. As has been shown (50), this activity can be accounted for by considering the superimposition of proton-promoted and ligand-promoted dissolution (eq 12). These proton and ligand effects are not fully independent, because ligand adsorption affects surface protonation (*see* Figure 7).

Fluoride and chloride are particularly nucleophilic and may even permeate the crystalline lattice. These ions have a pronounced effect on the structure and morphology of the oxide film, because they strongly influence nucleation, growth, and aging of polynuclear structures during hydrolysis of

Fe(III) (51); β-FeO(OH) polynuclear structures, stabilized by Cl⁻ ions, may be formed. The influence of organic anions on the crystallization of ferrihydrite has been studied by Cornell and collaborators (52, 53).

As already discussed, organic bidentates with oxygen donor groups that form mononuclear surface chelates (oxalate, salicylate, catechol, and citrate) are very dissolution-active. These surface ligand orbitals shift electron density toward the Fe(III) central ion and labilizes its binding to oxygen lattice atoms (trans effect). Oxalate, catechol, and salicylate can form mononuclear (bidentate) surface complexes (surface chelates), whereas steric effects require that the two carboxylate groups in iso- and terephthalate interact with different surface metal ions (47). The role of EDTA in influencing the dissolution of Fe(III) hydroxides is unclear. Although we observed (33) a pronounced enhancement of dissolution of lepidocrocite over a wide pH range, Rubio and Matijević (54) observed an inhibition of dissolution with β-FeOOH.

If these weak-field ligands are readily oxidizable (e.g., ascorbate), they initiate reductive dissolution (eq 14). Oxalate and EDTA in the presence of Fe(III) are thermodynamically metastable. Fe(II) complexed to these ions can reductively catalyze the dissolution of Fe(III) hydroxide (24).

Oxidants. Oxidizing substances added or present in the water restore high electrode potential and prevent the reductive dissolution of the passive oxide film. Oxidants that are specifically adsorbed, such as chromate and nitrite, are especially efficient. Both may form binuclear surface complexes.

Reductants. Reductants can break down passive oxide films because they favor reductive dissolution and lower the electrode potential. Naturally occurring organic substances, including fulvic and humic acids and phenols, can reduce Fe(III) (hydr)oxides. Especially detrimental are H_2S and S(-II) compounds. S(-II) has a strong affinity to Fe(III) (Figure 3) (55).

Inorganic Oxoanions. Oxoanions like phosphate, silicates, borate, molybdate, vanadate, arsenate, wolframate, and pertechnate are efficient inhibitors. These oxoanions are all specifically adsorbed, but in contrast to many organic bidentates, they tend to form binuclear or multinuclear surface complexes. By blocking surface functional sites, they prevent access of H^+ and dissolution-active ligands and mitigate surface reactivity. At higher concentrations these oxoanions may form surface clusters. Figure 11 illustrates that borate and phosphate on the surface of goethite prevent its reductive dissolution by H_2S (pH 5).

Chromium. So-called stainless steels contain in their alloys elements (such as Cr, Ni, Co, and Mo) that eventually become incorporated into the passive oxide films. Cr(III), as shown by EXAFS studies with hydrous ferric oxide and goethite (18), forms inner-sphere surface complexes in the form of multinuclear clusters. When Cr(III) is coprecipitated with Fe(III), isomorphic substitution of Cr(III) for Fe(III) occurs. Cr(III), characterized by very strong binding to oxygen atoms (water-exchange rate $k_{-w} = 10^{-5.6}/s$), decreases the reactivity of the oxide structure. Similar effects may be expected for Co and Ni (k_{-w} for Co(III) = $10^{-1}/s$; Co(II) = $10^{-5.3}/s$; and Ni(II) = $10^{-4.2}/s$). Mo plausibly adds polymolybdate interligating with iron corners to the oxide film.

Cations. Cations like Zn^{2+} form binuclear complexes and reduce surface protonation. Such ions can also form ternary surface complexes with inhibitors.

Acknowledgments

Two Ph.D. theses in my laboratory were instrumental in initiating some of the research described here: C. P. Huang (1970, now at the University of Delaware) quantified surface complex formation with H^+ and cations (by using the diffuse double-layer model for electrostatic correction). Laura Sigg (1979, EAWAG, Swiss Federal Institute of Technology) established much of the underpinning groundwork in the theory of ligand-exchange equilibria. I am also indebted to many other doctoral students and colleagues. Above all, I would like to acknowledge the great influence of James J. Morgan (California Institute of Technology, Pasadena) on concepts of aquatic chemical processes; Charles R. O'Melia (Johns Hopkins University, Baltimore) on the theory of particles and colloids in natural systems; Paul W. Schindler (University of Berne) on the ideas of surface coordination; and Garrison Sposito (University of California at Berkeley) on the surface chemistry of soils. Finally, it is a pleasure to acknowledge the Swiss National Science Foundation, which has supported our research on surface coordination chemistry for the past 20 years.

References

1. Stumm, W. *Chemistry of the Solid–Water Interface;* Wiley Interscience: New York, 1992.
2. Sposito, G. *The Surface Chemistry of Soils;* Oxford University: New York, 1984.
3. Sposito, G. *The Chemistry of Soils;* Oxford University: New York, 1989.
4. Schindler, P. W.; Stumm, W. In *Aquatic Surface Chemistry;* Stumm, W., Ed.; Wiley Interscience: New York, 1987; pp 83–110.
5. Dzombak, D. A.; Morel, F. M. M. *Surface Complexation Modeling; Hydrous Ferric Oxide;* Wiley Interscience: New York, 1990.
6. Sigg, L.; Stumm, W. *Colloids Surf.* **1981**, *2*, 101–117.
7. Smith, R. M.; Martell, A. E. *Critical Stability Constants;* Plenum: New York, 1976.
8. Kummert, R.; Stumm, W. *Colloid Interface Sci.* **1980**, *75*, 373–385.
9. Van Cappellen, P.; Charlet, L.; Stumm, W.; Wersin, P. *Geochim. Cosmochim. Acta* **1993**, *57*, 3505.

10. Schott, J.; Berner, R. A. *The Chemistry of Weathering;* NATO ASI Series C; Reidel: Dordrecht, The Netherlands, 1985; Vol. 149, pp 35–53.
11. Torrents, A.; Stone, A. T. *Environ. Sci. Technol.* **1991**, *25*, 143–149.
12. Motschi, H. In *Aquatic Surface Chemistry;* Stumm, W., Ed.; Wiley Interscience: New York, 1987; pp 111–124.
13. Tejedor-Tejedor, I.; Yost, E.; Anderson, M. *Langmuir* **1990**, *6*, 979–987.
14. Hayes, K. F.; Roe, A. L.; Brown, G. E., Jr.; Hodgson, K. O.; Leckie, J. O.; Parks, G. A. *Science (Washington, D.C.)* **1987**, *238*, 783–786.
15. Brown, G. E., Jr.; Parks, G. A.; Chisholm-Brause, C .J. *Chimia* **1989**, *43*, 248–256.
16. Brown, G. E., Jr. In *Mineral-Water Interface Geochemistry;* Hochella, M. F., Jr.; White, A. F., Eds.; American Mineralogical Society: Washington, DC, 1990; pp 309–363.
17. Manceau, A.; Charlet, L. J. *Colloid Interface Sci.* **1992**, *148*, 425.
18. Charlet, L.; Manceau, A. J. *Colloid Interface Sci.* **1992**, *148*, 443.
19. Chisholm-Brause, C. J.; Brown, G. E., Jr.; Parks, G. A. *Physica B* **1989**, *158*, 646–648.
20. Wieland, E.; Wehrli, B.; Stumm, W. *Geochim. Cosmochim. Acta* **1988**, *52*, 1969–1981.
21. Furrer, G.; Stumm, W. *Geochim. Cosmochim. Acta* **1986**, *50*, 1847–1860.
22. Blum, A. E.; Lasaga, A. C. In *Aquatic Surface Chemistry;* Stumm, W., Ed.; Wiley Interscience: New York, 1987; pp 255–292.
23. Stumm, W.; Wollast, R. *Rev. Geophys.* **1990**, *28/1*, 53–69.
24. Suter, D.; Banwart, S.; Stumm, W. *Langmuir* **1991**, *7*, 809–813.
25. Sulzberger, B. *Chemistry of the Solid–Water Interface;* Wiley Interscience: New York, 1992; Chapter 10.
26. Lakind, J. S.; Stone, A. T. *Geochim. Cosmochim. Acta* **1989**, *53*, 961-971.
27. Dos Santos Afonso, M.; Stumm, W. *Langmuir* **1992**, *8*, 1671.
28. Plummer, L. N.; Wigley, T. M. L.; Parkhurst, D. L. *Am. J. Sci.* **1978**, *278*, 179–216.
29. Wehrli, B.; Stumm, W. *Geochim. Cosmochim. Acta* **1989**, *53*, 69–77.
30. Wehrli, B. In *Aquatic Chemical Kinetics;* Stumm, W., Ed.; Wiley Interscience, New York, 1990; pp 311–336.
31. Stone, A. T. *Geochim. Cosmochim. Acta* **1987**, *51*, 919–925.
32. Grandstaff, D. E. In *Rates of Chemical Weathering of Rocks and Minerals;* Colman, S. M.; Dethier, D. P., Eds.; Academic: Orlando, FL, 1986; pp 41–59.
33. Bondietti, G.; Sinniger, J.; Stumm, W. *Colloids Surf.* **1993**, *79*, 157.
34. Parfitt, R.; Smart, R. J. *Chem. Soc. Faraday Trans.* **1977**, *173*, 796.
35. Sposito, G.; DeWitt, J.; Neal, R. *Soil Sci. Soc. Am. J.* **1988**, *52*, 947.
36. Waychunas, G.; Rea, B.; Fuller, C.; Davis, J. In *X-Ray Absorption Fine-Structure VI;* Hasnain, S. S., Ed.; Ellis Harwood: Chichester, England, 1990.
37. Biber, M. V.; Dos Santos Afonso, M.; Stumm, W. *Geochim. Cosmochim. Acta* **1994**, *58*, 1994.
38. Berner, R. A.; Lasaga, A. C. *Sci. Am.* **1989**, *260*, 74–81.
39. Drever, J. I. *Geochim. Cosmochim. Acta* **1994**, *58*, 2325.
40. Chou, L.; Wollast, R. *Am. J. Sci.* **1985**, *285*, 963–993.
41. Brady, P. V.; Walther, J. V. *Geochim. Cosmochim. Acta* **1989**, *53*, 2823–2830.
42. Stumm, W.; Wollast, R. *Rev. Geophys.* **1990**, *28/1*, 53–69.
43. Wollast, R. In *Aquatic Chemical Kinetics: Reaction Rates of Processes in Natural Waters;* Stumm, W., Ed.; Wiley & Sons: New York, 1990; pp 431–445.
44. Stumm, W.; Sulzberger, B. *Geochim. Cosmochim. Acta* **1992**, *56*, 3233.
45. Biber, M.; Stumm, W. *Environ. Sci. Technol.* **1994**, *28*, 763.
46. Blesa, M. A.; Marinovich, H. A.; Baumgartner, E. C.; Maroto, A. J. G. *Inorg. Chem.* **1987**, *26*, 3713–3717.

47. Moser, J.; Punchiheva, S.; Infelta, P.; Grtzel, M. *Langmuir* **1991,** *7,* 3012.
48. Tunesi, I.; Anderson, M. *Langmuir* **1992,** *8,* 487.
49. Pou, T. E.; Murphy, P. J.; Young, V.; Bockris, J. O. M. *J. Electrochem. Soc.* **1984,** *131,* 1243.
50. Grauer, R.; Stumm, W. *Colloid. Polymer. Sci.* **1982,** *260,* 959-970.
51. Schneider, W.; Schwyn, B. In *Aquatic Surface Chemistry;* Stumm, W., Ed.; Wiley and Sons: New York, 1987; pp 167–196.
52. Cornell, R. M.; Schwertmann, U. *Clay Miner.* **1979,** *27,* 402–410.
53. Cornell, R. M.; Giovanoli, R.; Schneider, W. *J. Chem. Technol. Biol.* **1989,** *46,* 115–134.
54. Rubio J.; Matijevic, E. *J. Colloid Interface Sci.* **1979,** *68,* 408.
55. Luther, G. W., III *Geochim. Cosmochim. Acta* **1987,** *51,* 3193–3199.

RECEIVED for review October 23, 1992. ACCEPTED revised manuscript May 24, 1993.

Adsorption as a Problem in Coordination Chemistry

The Concept of the Surface Complex

Garrison Sposito

Department of Environmental Science, Policy and Management, University of California, Berkeley, CA 94720–3110

> *The interfacial aqueous coordination chemistry of natural particles, in particular their surface complexation reactions, owes much of its development to the research of Werner Stumm. Beginning with the tentative interpretation of specific adsorption processes in terms of chemical reactions to form inner-sphere surface complexes, his seminal questions spawned a generation of research on the detection and quantitation of these surface species. The application of noninvasive spectroscopy in this research is exemplified by electron spin resonance and extended X-ray absorption fine structure studies. These studies, in turn, indicate the existence of a rich variety of surface species that transcend the isolated surface complex in both structure and reactivity, thereby stimulating future research in molecular conceptualizations of the particle–water interface.*

What in water did Bloom, waterlover, drawer of water, watercarrier ... admire? Its universality: its democratic equality and constancy to its nature in seeking its own level: its vastness in the ocean of Mercator's projection: its ... capacity to dissolve and hold in solution all soluble substances including millions of tons of the most precious metals: its slow erosions of peninsulas and islands, its persistent formation of homothetic islands, peninsulas and downward-tending promontories: its alluvial deposits: its weight and volume and density: its imperturbability in lagoons, atolls, highland tarns: its ... properties for cleansing, quenching thirst and fire, nourishing vegetation: its infallibility as paradigm and paragon: its metamorphoses as vapour, mist, cloud, rain, sleet, snow, hail: its ... variety of forms in loughs and bays and gulfs and bights and guts and lagoons and atolls and archipelagos and sounds and fjords and minches and tidal estuaries and arms of sea .... (*1*)

How often have the fine sons of Ireland been drawn to Zürich! The quotation from Joyce's *Ulysses* (*1*) that opens this chapter serves as an implicit reminder of this phenomenon and as a summary of the broad range of aqueous systems whose aesthetic qualities, geologic setting, and chemical behavior have attracted the interest of Werner Stumm during more than four decades of his scientific career. This chapter will not review the many successes of those four decades in all their details and ramifications; that task can be attempted only through the entire contents of this volume. Instead, my focus will be on aquatic surface chemistry, the subdiscipline that treats reactions at interfaces between natural colloids and the waters that bathe them. But (thanks in no small measure to the prolific research of Professor Stumm himself) even this subdisciplinary focus is too broad to cover in a single chapter.

It will not suffice even to focus on certain broad classes of surface reactions in natural waters or on selected aspects of those reactions (e.g., descriptions of their equilibria or of their kinetics), because whole books have been devoted to such topics. Therefore, this chapter in honor of Werner Stumm is about a single concept with which he has been identified closely over the past 20 years: the concept of the surface complex.

A surface complex can be defined simply as the stable molecular unit formed out of the reaction between a chemical species in aqueous solution and a functional group exposed at the surface of a solid (*2*). Perhaps the simplest example of surface complexation—and that most relevant to the work of Professor Stumm—is the binding of an ionized surface hydroxyl group to a metal cation. This molecular entity plays a central role in a wide variety of natural processes in water, ranging in scope from purification to biogeochemical cycling and in spatial scale from local to global (*3*).

The development of the surface complex in aquatic chemistry has been influenced markedly by the contributions of Werner Stumm. Fortunately, his many technical papers do not have to be perused to trace the evolution of his thinking about surface complexes. He has preserved much of his seminal thought in a small set of invited conference papers, published at approximately equal intervals over a 20-year period. These heuristic articles, whose provenance grows more precious by the day, can serve to introduce the concept of the surface complex, its picturesque and mathematical descriptions, and its experimental detection.

The Articles in Croatica Chemica Acta

The conceptual development of the surface complexation mechanism for adsorption processes, as fostered by Werner Stumm, is recorded in a remarkable series of six articles published between 1970 and 1990 in *Croatica Chemica Acta* (4–9). Each of these articles, based on lectures presented at summer conferences sponsored by the Ruder Bošković Institute, reflects the ever-broadening and deepening perspective of Stumm's ideas to match the growing success of his own *Weltanschauung* concerning surface chemical reactions (Table I). This series of position papers should be read and re-read by all students of aquatic surface chemistry, not—to borrow the pert phrase of Clifford Truesdell (10)—"to decorate a paper of their own by an early reference nor to write a history, but in search of understanding and method, revealed by the speech of giants untranslated by pygmies."

Cationic Surface Complexes. The first paper of the series (4), in a section entitled "Preliminary Approach to the Interfacial Coordination Chemistry of Hydrous Oxides," addressed the mechanism of cation adsorption by hydroxylated mineral surfaces. Here the term *coordination* was reserved for complex formation between cations and surface functional groups through bonding that "can be either electrostatic or covalent, or a mixture of both." The dichotomous qualification is much in the spirit of the classic Stern (11) picture of strong adsorption.

Chemical reactions were sketched for Brønsted acid phenomena and for metal adsorption on oxides, but the chief concern was how mass law considerations could be applied to functional groups attached to a solid surface. This concern was addressed by appeal to the well-known reactions of polyelectrolytes in aqueous solutions. For example, in the simple case of deprotonation reactions, the conditional dissociation constant (K_c)

Table I. The *Croatica Chemica Acta* Series of Articles

Title	Year
"Specific Chemical Interaction Affecting the Stability of Dispersed Systems"[a]	1970
"Interaction of Metal Ions with Hydrous Oxide Surfaces"	1976
"A Ligand Exchange Model for the Adsorption of Inorganic and Organic Ligands at Hydrous Oxide Surfaces"	1980
"The Role of Surface Coordination in Precipitation and Dissolution of Mineral Phases"	1983
"Surface Complexation and Its Impact on Geochemical Kinetics"	1987
"The Coordination Chemistry of the Oxide–Electrolyte Interface; The Dependence of Surface Reactivity (Dissolution, Redox Reactions) on Surface Structure"	1990

[a] Science Citation Index Classic, 1990.

$$K_c = a_H \frac{\alpha}{(1-\alpha)} \tag{1}$$

is related to a corresponding thermodynamic equilibrium constant (K_{int}) by the expression:

$$K_c = K_{int} \exp\left(\frac{F\psi_s}{RT}\right) \tag{2}$$

where a_H is aqueous proton activity, α is the mole fraction of deprotonated functional groups, ψ_s is the electrostatic potential difference between the solid surface and the bulk aqueous solution, F is the Faraday constant, R is the molar gas constant, and T is absolute temperature.

Equation 2 portrays a *factorization* of K_c into a part (K_{int}) representing the Gibbs energy change for dissociation of a proton complex and a part (the Boltzmann factor in ψ_s) representing an electrostatic constraint on proton dissociation arising from the close proximity of ionized surface functional groups. The Boltzmann factor reflects the increasing difficulty of deprotonation as α increases (and as ψ_s ostensibly becomes more negative). Thus, the equilibrium constant, K_{int}, equals the conditional dissociation constant extrapolated to the conditions for an uncharged surface ($\alpha = 0$ and, evidently, $\psi_s = 0$). Stumm et al. (*4*) went on to suggest how the potential ψ_s could be modeled approximately and how K_{int} could be estimated by graphical extrapolation of experimental proton titration data. The surface complexation of metals was discussed only in the context of its influence on these kinds of data and with respect to the similarity of oxide behavior to that of polymeric adsorbents (polyelectrolytes and synthetic cation exchangers).

Chemical Modeling of Metal Adsorption. The second paper of the series (*5*) addressed the still-unresolved issues of the precise meaning of surface complex and the determination of the potential ψ_s. Chemical reactions and conditional equilibrium constants were written out explicitly to describe the average acid–base behavior of a hydroxylated surface, that is,

$$SOH_2^+(s) \longrightarrow SOH(s) + H^+(aq) \tag{3a}$$

$$SOH(s) \longrightarrow SO^-(s) + H^+(aq) \tag{4a}$$

and

$$K_{1c} = \frac{x_{SOH} a_H}{x_{SOH_2^+}} \tag{3b}$$

$$K_{2c} = \frac{x_{SO^-} a_H}{x_{SOH}} \tag{4b}$$

where SOH(s) represents 1 mol of acidic surface hydroxyl groups and x is a mole fraction. [The notation here differs somewhat from that found in Stumm et al. (5).]

Equation 2 was invoked again to relate K_{1c} and K_{2c} to the corresponding thermodynamic equilibrium constants, but this time it was stated flatly that "there is no direct way to obtain ψ_s theoretically or experimentally" (5). The well-known technique of estimating the thermodynamic equilibrium constants for the reactions in equations 3a and 4a from a linear extrapolation (to the ordinate axis) of graphs of log K_{1c} (and of log K_{2c}) against $x_{SOH_2^+}$ (and x_{SO^-}) was illustrated in detail [see also Schindler and Gamsjäger (12)]. The authors concluded their example with the statement that "this linear extrapolation is justified because in the presence of an inert electrolyte (ionic strength $I = 0.1$), the charge, at low charge densities, is nearly proportional to the potential between the surface and the solution (approximately constant capacitance)." Thus, the inspiration was provided for Westall and Hohl (13) to dub this theoretical approach the "constant capacitance model", a name adopted subsequently in the rest of the *Croatica Chemica Acta* series.

Chemical reactions for the adsorption of metal cations (M^{m+}) by hydroxylated surfaces also were proposed by Stumm et al. (5) in the second paper:

$$SOH(s) + M^{m+}(aq) \longrightarrow SOM^{(m-1)+}(s) + H^+(aq) \tag{5a}$$

$$2SOH(s) + M^{m+}(aq) \longrightarrow (SO)_2 M^{(m-2)+}(s) + 2H^+(aq) \tag{6a}$$

with the corresponding conditional equilibrium constants

$$K_{1c}^* = \frac{x_{SOM} a_H}{x_{SOH} a_M} \tag{5b}$$

$$K_{2c}^* = \frac{x_{(SO)_2M} a_H^2}{x_{SOH}^2 a_M} \tag{6b}$$

where a_M is an aqueous free metal cation activity. Methodologies were outlined for estimating the thermodynamic equilibrium constants that represent the reactions in equations 5a and 6a [see also Schindler et al. (14) and Hohl and Stumm (15)].

Stumm et al. (5) ended their paper with a variety of remarks that, taken as a whole, implied that the adsorbed metal products in equations 5a and 6a are inner-sphere surface complexes. Their suggestion reflected observations of specific metal cation adsorption and comparisons of aqueous metal complexes with the corresponding surface complexes. They cautioned, however, that this kind of interpretation could not be made unequivocally without direct molecular evidence for inner-sphere complex formation.

Anion Adsorption. The third paper of the series (6) proposed chemical reactions analogous to equations 5a and 6a for the adsorption of an anion (L^{l-}) by a hydroxylated surface:

$$SOH(s) + L^{l-}(aq) \longrightarrow SL^{(l-1)-}(s) + OH^-(aq) \tag{7}$$

$$2SOH(s) + L^{l-}(aq) \longrightarrow S_2L^{(l-2)-}(s) + 2OH^-(aq) \tag{8}$$

Conditional equilibrium constants analogous to those in equations 5b and 6b were also defined, with a_{OH} instead of a_H, and a_L instead of a_M. Graphical methods of estimating reaction stoichiometry and thermodynamic equilibrium constants were illustrated similarly to what was shown for metal cation adsorption [see also Kummert and Stumm (16) and Sigg and Stumm (17)].

The significant conceptual advance in reference 6 was the identification of the ligand-exchange mechanism (i.e., exchange of OH for L) with inner-sphere surface complex formation. The positive correlation between the equilibrium constants for aqueous inner-sphere complexes of Fe(III) or Al(III) and those for surface complexes on α-FeOOH or γ-Al$_2$O$_3$ involving a variety of inorganic and organic ligands led Stumm et al. (6) to remark, "Since the solute complexes ... are 'inner sphere' complexes, we may infer that the surface complexes formed are of the inner sphere type." Elsewhere in the paper, the generic concept of specific adsorption was used synonymously with ligand exchange and, therefore, with inner-sphere surface complexation.

Statistical Thermodynamics of the Constant Capacitance Model

Although the first three articles in the *Croatica Chemica Acta* series provided a conceptual framework for the constant capacitance model, a number of theoretical loose ends remained. One was the vexing problem of how best to estimate model parameters from experimental data. This surprisingly complicated issue was reviewed critically by Westall and Hohl (13) and, more recently, by Hayes et al. (18) and Goldberg (19), so it will not be discussed in this review.

Another point in need of clarification is the suggested linear relationship between the electrostatic potential ψ_s and the surface charge density (13),

$$\sigma_p = C\psi_s \qquad (9)$$

where σ_p is the total net particle surface charge density and C is a differential capacitance density. Equation 9 is needed for closure in the model relation (eq 2) between conditional and thermodynamic equilibrium constants, because ψ_s cannot be measured. A fundamental molecular basis for equation 9 can be derived through a statistical thermodynamic analysis of adsorption that results in surface complex formation (20–22).

Molecular Derivation of the Model. The molecular theory of surface complexes is a special case of a multispecies lattice model. The surface complexes thus are interpreted as molecular species immobilized (relative to the 10-ps time scale for diffusive motion of an ion in aqueous solution) on an array of M sites that represent surface functional groups. If there are two different surface species on the sites (e.g., a protonated and an unprotonated surface hydroxyl group) and if each site has z nearest neighbors, then any distribution of the two species (call them A and B) over the sites must satisfy the general conditions (23)

$$zN_A = 2N_{AA} + N_{AB} \qquad (10a)$$

$$zN_B = 2N_{BB} + N_{AB} \qquad (10b)$$

$$N_A + N_B = M \qquad (10c)$$

where N_A is the total number of A species, N_{AA} is the number of nearest-neighbor A pairs (with analogous definitions for N_B and N_{BB}), and N_{AB} is the number of nearest-neighbor AB pairs. It is assumed that each species binds to just one site. If the array of sites is not regular, z can be interpreted as the average number of nearest neighbors of a site. Generalization to include multisite binding or binding of more than two species is straightforward, if tedious (20).

Sposito (20, 22) derived equations that lead to the calculation of the chemical potentials of surface species immobilized on a lattice of sites. The results can be expressed in terms of the relationship between K_c and K_{int} for the prototypical chemical reaction

$$A(aq) + B(s) \longrightarrow A(s) + B(aq) \qquad (11)$$

in which species A replaces species B on the surface. For example, if $A = H^+$ and $B = M^{m+}$, then equation 5a is a special case of equation 11; the same is true for equation 7 if $A = L^{l-}$ and $B = OH^-$. The statistical thermodynamic relation between $\ln K_c$ and $\ln K_{int}$ for this reaction is [cf. Sposito (22)]

$$\ln K_c = \ln K_{int} + \left(\frac{\partial \ln q_{AB}}{\partial N_a}\right)_M \quad (12)$$

where

$$K_c = \frac{x_A a_B}{x_B a_A} \quad (13)$$

and

$$q_{AB} \equiv \frac{N_A! N_B!}{M!} \sum_{N_{AB}} g(N_A, M, N_{AB}) \exp\left(\frac{N_{AB}\epsilon}{2RT}\right) \quad (14)$$

The function $g(N_A, M, N_{AB})$ is the number of ways that N_A species can be distributed on M sites to provide N_{AB} nearest-neighbor AB pairs. It is subject to the combinatorial constraint (23)

$$\sum_{N_{AB}} g(N_A, M, N_{AB}) = \frac{M!}{N_A! N_B!} \quad (15)$$

where the right side is the total number of ways that N_A indistinguishable species can be distributed on M sites, regardless of the value of N_{AB}. The sum over N_{AB} in equations 14 and 15 includes all possible values of this variable for the M sites. The function ϵ in equation 14 is the energy required to destroy two AB pairs and thereby create AA-and-BB pairs; or conversely, $-\epsilon$ is the energy required to create two AB pairs from an AA-and-BB set of pairs. For this reason, it appears in the Boltzmann factor that weights each value of the purely combinatorial factor g in equation 14.

The evaluation of q_{AB} is a formidable task if done exactly, but the constant capacitance model (as well as other surface complexation models) is the special case that results from equation 12 by a simple approximation (22):

$$q_{AB} \approx \left[\exp\left(\frac{z\epsilon}{2RT}\right)\right]^{N_A N_B / M} \quad (16)$$

This approximation has been deduced from equation 14 in three distinct ways in the literature of statistical thermodynamics:

1. replacing $\overline{N}_{AB}$ directly in equation 14 with an average value for the ideal system, ($MB\overline{N}_{AB}^{id}$, computed for a random distribution of the species A and B over the M sites (22):

$$N_{AB} \longrightarrow \overline{N}_{AB^{id}} \equiv \frac{zN_A N_B}{M} \quad (17a)$$

2. expanding the Boltzmann factor in equation 14 to first order in a MacLaurin series in $\epsilon/2RT$ (24):

$$q_{AB} \approx \frac{N_A! N_B!}{M!} \sum_{N_{AB}} g(N_A, M, N_{AB}) \left(1 + \frac{N_{AB}\epsilon}{2RT}\right)$$

$$= 1 + \left(\frac{\overline{N}_{AB}^{id}\epsilon}{2RT}\right) \approx \exp\left(\frac{\overline{N}_{AB}^{id}\epsilon}{2RT}\right) \quad (17b)$$

3. expanding q_{AB} to first order in MacLaurin series in both z^{-1} and ϵ (20, 25).

After the substitution of equation 16 into equation 12, the expression:

$$\ln K_c = \ln K_{int} - \frac{z\epsilon x_A}{RT} \quad (18)$$

is obtained (22), where $x_A = N_A/M$. Equation 18 indicates that a graph of $\ln K_c$ against the mole fraction x_A is a straight line with y intercept equal to $\ln K_{int}$, in agreement with the prescription of Stumm et al. (4, 5). Taking as an example the identifications $A = H^+$ and $B = SOH(s)$ (species B exists only in the solid phase in this case) one would plot $\ln K_c (= -\ln K_{1c}$ in eq 3b) against x_{SOH} to obtain an estimate of $\ln K_{int}$ by extrapolation. Moreover, a comparison of equations 2, 9, and 18 shows that, in this example, the capacitance density parameter C is to be identified through the relationship (22)

$$F\, SOH_T = Cz\epsilon \quad (19)$$

where SOH_T is M times the Avogadro constant divided by the adsorbent surface area, that is, the maximum moles of protonatable SOH per unit area ($F\, SOH_T$ is the corresponding maximum net proton surface charge density). Thus C is the ratio of the maximum surface charge density to the maximum surface interaction energy developed when all z nearest neighbors of a surface species are of its own kind.

Molecular Interpretation of the Model. A comparison of the three ways to derive equation 16 is instructive in understanding the molecular significance of equation 18. In method 1, all N_{AB} values are set equal to the total number of nearest neighbors of species A (z_{N_A}) times the fraction of total sites occupied by species B (N_B/M). This result would be strictly true if $\epsilon = 0$ (*ideal* mixing of A and B on the sites), such that no advantage accrued to any particular nearest-neighbor association. In the method 1 approximation, even when $\epsilon \neq 0$, the fraction of species B overall—the long-range order—still determines the short-range ordering of surface species into AB pairs. In method 2, the value of ϵ is supposed to be very small in relation to RT, such that ideal mixing of A and B occurs and the exponential in equation 14 can be expanded safely to first order. Then the purely combinatorial average of N_{AB}

$$\left(\frac{N_A!N_B!}{M!}\right) \sum_{N_{AB}} g(N_A, M, N_{AB}) N_{AB} = \frac{zN_A N_B}{M} \qquad (20)$$

can be calculated at once. In method 3, a small ϵ and a large z are assumed in tandem to facilitate the expansion of q_{AB}. This last approach is equivalent to the mathematical operation of letting z approach infinity while ϵ approaches zero in such a way that the product $z\epsilon$ remains finite. This is the *van der Waals* limit, interpreted physically as the result of placing each surface species in an average potential energy field determined collectively by all of its neighbors on other surface sites (*20*).

Comparative analysis shows that equation 18 represents a system in which the collective sum of very weak, short-range coulombic interactions is equivalent to a single, long-range average interaction that perturbs the system described by the reaction in equation 11 from ideal mixing only to the extent of causing a factorization of K_c into K_{int} and a Boltzmann factor in the average interaction (cf. eq 2). The short-range ordering in the system remains random despite the coulombic interaction. Clearly, this situation is realistic only when the actual surface interactions are few (i.e., when the mole fractions of the species that contribute to surface charge are small). Hence, it is appropriate to invoke equations 2 and 9 as a means of representing K_c for extrapolation to zero surface charge, but their use over the whole range of surface composition remains problematic and a matter for future research on surface complexation models.

Nature of the Surface Complexes. The constant capacitance model assumes an inner-sphere molecular structure for surface complexes formed in reactions like equation 5a or 7. But this structure does not manifest itself explicitly in the composition dependence of K_c; everything molecular is "buried" in K_{int}, which is an adjustable parameter. This encapsulating characteristic of the model was revealed dramatically by Westall and Hohl (*13*), who showed that five different surface speciation models, ranging from the Gouy–Chapman theory to the surface complex approach, could fit proton adsorption data on Al_2O_3 equally well, despite their mutually contradictory underlying molecular hypotheses [*see also* Hayes et al. (*19*)]. They concluded that "... no model will yield an unambiguous description of adsorption ...". To this conclusion one may add that no model *should* provide such a description, so long as goodness-of-fit to adsorption data is the sole criterion of model accuracy.

Recognizing this dilemma, Stumm et al. (6) added this sage advice in the third article of the *Croatica Chemica Acta* series:

> Thus, all the models may be viewed as being of the correct mathematical forms to represent the data but are not necessarily an accurate physical description at the interface. In other words, all models can be used to describe experimental data over the range of experimental data; the "intelligence" of the data, on the other hand is not sufficient to gain insight into the physical nature of the interface.

Johnston and Sposito (26) arrived at similar conclusions in a review of approaches to soil surface speciation and then went on to suggest an obvious alternative to the endless archiving of adsorption data:

> These models [also] have been applied successfully to soil colloids to bring matters full circle; but, like their predecessors, they rely solely on prior molecular concepts and are tested only by goodness-of-fit to adsorption data. Since the model assumptions are so different and the models so plausible, one is left to wonder what physical truth they bear. One fears that the fog will lift only to reveal a Tower of Babel.

> It is inevitable that methodologies not equipped to explore molecular structure will produce ambiguous results in the study of surface speciation. The method of choice for investigating molecular structures is spectroscopy. Surface spectroscopy, both optical and magnetic, is the way to investigate surface species, and thus to verify directly the molecular assumptions in surface speciation models. When the surface species are detected they need not be divined from adsorption data, and the choice of a surface speciation model from the buffet of available software becomes a matter unrelated to goodness-of-fit.

Spectroscopic Probes of Surface Species

The molecular spectroscopy of adsorbed cations and anions has two principal subdivisions: (1) invasive methods (such as X-ray photoelectron or secondary-ion mass spectrometry) that require sample desiccation and high vacuum and (2) noninvasive methods that require little or no alteration of a sample from its received condition. Invasive methods have an important role to play in the characterization of solid surfaces (27), but to use them for resolving surface speciation on particles in aqueous systems simply begs the question.

Noninvasive surface spectroscopies can be applied in the presence of liquid water; most of them involve the input and detection of photons. The best known examples are nuclear magnetic resonance, electron spin resonance, Raman, Fourier transform infrared, UV–visible fluorescence, X-ray absorption, and Mössbauer spectroscopies, although Brown (28) enumerated many others that are available to detect adsorbed ions. These methods, some of which are listed in Table II along with citations of illustrative applications, can be used both noninvasively and in conjunction with in situ probes.

Thus, a homologous adsorptive probe can be introduced into a sample without significantly affecting the type and distribution of surface species while permitting or enhancing the use of a noninvasive spectroscopic technique. A prototypical example of this approach is the introduction of a few mole percent of Cu^{2+} onto the surface of a hydrated clay mineral bearing adsorbed Ca^{2+} to examine surface speciation with electron spin resonance spectroscopy (33). Similarly, $^{113}Cd^{2+}$ can be doped onto the surface of a Ca-saturated clay mineral to facilitate nuclear magnetic resonance spectroscopy.

The intent of this chapter is not to survey noninvasive surface spectroscopy but to illustrate briefly how it is applied to resolve the Stummian issue of whether inner-sphere surface complexes form. For this purpose, the application of electron spin resonance (ESR), electron nuclear double resonance (ENDOR), and electron spin echo envelope modulation (ESEEM) spectroscopies to elucidate metal cation speciation and the use of extended X-ray absorption fine structure (EXAFS) spectroscopy to detect surface anion species will be described. Emphasis will be on the interpretation of spectra. Sample preparation and instrumentation details were reviewed in recent volumes edited by Hawthorne (55) and Perry (27). Because the constant capacitance model was developed in the context of adsorption by hydrous oxides, these

Table II. Noninvasive Methodologies for Investigating Surface Complexes

Acronym	Methodology	Application References
NMR	Nuclear magnetic resonance	29–32
ESR	Electron spin resonance	33, 34
ENDOR	Electron nuclear double resonance	35
ESEEM	Electron spin-echo envelope modulation	36, 37
RAMAN	Raman	38, 39
FTIR	Fourier transform infrared	39–43
IFQ	Interfacial fluorescence quenching	44
EXAFS	Extended X-ray absorption fine structure	45–48
XANES	X-ray absorption near-edge structure	28, 49
XRD	X-ray diffraction	50
SMOSS	Surface Mössbauer	51
NS	Neutron scattering (elastic and inelastic)	52
STM	Scanning tunneling microscopy	53
AFM	Atomic force microscopy	53, 54

adsorbents also will be the focus in the discussion of surface spectroscopy. Neither this model nor the spectroscopic methods, however, are limited to inorganic hydroxylated surfaces.

Electron Spin Resonance Spectroscopy of Adsorbed Metals

The basic principles of ESR spectroscopy were reviewed lucidly in the context of surface speciation applications by McBride (*33*, *56*). A comprehensive introduction oriented toward the use of ESR methods in mineral geochemistry was published by Calas (*57*). Senesi (*58*) did the same for organic geochemistry. Fundamentally, ESR spectroscopy detects chemical species with unpaired electrons. With respect to adsorbed metals, investigation is limited to to paramagnetic transition elements in certain oxidation states [e.g., V(IV), Cr(III), Mn(II), Fe(III), and Cu(II)], either as principal surface species or as in situ molecular probes of surface environments.

The line shape of an ESR spectrum is generated by transitions between electron spin states induced when a sample, after placement in a static magnetic field, is irradiated with microwave (X band) photons. The detailed structure of the spectrum depends on the local molecular environment (e.g., coordinating ligands and bond covalency) of the electrons undergoing the transitions, including the coupling of their spin magnetism with that of the nucleus of the atom in which they reside.

These transitions are illustrated in Figure 1 for the unpaired electron in Cu^{2+}, which will be the exemplar metal in this discussion. When an applied magnetic field $\mathbf{B}_0$ lifts the twofold degeneracy of the spin state $|½\rangle$, resonance transitions can occur when induced by photons of frequency v. Additional hyperfine structure in the resulting ESR spectral line shape can be present because of interaction between the electron spin and the ^{63}Cu (or ^{65}Cu) nuclear spin ($I = 3/2$, fourfold degeneracy). The line shape for Cu^{2+} in its common tetragonal (i.e., distorted octahedral, with elongation along the tetrad symmetry axis) coordination to oxygen ligands thus comprises a quadruplet (Figure 1) of closely spaced peaks, if the tetrad symmetry axis of the complex is either parallel or perpendicular to the direction of $\mathbf{B}_0$ over the time scale of the transitions (ca. 100 ps). The reason for this latter condition is that the spectroscopic *g* tensor (Figure 1) has different components corresponding to directions along ($g_{\|}$) or at right angles ($g_{\perp}$) to the tetrad symmetry axis. In a powder sample, however, a broad range of orientations of the symmetry axis relative to $\mathbf{B}_0$ is likely, and therefore, the resonance condition in Figure 1 will occur over a range of $\mathbf{B}_0$ values (for a fixed v): beginning with $hv/g_{\|}\beta$ and ending with $hv/g_{\perp}\beta$ [$g_{\|} > g_{\perp}$ for a tetragonal complex elongated along its tetrad symmetry axis (*33*)]. In this case, the detection of the orientationally broadened peaks and measurement of the components of the *g* tensor are facilitated by recording the derivative of the line shape with respect to $\mathbf{B}_0$ while the magnetic field is varied over the appropriate range. The derivative

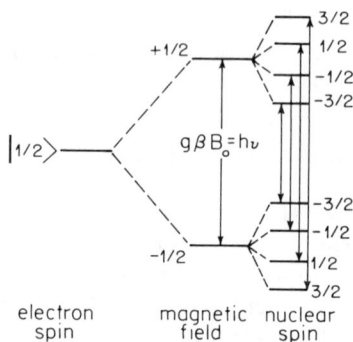

Figure 1. The removal of degeneracy in the electron spin state $|\frac{1}{2}\rangle$ by an applied magnetic field $\mathbf{B}_o$ and by coupling with a nuclear spin state (I = 3/2). Allowed "spin–flip" transitions are shown by arrows.

spectrum will exhibit characteristic "wiggles" wherever peaks or shoulders appear in the ESR absorption spectrum.

Additional information about the molecular environment within ca. 0.5 nm of a paramagnetic species over the time scale of 10^2 to 10^4 ps can be obtained with ENDOR and ESEEM techniques. The instrumentation and the analysis of spectra in ENDOR for transition metal complexes were reviewed in a monograph by Schweiger (59). The ENDOR spectrum is created by inducing ESR transitions with microwave photons while NMR transitions are induced with radio-wave photons. The ESR line shape is thus perturbed in the radio frequency range by electron nuclear spin coupling, such that pairs of ENDOR peaks appear at positions determined by the frequency of the NMR transition and the strength of the electron nuclear spin interaction. Model simulations of peak shape and separation yield information about metal–ligand bond lengths and stereochemistry.

A typical example of an ENDOR application is proton ENDOR (59), in which the coupling between an unpaired electron and vicinal ^{1}H nuclei is exploited. Motschi (60) reviewed applications for transition metal surface complexes. The ESEEM method was described by Kevan (61). In this technique, microwave pulses are applied to a paramagnetic species to generate an echo whose amplitude, as a function of the time interval between pulses, is modulated by dipolar hyperfine interactions between the electron and neighboring magnetic nuclei (e.g., protons). The modulation pattern is sensitive to the type and number of nearest-neighbor nuclei and to their distance from the paramagnetic species. By simulation of the modulation pattern, one can estimate the coordination number of the species in terms of the nuclei. Fourier transformation of the modulation pattern produces a spectrum whose principal peaks will correspond to the characteristic Larmor frequencies of the participating nuclei and permits their identification as near neighbors.

The $Cu(H_2O)_6^{2+}$ Solvation Complex. Spectral information from the ESR, ENDOR, and ESEEM techniques can be illustrated with data from the tetragonal Cu^{2+} solvation complex, whose molecular structure is sketched in the center of Figure 2. Elongation of the complex along its tetrad symmetry axis (taken as the z axis) is apparent, and this distortion of the octahedral coordination is accompanied by a partial removal of the degeneracy of the d orbitals for the unpaired electron (lower right in Figure 2). The components of the g tensor, $g_\parallel$ ($\equiv g_{zz}$) and $g_\perp$ ($\equiv g_{xx} = g_{yy}$), are determined by the energy separations, Δ_1, between the $d_{x^2-y^2}$ and d_{xy} orbitals, and Δ_2, between the $d_{x^2-y^2}$ and d_{xz}, d_{yz} orbitals, respectively (33). [The parameter ζ is the spin-orbit coupling constant; see, e.g., Calas (57).] In the tetragonal complex, $\Delta_1 < \Delta_2$ and $g_\parallel > g_\perp$. The values of the g tensor components are given at the upper right in Figure 2, but the (derivative) ESR spectrum at 25 °C (34) shows only a single, broad feature corresponding to $g_{iso} = (g_\parallel + 2g_\perp)/3$. This "washing out" of the g-tensor anisotropy occurs because displacements of the tetrad symmetry axis are rapid on the ESR time scale. Only if the displacement time

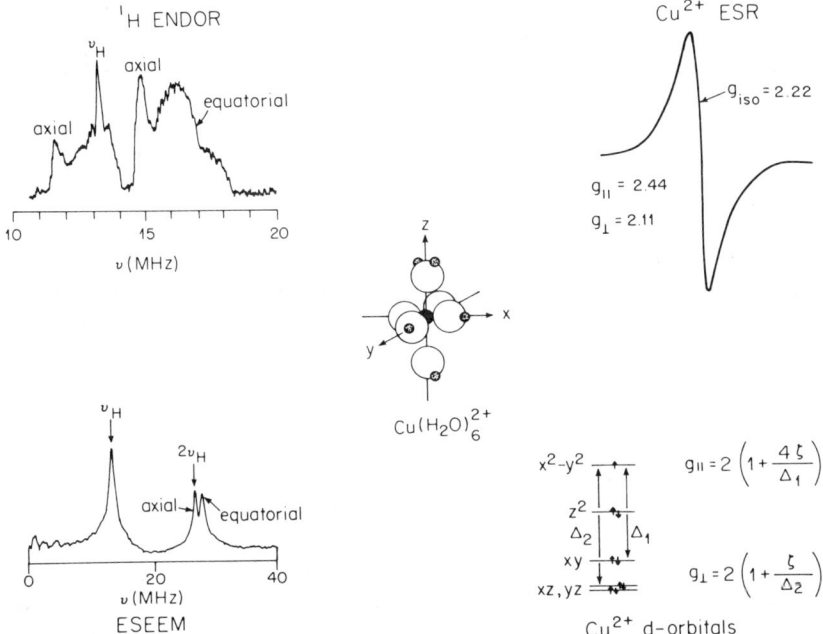

Figure 2. The Cu^{2+} solvation complex (center) with corresponding d orbital energy levels, showing the transitions and orbital level splittings (Δ_1, Δ_2) that yield ESR absorptions at $g_\parallel$ and $g_\perp$ (lower right). The broad ESR (derivative) spectrum (upper right) at 25 °C can be resolved into two component spectra ($g_\parallel = 2.44$, $g_\perp = 2.11$) at liquid He temperatures. Proton ENDOR (upper left) and ESEEM (lower left) spectra help to establish the detailed structure of the complex.

scale increases above 100 ps will the anisotropy be resolved, and this happens if solutions are examined at liquid helium temperatures (*34*).

The proton ENDOR spectrum of the solvation complex at 6 K (*35*) appears at the upper left in Figure 2. Besides the ^{1}H Larmor frequency (at 13.3 MHz), there are peaks at 11.7 and 15.0 MHz, arising from the two axially coordinated water molecules, and a broad spectral feature ending at 18.5 MHz that has been assigned to the four equatorial water molecules. Measurement of the spectral hyperfine coupling constants (*35*) leads to estimates of 0.2 and 0.23 nm, respectively, for the Cu–O distances for equatorially and axially coordinated water molecules. The ESEEM spectrum of the complex at 10 K (*37*), shown at the lower left in Figure 2, exhibits the ^{1}H Larmor peak and a peak at $2v_H$, arising as a combination peak from axially coordinated water molecules plus water molecules outside the solvation complex in the bulk liquid. For the equatorially coordinated water molecules, this sum peak occurs at $(2v_H + 1)$ MHz, but it can be resolved without difficulty. Direct fitting of the modulation pattern from ESEEM studies of $Cu(H_2O)_6^{2+}$ confirms the Cu–O distances and the coordination number of six (*36*).

Adsorbed Cu^{2+} on δ-Al_2O_3. The (derivative) ESR spectrum of Cu^{2+} adsorbed on δ-Al_2O_3 at 25 °C (*34*), shown at the upper right in Figure 3, reflects the four-peak hyperfine structure at $g_\parallel$ and a broad, unresolved peak at $g_\perp$. This result implies that a surface complex has formed whose tetrad axis is not undergoing rapid reorientation on the ESR time scale. Because $(g_\parallel - 2.00)/(g_\perp - 2.00) = 4.2 > 4$, the surface complex has tetragonal symmetry (*34*). Moreover, both components of the g tensor are smaller than those of the solvation complex; therefore, the d-orbital energies are more widely separated in the surface complex than in the solvation complex (lower right in Figure 3). This gain in ligand field stabilization suggests that oxygen ligands are coordinated more strongly to Cu^{2+} in the surface complex than in the solvation complex (*33*) and that this strengthening occurs through equatorial bonds (*34*).

The proton ENDOR spectrum at 6 K for surface-complexed Cu^{2+} (*35*), shown at the upper left in Figure 3, has features arising from axially and equatorially coordinated water molecules, but the axial peaks are now at 12.2 and 14.8 MHz. The axial Cu–O distance is thus shifted to 0.26 nm while the equatorial Cu–O distance remains at 0.2 nm (*35*). The surface complex is evidently more elongated along the tetrad symmetry axis than the solvation complex. The ESEEM spectrum of the surface Cu^{2+} complex at 10 K (*37*) appears at the lower left in Figure 3. The ^{1}H peaks arising from solvation and bulk water molecules appear in this spectrum, but a new feature is the strong ^{27}Al peak at the appropriate Larmor frequency (3.4 MHz). This peak is direct evidence that adsorbed Cu^{2+} is close to the δ-Al_2O_3 surface. Unfortunately, it is not equally clear that the surface complex is an inner-sphere complex, because only subtle effects on the ^{27}Al peaks arise from desolvating Cu^{2+} and

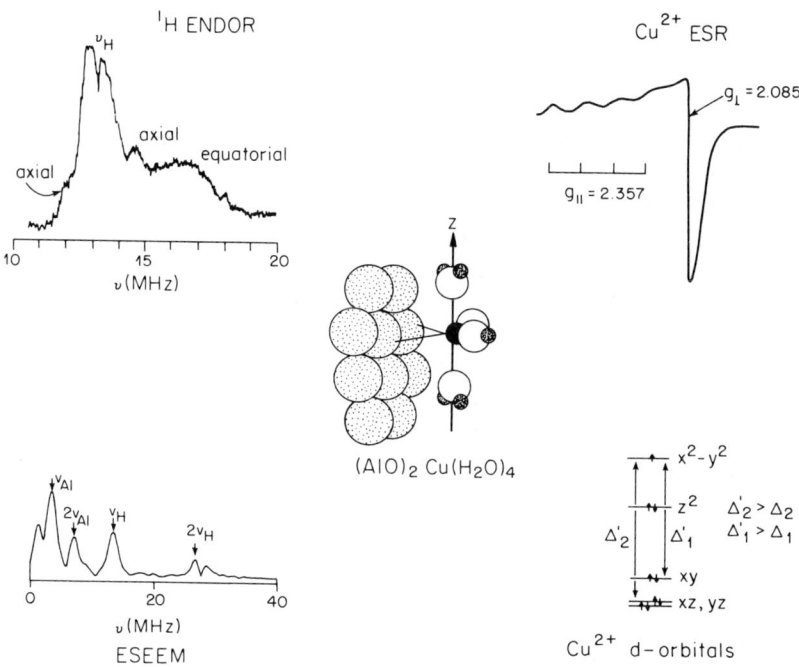

Figure 3. The Cu^{2+} surface complex on δ-Al_2O_3 (center), showing the corresponding d-orbital transitions (lower right), ESR (derivative) spectrum (upper right), proton ENDOR spectrum (upper left), and ESEEM spectrum (lower left).

moving it closer to the oxygen ions in the δ-Al_2O_3 surface (37). Motschi (34) suggested, on the basis of comparative ESR and ^{1}H ENDOR studies of ternary Cu^{2+} surface complexes on δ-Al_2O_3, that inner-sphere coordination, as illustrated in the center of Figure 3, in fact occurs.

Extended X-ray Absorption Fine Structure Spectroscopy of Adsorbed Anions

The use of X-ray absorption spectroscopy and, in particular, EXAFS spectroscopy to investigate surface speciation was reviewed in depth by Brown et al. (15), Brown (28), and Charlet and Manceau (62). This experimental methodology uses the photons from high-intensity, monochromatic synchrotron radiation to produce an absorption spectrum in the X-ray range of wavelengths. The spectral range covered includes the absorption edge of the chemical element whose surface configuration is of interest. Thus, the method is element-specific and can give information about the molecular-coordination environment of different elements in an interfacial region simply by tuning the spectrometer to the desired absorption edges. The EXAFS portion of the spectrum refers to incident photon energies up to 800–1000 eV beyond the absorption edge of an element (28, 62).

The mechanism by which an EXAFS spectrum is produced involves the ejection of excess-energy photoelectrons from an atomic absorber of beyond-edge X-ray photons (Figure 4). These electrons are singly scattered from the atoms that are first and second nearest neighbors of the absorber and thereby generate backscattered waves that can interfere with the ejected photoelectron wave to modulate the X-ray absorption spectrum. This modulation feature constitutes the EXAFS portion of the spectrum. Evidently, it contains implicit information about the number and type of near-neighbor scattering atoms, as well as their positions relative to the absorber atom. Brown (28) and Charlet and Manceau (62) summarized the relevance of EXAFS spectra to surface speciation:

1. Synchrotron-based EXAFS can be used to study most chemical elements in solid, liquid, or gas phases at concentrations as low as millimoles per cubic meter. The high intensity of synchrotron radiation allows the study of very small or dilute samples under conditions of varying temperature or pressure and in controlled environments, including the presence of liquid water. Thus, the method is noninvasive and can be used with in situ molecular probes.

2. The EXAFS spectrum gives information, on the time scale of 10^{-4} ps, concerning only the two or three closest shells of neighbors around an absorbing atom (≤ 0.6 nm) because of the small photoelectron mean free path in most materials.

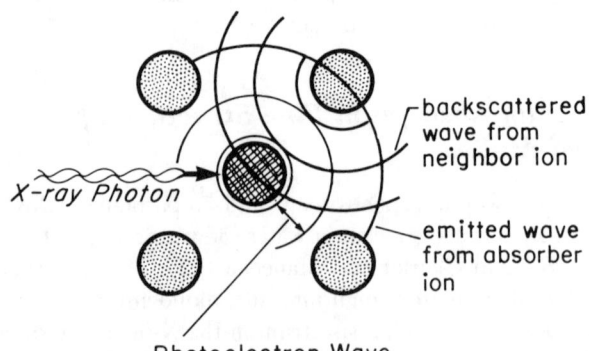

Figure 4. The fundamental photoelectron interference process in EXAFS spectroscopy (28).

3. If the difference in atomic number between the absorber element and the backscattering element is >10 and if only one kind of element backscatters, EXAFS spectra can be analyzed readily to provide local structural data on adsorbed species. However, because the electron mean free path, thermal and static disorder parameters (Debye–Waller factors), and coordination number for an absorber environment cannot be determined a priori with sufficient accuracy, EXAFS data for suitable reference compounds of known molecular structure must be used to help interpret the EXAFS spectrum for an interfacial region.

4. For most systems, EXAFS spectra can be analyzed to yield average interatomic distances accurate to 2 pm and average coordination numbers accurate to 10–20%, if systematic errors have been minimized in both the experiment and data analysis, and if static and thermal disorder are both small.

Adsorbed Selenium Anions. The principal anionic species of Se in natural waters are biselenate ($HSeO_4^-$) and selenite (SeO_3^{2-}). Hayes et al. (46) reported K-edge, fluorescence-yield Se EXAFS spectra for aqueous solutions (25 mol of Se per m^3) of these two species, as well as for goethite (α-FeOOH) suspensions in which the aqueous solutions of Se served as supporting electrolytes (Figure 5). The EXAFS spectra of selenate in solution and in suspension media (top of Figure 5) were identical and could be modeled accurately by a structure comprising simply one Se absorber and four nearest-neighbor O backscattering ions at 0.165 nm from the Se absorber. Thus, selenate was adsorbed as a *solvated* species, either in an outer-sphere surface complex or in the diffuse ion swarm near the charged surface. Because the EXAFS time scale is <<1 ps, both species are static and cannot be distinguished by their motional effects on the spectra.

The EXAFS spectra of selenite in the two media differed, on the other hand. In the presence of goethite, the spectrum exhibited features (marked by arrows in the lower part of Figure 5) indicative of more than one type of backscattering atom. This spectrum could be modeled by a structure comprising one Se absorber, three O backscattering ions at 0.170 nm from the absorber, and two Fe backscattering ions at 0.338 nm from the absorber. These structural data point to adsorbed selenite in an inner-sphere surface complex on goethite. Detailed consideration of the interionic distances and the crystal structure of goethite indicates that the surface complex is binuclear bidentate (46).

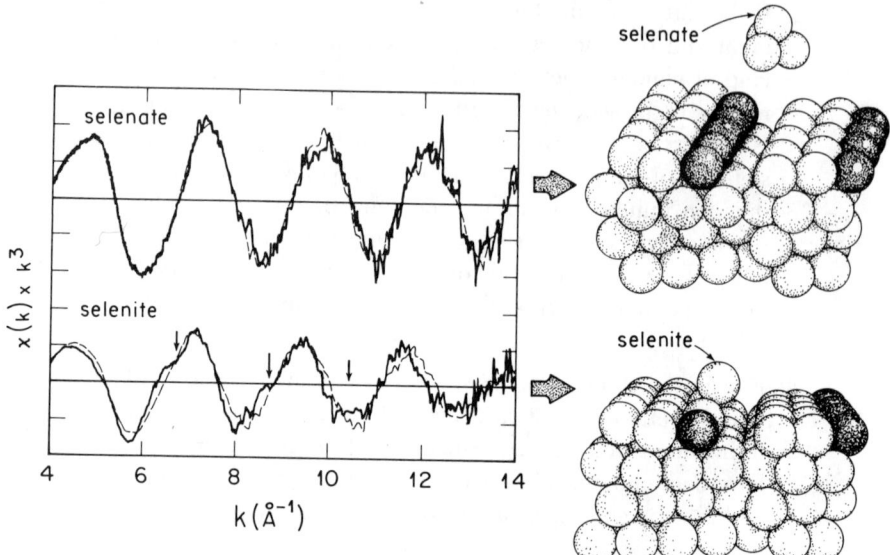

Figure 5. Evidence for surface species (right) in the normalized, background-subtracted, k^3-weighted EXAFS spectra (left) of selenate and selenite ions reacted with goethite. The EXAFS spectra of selenate or selenite ions in aqueous solution are shown as dashed curves, with vertical arrows denoting contributions from Fe(III) in the selenite spectrum. Reactive surface OH groups on goethite are shown in black.

Beyond the Surface Complex

These illustrative examples of ESR and EXAFS spectroscopy applied to detect surface complexes demonstrate the typical use of noninvasive methods but also expose what remains problematic about the information provided by these methods. Stumm et al. (4) wisely retained the Stern dichotomy for strongly adsorbed metals and considered surface complexes that contain both solvated and desolvated metal ions. This dichotomy persists in spectroscopic data of metal adsorption on hydroxylated surfaces. Often, they can give clear evidence for immobilization of a metal ion on a surface within the time scale of the spectroscopic method, but not necessarily for complete desolvation of the ion to form an inner-sphere complex with surface hydroxyl groups.

This dilemma emphasizes the two essential prerequisites for a successful molecular probe of surface speciation: (1) it must be able to distinguish species that are stationary at an interface for longer than 10 ps from those that are not, and (2) it must be able to disaggregate the behavior of oxygen atoms bonding to an adsorbed species into contributions from complexing surface groups and those from coordinated water molecules. The ESR techniques succeed with the time scale criterion (diffuse-ion species are mobile, whereas surface complexes are static) better than with O–ligand discrimination; the inverse is true for the EXAFS method. For this reason a complete characterization of surface species usually requires the application of several of the spectroscopic methods listed in Table II to the same sample (26, 51).

Reviews (9, 63, 64) of the reactions between hydroxylated mineral surfaces and aqueous solutions brought out the richness of variety found in surface phenomena involving natural particles. Isolated surface complexes, the principal topic of this chapter, are expected when reaction times are short and the adsorbate content is low [Figure 6, inspired by Schindler and Stumm (63)]. Thus, surface complexes occupy a reasonably well-defined domain in the tableau of reaction time scale versus sorbate concentration. Localized clusters of adsorbate (47, 48, 65, 66) that contain two or more adsorbate ions bonded together can form if the amount sorbed is increased by accretion or by the direct adsorption of polymeric species (multinuclear surface complexes). Surface clusters can erase the hyperfine structure in the ESR spectrum of an immobilized adsorbate (33, 67) or produce new second-neighbor peaks from ions like the absorber in its EXAFS spectrum (47, 66).

Surface nuclei are to be distinguished structurally from mere surface clusters. For surface nuclei, accretion and rearrangement of constituent ions are needed to present a kernel on which a surface precipitate can grow successfully (7). A case in point is the formation of calcium phosphate nuclei on the surface of calcite after rearrangement of adsorbed phosphate clusters (68). The transition from surface complexes to clusters to precipitate was reviewed in detail by Charlet and Manceau (62). They stressed the important interre-

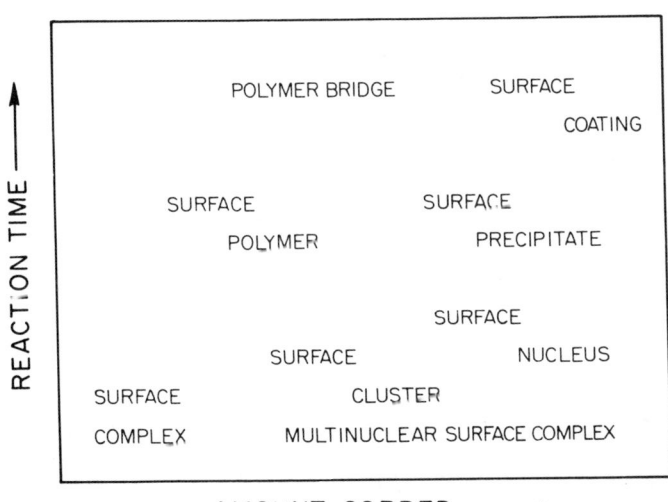

Figure 6. A tableau of surface reaction domains differentiated by reaction time and amount sorbed (63).

lations among reaction kinetics, adsorbate concentration, and epitaxy in this transition, which can be characterized at a macroscopic level by the appearance of an inflection in a log–log plot of a sorption isotherm (Figure 7). Farley et al. (69) developed a successful phenomenological model of this kind of curve (48, 70, 71).

If the three-dimensional structure of a surface precipitate is precluded from development because of unfavorable conditions relating to either the sorbate concentration or the requirements of epitaxial growth, surface polymeric structures may evolve instead whose configurations result from the more or less random sequential attachment of aqueous species to solid accretions rooted sporadically on an adsorbent surface. Thus, a surface polymer would differ from a surface precipitate both in its less-organized external morphology and in its ability to cover the surface on which it grows. Ultimately, surface polymers may attach their protruding meristems to vicinal particles other than the one to which they are anchored and thereby form interparticle bridges that figure in colloid aggregation processes (72).

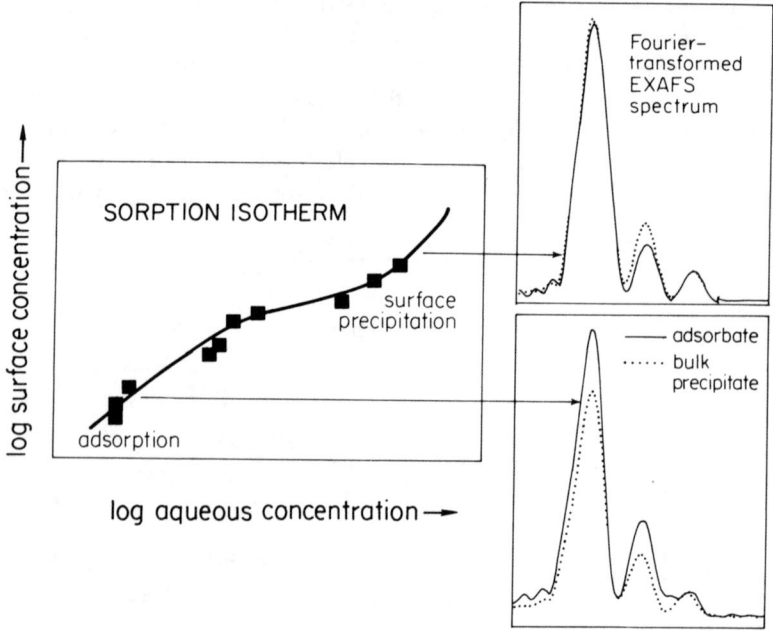

Figure 7. A log–log plot of a sorption isotherm, with an inflection indicating the transition from adsorption to surface precipitation processes. On the right, illustrative Fourier transformed EXAFS spectra for the adsorbate (solid curve) are compared with that for a precipitate (dotted curve) to show the adsorption → precipitation transition. Data are from Charlet and Manceau (48) for Cr(III) sorbed on hydrous ferric oxide.

This effect can be distinguished from the new mineral–water interface created ultimately by a surface precipitate, whose lateral growth and thickness succeed in entirely coating the original adsorbent surface (62). Atomic force microscopy (AFM) and scanning tunneling microscopy (STM) may be useful in identifying the morphological differences between surface polymers and surface precipitates at the molecular level. These structures have a life of their own and may bear only the faintest imprint of the coordination chemistry from which they were born as isolated surface complexes.

Epilog

Few of us are fortunate enough to have our professional lives touched by contact with a man of the personal charm and exploratory genius of Werner Stumm. Our intellectual debt to him is enormous, both for the way in which he has changed the science of aquatic chemistry and for the inspiring facets of his generous nature. To this we may add the gratitude of our fellow human beings on this planet for his crowning achievements in helping to provide a scientific basis for the global effort to protect our precious endowment of water resources throughout the millennia to come.

Acknowledgments

Gratitude is expressed to James J. Morgan for providing the inspiration to use the quotation from Joyce's *Ulysses* and to Laurent Charlet for providing preprints of references 48, 49, and 62. Thanks also to two anonymous referees for helpful criticism and to Terri DeLuca for her excellent typing of the manuscript. The preparation of this review was supported in part by NSF Grant No. EAR-9206052.

References

1. Joyce, J. *Ulysses: The Corrected Text;* Vintage Books: New York, 1986; p 549. (Quoted from with permission from Random House on pp. 33 and 34.)
2. Sposito, G. *Chimia* **1989**, *43*, 160–176.
3. Stumm, W.; Morgan, J. J. *Aquatic Chemistry;* John Wiley: New York, 1981; Chapters 9–11]
4. Stumm, W.; Huang, C. P.; Jenkins, S. R. *Croat. Chem. Acta* **1970**, *42*, 223–245.
5. Stumm, W.; Hohl, H.; Dalang, F. *Croat. Chem. Acta* **1976**, *48*, 491–504.
6. Stumm, W.; Kummert, R.; Sigg, L. *Croat. Chem. Acta* **1980**, *53*, 291–312.
7. Stumm, W.; Furrer, G.; Kunz, B. *Croat. Chem. Acta* **1983**, *56*, 593–611.
8. Stumm, W.; Wehrli, B.; Wieland, E. *Croat. Chem. Acta* **1987**, *60*, 429–456.
9. Stumm, W.; Sulzberger, B.; Sinniger, J. *Croat. Chem. Acta* **1990**, *63*, 277–312.
10. Truesdell, C. *Essays in the History of Mechanics;* Springer-Verlag: New York, 1968; Foreword.
11. Stern, O. *Z. Elektrochem.* **1924**, *30*, 508–516.

12. Schindler, P. W.; Gamsjä:ger, H. *Kolloid Z. Z. Polym.* **1972,** *250,* 759–763.
13. Westall, J.; Hohl, H. *Ad. Colloid Interface Sci.* **1980,** *12,* 265–294.
14. Schindler, P. W.; Fü:rst, B.; Dick, R.; Wolf, P. U. *J. Colloid Interface Sci.* **1976,** *55,* 469–475.
15. Hohl, H.; Stumm, W. *J. Colloid Interface Sci.* **1976,** *55,* 281–288.
16. Kummert, R.; Stumm, W. *J. Colloid Interface Sci.* **1980,** *75,* 373–385.
17. Sigg, L.; Stumm, W. *Colloids Surfaces* **1980–1981,** *2,* 101–117.
18. Hayes, K. F.; Redden, G.; Leckie, J. O. *J. Colloid Interface Sci.* **1991,** *142,* 448–469.
19. Goldberg, S. *Ad. Agronomy* **1992,** *47,* 233–329.
20. Sposito, G. *J. Colloid Interface Sci.* **1983,** *91,* 329–340.
21. Sposito, G. *The Surface Chemistry of Soils;* Oxford University: New York, 1984; Chapter 5.
22. Sposito, G. In *Mineral–Water Interface Geochemistry;* Hochella, M. F.; White, A. F., Eds.; Mineralogical Society of America: Washington, DC, 1990; Chapter 5.
23. Hill, T. L. *An Introduction to Statistical Thermodynamics;* Addison-Wesley: Reading, MA, 1960; Chapters 14, 20.
24. Fowler, R. H.; Guggenheim, E. A. *Statistical Thermodynamics;* Cambridge University: London, 1949; pp 574–576.
25. Guggenheim, E. A. *Proc. Royal Soc. (London)* **1944,** *A183,* 213–227.
26. Johnston, C. T.; Sposito, G. In *Future Developments in Soil Science Research;* Boersma, L. L., Ed.; Soil Science Society of America: Madison, WI, 1987; pp 89–99.
27. Perry, D. L. *Instrumental Surface Analysis of Geologic Materials;* VCH Publishers: New York, 1990; Chapters 3–10.
28. Brown, G. E. In *Mineral–Water Interface Geochemistry;* Hochella, M. F.; White, A. F., Eds.; Mineralogical Society of America: Washington, DC, 1990; Chapter 8.
29. Bank, S.; Bank, J. F.; Ellis, P. D. *J. Phys. Chem.* **1989,** *93,* 4847–4855.
30. Weiss, C. A.; Kirkpatrick, R. J.; Altaner, S. P. *Geochim. Cosmochim. Acta* **1990,** *54,* 1655–1669.
31. Laperche, V.; Lambert, J. F.; Prost, R.; Fripiat, J. J. *J. Phys. Chem.* **1990,** *94,* 8821–8831.
32. Lambert, J. F.; Prost, R.; Smith, M. E. *Clays Clay Miner.* **1992,** *40,* 253–261.
33. McBride, M. B. In *Instrumental Surface Analysis of Geologic Materials;* Perry, D. L., Ed.; VCH Publishers: New York, 1990; Chapter 8.
34. Motschi, H. *Colloids Surf.* **1984,** *9,* 333–347.
35. Rudin, M.; Motschi, H. *J. Colloid Interface Sci.* **1984,** *98,* 385–393.
36. Brown, D. R.; Kevan, L. *J. Am. Chem. Soc.* **1988,** *110,* 2743–2748.
37. Möhl, W.; Schweiger, A.; Motschi, H. *Inorg. Chem.* **1990,** *29,* 1536–1543.
38. Johnston, C. T.; Sposito, G.; Bocian, D. F.; Birge, R. R. *J. Phys. Chem.* **1984,** *88,* 5959–5964.
39. Johnston, C. T. In *Instrumental Surface Analysis of Geologic Materials;* Perry, D. L., Ed.; VCH Publishers: New York, 1990; Chapter 5.
40. Tejedor-Tejedor, M. I.; Anderson, M. A. *Langmuir* **1986,** *2,* 203–210.
41. McBride, M. B. *Soil Sci. Soc. Am. J.* **1987,** *51,* 1466–1472.
42. McBride, M. B.; Wesselink, L. G. *Environ. Sci. Technol.* **1988,** *22,* 703–708.
43. Tejedor-Tejedor, M. I.; Anderson, M. A. *Langmuir* **1990,** *6,* 602–611.
44. Traina, S. J. *Adv. Soil Sci.* **1990,** *14,* 167–190.
45. Brown, G. E.; Parks, G. A.; Chisholm-Brause, C. J. *Chimia* **1989,** *43,* 248–256.
46. Hayes, K. F.; Roe, A. L.; Brown, G. E.; Hodgson, K. O.; Leckie, J. O.; Parks, G. A. *Science (Washington, DC)* **1987,** *238,* 783–786.

47. Chisholm-Brause, C. J.; Hayes, K. F.; Roe, A. L.; Brown, G. E.; Parks, G. A.; Leckie, J. O. *Geochim. Cosmochim. Acta* **1990**, *54*, 1897–1909.
48. Charlet, L.; Manceau, A. *J. Colloid Interface Sci.* **1992**, *148*, 443–458.
49. Manceau, A.; Charlet, L. *J. Colloid Interface Sci.* **1992**, *148*, 425–442.
50. MacEwan, D. M. C.; Wilson, M. J. In *Crystal Structures of Clay Minerals and Their X-ray Identification*; Brindley, G. W.; Brown, G., Eds.; Mineralogical Society: London, 1980; Chapter 3.
51. Rea, B. A. Davis, J. A.; Waychunao, G. A. *Clays Clay Miner.* **1994**, *42*, 23–34.
52. Hall, P. L. In *Advanced Techniques for Clay Mineral Analysis*; Fripiat, J. J., Ed.; Elsevier: Amsterdam, The Netherlands, 1982; Chapter 3.
53. Hochella, M. F. In *Mineral–Water Interface Geochemistry*; Hochella, M. F., White, A. F., Eds.; Mineralogical Society of America: Washington, DC, 1990; Chapter 3.
54. Hartman, H.; Sposito, G.; Yang, A.; Manne, S.; Gould, S. A. C.; Hansma, P. K. *Clays Clay Miner.* **1990**, *38*, 337–342.
55. Hawthorne, F. C. *Spectroscopic Methods in Mineralogy and Geology*; Mineralogical Society of America: Washington, DC, 1988.
56. McBride, M. B. In *Geochemical Processes at Mineral Surfaces*; Davis, J. A.; Hayes, K. F., Eds.; ACS Symposium Series 323; American Chemical Society: Washington, DC, 1986; Chapter 17.
57. Calas, G. In *Spectroscopic Methods in Mineralogy and Geology*; Hawthorne, F. C., Ed.; Mineralogical Society of America: Washington, DC, 1988; Chapter 12.
58. Senesi, N. *Anal. Chim. Acta* **1990**, *232*, 51–75.
59. Schweiger, A. *Electron Nuclear Double Resonance of Transition Metal Complexes with Organic Ligands*; Springer-Verlag: Berlin, Germany, 1982.
60. Motschi, H. In *Aquatic Surface Chemistry*; Stumm, W., Ed.; John Wiley: New York, 1987; Chapter 5.
61. Kevan, L. In *Time Domain Electron Spin Resonance*; Kevan, L.; Schwartz, R., Eds.; John Wiley: New York, 1979; Chapter 8.
62. Charlet, L.; Manceau, A. In *Characterization of Environmental Particles*; Buffle, J.; van Leeuwen, H. P., Eds.; Lewis Publishers: Chelsea, MI, 1993; Vol. II, Chapter 3.
63. Schindler, P. W.; Stumm, W. In *Aquatic Surface Chemistry*; Stumm, W., Ed.; John Wiley: New York, 1987; Chapter 4.
64. Schindler, P. W.; Sposito, G. In *Interactions at the Soil Colloid–Soil Solution Interface*; Bolt, G. H.; De Boodt, M. F.; Hayes, M. H. B.; McBride, M. B., Eds.; Kluwer: Dordrecht, The Netherlands, 1991; Chapter 4.
65. McBride, M. B. *Soil Sci. Soc. Am. J.* **1979**, *43*, 693–698.
66. Chisholm-Brause, C. J.; O'Day, P. A.; Brown, G. E.; Parks, G. A. *Nature (London)* **1990**, *348*, 528–530.
67. Wersin, P.; Charlet, L.; Karthein, R.; Stumm, W. *Geochim. Cosmochim. Acta* **1989**, *53*, 2787–2796.
68. Sposito, G. In *Geochemical Processes at Mineral Surfaces*; Davis, J. A.; Hayes, K. F., Eds.; ACS Symposium Series 323; American Chemical Society: Washington, DC, 1986; Chapter 11.
69. Farley, K. J.; Dzombak, D. A.; Morel, F. M. M. *J. Colloid Interface Sci.* **1985**, *106*, 226–242.
70. Dzombak, D. A.; Morel, F. M. M. *J. Colloid Interface Sci.* **1986**, *112*, 588–598.
71. Comans, R. N. J.; Middleburg, J. J. *Geochim. Cosmochim. Acta* **1987**, *51*, 2587–2591.
72. Hansmann, D.; Anderson, M. A. *Environ. Sci. Technol.* **1985**, *19*, 544–551.

RECEIVED for review October 23, 1992. ACCEPTED revised manuscript April 27, 1993.

3

Ion Exchange

The Contributions of Diffuse Layer Sorption and Surface Complexation

David A. Dzombak[1] and Robert J. M. Hudson[2]

[1] Department of Civil and Environmental Engineering, Carnegie Mellon University, Pittsburgh, PA 15213
[2] Institute of Marine Sciences, University of California, Santa Cruz, CA 95064

Models for ion exchange on soils, sediments, and aquatic particles remain largely empirical because of the heterogeneity of these materials and the complexity of the phenomena that constitute ion exchange—specific adsorption at particle surfaces and nonspecific electrostatic (diffuse layer) sorption. Although empirical models can fit data for narrow ranges of calibration, they provide little insight into the microscale physical and chemical processes involved. A lack of the computational tools needed to develop general models incorporating both diffuse layer and surface sorption hindered past attempts to develop physicochemical models for ion exchange. Surface complexation models, which have emerged as powerful tools for describing specific sorption onto reactive mineral surfaces, may be extended to represent ion exchange by using the Gouy–Chapman theory to determine the contribution of diffuse-layer sorption to the overall sorption of ionic species. Distinguishing between surface complexation and diffuse-layer sorption provides insights into ion-exchange phenomena and the physicochemical basis of empirical exchange equations.

ION EXCHANGE AT THE MINERAL–WATER INTERFACE has been studied extensively because of its importance in soil chemistry (1, 2) and its usefulness in chemical processes such as water demineralization (3). The scientific literature on ion exchange is vast, encompassing many experimental and theoretical developments over the past 150 years. Since the development of synthetic ion-exchange resins in the 1940s, research on technological uses of ion exchange has shifted to synthetic resins (3, 4). In soil and aquatic chemistry research, however, interest in ion exchange at the surfaces of minerals has continued and accelerated. Cation exchange has been of foremost interest in environmental chemistry, because most natural particles carry a negative charge in the pH range of aquatic systems.

The process of ion exchange is usually conceptualized as the reversible exchange of electrolyte counterions in the diffuse layer near a charged surface or in a separate phase, as for Donnan equilibria. Ion exchange is considered to be a stoichiometric process; that is, every ion removed from solution by electrostatic attraction to the charged surface is replaced by an equivalent amount of a similarly charged ionic species displaced from the interfacial region. The notion of ion exchange as a wholly electrostatic process is convenient for qualitative interpretation of ion-exchange data and construction of simple quantitative models to fit such data. It is widely recognized, however, that most ions chemisorb to varying degrees at reactive sites on mineral surfaces. This tendency generates significant differences in chemical activities of the ions in the sorbed phase.

Because the Gouy–Chapman theory for electrostatic (diffuse-layer) sorption does not yield tractable analytical solutions for mixed electrolytes, and because the specific surface interactions possible in ion-exchange systems are so complex, interactions of ions with charged surfaces in soil–water and other natural aquatic systems are usually described with semiempirical ion-exchange equations. General thermodynamic treatments of ion exchange that use composition-specific activity coefficients to describe the sorbed phase have also been developed (5–7). While the purely thermodynamic approaches are applicable to a wide range of conditions, they provide little physical insight and do not permit extrapolation from binary to higher order exchange systems.

Specific physicochemical models (namely, diffuse-layer theory for electrostatic sorption and surface complexation theory for chemisorption) have been developed to describe chemical and electrostatic sorption–desorption of ions on mineral surfaces. However, few attempts have been made to apply these theories to model ion-exchange phenomena in mineral–water systems. Bolt and co-workers (8–10) investigated in detail the application of the Gouy–Chapman diffuse-layer theory to ion-exchange processes. Their work demonstrated that consideration of electrostatic sorption alone is not sufficient to explain ion-exchange data and that chemisorption (or "specific" sorption) needs to be included in ion-exchange models. The lack of adequate models and computational tools to describe specific sorption hindered attempts to develop general physicochemical models for ion exchange.

Spurred by the work of Stumm (e.g., 11–13), surface complexation modeling has emerged as a powerful tool for describing chemical sorption of ions onto reactive mineral surfaces. In surface complexation models, ions and individual functional groups on the surface are considered to react to form coordination complexes and ion pairs. These models consider electric field effects on surface sorption via the Gouy–Chapman theory, but they do not distinguish sorption in the diffuse layer from total sorption. The former is usually a small fraction of the total for ions that exhibit significant chemisorption.

Surface complexation models can be extended to include diffuse-layer sorption. This approach permits their application in modeling the sorption of ions (such as monovalent electrolyte ions) that exhibit weak specific sorption. The generality of such an extended surface complexation approach together with the mathematical power of modern chemical speciation models offers the potential for accurate physicochemical modeling of ion exchange, as foreseen by Bolt (9) and others (e.g., 14, 15). Such models, although undoubtedly difficult to apply to complex natural systems, would link the soil science approaches to ion-exchange modeling with the diffuse-layer and surface complexation theory favored in the aquatic chemistry literature.

This chapter presents an approach to the physicochemical modeling of ion exchange. By using general chemical equilibrium models as a mathematical framework, we synthesize existing models in a way that accounts explicitly for both nonspecific, counterion sorption in the diffuse layer and chemisorption via ion pairing and complexation at surface sites. Application of the physicochemical model to example data sets illustrates the relative contributions in the diffuse layer and on the surface to overall sorption. The extensive data requirements and other impediments to applying such a physicochemical model to describe or predict ion exchange in a complex aquatic or soil–water system are also discussed.

A review of some leading semiempirical models precedes examination of physicochemical modeling of ion exchange. Such models will likely be used for the foreseeable future to describe ion-exchange phenomena in complex systems. Thus, they represent the reference point for development of improved models. Methods of incorporating the semiempirical ion-exchange equations in general chemical equilibrium models are also described.

Ion-Exchange Processes and Data

Ion exchange is usually conceived of as the exchange of counterions in the diffuse and Stern layers of charged surfaces in aqueous suspension. For example, the net negative charge on a clay platelet in water is counteracted predominantly by Na^+ in an NaCl electrolyte solution. However, if the clay particles with sodium counterions are placed in a $CaCl_2$ electrolyte solution, the Ca^{2+} ions will displace the Na^+ ions on an equivalent-for-equivalent basis (Figure 1). That is, Na^+ counterions sorbed on the surface will be exchanged for Ca^{2+} counterions. The total exchangeable equivalents of cationic charge (in mequiv/100 g) under a particular experimental condition is referred to as the "cation-exchange capacity."

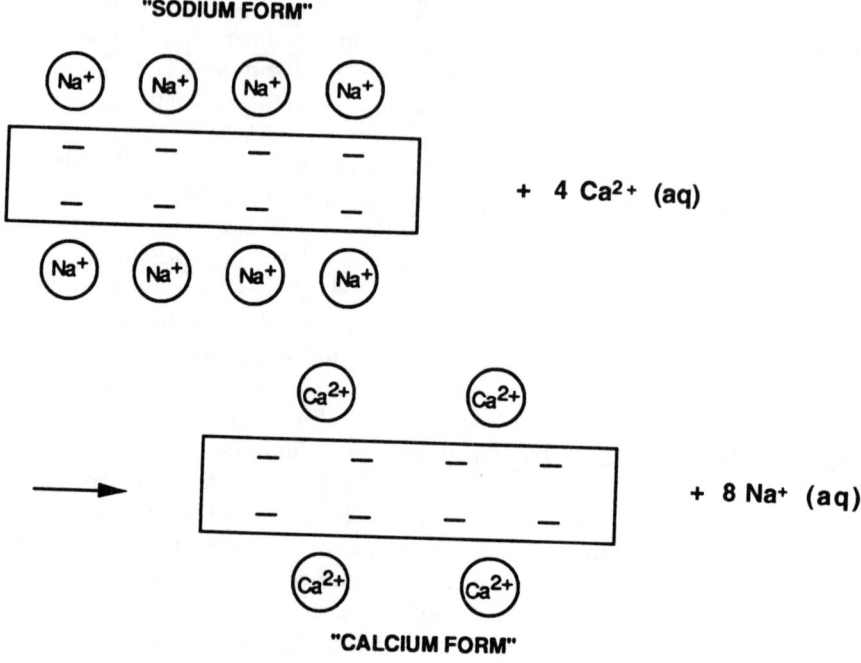

Figure 1. Schematic representation of Na^+–Ca^{2+} exchange on a clay platelet.

The experiment described, in which the dominant electrolyte ion is switched, is basically the method used to measure the exchange characteristics of a particular soil, sediment, or aquatic particle. Batch techniques are used most commonly, but ion exchange is also sometimes measured via experiments involving continuous flow through packed columns (16, 17). In both techniques, changes in the aqueous and/or solid-phase concentrations of the competing ions are measured to provide information about ion uptake or release by the solid phase.

The ion-exchange properties of a particular solid material are represented by its cation-exchange capacity (CEC) or anion-exchange capacity (AEC) relative to a reference electrolyte, and by its ion-exchange isotherms. Ion-exchange isotherms are plots of equilibrium concentration(s) of ions in the exchanger (solid) phase versus equilibrium concentration(s) of ions in the solution phase for a particular pair of exchangeable ions in an aqueous suspension at fixed temperature and pressure. The sorbed and dissolved concentrations may be expressed as mole or equivalent fractions. As illustrated in Figure 2, which shows three isotherms (18) for sodium–calcium exchange on a clay soil, clays exhibit a strong preference for calcium relative to sodium; large amounts of Na^+ in the solution phase are required to effect significant sorption of Na^+. A qualitative explanation of the preference, or selectivity, of the sur-

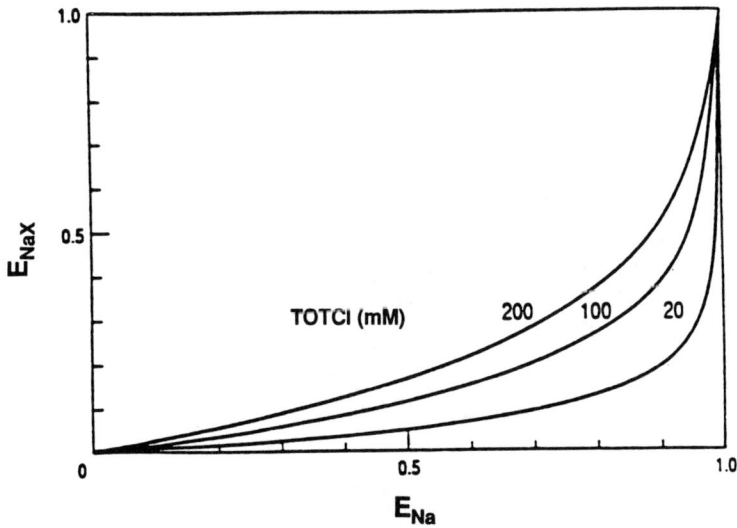

Figure 2. Isotherms for Na^+–Ca^{2+} exchange on Brucedale clay soil for various TOTCl concentrations (TOTNa and TOTCa varied). E_{NaX} and E_{Na} are the equivalent fractions of Na^+ in the exchanger and in the bulk solution, respectively. (Data are from reference 18.)

face for calcium relative to sodium is the greater positive charge associated with the calcium ion and hence the greater energy of attraction to a negatively charged surface.

Electrostatic attraction–repulsion is clearly important in ion exchange, but available data make it clear that chemisorption at the surface can play a significant role as well. Soils exhibit selectivities for certain ions (e.g., $Cs^+ > Rb^+ > K^+ \approx NH_4^+ > Na^+ > Li^+$), indicating significant energies of interaction at the surface typical of energies associated with ion pairing and/or covalent bonding (19). The formation of such bonds requires the close approach of an ion to the surface and partial loss of its water of solvation. An ion's ability to participate in inner-sphere bonding (which requires partial desolvation of the ion) and in outer-sphere bonding (which involves retention of its solvation sphere) depends on its ionic radius and charge, the strength of water coordination, and other ionic properties.

A schematic diagram that summarizes different mechanisms of ion binding at the mineral–water interface is shown in Figure 3. A silicon oxide surface, which bears a negative charge at all pH values greater than 3, is used for illustrative purposes. The negative charge of the surface is counteracted by calcium ions from the electrolyte solution. Electrostatic attraction results in the formation of a diffuse layer of nonspecifically sorbed calcium ions in the water adjacent to the oxide surface. As indicated in Figure 3, some of the counterions may approach the surface more closely to form weak outer-sphere

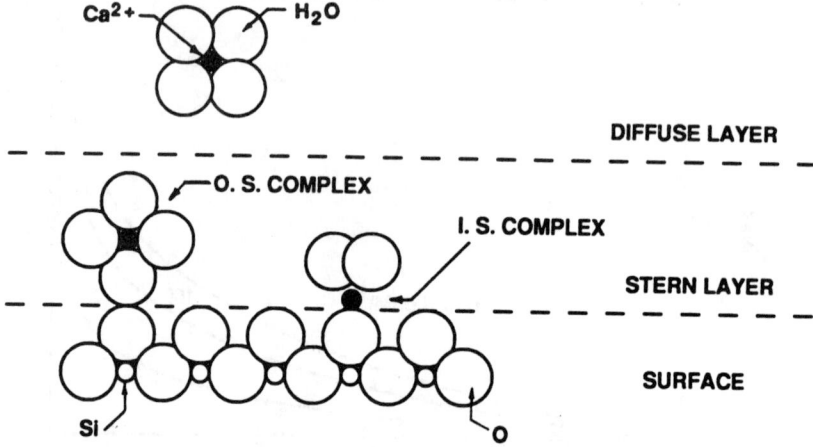

Figure 3. Schematic representation of mechanisms responsible for ion sorption on charged mineral surfaces. Key: O.S., outer sphere; I.S., inner sphere.

surface complexes (primarily through electrostatic binding or ion pairing) or stronger inner-sphere surface complexes (primarily through chemical bonding).

Other physical factors can influence the ion-exchange characteristics of certain kinds of solids. Primary among these factors are ion exclusion due to the size of structural pore spaces in porous charged solids, and physical swelling or shrinking of the solid phase as ions of different sizes and degrees of hydration move into and out of porous solids (4). Ion size exclusion and swelling–shrinking are of greatest interest with respect to ion-exchange resins but can also be important for certain minerals such as montmorillonite (4, 19). In models of ion exchange for soils, these factors are usually not considered (7, 9, 10, 19).

Ion-Exchange Equations

The exchange of ions at the solid–water interface is usually described by reactions in which equivalents of counterion charge are conserved. As an example of heterovalent exchange, the replacement of Na^+ by Ca^{2+} as the counterion adjacent to a negatively charged surface may be represented by

$$Ca^{2+} + 2NaX \longrightarrow CaX_2 + 2Na^+ \tag{1}$$

where X^- denotes an equivalent of exchange capacity of the solid adsorbent, and thus NaX and CaX$_2$ represent species in the exchanger phase. The corresponding mass law equation is given as

$$K_{\text{Na-Ca}} = \frac{\{\text{Na}^+\}^2\{\text{CaX}_2\}}{\{\text{Ca}^{2+}\}\{\text{NaX}\}^2} \quad (2)$$

where $K_{\text{Na-Ca}}$ is the thermodynamic equilibrium constant for the reaction and { } represent activities. Even in ion-exchange systems involving no specific ion–surface interactions (i.e., purely sorption in the diffuse layer), formulations such as equation 1 are convenient in that they help describe the charge and mass conservation associated with ion exchange. Such equations are thus useful for modeling the macroscopic behavior of ion interactions with charged solids, particularly with heterogeneous natural solids.

Because the activities of species in the exchanger phase are not well defined in equation 2, a simplified model—that of an ideal mixture—is usually employed to calculate these activities according to the approach introduced by Vanselow (20). Because of the approximate nature of this assumption and the fact that the mechanisms involved in ion exchange are influenced by factors (such as specific sorption) not represented by an ideal mixture, ion-exchange constants are strongly dependent on solution- and solid-phase characteristics. Thus, they are actually conditional equilibrium constants, more commonly referred to as selectivity coefficients. Both mole and equivalent fractions of cations have been used to represent the activities of species in the exchanger phase. Townsend (21) demonstrated that both the mole and equivalent fraction conventions are thermodynamically valid and that their use leads to solid-phase activity coefficients that differ but are entirely symmetrical and complementary.

The mole fraction convention is employed in the Vanselow equation (20)

$$K^V_{\text{Na-Ca}} = \frac{\{\text{Na}^+\}^2 N_{\text{CaX}_2}}{\{\text{Ca}^{2+}\} N_{\text{NaX}}^2} \quad (3)$$

where $K^V_{\text{Na-Ca}}$ is the Vanselow selectivity coefficient. The mole fractions of cations in the exchanger phase, N_{CaX_2} and N_{NaX}, are defined by

$$N_{\text{CaX}_2} = \frac{[\text{CaX}_2]}{[\text{CaX}_2] + [\text{NaX}]} \quad (4)$$

$$N_{\text{NaX}} = \frac{[\text{NaX}]}{[\text{CaX}_2] + [\text{NaX}]} \quad (5)$$

where [ ] represent molar concentrations.

In the Gaines–Thomas equation (22), equivalent fractions are employed to represent the activities of ions in the exchanger phase, that is,

$$K^{GT}_{Na-Ca} = \frac{\{Na^+\}^2 E_{CaX_2}}{\{Ca^{2+}\} E_{NaX}^2} \tag{6}$$

where K^{GT}_{Na-Ca} is the Gaines–Thomas selectivity coefficient. The equivalent fractions of cations in the exchanger phase, E_{CaX_2} and E_{NaX}, are defined by

$$E_{CaX_2} = \frac{2[CaX_2]}{2[CaX_2] + [NaX]} \tag{7}$$

$$E_{NaX} = \frac{[NaX]}{2[CaX_2] + [NaX]} \tag{8}$$

Equations 3 and 6, and some closely related empirical variants, have been used to fit many isotherms for monovalent–divalent exchange (23). For example, a two-parameter fit to isotherm data can be obtained by using the Rothmund–Kornfeld equation

$$K^{RK}_{Na-Ca} = \frac{\{Na^+\}^2 E_{CaX_2}^{2/n}}{\{Ca^{2+}\} E_{NaX}^{2/n}} \tag{9}$$

where K^{RK}_{Na-Ca}, the Rothmund–Kornfeld selectivity coefficient, and $1/n$, the mass law exponent, are empirical parameters specific to the exchange reaction under study (19). A special case ($n = 1$) of the Rothmund–Kornfeld equation yields the venerable Gapon equation

$$K^{G}_{Na-Ca} = (K^{RK}_{Na-Ca})^{0.5} = \frac{\{Na^+\} E_{Ca_{0.5}X}}{\{Ca^{2+}\}^{0.5} E_{NaX}} \tag{10}$$

where K^{G}_{Na-Ca} is the Gapon selectivity coefficient. The Gapon equation follows from conceptualizing the exchange reaction as a 1:1 interaction of monovalent negative charges and equivalents of cationic charge:

$$NaX + 0.5Ca^{2+} = Ca_{0.5}X + Na^+ \tag{11}$$

The Gapon equation is widely recognized as empirical in nature and thermodynamically dubious (e.g., see references 7 and 24) but has nonetheless often been used successfully to fit cation-exchange data. We will demonstrate later that the Gapon equation can indeed describe monovalent–divalent exchange under conditions in which sorption in the diffuse layer is minor in comparison with chemisorption.

Ion Exchange in Chemical Equilibrium Models

Ion-exchange equations such as those presented have been incorporated in chemical equilibrium models to account for interactions of ions with charged solids in aquatic systems. Models adapted in this way include general aqueous speciations models (25, 26) and solute transport models with sorption equilibria (e.g., 27–29). In using ion-exchange equations, it is important to remember that values of selectivity coefficients correspond to a relatively narrow set of conditions and that their use is proper only for calculations within the range of calibration conditions. Ion-exchange reactions can be included in chemical equilibrium models by including one of the exchanger species and the electrolyte ions as components. Other relevant exchanger species are then represented in terms of these components by using the empirical ion-exchange equations. The general formulation of chemical equilibrium problems is described in detail elsewhere (30, 31). For a particular aqueous system, a set of components is selected such that each of the chemical species present in the system can be expressed as a product of a reaction involving only the components.

One approach to including ion-exchange reactions in chemical equilibrium models is to break up the exchange reactions into half reactions by making X a component. If we consider the reaction in equation 1 as an example, the two half reactions would be

$$NaX \longrightarrow Na^+ + X^- \qquad \frac{1}{K_{NaX}} \qquad (12)$$

$$Ca^+ + 2X^- \longrightarrow CaX_2 \qquad K_{CaX_2} \qquad (13)$$

which yield the net exchange reaction in equation 1 upon combination. In these half reactions, X^- represents one equivalent of the negative surface charge that must be counteracted by cations, and K_{NaX} and K_{CaX} are equilibrium constants. Because the surface charge is always neutralized, free X^- never exists. Rather, X^- is used here as a fictitious species in the same way that the free electron is used in oxidation–reduction half reactions to account for the electrode potential in an electrochemical cell (30). Physical interpretations of the activity of X^- are model-dependent. As shown in Appendix 1 for a reference half reaction and the case of the Donnan model (32), the X^- activity is related to the electrical potential in the exchanger relative to the bulk solution according to

$$\{X^-\}^Z = \exp\left(\frac{-ZF\Psi}{RT}\right) \qquad (14)$$

where Z is the charge of the cation in the reference half reaction, Ψ is the electrical potential of the exchanger relative to bulk solution, F is the Faraday constant (96,485 C/mol), R is the gas constant (8.314 J/mol per K or V·C/mol per K), and T is the absolute temperature (K). The total concentration of the component X^-, or TOTX, is the measured CEC.

$$\text{TOTX} = [\text{NaX}] + 2[\text{CaX}_2] + [\text{HX}] + \cdots \quad (15)$$

In equation 15 co-ion exclusion also contributes slightly and, for completeness, should be considered. With the introduction of X^- as a component, it is a straightforward matter to incorporate cation-exchange equations into general chemical equilibrium models. All of the exchanger species can be defined in terms of X^- and one solution component (e.g., $\{\text{NaX}\} = K_{\text{NaX}} \{\text{Na}^+\} \{X^-\}$).

If the equivalent fraction convention for exchanger phase activities is adopted (i.e., the Gaines–Thomas approach), the molar concentrations are proportional to the equivalent fractions, and it can easily be shown that the half reaction constants and the binary-exchange constants (selectivity coefficients) are related by

$$K_{\text{Na-Ca}}^{\text{GT}} = \frac{2\text{TOTX}\, K_{\text{CaX}_2}}{K_{\text{NaX}}^2} \quad (16)$$

For monovalent–monovalent exchange, a simpler relationship pertains.

$$K_{\text{Na-K}}^{\text{GT}} = \frac{K_{\text{KX}}}{K_{\text{NaX}}} \quad (17)$$

With an arbitrary definition of K_{NaX} as equal to unity, thus establishing a reference half reaction, the equilibrium constant for any other half reaction can be determined from measured selectivity coefficients. The Gapon equation can be readily implemented in this manner. Implementation of the Vanselow equation, however, requires modification of the general equilibrium models to account for the more complex dependence of mole fractions on the molar concentrations. An example ion-exchange calculation using the half reaction approach to represent the Gapon equation is presented in Appendix 2.

Ion-exchange reactions thus can be incorporated in general chemical equilibrium models, and solid–water partitioning can be taken into account in calculating speciation for complex systems. We used this approach to formulate a soil–water chemical equilibrium model that has been incorporated in a comprehensive model of nutrient cycling in forest soils (29). A listing of some of the important reactions, including ion-exchange reactions formulated by using the Gaines–Thomas convention, is given in List I and the corresponding tableau (30) is provided as Table I. As a practical matter, selection of the

List I. Partial Chemical Model for Ion Exchange in a Soil–Water System

Species: H^+, OH^-, $H_2CO_3^*$, HCO_3^-, CO_3^{2-}, Na^+, Ca^{2+}, Al^{3+}, $AlOH^{2+}$, $Al(OH)_2^+$, NaX, CaX_2, AlX_3

Reactions:
$$H_2O \rightleftharpoons H^+ + OH^- \qquad K_w$$
$$H^+ + HCO_3^- \rightleftharpoons H_2CO_3^* \qquad K_{a1}^{-1}$$
$$HCO_3^- \rightleftharpoons CO_3^{2-} + H^+ \qquad K_{a2}$$
$$Al^{3+} + H_2O \rightleftharpoons AlOH^{2+} + H^+ \qquad K_{1,Al}$$
$$Al^{3+} + 2H_2O \rightleftharpoons Al(OH)_2^+ + 2H^+ \qquad K_{2,Al}$$
$$H^+ + X^- \rightleftharpoons HX \qquad K_{HX}$$
$$Na^+ + X^- \rightleftharpoons NaX \qquad K_{NaX}$$
$$Ca^{2+} + 2X^- \rightleftharpoons CaX_2 \qquad K_{CaX_2}$$
$$Al^{3+} + 3X^- \rightleftharpoons AlX_3 \qquad K_{AlX_3}$$

Mass law equations:
$$[OH^-] = [H^+]^{-1} K_w$$
$$[H_2CO_3^*] = [H^+][HCO_3^-] K_{a1}^{-1}$$
$$[CO_3^{2-}] = [H^+]^{-1} [HCO_3^-] K_{a2}$$
$$[AlOH^{2+}] = [Al^{3+}][H^+]^{-1} K_{1,Al}$$
$$[Al(OH)_2^+] = [Al^{3+}][H^+]^{-2} K_{2,Al}$$
$$[HX] = [H^+][X^-] K_{HX}$$
$$[NaX] = [Na^+][X^-] K_{NaX}$$
$$[CaX_2] = [Ca^{2+}][X^-]^2 K_{CaX_2}$$
$$[AlX_3] = [Al^{3+}][X^-]^3 K_{AlX_3}$$

Mole balance equations:
$$TOTH = [H^+] - [OH^-] + [H_2CO_3^*] - [CO_3^{2-}] - [AlOH^{2+}]$$
$$\quad - 2[Al(OH)_2^+] + [HX]$$
$$TOTNa = [Na^+] + [NaX]$$
$$TOTCa = [Ca^{2+}] + [CaX_2]$$
$$TOTHCO_3 = [H_2CO_3^*] + [HCO_3^-] + [CO_3^{2-}]$$
$$TOTAl = [Al^{3+}] + [AlOH^{2+}] + [Al(OH)_2^+] + [AlX_3]$$
$$TOTX = [NaX] + 2[CaX_2] + 3[AlX_3] = CEC$$

---

dominant exchanger species (e.g., CaX_2) rather than X^- as a component yields a more stable and more rapidly converging numerical solution.

Questions remain about how best to extrapolate from empirical binary ion-exchange data to the multicomponent exchange predominant in natural systems. Calibrating such empirical models by using measurements of solution and exchanger phase concentrations provides the most reasonable basis for representing small-to-moderate perturbations. An improved mechanistic understanding of the basis for cation-exchange selectivity is essential, however, before the accuracy of such extrapolations is known. Accurate description of

Table I. Tableau for Partial Chemical Model of List I

Species	H^+	HCO_3^-	Na^+	Ca^{2+}	Al^{3+}	X^-	K
H^+	1						1
OH^-	-1						K_w
$H_2CO_3^*$	1	1					K_{a1}^{-1}
HCO_3^-		1					1
CO_3^{2-}	-1	1					K_{a2}
Na^+			1				1
Ca^{2+}				1			1
Al^{3+}					1		1
$AlOH^{2+}$	-1				1		$K_{1,Al}$
$Al(OH)_2^+$	-2				1		$K_{2,Al}$
HX	1					1	K_{HX}
NaX			1			1	K_{NaX}
CaX_2				1		2	K_{CaX2}
AlX_3					1	3	K_{AlX3}
	TOTH	TOTHCO$_3$	TOTNa	TOTCa	TOTAl	TOTX	

NOTE: Components are listed on the top row (H$_2$O is a component, but not shown); species are listed in the first column. Mass law equations are given across rows; mole balance equations are given down columns. Additional information on the formulation of a chemical equilibrium problem in a tableau is given in Morel and Hering (30).

ion exchange in mechanistic models remains difficult, but such models can provide insight not afforded by empirical models.

Electrostatic Sorption in Ion Exchange

The two main processes involved in sorption of an ion onto a charged solid in an aqueous system are (1) nonspecific electrostatic attraction to the charged surface and (2) chemical bonding at discrete sites on the surface (including ion pairing or outer-sphere complex formation). Because of the balance of electrostatic attraction and thermal excitation forces, electrostatically sorbed ions are distributed throughout the interfacial region known as the diffuse layer (see Figure 4). This diffuse-layer sorption is well described by the Gouy–Chapman theory. The nonspecific contribution to sorption is most important for major electrolyte ions, as these comprise nearly all of the counterion charge adjacent to charged particles in aquatic systems. For ions present at minor-to-trace concentrations, surface sorption dominates solid–water partitioning.

Substantial efforts have been made to develop physicochemical models for ion exchange based on the Gouy–Chapman diffuse-layer theory (e.g., 9, 10). This work not only has provided insight into the role of diffuse-layer sorption in the ion-exchange process but also has pointed to the need to consider other factors, especially specific sorption at the surface. Consideration of specific sorption enables description of the different tendencies of ions to

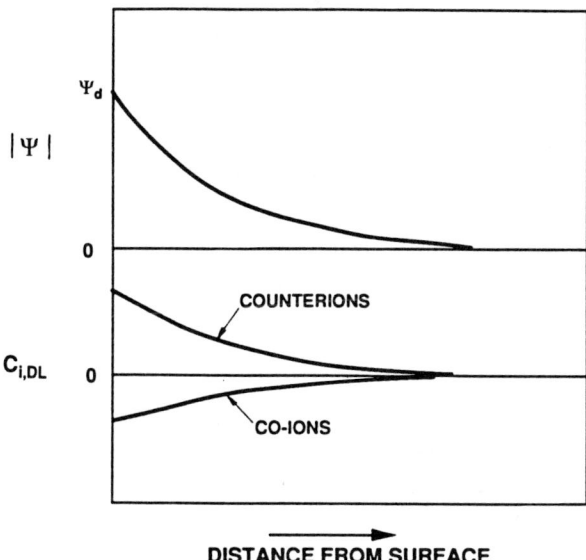

Figure 4. Schematic plots of electrical potential, counterion concentration, and co-ion concentration in the diffuse layer versus distance from a charged surface.

form bonds or ion pairs at the surface and also mitigates the problem of predicting impossibly high counterion densities very close to the surface. Such predictions occur because the Gouy–Chapman theory approximates ions as point charges.

The application of the Gouy–Chapman theory for describing ion exchange can be illustrated by deriving an exchange isotherm. Consider that the surface excess (in moles per square meter) of an electrostatically sorbed ion i is given by

$$\Gamma_i = \int_0^\infty (C_{i,x} - C_{i,0})(10^3 \text{ L/m}^3) \, dx \quad (18)$$

where $C_{i,x}$ is the molar concentration of ion i at distance x from the surface, and $C_{i,0}$ is the molar concentration of i in bulk solution (refer to Figure 4). The concentration in the diffuse layer of ion i with charge Z_i can be related to the bulk concentration through the Boltzmann equation

$$C_{i,x} = C_{i,0} \exp\left[\frac{-Z_i F \Psi(x)}{RT}\right] \quad (19a)$$

or, at low potential ($\Psi < 25$ mV), where the exponential term can be linearized,

$$C_{i,x} = C_{i,0} \left[1 - \frac{Z_i F \Psi(x)}{RT} \right] \tag{19b}$$

This expression assumes that the activity coefficients of an ion in the diffuse layer and in solution are the same. For low potentials, the Gouy–Chapman theory yields

$$\Psi(x) = \Psi_d \exp(-\kappa x) \tag{20}$$

$$\sigma_d = \epsilon \epsilon_0 \kappa \Psi_d \tag{21}$$

where Ψ_d and σ_d are the potential (V) and charge density (C/m²) at the surface, respectively, ϵ_0 (8.854 × 10⁻¹² C/V·m) is the dielectric permittivity of free space; ϵ is the dielectric constant of water (dimensionless); and κ is the inverse diffuse layer thickness (m⁻¹).

$$\kappa = \left[\left(\frac{F^2}{\epsilon \epsilon_0 RT} \right) (\Sigma Z_i^2 C_{i,0})(10^3 \text{ L/m}^3) \right]^{0.5}$$

$$\kappa = \left[\left(\frac{2F^2 I}{\epsilon \epsilon_0 RT} \right) (10^3 \text{ L/m}^3) \right]^{0.5} \tag{22}$$

The parameter I represents the ionic strength of the system (mol/L). Combining equations 18–21 yields the surface excess or deficit of an electrolyte ion, Γ_i (mol/m²), under the condition of low surface potential:

$$\Gamma_i = C_{i,0} (10^3 \text{ L/m}^3) \int_0^\infty \left[-Z_i F \Psi_d \frac{\exp(-\kappa x)}{RT} \right] dx$$

$$\Gamma_i = -C_{i,0} (10^3 \text{ L/m}^3) \frac{Z_i F \sigma_d}{\epsilon \epsilon_0 RT \kappa^2} \tag{23}$$

For a positive surface charge, positively charged ions will have a negative surface excess corresponding to expulsion of ions from the diffuse layer. Likewise, negatively charged ions near negatively charged surfaces will experience negative sorption in the diffuse layer.

Now assume that Na^+ – Ca^{2+} exchange involves only electrostatic sorption in the diffuse layer. According to the Gouy–Chapman theory, the surface excesses (mol/m²) of these two ions are given by

$$\Gamma_{Na} = -[Na^+](10^3 \text{ L/m}^3) \frac{F\sigma_d}{\epsilon\epsilon_0 RT\kappa^2} \quad (24)$$

$$\Gamma_{Ca} = -2[Ca^{2+}](10^3 \text{ L/m}^3) \frac{F\sigma_d}{\epsilon\epsilon_0 RT\kappa^2} \quad (25)$$

These expressions can be inserted in the definitions of the equivalent fractions of sorbed Na^+ and Ca^{2+} to obtain

$$\frac{E_{CaX_2}}{E_{NaX}} = \frac{2\Gamma_{Ca}/(2\Gamma_{Ca} + \Gamma_{Na})}{\Gamma_{Na}/(2\Gamma_{Ca} + \Gamma_{Na})} = \frac{4[Ca^{2+}]}{[Na^+]} \quad (26)$$

which is an ion-exchange mass law expression for Na^+–Ca^{2+} exchange on a low-potential surface. This exchange equation and the corresponding isotherm (Figure 5) indicate a constant, weak selectivity in favor of Ca^{2+}. Equation 26 lacks the second-order dependence on $[Na^+]$ of equations 3 and 6, both of which can fit monovalent–divalent exchange data, because at low potential the surface excess of each ion is simply proportional to its charge number (see eq 23). At higher potentials, which are examined later, the Gouy–Chapman theory predicts a more realistic selectivity for heterovalent exchange. However, because no surface sorption and no ionic properties other than charge are considered, many empirical observations (e.g., the selectivity sequence for monovalent ion sorption on clays) cannot be explained. For example, no selectivity would be expected for homovalent exchange, and thus the observed

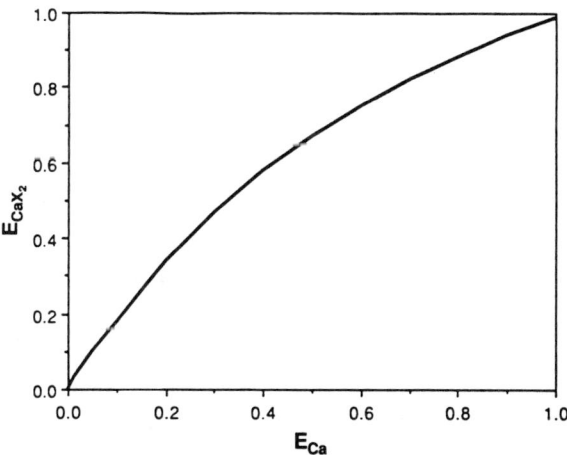

Figure 5. Isotherm predicted from the Gouy–Chapman theory (with low-potential approximation) for Na^+–Ca^{2+} exchange on a charged solid.

differences between K^+–Ca^{2+} and Na^+–Ca^{2+} exchange isotherms cannot be predicted.

As seen in equation 23 for the low-potential case, the concentrations of excess ions in the diffuse layer are related to the surface potential, the bulk solution concentrations of the individual ions, and the overall composition of the bulk solution phase. Equation 23 may be rewritten as

$$C_{i,\mathrm{DL}} = as\Gamma_i = g_i\, C_{i,0} \tag{27}$$

where

$$g_i = -as\ (10^3\ \mathrm{L/m^3})\,\frac{Z_i F \sigma_\mathrm{d}}{\epsilon \epsilon_0 RT \kappa^2} \tag{28}$$

and $C_{i,\mathrm{DL}}$ is the excess or deficit concentration (mol/L) of ion i in the diffuse layer, a is the specific surface area (m²/g) of the charged solid, and s is the mass concentration (g/L) of the solid. $C_{i,\mathrm{DL}}$ will be positive for counterions and negative for co-ions. Equation 27, which is based on the Gouy–Chapman theory, accounts for a number of factors that influence ion concentrations in the diffuse layer. However, it does not account for ion–ion interactions in the diffuse layer, which may become important at high electrolyte concentrations.

Borkovec and Westall (33) derived a general expression for g_i —an integral equation that requires numerical solution except in the case of symmetrical electrolytes—that is applicable to all potentials. The general analytical solution for g_i is given (33) by

$$g_i = \left(\frac{as}{F}\right)\left(\frac{\epsilon\epsilon_0 RT}{2}\right)^{0.5} \mathrm{sgn}(P-1) \int_1^P \frac{X^{Z_i}-1}{[X^2 \Sigma C_{j,\mathrm{B}}(X^{Z_j}-1)]^{0.5}}\, dX \tag{29}$$

where $P = \exp(-F\Psi_\mathrm{d}/RT)$ is the dimensionless potential, $C_{j,\mathrm{B}}$ is the bulk solution concentration of species j, and the summation runs over all bulk solution species.

Physicochemical Modeling of Ion Exchange

Several physicochemical models of ion exchange that link diffuse-layer theory and various models of surface adsorption exist (9, 10, 14, 15). The difficulty in calculating the diffuse-layer sorption in the presence of mixed electrolytes by using analytical methods, and the sometimes over simplified representation of surface sorption have hindered the development and application of these models. The advances in numerical solution techniques and representations of surface chemical reactions embodied in modern surface complexation models (*11–13, 34–41*) show promise as a foundation for the further development of physicochemical models of ion exchange.

In surface complexation modeling, chemisorption of ions on mineral surfaces is described by assuming reactions analogous to those that occur among solutes. Reactive surface sites are represented as independent reactant species. Surface hydroxyl groups, for example, are represented by $\equiv SOH^0$, where $\equiv S$ indicates a surface metal atom having multiple bonds in the bulk solid phase. With this notation, the coordination of an ion by a surface hydroxyl group may be described by

$$\equiv SOH^0 + M^{2+} = \equiv SOM^+ + H^+ \quad (30)$$

where M^{2+} is a divalent cation. Mass law expressions corresponding to such reactions have been employed with success to describe equilibrium sorption of ions at mineral surfaces.

Similar reactions can be written for the specific sorption of cations and anions at other kinds of surface sites. Here, we represent ion binding at permanent charge sites by the reaction

$$\equiv S_p^- + M^{Z+} = \equiv S_p-M^{Z-1} \quad (31)$$

$\equiv S_p^-$ represents a site of fixed charge arising from isomorphous substitution or other structural defects. Because the intrinsic equilibrium constants for equations 30 and 31 reflect solute concentrations at the surface of the sorbent, which depend in turn on the surface potential, a coulombic term must be included in the mass law expression

$$K_{S_p-M}^{int} \exp\left(\frac{-\Delta Z\, F\Psi_d}{RT}\right) = \frac{[\equiv S_p-M^{Z-1}]}{[\equiv S_p^-](M^{Z+})} \quad (32)$$

where ΔZ is the net change in charge number of the surface species, Ψ_d is the surface potential, and $K_{S_p-M}^{int}$ is the intrinsic (chemical interaction) equilibrium constant for sorption of cation M^{Z+}. The coulombic term is incorporated in general chemical equilibrium models by defining a new component, *P* (*30, 31*). To calculate the surface potential and hence *P*, a molecular model for the geometry and location of ions at the mineral–water interface must be invoked. Various molecular models have been used for this purpose, resulting in a number of closely related but somewhat different surface complexation models (*35–37*).

Current surface complexation models were developed with a focus on minor and trace ions and hence do not consider sorption in the diffuse layer. Even the triple-layer model (*34*), which can include electrolyte sorption as outer-sphere complexes, does not consider sorption in the diffuse layer. To the extent that diffuse-layer sorption contributes to total sorption, its contribution is included with specific sorption and modeled as such.

Surface complexation models can be extended to account explicitly for electrostatic sorption by calculating excess counterion concentrations in the diffuse layer in addition to specific sorption. Counterions in the diffuse layer (e.g., Ca^{2+}_{DL}) can then be treated as distinct from those in bulk solution (e.g., Ca^{2+}) and those that are specifically sorbed (e.g., $\equiv S_p-Ca^+$). The total sorption is given by the sum of the concentrations of specifically sorbed and electrostatically sorbed species:

$$\Gamma_{Ca} = \left(\frac{1}{as}[Ca^{2+}_{DL}] + [\equiv SO-Ca^+] + [\equiv S_p-Ca^+]\right) \tag{33}$$

A schematic representation of the locations of these species relative to a charged mineral surface is given in Figure 6.

To demonstrate the physicochemical modeling of ion exchange via an extended surface complexation model, we consider an aqueous system containing a mineral solid that bears fixed- and variable-charge sites. As seen in Figure 6, the mineral–water interface is represented by a simple two-layer

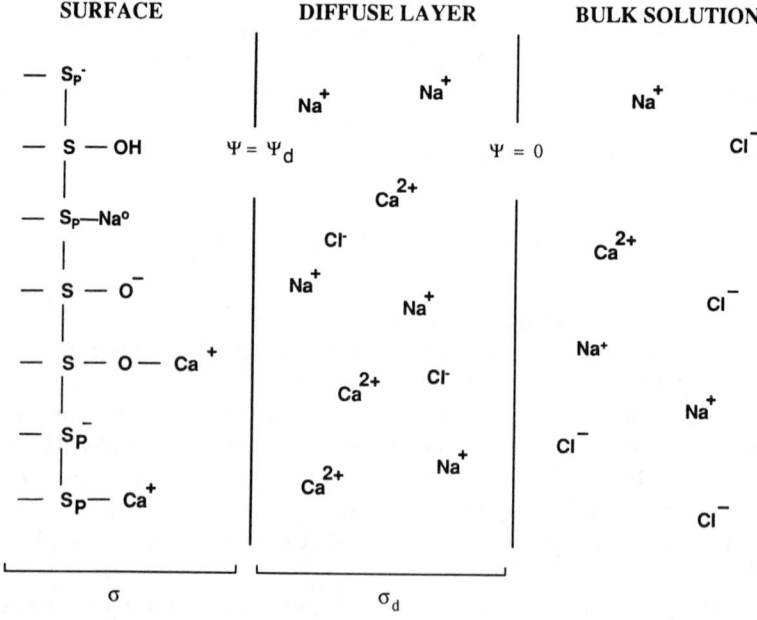

Figure 6. Schematic representation of electrostatic sorption and surface complexation involved in Na^+–Ca^{2+} exchange at the mineral–water interface. $\equiv SOH^0$ represents a surface hydroxyl (variable-charge) site; $\equiv S_p^-$ represents a site of fixed (permanent) negative charge.

model—one surface layer and one diffuse layer of counterions. The model also includes calcium and sodium sorbed specifically at both variable- and fixed-charge surface sites. Surface species reflecting one-to-one binding are shown. Such surface complexes would be expected for clay minerals with a relatively low density of fixed charge sites (typically about 7–15 ° between sites). The equations given in List II may be used to define the equilibrium composition of the system. The surface excess or deficit of each ion in the diffuse layer is represented explicitly as a species.

The system of equations in List II may be summarized as a tableau (30), as is done in Table II. Different values are required for the mass law exponents and mole balance coefficients for some species. The mole balance equation for the coulombic component P imposes the electroneutrality constraint for the solid–water interface. The charge–potential relationship given by the Gouy–Chapman theory, which is incorporated explicitly in surface complexation models through component P (31), is included implicitly through the g_i factors in the tableau. This model is based on the elegant approach of Borkovec and Westall (33) for incorporation of diffuse-layer sorption in a chemical equilibrium model. Solutions to the model equations were obtained by incorporating the tableau of Table II in the chemical equilibrium program MICROQL (42, 43). A simplification of Borkovec and Westall's (33) general approach to solving for the surface potential was employed. Ion concentrations in bulk solution were fixed so that $\equiv S_p^-$ and P were left as the only unknown components. Equilibrium compositions for different combinations of Na^+, Ca^{2+}, and Cl^- concentrations were then obtained by calculating the speciation with P as a fixed activity component and iterating on P until the electroneutrality constraint for the surface and diffuse layer was satisfied. The integrals defining g_i (eq 29) were integrated by using a logarithmic transformation of X and the extended midpoint rule (44). With this approach, no modification of the Jacobian matrix in the Newton–Raphson solution technique of MICROQL was needed.

We used the model to analyze the empirical isotherms obtained by Bond and Phillips (17) for Na^+–Ca^{2+} and K^+–Ca^{2+} exchanges on a clay subsoil. Properties of the soil and the experimental conditions are summarized in List III. Because of the incompleteness of information about the soil and the soil surface properties, our model "fits" must be regarded as qualitative investigations of the character of ion exchange on these clays. Model parameter values and assumptions are found in List IV. To simplify the calculations, we assumed that fixed-charge sites dominate the surface of the soil [i.e., $TOT(\equiv S_p) = CEC$ and $TOT(\equiv SOH) = 0$]. More generally, the contributions of variable-charge sites to CEC (or AEC), obtainable by fitting pH titration data, must be taken into account.

Bond and Phillips (17) presented their exchange isotherm data in reference to the ratio of monovalent ion activity to divalent ion activity (e.g., $\{Na^+\}/\{Ca^{2+}\}^{0.5}$) for the concentrations in the solution phase. The activity ratio,

List II. Chemical Model for Na^+–Ca^{2+} Exchange in an Aqueous Suspension of a Solid with Variable- and Fixed-Charge Sites

Species: H^+, OH^-, Na^+, Ca^{2+}, Cl^-, $\equiv SOH_2^+$, $\equiv SOH^0$, $\equiv SO^-$, $\equiv S_p^-$, Na_{DL}^+, H_{DL}^+, OH_{DL}^-, CA_{DL}^{2+}, Cl_{DL}^-, $\equiv SO\text{–}Ca^+$, $\equiv S_p\text{–}Ca^+$, $\equiv SO\text{–}Na^0$, $\equiv S_p\text{–}Na^0$

Reactions:
$$H_2O \rightleftarrows H^+ + OH^- \qquad K_w$$
$$\equiv SOH_2^+ \rightleftarrows \equiv SOH^0 + H^+ \qquad K_{a1}^{int}\, P^{-1}$$
$$\equiv SOH^0 \rightleftarrows \equiv SO^- + H^+ \qquad K_{a2}^{int}\, P^{-1}$$
$$\equiv SOH^0 + Ca^{2+} \rightleftarrows \equiv SO\text{–}Ca^+ + H^+ \qquad K_{SO-Ca}^{int}\, P$$
$$\equiv S_p^- + Ca^{2+} \rightleftarrows \equiv S_p\text{–}Ca^+ \qquad K_{Sp-Ca}^{int}\, P^2$$
$$\equiv SOH^0 + Na^+ \rightleftarrows \equiv SO\text{–}Na^0 + H^+ \qquad K_{SO-Na}^{int}$$
$$\equiv S_p^- + Na^+ \rightleftarrows \equiv S_p\text{–}Na^0 \qquad K_{Sp-Na}^{int}\, P$$

Mass law equations:
$$[OH^-] = [H^+]^{-1} K_w$$
$$[\equiv SOH_2^+] = [H^+][\equiv SOH^0] P (K_{a1}^{int})^{-1}$$
$$[\equiv SO^-] = [H^+]^{-1}[\equiv SOH^0] P^{-1} K_{a2}^{int}$$
$$[\equiv SO\text{–}Ca^+] = [H^+]^{-1}[\equiv SOH^0][Ca^{2+}] P\, K_{SO-Ca}^{int}$$
$$[\equiv S_p\text{–}Ca^+] = [\equiv S_p^-][Ca^{2+}] P^2\, K_{Sp-Ca}^{int}$$
$$[\equiv SO\text{–}Na^0] = [H^+]^{-1}[\equiv SOH^0][Na^+]\, K_{SO-Na}^{int}$$
$$[\equiv S_p\text{–}Na^0] = [\equiv S_p^-][Na^+]\, K_{Sp-Na}^{int}\, P$$

Diffuse layer sorption:
$$[Na_{DL}^+] = g_{Na}[Na^+]$$
$$[Ca_{DL}^{2+}] = g_{Ca}[Ca^{2+}]$$
$$[Cl_{DL}^-] = g_{Cl}[Cl^-]$$
$$[H_{DL}^+] = g_H[H^+]$$
$$[OH_{DL}^-] = g_{OH}[OH^-]$$

More balance equations:
$$TOTH = [H^+] - [OH^-] + [\equiv SOH_2^+] - [\equiv SO^-] - [\equiv SO\text{–}Ca^+] - [\equiv SO\text{–}Na^0] + [H_{DL}^+] - [OH_{DL}^-]$$
$$TOT[\equiv SOH] = [\equiv SOH_2^+] + [\equiv SOH^0] + [\equiv SO^-] + [\equiv SO\text{–}Ca^+] + [\equiv SO\text{–}Na^0]$$
$$TOT[\equiv S_p^-] = [\equiv S_p^-] + [\equiv S_p\text{–}Ca^+] + [\equiv S_p\text{–}Na^0]$$
$$TOTNa = [Na^+] + [\equiv SO\text{–}Na^0] + [\equiv S_p\text{–}Na^0] + [Na_{DL}^+]$$
$$TOTCa = [Ca^{2+}] + [\equiv SO\text{–}Ca^+] + [\equiv S_p\text{–}Ca^+] + [Ca_{DL}^{2+}]$$
$$TOTCl = [Cl^-] + [Cl_{DL}^-]$$

Electrical double layer neutrality:
$$\sigma_d + \sigma_{DL} = (F/as)TOTP = (F/as)\{[\equiv SOH_2^+] - [\equiv SO^-] + [\equiv SO\text{–}Ca^+] - [\equiv S_p^-] + [\equiv S_p\text{–}Ca^+] + [H_{DL}^+] - [OH_{DL}^-] + [Na_{DL}^+] + 2[Ca_{DL}^{2+}] - [Cl_{DL}^-]\} = 0$$

NOTE: $P = \exp(-F\Psi_d/RT)$.

Table II. Tableau for 0Chemical Model of List II

Species	H^+	Na^+	Ca^{2+}	Cl^-	$\equiv SOH^0$	$\equiv S_p^-$	P	K
H^+	$1, 1 + g_H$						$0, g_H$	1
OH^-	$-1, -1 - g_{OH}$						$0, -g_{OH}$	K_w
Na^+		$1, 1 + g_{Na}$					$0, g_{Na}$	1
Ca^{2+}			$1, 1 + g_{Ca}$				$0, 2g_{Ca}$	1
Cl^-				$1, 1 + g_{Cl}$			$0, -g_{Cl}$	1
$\equiv SOH_2^+$	1				1		1	$(K_{a1}^{int})^{-1}$
$\equiv SOH^0$					1			1
$\equiv SO^-$	-1				1		-1	K_{a2}^{int}
$\equiv SO-Ca^+$	-1		1		1		1	K_{SO-Ca}^{int}
$\equiv SO-Na^0$	-1	1			1			K_{SO-Na}^{int}
$\equiv S_p^-$						1	$0, -1$	1
$\equiv S_p-Ca^+$			1			1	$2, 1$	K_{Sp-Ca}^{int}
$\equiv S_p-Na^0$		1				1	$1, 0$	K_{Sp-Na}^{int}
	TOTH	TOTNa	TOTCa	TOTCl	TOT($\equiv$SOH)	TOT($\equiv S_p$)	O	

NOTE: Components are on the top row (H_2O is a component but not shown); species are in the first column. Mass law equations are given across rows; mole balance equations are given down columns. Where two numbers are given, the first is the mass alw exponent and the second is the mole balance coefficient. $P = \exp(-F\Psi_d/RT)$ is the coulombic term. Additional information on formulation of a chemical equilibrium problem in a tableau is given in Morel and Hering (reference 30).

List III. Soil Properties and Experimental Conditions for Na^+–Ca^{2+} and K^+–Ca^{2+} Exchange Data

Brucedale clay soil [> 60% clay]

Clay composition: 30–40% illite
30–40% kaolinite
20–30% interstratified clay

CEC = 22 mequiv/100 g (sum of exchangeable Na^+, K^+, Mg^{2+}, Ca^{2+})

Exchangeable Ca^{2+} = 0.214 equiv/kg

Soil concentration in batch experiments = 5 g/25 mL = 200 g/L

Soil in Ca-form at beginning of each batch experiment

Soil Ca = 0.214 equiv/kg × 0.2 kg/L = 0.043 equiv/L.
Additional Ca as $CaCl_2$ added in some experiments.

SOURCE: Data are taken from reference 17.

which may be derived by taking the square root of equation 6, is a natural variable for fitting data with empirical ion-exchange equations. The experimental Na^+–Ca^{2+} exchange data, shown in Figure 7, were first fitted by considering diffuse-layer sorption only. The Gouy–Chapman high-potential isotherm is sensitive to surface site density (= CEC/a). At values greater than about 2 μequiv/m², the high surface potential leads to greater sorption of Ca^{2+} than is observed, whereas the opposite is true at low site densities. Because the high CEC of the soil suggests that its character is closer to that of illite than to that of kaolinite, we adopted a high specific surface area, 100 m²/g, in the range of those reported for illite (List IV). This value also has the virtue of matching the empirical isotherm relatively well (Figure 7).

The fit of the Na^+–Ca^{2+} isotherm with diffuse-layer sorption alone predicts surface potentials in the range of −70 to −120 mV. These potentials imply surface concentrations of Ca^{2+} or Na^+ in excess of 10 M over much of the isotherm. Although such concentrations may be physically possible, it would seem likely that surface ion pairing occurs at such high concentrations. In fact, some specific binding of Na^+ is expected on the basis of observations of Na^+ ion pairing in solution and estimates of ion pairing at charged mineral surfaces (14, 15, 45). To illustrate the effect of this process on ion exchange, we introduced the surface reactions yielding $\equiv S_p$–Na^0 and $\equiv S_p$–Ca^+ into the model (List II). We assumed that calcium interacts with only a single site, because the site spacing, typically 1.0–1.5 nm in clays (9), is much greater than the ionic radius of the Ca^{2+} ion, 0.1 nm (46). Adopting the surface complexation ("inner-sphere, surface ion pair formation") constant calculated for Na^+ on montmorillonite by Shainberg and Kemper (15, 45) as a starting point for our fitting exercise, we adjusted K^{int}_{Sp-M} for formation of $\equiv S_p$–Ca^+

List IV. Parameter Values and Assumptions Used in Physiocochemical Model for Ion Exchange

a = specific surface area of Brucedale clay soil = 100 m²/g

This value was assumed on the basis of the high CEC of the Brucedale clay soil (220 μequiv/g), which indicates that the clay mixture in the soil has properties more like illite (a = 90–130 m²/g, CEC = 200–400 μequiv/g) than like kaolinite (a = 10–20 m²/g, CEC = 20–60 μequiv/g). The ranges of clay properties are taken from Talibudeen (46).

s = mass concentration of soil in water = 200 g/L

[Cl⁻] was fixed at 0.1 M; TOTCl ranged from 0.096 M to 0.098 M.

[Ca²⁺], [Na⁺], and [K⁺] were varied to obtain the range of activity ratios shown in Figures 7 and 9, while maintaining electroneutrality in the bulk solution. The resulting range of TOTCa was 0.0044–0.14 equiv/L, and the TOTNa and TOTK range was 0.002–0.14 equiv/L.

TOT($\equiv S_p^-$) = CEC = (0.022 equiv/100 g)(200 g/L) = 0.044 equiv/L

Surface complexation constants:

K_{Sp-Na}^{int} = 0.2 M⁻¹   Fixed at the value for surface complexation of Na⁺ on montmorillonite calculated by Kemper and Shainberg (45).

K_{Sp-Ca}^{int} = 1.0 M⁻¹   Adjusted to fit the Na⁺–Ca²⁺ exchange data of Bond and Phillips (17), with K_{Sp-Na}^{int} fixed.

K_{Sp-K}^{int} = 3.6 M⁻¹   Adjusted to fit the K⁺–Ca²⁺ exchange data of Bond and Phillips (17), with K_{Sp-Ca}^{int} fixed at value determined in fitting Na⁺–Ca²⁺ exchange data.

---

until the predicted and observed Na⁺ sorption values were similar. The relative values of the two surface complexation constants, 0.2 and 1.0 M⁻¹, are consistent with the greater electrostatic attraction of the divalent ion for the negative site, as is observed for Na⁺–anion and Ca²⁺–anion pairs in solution (47).

In the fit associated with consideration of specific sorption, the formation of $\equiv S_p$–Na⁰ and $\equiv S_p$–Ca⁺ reduces the negative surface potential relative to the no-surface-complexation case by about 25 mV over the entire isotherm. The result is a decrease in the predicted surface concentrations of sodium (<4 M) and calcium (<1.5 M) and their accumulation in the diffuse layer.

Concentrations of the major solution and surface species corresponding to the calculated isotherms in Figure 7 are shown on Figures 8a (no surface complexation) and 8b (surface complexation of Na⁺ and Ca²⁺). Without surface complexation (Figure 8a), Na⁺ dominates sorption in the diffuse layer at high TOTNa concentrations, although Ca²⁺ quickly becomes dominant as TOTNa decreases. With specific sorption, the surface complex of Na⁺ and its diffuse-layer species constitute roughly similar fractions of the total sorbed

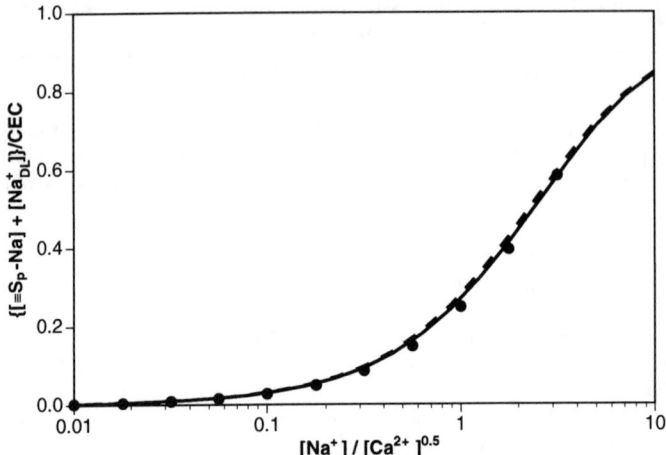

Figure 7. Composite isotherm for Na^+–Ca^{2+} exchange on Brucedale clay soil for various TOTNa and TOTCa concentrations. [Data (representative points shown) are from reference 17.] Curves are fits obtained with physicochemical ion-exchange model considering diffuse-layer sorption alone (– – –) and combined diffuse-layer sorption plus surface complexation of Na^+ and Ca^{2+} (solid line). See Lists III and IV for model parameters.

Na^+, whereas for Ca^{2+}, specific sorption dominates partitioning to the solid phase.

Bond and Phillips (17) also reported K^+–Ca^{2+} exchange isotherms for the same soil. In accordance with exchange data reported in the literature for many other soils, K^+ was observed to sorb more strongly on the test soil than did Na^+ in the presence of Ca^{2+}, as is evident by comparison of the Na^+–Ca^{2+} and the K^+–Ca^{2+} exchange isotherms in Figures 7 and 9. The stronger sorption of K^+ relative to Na^+ indicates a greater degree of surface complexation of K^+.

We fitted the K^+–Ca^{2+} exchange isotherms of Bond and Phillips with a model mathematically identical to that used to fit the Na^+–Ca^{2+} exchange data (Table II). Because the same soil was involved, we used the same value for K^{int}_{Sp-Ca} but adjusted the value for K^{int}_{Sp-K} (3.6 M^{-1}) was greater than both K^{int}_{Sp-Na} and K^{int}_{Sp-Ca}. The K^{int}_{Sp-K} value is slightly higher than the value of 2.0 M^{-1} estimated by Kemper and Shainberg (45) but is consistent with the high affinity of illite for K^+ (46). The necessity of including specific sorption of K^+ is clearly indicated by the poor fit of the model for diffuse-layer sorption only (dashed curve in Figure 9). The diffuse-layer sorption represented by this curve is the same as that predicted for Na^+ in the absence of specific sorption, a consequence of the Gouy–Chapman theory accounting only for ionic charge. Calculated concentrations of the major solution and surface species for the

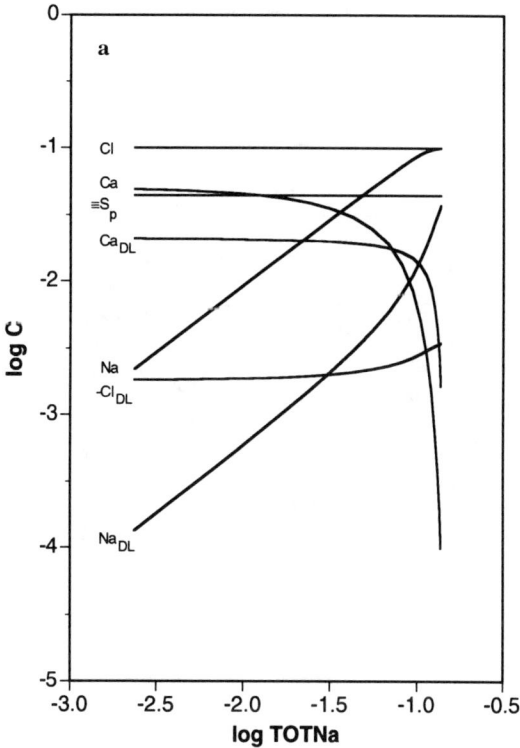

Figure 8. Predicted concentrations of major surface and solution species in monovalent–divalent exchange model fits of Figures 7 and 9. The diffuse-layer species represent the excess (K^+, Na^+, Ca^{2+}) or deficit (Cl^-) of ions relative to the bulk solution. Part a: log C versus log TOTNa for fit of Na^+–Ca^{2+} exchange data (Figure 7) without consideration of surface complexation. Part b: log C versus log TOTNa for fit of Na^+–Ca^{2+} exchange data (Figure 7) with surface complexation of Na^+ and Ca^{2+}. Part c: log C versus log TOTK for fit of K^+–Ca^{2+} exchange data (Figure 9) with surface complexation of K^+–Ca^{2+}. Continued on next page.

K^+–Ca^{2+} isotherm (Figure 8c) show that the K^+ surface complex is the predominant sorbed species.

Physicochemical models of partitioning at the solid–water interface, such as that used here to model ion exchange, require detailed knowledge about the particles. The surface properties of the mineral phases present, as well as equilibrium constants for ion binding to both fixed and variable charge sites associated with each phase, are required. These data requirements and the uncertainty about modeling sorption in mixtures of minerals (e.g., 48–50) make such models difficult to apply to complex natural systems. This is especially the case for modeling solute transport in soil–water systems, which

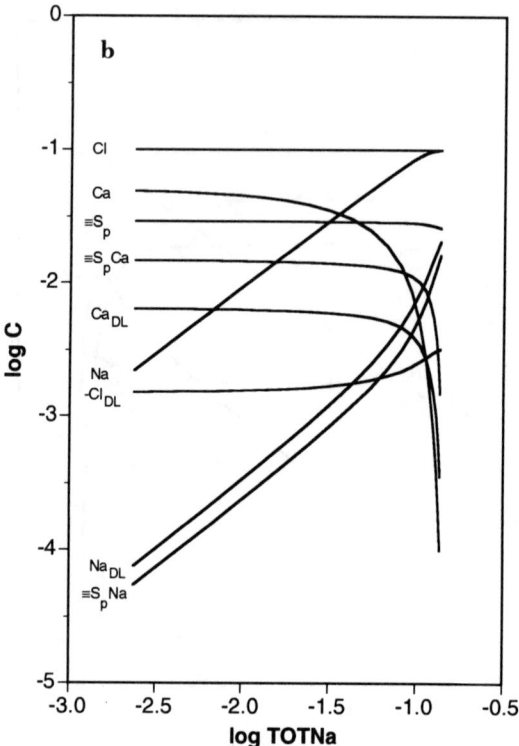

Figure 8.—Continued.

are characterized by substantial heterogeneity (29, 51, 52). Nevertheless, for a simple, well-characterized aqueous suspension, physicochemical models of ion exchange that consider diffuse-layer sorption and surface complexation can provide insights into solute partitioning that cannot be obtained from an empirical model.

Insights into Exchange Equations

The physicochemical ion-exchange model and example applications presented can be used to gain understanding about the suitability of different exchange equations for different systems. Again, we consider the case of monovalent–divalent ion exchange. In Bruggenwert and Kamphorst's survey (23) of experimental information on cation exchange in soil systems, most Na^+–Ca^{2+} exchange data are reported as Gaines–Thomas selectivity coefficients, whereas the majority of K^+–Ca^{2+} selectivity coefficients are of the Gapon (Rothmund–Kornfeld) form. Similarly, Bond and Phillips (17) fitted their Na^+–Ca^{2+} data with the Gaines–Thomas equation but their K^+–Ca^{2+} data with the Roth-

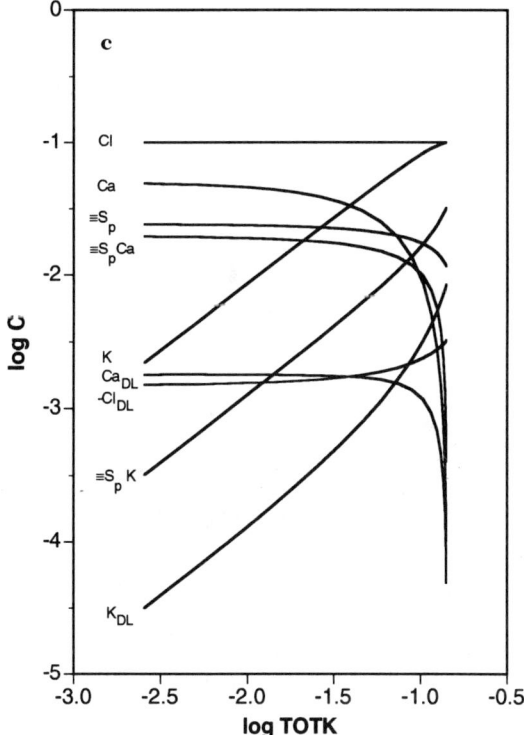

Figure 8.—Continued.

mund–Kornfeld equation. The difference in behavior of the Gaines–Thomas and Gapon equations may be seen in Figure 10, where both isotherms have been fitted to the K^+–Ca^{2+} data of Bond and Phillips. As seen there, the Gapon equation is almost capable of fitting the Bond and Phillips K^+–Ca^{2+} exchange data, but the Gaines–Thomas equation is clearly not appropriate.

In the context of the physicochemical model described, the Gapon equation reflects the strong influence of surface sorption, whereas the Gaines–Thomas equation is more useful where diffuse-layer sorption dominates. The reason for this behavior can be seen by considering a limiting case where specific sorption of both cations, as either outer- or inner-sphere complexes, is the predominant form of sorption in the exchanger phase. The relevant binary exchange equation can be derived by eliminating P from the mass law expressions for $[\equiv S_p\text{–}Ca^+]$ and $[\equiv S_p\text{–}Na^0]$ shown in List II:

$$\frac{[\equiv S_p\text{–}Ca^+]}{[\equiv S_p\text{–}Na^0]^2} = \frac{[Ca^{2+}]K^{int}_{S_p\text{–}Ca}}{[\equiv S_p^-][Na^+]^2(K^{int}_{S_p\text{–}Na})^2} \tag{34}$$

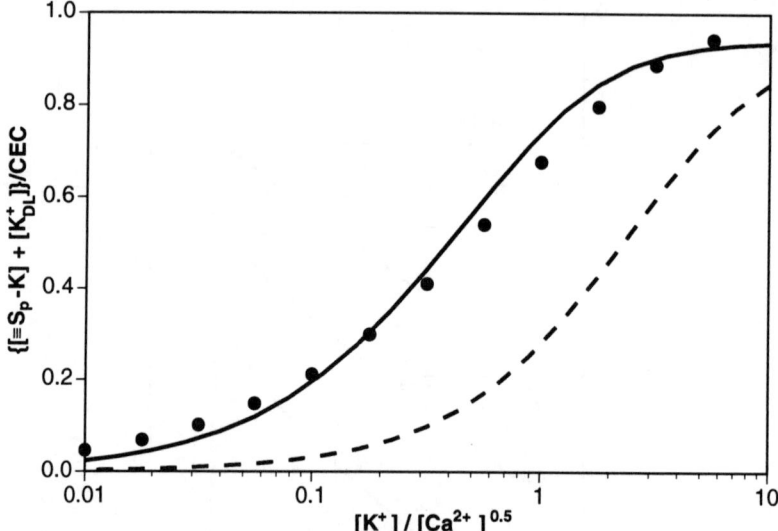

Figure 9. Composite isotherm for K^+–Ca^{2+} exchange on Brucedale clay soil for various TOTK and TOTCa concentrations. [Data (representative points shown) are from reference 17.] Curves are fits obtained with physicochemical ion-exchange model considering diffuse-layer sorption alone (– – –) and combined diffuse layer sorption plus surface complexation of K^+ and Ca^{2+} (solid line). See Lists III and IV for model parameters.

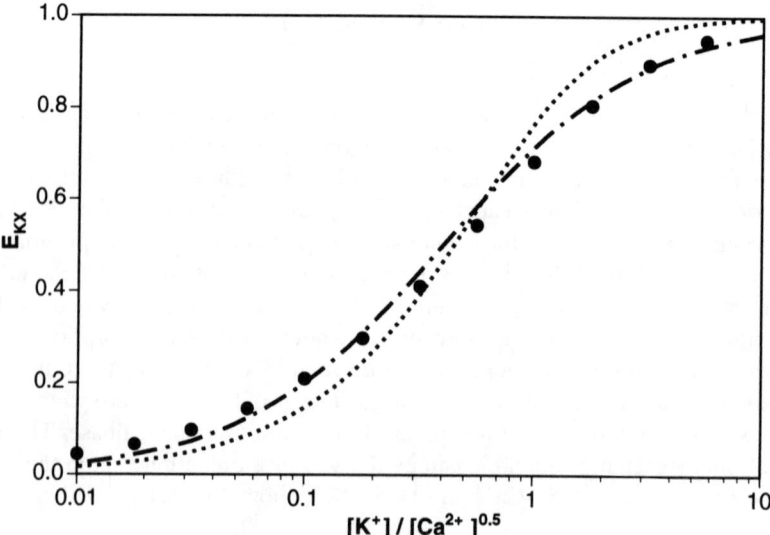

Figure 10. Empirical fits of K^+–Ca^{2+} exchange data from reference 17. Curves are fits obtained with Gapon isotherm equation (– – –), $K^G_{K-Ca} = 0.4$, and with Gaines–Thomas isotherm equation (.....), $K^{GT}_{K-Ca} = 0.4$.

When specific sorption predominates, the electroneutrality condition at the surface (List II) reduces to:

$$[\equiv S_p-Ca^+] \approx (\equiv S_p^-) \tag{35}$$

Use of this equality permits simplification of equation 32 to

$$\frac{[\equiv S_p-Ca^+]}{[\equiv S_p-Na^0]} = \frac{[Ca^{2+}]^{0.5}(K^{int}_{S_p-Ca})^{0.5}}{[Na^+](K^{int}_{S_p-Na})} \tag{36}$$

which can be readily converted to the Gapon equation (eq 10), with

$$K^G_{Na-Ca} = \frac{2(K^{int}_{S_p-Ca})^{0.5}}{K^{int}_{S_p-Na}} \tag{37}$$

This argument suggests that the Gapon equation results from the existence of equal concentrations of free and calcium-bound specific sites under these conditions. Such conditions may be obtained when high electrolyte concentrations or extensive surface sorption leads to low surface potentials. The stronger surface sorption of K^+ relative to Na^+ may account for the fact that Gapon-like behavior is more often observed in systems with potassium than in those with sodium. Although equations 34–36 suggest that the Gapon equation may have physical significance after all, equation 35 is not pertinent to ternary or higher-order exchange systems. These exceptions imply that the Gapon formulation is not generalizable to multicomponent systems.

Summary and Conclusions

The exchange of ions at the charged surfaces of solids in aqueous suspension is an important process in solute fate and transport in soil and aquatic systems. Ion exchange involves both nonspecific, electrostatic sorption in the diffuse layer and specific sorption attributable to ion pairing and complexation at surface sites. Despite the complexity of the process, simple stoichiometric reactions describing the transfer of equivalents of charge between the solution and the surface are commonly employed to model ion exchange. The corresponding mass law equations are used to describe macroscopic partitioning behavior of solute ions under particular conditions and are thus largely empirical. Because the equilibrium constants for these reactions are highly dependent on the nature of the solid and the conditions in solution, they are actually variable, conditional constants more commonly referred to as selectivity coefficients.

Ion-exchange reactions can be useful for describing partitioning in heterogeneous systems, provided that the selectivity coefficients are used only within the range of calibration conditions. These reactions may be incorporated in equilibrium chemical speciation models through use of exchange half reactions between each exchanging ion and the species X^-, a model construct representing a unit of surface charge to be counteracted.

The utility of semiempirical ion-exchange reactions in describing solid–water partitioning of ions in complex systems is evident from their extensive use, and physicochemical modeling is necessary to investigate the fundamental principles of the ion-exchange behavior of such systems. With sufficient knowledge about the properties of a charged solid in an aqueous system, a physicochemical model for ion exchange can be formulated by extending surface complexation models to account explicitly for electrostatic sorption in the diffuse layer. For each ion, the excess diffuse-layer concentration, calculated via the Gouy–Chapman theory, can be combined with the surface complex concentration to give overall sorption. The diffuse-layer sorption calculations can be incorporated in general chemical speciation models by representing the excess ions in the diffuse layer as distinct species. Although the contribution of diffuse-layer sorption to total sorption is small for many ions, it can be important for weakly sorbing electrolyte ions.

The utility of physicochemical models for ion exchange, such as that presented here, can be assessed only by application to a larger number of data sets. Such models can serve as a basis for extrapolating from binary to multicomponent exchange systems and for generating additional physical insights into exchange isotherms. Physicochemical models may also prove useful in analyzing exchange of complex ions. Systematic study of ion-exchange systems by using a consistent model may yield a limited set of model parameters that can be used to describe the behavior of a wide range of exchanging species, at least for simple systems.

A physicochemical sorption model requires detailed knowledge about the surface properties of the solid phase and the reactions occurring there, knowledge that is often difficult or impossible to acquire for natural particles in environmental systems. Nevertheless, research on sorption at the mineral–water interface has demonstrated that physicochemical models can be applied with success to describe solute partitioning in simple, well-characterized aqueous suspensions. It appears that such models can be used to describe ion exchange in simple systems as well. Insights gained from application of such models will aid interpretation of complicated phenomena observed in environmental systems and may help with construction of more robust semiempirical models. Although the heterogeneity of natural solids will require much additional effort to accommodate in physicochemical models of ion exchange, such efforts promise to advance our understanding, as the progress in knowledge and modeling of ion–humate interactions attests.

Acknowledgments

We thank Steven Gherini for his support and encouragement in the initiation of this work. We are also indebted intellectually to many others, especially Gerhard Bolt, Garrison Sposito, Werner Stumm, and John Westall, whose research and writings have helped us (try) to understand ion exchange and its relationship to mechanistic sorption models. This work was supported in part by the Electric Power Research Institute under a contract with Tetra Tech, Inc., Robert Hudson's former affiliation; and by grants from Duquesne Light Company, Chambers Development Company, the Donors of the Petroleum Research Fund administered by the American Chemical Society, and the National Science Foundation (PYI Grant No. BCS–9157086) to Carnegie Mellon University.

Appendix 1. Interpretation of {x} Component

Physical interpretation of the X^- species employed in ion-exchange half reactions is governed by the molecular model used to describe the ion-exchange process. If the ion exchanger is viewed as a porous gel, for example, in which counterions are distributed within the charged solid lattice, the ion concentrations in the exchanger phase may be related to the corresponding bulk solution concentrations by (32)

$$(A_{EX}^Z) = (A_B^Z) \exp\left(\frac{-ZF\Psi}{RT}\right) \quad (1.1)$$

where Ψ is the potential in the exchanger phase relative to bulk solution. Equation 1.1 describes the well-known Donnan equilibrium. The transfer of an ion from the bulk solution to the exchanger phase in the Donnan model could also be represented by

$$A^Z + ZX^- = AX_Z \quad (1.2)$$

where the species X^- represents a unit charge in the exchanger phase. The mass law expression corresponding to equation 1.2 is

$$K_{A-X} = \frac{\{AX_Z\}}{\{A^z\}\{X^-\}^Z} \quad (1.3)$$

In the context of the porous gel, or Donnan model, of the exchanger,

$$\frac{\{AX_Z\}}{\{A^Z\}} = \exp\left(\frac{-ZF\Psi}{RT}\right) \tag{1.4}$$

which, upon substitution in equation 1.3, yields

$$\{X^-\}^Z = \frac{\exp\left(\dfrac{-ZF\Psi}{RT}\right)}{K_{A-X}} \tag{1.5}$$

If equation 1.2 corresponds to the reference exchange half reaction in a system, then K_{A-X} may be defined as unity by convention and equation 1.5 indicates plainly that the component X^- represents the dimensionless exchanger phase potential.

Appendix 2. An Example Demonstrating the Use of Ion-Exchange Half Reactions

This example is based on an illustrative problem from Bolt et al. (53). Starting with a 100-g soil sample (dry weight) that is brought to water saturation, one finds a moisture content, W, of 60 cm^3 of water per 100 g of dry soil. The concentration of the ions present in the soil solution at this moisture content is determined in a small amount of filtrate obtained from a subsample:

$[Na^+] = 0.006$ N $= 0.006$ M
$[Ca^{2+}] = 0.001$ N $= 0.0005$ M
$[Cl^-] = 0.007$ N $= 0.007$ M

Other ions are present only in negligible amounts. The sample is percolated with a sufficient amount of 1 N NH_4NO_3 solution to ensure complete exchange. Analysis of the percolate yields the following total amounts of ions in the system (per 100 g of dry soil):

$T_{Na} = 3.40$ mequiv/100 g
$T_{Ca} = 22.12$ mequiv/100 g
$T_{Cl} = 0.28$ mequiv/100 g

Now the CEC, exchanger species concentrations, and exchange constant can be calculated:

$$CEC = \Sigma\, T_{cat} - \Sigma\, T_{an}$$

3. DZOMBAK & HUDSON   *Ion Exchange*

$$CEC = (3.40 + 22.12 - 0.28) \text{ mequiv}/100 \text{ g}$$
$$CEC = 25.24 \text{ mequiv}/100 \text{ g}$$

$$[NaX] = [T_{Na}-W[Na^+]] / W = (T_{Na}/W)-[Na^+]$$
$$[NaX] = (3.40 \text{ mequiv}/100 \text{ g})/(60 \text{ mL}/100 \text{ g})-0.006 \text{ N}$$
$$[NaX] = 0.0507 \text{ N}$$

$$[Ca_{0.5}X] = [T_{Ca}-W[Ca^{2+}]] / W = (T_{Ca}/W)-[Ca^{2+}]$$
$$[Ca_{0.5}X] = (22.12 \text{ mequiv}/100 \text{ g})/(60 \text{ mL}/100 \text{ g})-0.001 \text{ N}$$
$$[Ca_{0.5}X] = 0.3677 \text{ N}$$

$$K_{Na-Ca} = \frac{[Ca_{0.5}X][Na^+]}{[NaX][Ca^{2+}]^{0.5}} = \frac{(0.3677 \text{ N})(0.006 \text{ M})}{(0.0507 \text{ N})(0.0005 \text{ M})^{0.5}}$$

$$K_{Na-Ca} = 1.95 \approx 2$$

If to a sample of this system (containing 100 g of soil and 60 mL of solution), one adds 140 mL of a solution containing 9.6 mequiv of NaCl and 0.5 mequiv of $CaCl_2$, what is the equilibrium composition?

To set up the problem for input to a chemical equilibrium model, we choose Na^+, Ca^{2+}, and X as components and write the following reactions:

$$Na^+ + X^- = NaX \qquad K_{NaX} = 1$$
$$0.5 \text{ } Ca^{2+} + X^- = Ca_{0.5}X \qquad K_{Ca_{0.5}X} = 2$$

Combination of these two half reactions yields

$$K_{Na-Ca} = K_{Ca_{0.5}X}/K_{NaX} = 2$$

Now we need to calculate the total concentrations for each component.

$$TOTX = CEC = (25.24 \text{ mequiv}/100 \text{ g})(100 \text{ g}/200 \text{ mL}) = 0.126 \text{ N}$$

$$TOTNa = (0.0507 \text{ N} + 0.006 \text{ N})(60 \text{ mL}/200 \text{ mL}) + (9.6 \text{ mequiv}/200 \text{ mL})$$

$$TOTNa = 0.065 \text{ N}$$
$$TOTNa = 0.065 \text{ M}$$

$$TOTCa = (0.3677 \text{ N} + 0.001 \text{ N})(60 \text{ mL}/200 \text{ mL}) + (0.5 \text{ mequiv}/200 \text{ mL})$$

TOTCa = 0.1131 N

TOTCa = 0.0566 M

The appropriate mass balance equations are

$$\text{TOTNa} = [\text{Na}^+] + [\text{NaX}] = 0.065 \text{ M}$$

$$\text{TOTCa} = [\text{Ca}^{2+}] + 0.5\ [\text{Ca}_{0.5}X] = 0.0566 \text{ M}$$

$$\text{TOTX} = [\text{NaX}] + [\text{Ca}_{0.5}X] = 0.126 \text{ N}$$

where co-ion (Cl⁻) exclusion has been neglected in formulating TOTX as a simplification. We thus have five equations in five unknowns, which can be solved to obtain the solution to the equilibrium problem:

$[\text{Na}^+] = 0.0401$ M

$[\text{Ca}^{2+}] = 0.0063$ M

$[\text{NaX}] = 0.0254$ N $= 0.0254$ M

$[\text{Ca}_{0.5}X] = 0.1006$ N $= 0.0503$ M

This solution, which was obtained with MICROQL (42), agrees with that obtained graphically by Bolt et al. (53).

References

1. *Soil Chemistry. Vol. A, Basic Elements;* Bolt, G. H.; Bruggenwert, M. G. M., Eds.; Elsevier: Amsterdam, The Netherlands, 1978.
2. *Soil Chemistry. Vol. B, Physico-Chemical Models;* Bolt, G. H., Ed.; Elsevier: Amsterdam, The Netherlands, 1982.
3. Applebaum, S. B. *Demineralization by Ion Exchange;* Academic: Orlando, FL, 1968.
4. Helfferich, F. *Ion Exchange;* McGraw-Hill: New York, 1962.
5. Fletcher, P.; Townsend, R. P. *J. Chem. Soc. Faraday Trans. 2,* **1981,** 77, 965.
6. Chu, S. Y.; Sposito, G. *Soil Sci. Soc. Am. J.* **1981,** 45, 1084.
7. Sposito, G. *The Thermodynamics of Soil Solutions;* Oxford University: New York, 1981; Chapters 5, 6.
8. Bolt, G. H.; Page, A. L. *Soil Sci.* **1965,** 99, 357.
9. Bolt, G. H. *Neth. J. Agric. Sci.* **1967,** 15, 81.
10. Bolt, G. H. In *Soil Chemistry. Vol. B, Physico-Chemical Models;* Bolt, G. H., Ed.; Elsevier: Amsterdam, The Netherlands, 1982; pp 47–75.
11. Stumm, W.; Huang, C. P.; Jenkins, S. R. *Croat. Chem. Acta* **1970,** 42, 223.
12. Stumm, W.; Hohl, H.; Dalang, F. *Croat. Chem. Acta* **1976,** 48, 491.
13. Stumm, W.; Kummert, R.; Sigg, L. *Croat. Chem. Acta* **1980,** 53, 291.
14. Heald, W. R.; Frere, M. H.; de Wit, C. T. *Soil Sci. Soc. Am. Proc.* **1964,** 28, 622.
15. Shainberg, I.; Kemper, W. D. *Soil Sci.* **1967,** 103, 4.

16. *Methods of Soil Analysis. Part 2, Chemical and Microbiological Analysis,* 2nd ed.; Page, A. L.; Miller, R. H.; Keeney, D. R., Eds.; American Society of Agronomy: Madison, WI, 1982.
17. Bond, W. J.; Phillips, I. R. *Soil Sci. Soc. Am. J.* **1990**, *54*, 722.
18. Bond, W. J.; Phillips, I. R. *Soil Sci. Soc. Am. J.* **1990**, *54*, 636.
19. Sposito, G. *The Chemistry of Soils;* Oxford University: New York, 1989.
20. Vanselow, A. P. *Soil Sci.* **1932**, *33*, 95.
21. Townsend, R. *Phil. Trans. Royal Soc. London* **1984**, *311A*, 301.
22. Gaines, G. L.; Thomas, H. C. *J. Chem. Phys.* **1953**, *21*, 714.
23. Bruggenwert, M. G. M.; Kamphorst, A. In *Soil Chemistry. Vol. B, Physico-Chemical Models;* Bolt, G. H., Ed.; Elsevier: Amsterdam, The Netherlands, 1982; pp 141–203.
24. Sposito, G. *Soil Sci. Soc. Am. J.* **1977**, *41*, 1205.
25. Felmy, A. R.; Girvin, D. C.; Jenne, E. A. *MINTEQ: A Computer Program for Calculating Aqueous Geochemical Equilibria;* EPA–600/3–84–032; U.S. Environmental Protection Agency: Athens, GA, 1984; 84 pp.
26. Parkhurst, D. L.; Thorstenson, D. C.; Plummer, L. N. *PHREEQE: A Computer Program for Geochemical Calculations;* U.S. Geological Survey, WRI 80–96; U.S. Geological Survey: Reston, VA, 1980; 210 pp.
27. Valocchi, A. J.; Street, R. L.; Roberts, P. V. *Water Resour. Res.* **1981**, *17*, 1517.
28. Cederberg, G. A.; Street, R. L.; Leckie, J. O. *Water Resour. Res.* **1985**, *21*, 1095.
29. Liu, S.; Munson, R.; Johnson, D.; Gherini, S.; Summers, K.; Hudson, R.; Wilkinson, K.; Pitelka, L. *Tree Physiol.* **1991**, *9*, 173.
30. Morel, F. M. M.; Hering, J. G. *Principles and Applications of Aquatic Chemistry;* Wiley: New York, 1993.
31. Westall, J. C. In *Particulates in Water;* Kavanaugh, M. C.; Leckie, J. O., Eds.; ACS Advances in Chemistry 189; American Chemical Society: Washington, DC, 1980; pp 33–44.
32. Westall, J. C. In *Aquatic Surface Chemistry;* Stumm, W., Ed.; Wiley: New York, 1987; pp 3–32.
33. Borkovec, M.; Westall, J. C. *J. Electroanal. Chem.* **1983**, *150*, 325.
34. Davis, J. A.; James, R. O.; Leckie, J. O. *J. Colloid Interface Sci.* **1978**, *63*, 480.
35. Westall, J. C.; Hohl, H. *Adv. Colloid Interface Sci.* **1980**, *12*, 265.
36. Sposito, G. *The Surface Chemistry of Soils;* Oxford University: New York, 1984.
37. Dzombak, D. A.; Morel, F. M. M. *J. Hydraul. Eng.* **1987**, *113*, 430.
38. Dzombak, D. A.; Morel, F. M. M. *Surface Complexation Modeling: Hydrous Ferric Oxide;* Wiley: New York, 1990.
39. Sposito, G. *Chimia* **1989**, *43*, 169.
40. Davis, J. A.; Kent, D. B. *Rev. Mineral.* **1990**, *23*, 177.
41. Stumm, W. *Chemistry of the Solid–Water Interface;* Wiley: New York, 1992.
42. Westall, J. C. *MICROQL I. A Chemical Equilibrium Program in BASIC;* Technical Report; Swiss Federal Institute of Technology: Dübendorf, Switzerland, 1979; 42 pp.
43. Westall, J. C. *MICROQL II. Computation of Adsorption Equilibria in BASIC;* Technical Report; Swiss Federal Institute of Technology: Dübendorf, Switzerland, 1979; 35 pp.
44. Press, W. H.; Flannery, B.; Teukolsky, S. A.; Vetterling, W. T. *Numerical Recipes in C: The Art of Scientific Computing;* Cambridge University: New York, 1988.
45. Kemper, W. D.; Shainberg, I. *Soil Sci.* **1967**, *104*, 448.
46. Talibudeen, O. In *The Chemistry of Soil Processes;* Greenland, D. J.; Hayes, M. H. B., Eds.; Wiley: Chichester, England, 1981; pp 115–177.
47. Fuoss, R. M. *J. Am. Chem. Soc.* **1958**, *80*, 5059.

48. Honeyman, B. D.; Santschi, P. H. *Environ. Sci. Technol.* **1988,** 22, 862.
49. Anderson, P. R.; Benjamin, M. M. *Environ. Sci. Technol.* **1990,** 24, 1586.
50. Meng, X.; Letterman, R. D. *Environ. Sci. Technol.* **1993,** 27, 970.
51. Bolt, G. H.; Van Riemsdijk, W. H. In *Aquatic Surface Chemistry;* Stumm, W., Ed.; Wiley: New York, 1987; pp 127–164.
52. Dzombak, D. A.; Ali, M. A. *Water Pollut. Res. J. Can.* **1993,** 28, 7.
53. Bolt, G. H.; Bruggenwert, M. G. M.; Kamphorst, A. In *Soil Chemistry. Vol. A, Basic Elements;* Bolt, G. H.; Bruggenwert, M. G. M., Eds.; Elsevier: Amsterdam, The Netherlands, 1978; pp 58–60, 71.

RECEIVED for review August 17, 1992. ACCEPTED revised manuscript November 17, 1993.

4

Interaction of Organic Matter with Mineral Surfaces

Effects on Geochemical Processes at the Mineral–Water Interface

Janet G. Hering

Department of Civil and Environmental Engineering, University of California–Los Angeles, Los Angeles, CA 90024-1593

The interaction of organic matter with mineral surfaces influences mineral properties and reactivity. Organic matter–surface associations may either protect minerals from dissolution or increase rates of mineral weathering, depending on the structure of the adsorbing organic compound and the nature of the organic matter–surface interaction. The dissolved organic matter (DOM) in soils consists of a mixture of substances with widely varying structures and properties. Low-molecular-weight organic ligands have been identified in soil waters and in leachates of forest litter. Laboratory studies suggest that such compounds, although only minor components of soil DOM, may be effective at promoting the solubilization of metals in soils. In contrast, humic substances appear to have only slight effects on the dissolution of oxide and silicate minerals; both inhibition and acceleration of mineral weathering (depending on pH) have been observed. This variability in the effects of DOM components on mineral dissolution suggests that metal mobility in soils will depend on the source and composition of soil DOM.

MANY IMPORTANT GEOCHEMICAL PROCESSES occur at mineral–water interfaces. Iron and aluminum are mobilized in soils by the dissolution of oxide and silicate minerals or of amorphous oxide coatings on soil minerals (1, 2). The leaching of iron and aluminum immediately below the organic-rich layers of the upper soil column and their reprecipitation at depth results in the characteristic banded appearance of podzolic soils (3–5). The mobilization of aluminum in soils and the resulting increased aluminum concentrations in soil pore waters, groundwaters, and groundwater-fed lakes are of considerable concern because of aluminum toxicity to higher plants and fish (6–8).

When oxides in soils or sediments dissolve, substances adsorbed to the oxide surfaces will also be released into solution. Thus, for example, phosphate release into sediment pore waters accompanies the reductive dissolution of iron oxides in anoxic sediments. Release of phosphate into the overlying (oxic) water column is limited by phosphate adsorption on freshly precipitated amorphous iron oxides at the oxic–anoxic interface (9, 10). A similarly coupled cycle of phosphate and iron is observed in surface waters where photochemical reductive dissolution of iron oxides results in increased dissolved concentrations of ferrous iron and phosphate during the day (11).

Enhanced dissolution of aluminum oxides occurs at low pH and in the presence of organic ligands. The dissolution of reducible oxides, particularly manganese and iron oxides, is also facilitated by reductants. In soils, one effect of acid precipitation is increased mobilization of aluminum (12, 13). Significant concentrations of dissolved organic matter (DOM) may be present in water infiltrating through soils as a result of biological activity in the root zone.

Correlations between concentrations of dissolved metals (iron and/or aluminum) and of either DOM or organic acids in infiltrating waters suggest that DOM, or some component of DOM, facilitates metal solubilization and translocation in soils (14). The role of DOM in metal mobilization is also supported by the occurrence of organically complexed metals in soil solutions (3, 15, 16). Precipitation and transformations of soil minerals can be inhibited by DOM, which tends to stabilize metastable phases [such as amorphous aluminum hydroxide and pseudoboehmite (17), ferrihydrite (18), and octacalcium phosphate (19)] and to prevent formation of more crystalline phases.

Soil DOM comprises a wide variety of organic substances. Specific microbial or plant exudates are released as a consequence of metabolic activity of soil biota. Transformations of biogenic compounds, through partial degradation and polymerization, result in the formation of a structurally ill-defined mixture of humic substances, a term used to include both humic and fulvic acids (20, 21). Nonuniform effects of DOM components on metal mobilization may be anticipated because of the heterogeneity of their chemical structures and properties.

Laboratory studies of oxide dissolution in the presence of organic ligands (ligand-promoted dissolution) have shown that rates of oxide dissolution can be increased by organic matter–surface interactions; this enhancement depends on the structure of the organic ligand (22–24). The extension of these results to natural systems suggests that low-molecular-weight (LMW) organic acids, although not the major constituents of soil DOM, may nonetheless contribute significantly to the overall effect of DOM on metal mobilization in soils. The detailed interactions of humic substances with oxide surfaces are not known but may be deduced from observed effects of humic substances on dissolution.

In this chapter, possible roles of DOM constituents, particularly LMW organic acids and humic substances, in the solubilization and translocation of metals in soils are discussed. Competitive and synergistic interactions between these DOM constituents, based on laboratory studies of oxide dissolution, are proposed.

Surface Structure and Reactivity

Surface Chemistry of Oxide Minerals. Because of their coordinative unsaturation, the metal centers at the surface of oxide minerals interact strongly with water. Thus the oxide surface can be characterized as a polymeric oxo acid, and the specific adsorption of protons or hydroxide ions can be interpreted in terms of acid–base reactions at the oxide surface. According to the surface complexation model (25–27), surface charge is developed by reactions yielding charged surface species, such as protonated or deprotonated surface hydroxyl groups. In many soil minerals, permanent structural charge, arising from isomorphic substitution, also contributes to surface charge.

The effects of surface charge are evident in the stabilization of colloidal particles. Colloidal suspensions are destabilized (subject to rapid aggregation) at their point of zero charge (PZC), where there is no net surface charge on the particles. This minimum in colloidal stability coincides with the absence of electrophoretic mobility of the particles at the isoelectric point (IEP). In the absence of other specifically adsorbing species, the PZC of oxide particles corresponds to the PZNPC, point of zero net proton condition, where the concentrations of positively charged, protonated and negatively charged, deprotonated surface hydroxyl groups are equal (28).

Organic–Surface Interactions. The association of an organic compound with an oxide surface may be dominated by favorable organic matter–surface interactions, unfavorable organic matter–solvent interactions (hydrophobic exclusion), or favorable second-order interactions such as hydrophobic interactions of adsorbed organic molecules or coadsorption of metals and organics. The organic matter–surface interactions include both chemical interactions, which may be either specific (surface complexation) or nonspecific (hydrogen bonding or van der Waals attraction), and electrostatic interactions. The relative contributions of these types of interactions to the organic matter–surface association will vary, depending on the physicochemical characteristics of both the organic adsorbate and the surface, as well as on parameters such as pH and ionic strength (29, 30).

Direct spectroscopic evidence demonstrates the formation of inner-sphere complexes between organic ligands and surface metal centers, in which the organic ligand replaces a surface hydroxyl group. Formation of surface complexes and of metal–organic complexes in solution result in similar changes in the spectroscopic properties of the ligand, as determined by Fourier transform infrared, electron spin resonance (ESR), and fluorescence spectroscopy (31–33).

In the adsorption of organic ligands through surface complex formation, adsorption is limited by the number of exchangeable hydroxyl groups on the oxide surface. Surface saturation is commonly observed in the adsorption isotherms of LMW organic ligands on oxides (34–36), which may be described by the Langmuir equation

$$\{\equiv ML\} = \frac{KS_T[L]}{1 + K[L]} \tag{1}$$

where $\{\equiv ML\}$ is the ligand concentration at the surface, $[L]$ is the concentration of dissolved ligand, S_T is the total concentration of surface sites, and K is an apparent stability constant for the surface complex that includes a coulombic factor arising from the surface charge of the oxide. Because ligand adsorption influences the surface charge of oxides, the stability constant K will vary somewhat with surface coverage, an effect that is often not discernible in the adsorption isotherms (23). The extent of adsorption of organic ligands is influenced by pH and the presence of competing anionic adsorbates. The pH effects are due to ligand protonation (at low pH) and competition with hydroxide in the ligand-exchange reaction at the surface (at high pH). Other inorganic anions, such as phosphate and chromate, can compete with and decrease the adsorption of organic ligands (37–39).

Organic adsorbates that are more hydrophobic exhibit different adsorption behavior, particularly at higher concentrations. Long-chain fatty acids adsorb to oxide surfaces in part through surface complexation, as shown by electron spin resonance spectroscopy (32). At higher concentrations at the surface, however, favorable interactions between sorbed molecules (hemimicelle formation) appear to dominate and result in greater than monolayer adsorption (40, 41). Because humic substances (like the fatty acids) are amphiphilic, both surface complexation and hydrophobic interactions may be involved in the adsorption of humic substances on oxide surfaces.

Ligand-Promoted Dissolution. In some cases, organic matter–surface associations markedly increase oxide dissolution rates. The acceleration of oxide dissolution by organic ligands exhibits saturation kinetics; that is, the dissolution rate reaches a plateau at high concentration of dissolved ligand as the surface becomes saturated. This behavior is characteristic of surface-controlled reactions (in which the reaction of a surface-bound species is the rate-limiting step) and is consistent with the direct dependence of the dissolution rate on the concentration of the reactive species at the surface (22–24).

The proposed mechanism for the ligand-promoted dissolution of aluminum oxide involves three general steps: (1) ligand adsorption and surface complex formation, (2) slow detachment of a surface metal center (as a complex with the ligand), and (3) regeneration of the surface (shown schematically in Figure 1).

Transport of reactants or products between the bulk solution and the surface is considered to be rapid compared with surface reactions. For this mechanism, the rate law for dissolution is of the form

$$\text{rate} = \frac{d[M]}{dt} = k\{\equiv ML\} \qquad (2)$$

where $\{\equiv ML\}$ is the concentration of the surface complex, $d[M]/dt$ is the change in the concentration of dissolved metal with time, and k is a rate constant. Steady-state dissolution kinetics (i.e., a constant dissolution rate) has been observed for aluminum oxide with a series of organic ligands (22). This kinetic behavior indicates, as does more direct spectroscopic evidence (33), that a constant ligand concentration at the surface is maintained over the course of the dissolution.

Not all organic ligands are equally effective at promoting oxide dissolution. Surface reactivity, specifically the tendency for the complexed surface metal center to detach from the crystal lattice, depends on the structure of the surface complex. Rates of oxide dissolution are markedly enhanced in the presence of bidentate organic ligands that form chelate complexes with surface metal centers. The relative enhancement by different bidentate ligands has been related to the ring size of the surface chelate; rate constants for ligand-promoted dissolution (i.e., the rate constant k in eq 2) decrease with increasing size of the chelate ring from a five-membered ring for oxalate and six-membered ring for malonate to seven-membered ring for succinate (Figure 2a). In contrast, monodentate ligands, such as benzoate, have little effect on oxide dissolution and can competitively inhibit dissolution by bidentate ligands. The decreasing effectiveness of some bidentate ligands in promoting dissolution at low pH has been attributed to opening of the chelate ring (Figure 2b). Thus the effect of organic matter–surface interactions on surface reactivity may be explained in terms of the specific nature of organic matter–surface interactions, in particular the formation of bidentate surface complexes (22).

When oxide minerals are exposed to a mixture of organic substances, such as soil DOM, the composition of the mixture will determine the extent to which oxide dissolution is enhanced. Thus, to evaluate the role of soil DOM in the dissolution of soil minerals, the effectiveness of individual DOM components in promoting dissolution and the interactions of various components must be considered. The following discussion will examine the possible roles of two types of soil DOM components, humic substances and LMW organic

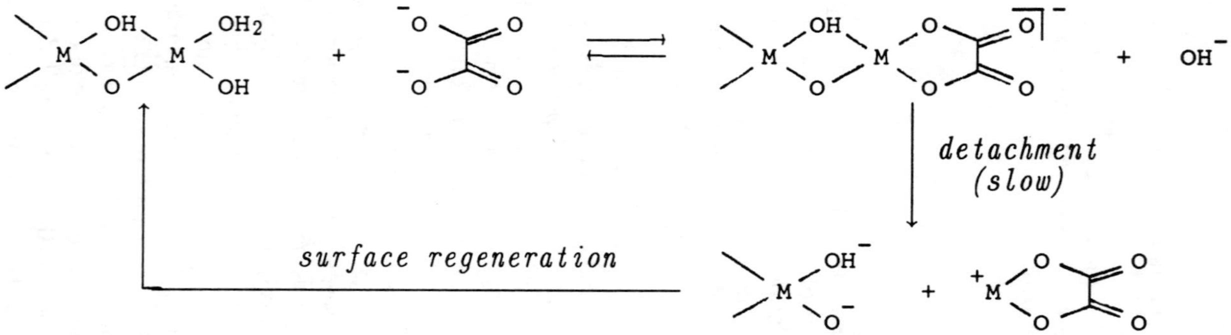

Figure 1. *Schematic diagram of oxalate-promoted dissolution of an oxide mineral for* $M = Al(III)$ *or* $Fe(III)$ *based on the surface-controlled dissolution model. (Reproduced with permission from reference 22. Copyright 1986 Pergamon Press.)*

4. HERING *Interaction of Organic Matter with Mineral Surfaces*   101

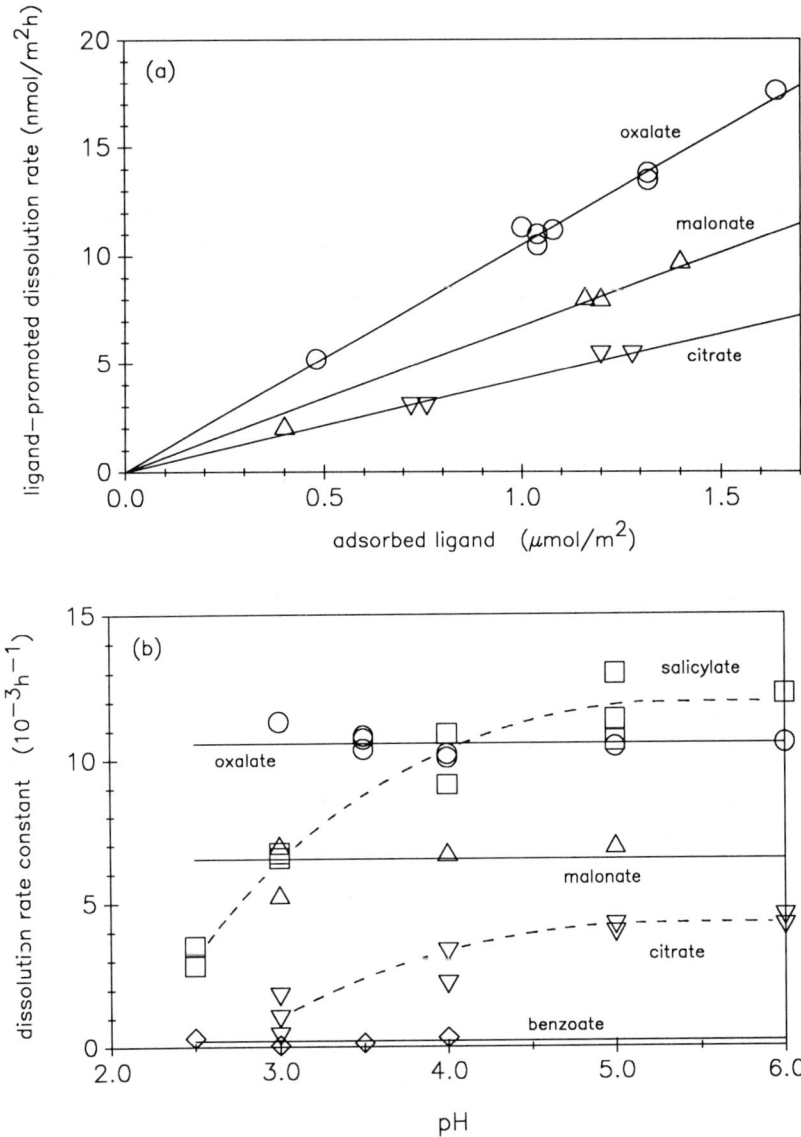

Figure 2. Ligand-promoted dissolution of δ-Al_2O_3 (2.2 g/L). Part a: Dissolution rates as a function of adsorbed ligand concentrations for a series of organic ligands. Part b: Dissolution rate constants as a function of pH. Symbols: (○) oxalate, (△) malonate, (▽) citrate, (□) salicylate, and (◇) benzoate. (Adapted with permission from reference 22. Copyright 1986 Pergamon Press.)

acids, in the dissolution of soil minerals and in the stabilization of dissolved metals in soil waters.

The Role of Humic Substances in Complexation, Adsorption, and Dissolution

The preceding discussion relies on an analogy between complexation in solution and complexation at the mineral surface, a fundamental tenet of the surface complexation model (27). Strong complexation of metals in solution by humic substances is well-documented (16, 42–44). Thus surface complex formation is a likely mechanism for the adsorption of humic substances on oxide surfaces.

Humic substances are polymeric organic acids of variable composition; acidity (typically 10–20 mequiv/g of C) is contributed by carboxyl and phenolic functional groups. Terrigenous humic substances are derived in large part from degradation of higher plant material and retain a high degree of aromaticity from their biogenic precursors. Metal complexation by humic substances is attributed to chelation by proximate acidic functional groups (21). The molecular weights of humic substances range from approximately 1000 Da for fulvic acids to 25,000 Da or higher for humic acids (45). The structural complexity of humic substances is reflected in their acid–base and metal complexation chemistry and in their adsorption behavior.

Adsorption of Humic Substances on Oxide Surfaces. The extent of adsorption of humic substances on goethite, aluminum oxide, and kaolinite decreases with increasing pH (46–51); this general feature of anion adsorption is also exhibited by LMW organic acids (27, 29, 34–36, 52). Decreased adsorption of phosphate in the presence of humic substances has been attributed to competition for binding sites (53, 54). High-molecular-weight fractions of humic substances are preferentially adsorbed (55); incomplete adsorption of sedimentary DOM on aluminum oxide has been taken to indicate that some DOM components are negligibly adsorbed (56). Isotherms for humate adsorption on oxides (in the absence of divalent cations) generally exhibit surface saturation and have been described in terms of surface complex formation (46–48, 56).

The effects of calcium and magnesium on the adsorption of humic substances have not been entirely explained. Humate adsorption on goethite is increased in the presence of calcium and magnesium, and clear saturation is not observed. These effects have been attributed to coadsorption of the divalent cations (47, 48) and may be related to the ability of calcium and magnesium to coagulate some fraction of humic substances (57, 58).

Adsorption of humic substances on oxides markedly affects particle surface charge and colloidal stability. In these effects, a contrast may be noted between humic substances and LMW organic acids. Decreasing stability of colloidal hematite (initially positively charged at pH 5–6) with adsorption is observed for both humic substances and LMW organic acids, which is consistent with the effect of adsorbed organics on surface charge. Colloidal stability reaches a minimum value at the isoelectric point of the oxide. Further increase in the organic concentration results in a marked restabilization of the colloids for humic substances and fatty acids but not for the LMW organic ligands phthalate and oxalate. For phthalate, the development of negative surface charge (charge reversal) and colloidal restabilization appear to be limited by the adsorption capacity of the surface. In contrast, surface saturation is not observed in adsorption of long-chain fatty acids, which appears to be strongly influenced by hydrophobic interactions (59, 60). The parallel between the effects of humic substances and long-chain fatty acids on colloidal stability suggests that the hydrophobicity of humic substances may also contribute to their adsorption, particularly at high sorbate concentrations.

Effects of Humic Substances on Mineral Dissolution. Although humic substances appear to adsorb to oxide surfaces at least in part through surface complex formation, they have only slight effects on the dissolution of oxide and silicate minerals in laboratory studies. Both inhibition (at pH 4) and acceleration (at pH 3) of aluminum oxide dissolution have been observed (61), but in neither case was the effect dramatic (Table I). Kaolinite dissolution at pH 4.2 was also observed to be only minimally affected by humic substances (62).

Some information on the nature of humate–surface interactions may be derived from the absence of any significant enhancement of mineral dissolution by humic substances. The contrasting effects of bidentate, LMW organic ligands (e.g., oxalate, phthalate, and salicylate) and humic substances indicate that, in this regard, the simple, LMW organic ligands are not appropriate models for the more structurally complex geopolymers. This difference may arise from the polyfunctionality of the humic substances.

Humic substances and LMW organic ligands share common chelating functional groups, and it is likely that isolated segments of humate molecules

Table I. Effect of Humic Substances on the Dissolution of Aluminum Oxide

pH	DOC (mg/L)	Normalized Dissolution Rate[a]
3	1.5	1.08
3	7.5	1.13
4	7.5	0.83

[a] Ratio of dissolution rates observed in the presence and absence of humic substances.
SOURCE: Data are taken from reference 61.

form bidentate surface complexes comparable with those formed by LMW organic ligands. The larger, polyfunctional humate molecules may, however, bridge across metal centers on the oxide surface, to form multinuclear surface complexes (*61, 62*). Formation of multinuclear surface complexes would be unlikely to facilitate, and might even inhibit, dissolution, because detachment of two (or more) metal centers would be required.

A similar argument has been advanced to explain the accelerating effect of phosphate on the dissolution of crystalline iron oxides (goethite, hematite, maghemite, and lepidocrocite) by the multidentate ligand ethylenediaminetetraacetic acid (EDTA). In this case, it has been suggested that competitive adsorption of phosphate may shift the complex formed by EDTA at the surface from binding at two surface metal centers to binding (as a bidentate complex) at only one metal center, which is then more favorable for dissolution of the oxide (*63*).

Although some caution must be exercised in extrapolating from laboratory studies to natural systems, the laboratory evidence does not support a major, direct role of humic substances in metal mobilization in soils. Variations in weathering rates between laboratory and natural systems may arise from the preferential reaction of more easily weathered phases (e.g., amorphous oxide coatings) or from hydrologic control (transport limitation) of dissolution reactions in soils (*64*). Nonetheless, it is likely that the relative efficiencies of humic substances and LMW (bidentate) organic ligands in promoting dissolution in the laboratory and in natural systems will be similar. Then, the importance of ligand-promoted dissolution as a contributing mechanism for metal mobilization in soils may well be governed by the abundance of LMW organic ligands.

Metal Mobilization in Soils: Direct Influence of Individual DOM Components

Occurrence of Low-Molecular-Weight Organic Ligands.

Although LMW organic acids are only minor constituents of soil DOM, these compounds have been identified both in soil solutions and in leachates of forest litter and soils. Concentrations of LMW organic acids up to 1.2 mM have been measured in leachates of forest litter and soil, corresponding to up to 5% of the DOM in leachate (*65*). Oxalate concentrations as high as 1 mM have been measured in soil solutions (*66*).

Concentrations of LMW organic acids are high enough in some soils that salts of the acids are precipitated. Weddellite and whewellite (hydrated calcium oxalate) have been observed as coatings on fungal hyphae in forest soils (*67*). Oxalate, often one of the more abundant of the identifiable LMW organic acids, significantly enhances oxide dissolution in laboratory studies (*22, 23*).

The Role of Naturally Occurring Low-Molecular-Weight Organic Ligands in Dissolution.

Laboratory studies have demonstrated the effectiveness of bidentate organic ligands (at high concentrations) in promoting mineral dissolution. As already discussed, LMW organic ligands are present, at least in some soils, at concentrations sufficient to effect oxide dissolution. Leaching studies suggest that naturally occurring organic substances do promote metal mobilization in soils. When soils were leached with overlying layers of forest litter from several sources, a correlation was observed between dissolved organic carbon (DOC) and Al concentrations in leachate, as shown in Figure 3. LMW organic acids, including oxalic acid, were identified in the leachates (65).

The effect of microbial activity on ligand-promoted dissolution suggests that biologically labile DOM components are most important in facilitating dissolution. Table II shows that the enhancement of plagioclase dissolution by stream waters could be largely eliminated by inoculation with microorganisms. A similar effect was observed for feldspar dissolution by citrate. These effects, attributed to biodegradation of the reactive components, suggest that biologically labile constituents of DOM (such as LMW organic acids) rather than refractory humic and fulvic substances were responsible for enhancement of dissolution (68).

Chemical weathering of silicates was observed in a microcosm study conducted in situ in a petroleum-contaminated aquifer. Enhanced mineral weathering was attributed to the colonization of mineral surfaces by indigenous

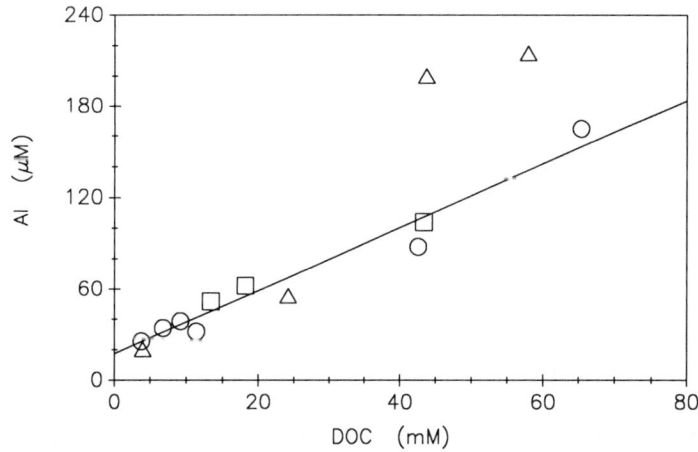

Figure 3. Correlation of dissolved Al and DOC in leachate of soil with overlying forest litter from different sources. Symbols: ($\bigcirc$, $\square$) soils from a single area collected on different dates (data used for linear regression); ($\triangle$) soil from different area (data not used in regression). Samples were leached with pH 3 HNO_3; final pH of leachates ranged from 5.3 to 5.7. (Data are taken from reference 65.)

Table II. Effect of Microbial Inoculum on Weathering Rates at pH 5.1

Mineral	Medium	Normalized Dissolution Rate[a]	
		Without Inoculum	With Inoculum
Plagioclase	Stream water (DOC, 24 mg/L)	3.5	1.2
K-feldspar	Citrate (5 mM)	2	0.9

[a] Ratio of dissolution rates under specified conditions to rates in distilled water at pH 5.1.
SOURCE: Data are taken from reference 68.

bacteria and the release of LMW organic acids accompanying hydrocarbon metabolism (69).

Indirect Effects of Humic Substances on Metal Mobilization in Soils

Even if humic substances do not directly promote mineral dissolution, they may still influence the mobilization of metals in soils. A possible indirect effect of humic substances is inhibition of ligand-promoted dissolution resulting from competitive adsorption of humic substances and LMW organic ligands. The decrease in rates of (bidentate) ligand-promoted dissolution of aluminum oxides in the presence of monodentate organic ligands has been attributed to such competitive adsorption (22). Adsorption of inorganic anions (such as phosphate and chromate) can also block adsorption of organic ligands (37–39); rates of oxalate-promoted dissolution of goethite have been shown to decrease with increasing chromate concentrations (70). Because rates of ligand-promoted dissolution depend on ligand concentrations at the surface, the blocking of surface sites by adsorption of humic substances would inhibit dissolution.

Humic substances may also facilitate the translocation of metals in soils, however, by stabilizing metals in solution as (dissolved) metal–humate complexes. Although the initial release of metals from soil minerals by dissolution would not be directly affected by humic substances, the migration of metals through the soil column could be enhanced. In this case, mineral dissolution promoted by LMW organic ligands would yield the corresponding metal complex with the LMW organic ligand in the soil solution; the metal–humate complex would then be formed at the expense of this initial complex. Formation of the humate–metal complex could occur through an abiotic metal-exchange reaction in which the humic substance displaces the LMW organic ligand. Alternatively, microbial activity might result in degradation of the LMW organic ligand, thus allowing formation of the metal–humate complex. Field observations of maximum dissolution immediately below the organic (O) horizon suggest a synergistic effect of LMW organics and humic substances on metal mobilization in soils (Johnsson, P., Stanford University, personal communication).

Conclusions

Increased mobility of metals in soils may be the result of the enhanced dissolution of soil minerals, the stabilization of dissolved metals in soil waters, or a combination of these processes. Direct involvement of soil DOM in metal mobilization is suggested by leaching studies with forest litter and soils and by field observations. Neither the components of DOM responsible for the enhancement of metal mobility nor the mechanism of their action has been unambiguously identified.

Laboratory studies provide some insight into the possible role of DOM constituents in metal mobilization in soils. These studies suggest that LMW organic ligands, specifically bidentate ligands, are more likely to promote mineral dissolution than are humic substances. LMW organic acids have been identified in soil waters and in leachates of forest litter at concentrations sufficient to effect oxide dissolution.

Humic substances, although a major fraction of soil DOM, do not facilitate oxide dissolution in laboratory studies. The mobilization of metals in soils may, however, be indirectly affected by humic substances. Solubilization of metals may be inhibited by competitive adsorption of humic substances and LMW organic ligands. The opposite effect—an enhancement of metal mobility—might result from the stabilization of dissolved metals as humate complexes.

Metal mobility in soils is governed by interfacial processes, such as dissolution. The role of DOM in such processes will be determined by the nature of organic matter–surface associations. The surface complexation model provides a conceptual framework for estimating the contributions of specific DOM components, particularly LMW organic ligands, to the mobilization of metals in soils. With this framework, the effects of humic substances on mineral dissolution can be interpreted to provide some insight into humate–surface interactions.

Acknowledgments

I thank W. Stumm (EAWAG) for his encouragement and support. I have benefited greatly from his exceptional insight into the chemistry of the mineral–water interface. I also thank P. Johnsson (Stanford University), W. Fish (Oregon Graduate Institute), and M. Ochs (EAWAG) for sharing preliminary data and for helpful discussions of this work and D. Dzombak (Carnegie Mellon University) for his comments on an earlier draft of this manuscript. This work was supported in part by the National Science Foundation (BCS-92-58431).

References

1. Giovanoli, G.; Schnoor, J. L.; Sigg, L.; Stumm, W.; Zobrist, J. *Clay Clay Min.* **1988**, *36*, 521–529.
2. Lee, F. Y.; Yuan, T. L.; Carlisle, V. W. *Soil Sci. Soc. Am. J.* **1988**, *52*, 1411–1418.
3. Stobbe, P. C.; Wright, J. R. *Soil Sci. Soc. Am. Proc.* **1959**, *23*, 161–164.
4. McKeague, J. A.; Cheshire, M. V.; Andreaux, F.; Berthelin, J. In *Interactions of Soil Minerals with Natural Organics and Microbes;* Huang, P. M.; Schnitzer, M., Eds.; Soil Science Society of America: Madison, WI, 1986; pp 549–592.
5. Tan, K. H. In *Interactions of Soil Minerals with Natural Organics and Microbes;* Huang, P. M.; Schnitzer, M., Eds.; Soil Science Society of America: Madison, WI, 1986; pp 1–27.
6. Baker, J. P. In *Acid Rain/Fisheries;* Johnson, R. E., Ed.; American Fisheries Society: Bethesda, MD, 1982; pp 165–176.
7. Driscoll, C. T.; Schecher, W. D. In *Metal Ions in Biological Systems;* Sigel, H.; Sigel, A., Eds.; Marcel Dekker: New York, 1988; Vol. 24, pp 59–122.
8. Joslin, J. D.; Wolfe, M. H. *Soil Sci. Soc. Am. J.* **1989**, *53*, 274–281.
9. Mayer, L. M.; Liotta, F. P.; Norton, S. A. *Water Res.* **1982**, *16*, 1189–1196.
10. Chambers, R. M.; Odum, W. E. *Biogeochemistry* **1990**, *10*, 37–52.
11. Francko, D. A.; Heath, R. T. *Limnol. Oceanogr.* **1982**, *27*, 564–569.
12. Johnson, N. M.; Driscoll, C. T.; Eaton, J. S.; Likens, G. E.; McDowell, W. H. *Geochim. Cosmochim. Acta* **1981**, *45*, 1421–1437.
13. Huang, P. M.; Violante, A. In *Interactions of Soil Minerals with Natural Organics and Microbes;* Huang, P. M.; Schnitzer, M., Eds.; Soil Science Society of America: Madison, WI, 1986; pp 159–221.
14. Drever, J. I.; Blum, A. E. *Processes Controlling the Composition of Infiltrating Water in Forested Watersheds;* prepared for U.S. Department of the Interior, U.S. Geological Survey, Wyoming Water Research Center, University of Wyoming, Laramie, WY, 1984.
15. Driscoll, C. T.; van Breeman, N.; Mulder, J. *Soil Sci. Soc. Am. J.* **1985**, *49*, 437.
16. Stevenson, F. J.; Fitch, A. In *Interactions of Soil Minerals with Natural Organics and Microbes;* Huang, P. M.; Schnitzer, M., Eds.; Soil Science Society of America: Madison, WI, 1986; pp 29–58.
17. Huang, P. M.; Violante, A. In *Interactions of Soil Minerals with Natural Organics and Microbes;* Huang, P. M.; Schnitzer, M., Eds.; Soil Science Society of America: Madison, WI, 1986; pp 159–221.
18. Schwertmann, U.; Kodama, H.; Fischer, W. R. In *Interactions of Soil Minerals with Natural Organics and Microbes;* Huang, P. M.; Schnitzer, M., Eds.; Soil Science Society of America: Madison, WI, 1986; pp 223–250.
19. Grossl, P. R.; Inskeep, W. P. *Geochim. Cosmochim. Acta* **1992**, *56*, 1955–1961.
20. Schnitzer, M.; Khan, S. U. *Humic Substances in the Environment;* Marcel Dekker: New York, 1972.
21. *Humic Substances in the Soil, Sediment, and Water,* Aiken, G. R.; McKnight, D. M.; Wilson, R. L.; MacCarthy, P., Eds.; Wiley: New York, 1985.
22. Furrer, G.; Stumm, W. *Geochim. Cosmochim. Acta* **1986**, *50*, 1847.
23. Stumm, W.; Furrer, G. In *Aquatic Surface Chemistry;* Stumm, W., Ed.; Wiley-Interscience: New York, 1987; pp 197–219.
24. Stumm, W.; Wehrli, B.; Wieland, E. *Croat. Chem. Acta* **1987**, *60*, 429–456.

25. Schindler, P.; Furst, B.; Dick, R.; Wolf, P. J. *Colloid Interface Sci.* **1976,** *55,* 469–475.
26. Stumm, W.; Kummert, R.; Sigg, L. *Croat. Chem. Acta* **1980,** *53,* 291–312.
27. Schindler, P. W.; Stumm, W. In *Aquatic Surface Chemistry;* Stumm, W., Ed.; Wiley-Interscience: New York, 1987; p 83.
28. Stumm, W. *Chemistry of the Solid-Water Interface;* Wiley-Interscience: New York, 1992.
29. Westall, J. C. In *Aquatic Surface Chemistry;* Stumm, W., Ed.; Wiley-Interscience: New York, 1987; p 3.
30. Tipping, E. In *Organic Acids in Aquatic Ecosystems;* Perdue, E. M.; Gjessing, E. T., Eds.; Wiley: New York, 1990; p 209.
31. Zeltner, W. A.; Yost, E. C.; Machesky, M. L.; Tejedor-Tejedor, M. I.; Anderson, M. A. In *Geochemical Processes at Mineral Surfaces;* Davis, J. A.; Hayes, K. F., Eds.; ACS Symposium Series 323, American Chemical Society: Washington, DC, 1986; p 142.
32. McBride, M. B. *J. Colloid Interface Sci.* **1980,** *76,* 393–398.
33. Hering, J. G.; Stumm, W. *Langmuir* **1991,** *7,* 1567.
34. Parfitt, R. L.; Farmer, V. C.; Russell, J. D. *J. Soil Sci.* **1977,** *28,* 29–39.
35. Parfitt, R. L.; Fraser, A. R.; Russell, J. D.; Farmer, V. C. *J. Soil Sci.* **1977,** *28,* 40–47.
36. Kummert, R.; Stumm, W. *J. Colloid Interface Sci.* **1980,** *75,* 373–385.
37. Kafkafi, U.; Bar-Yosef, B.; Rosenberg, R.; Sposito, G. *Soil Sci. Soc. Am. J.* **1988,** *52,* 1585–1589.
38. Violante, A.; Columbo, C.; Buondonna, A. *Soil Sci. Soc. Am. J.* **1991,** *55,* 65–70.
39. Mesuere, K.; Fish, W. *Environ. Sci. Technol.* **1992,** *26,* 2365–2370.
40. Wakamatsu, T.; Fuerstenau, D. W. In *Adsorption from Aqueous Solution;* Weber, W. J.; Matijevic, E., Eds.; ACS Advances in Chemistry 79; American Chemical Society: Washington, DC, 1968; pp 161–172.
41. Yap, S. N.; Mishra, R. K.; Raghaven, S.; Fuerstenau, D. W. In *Adsorption from Aqueous Solution;* Tewari, P., Ed.; Plenum: New York, 1981; p 119.
42. McKnight, D. M.; Feder, G. L.; Thurman, E. M.; Wershaw, R. L.; Westall, J. C. *Sci. Tot. Environ.* **1983,** *28,* 65–76.
43. Cabaniss, S. E.; Shuman, M. S. *Geochim. Cosmochim. Acta* **1988,** *52,* 194.
44. Buffle, J. *Complexation Reactions in Aquatic Systems: An Analytical Approach;* Ellis Horwood: Chichester, England, 1988.
45. Beckett, R.; Zue, Z.; Giddings, J. C. *Environ. Sci. Technol.* **1987,** *21,* 289–295.
46. Parfitt, R. L.; Fraser, A. R.; Farmer, V. C. *J. Soil Sci.* **1977,** *28,* 289–296.
47. Tipping, E. *Chem. Geol.* **1981,** *33,* 81–89.
48. Tipping, E. *Geochim. Cosmochim. Acta* **1981,** *45,* 191–199.
49. Davis, J. A. *Geochim. Cosmochim. Acta* **1982,** *46,* 2381.
50. Ochs, M. Ph.D. Thesis, Swiss Federal Institute of Technology, Zurich, Switzerland, 1991.
51. Schultess, C. P.; Huang, C. P. *Soil Sci. Soc. Am. J.* **1991,** *55,* 34–42.
52. Mesuere, K.; Fish, W. *Environ. Sci. Technol.* **1992,** *26,* 2357–2364.
53. Sibanda, H. M.; Young, S. D. *J. Soil Sci.* **1986,** *37,* 197–204.
54. Fontes, M. R.; Weed, S. B.; Bowen, L. H. *Soil Sci. Soc. Am. J.* **1992,** *56,* 982–990.
55. Davis, J. A.; Gloor, R. *Environ. Sci. Technol.* **1981,** *15,* 1223–1229.
56. Davis, J. A. In *Contaminants and Sediments;* Baker, R. A., Ed.; Ann Arbor Science: Ann Arbor, MI, 1980; Vol. 2, pp 279–304.
57. Ong, H. L.; Bisque, R. E. *Soil Sci.* **1968,** *106,* 220–224.
58. Sholkovitz, E. R.; Copland, D. *Geochim. Cosmochim. Acta* **1981,** *45,* 181–189.

59. Liang, L.; Morgan, J. J. In *Chemical Modeling of Aqueous Systems II;* Melchior, D. C.; Bassett, R. L., Eds.; ACS Symposium Series 416; American Chemical Society: Washington, DC, 1990; pp 293–308.
60. Liang, L.; Morgan, J. J. *Aquatic Sci.* **1990**, *52*, 32.
61. Ochs, M.; Brunner, I.; Stumm, W.; Cosovic, B. *Water Air Soil Poll.* **1993**, *68*, 213.
62. Chin, P. K. F.; Mills, G. L. *Chem. Geol.* **1991**, *90*, 307–317.
63. Borggaard, O. K. *Clays Clay Miner.* **1991**, *39*, 324–328.
64. Schnoor, J. L. In *Aquatic Chemical Kinetics;* Stumm, W., Ed.; Wiley-Interscience: New York, 1990; pp 475–504.
65. Pohlman, A. A.; McColl, J. G. *Soil Sci. Soc. Am. J.* **1988**, *52*, 265.
66. Fox, T. R.; Comerford, N. B. *Soil Sci. Soc. Am. J.* **1990**, *54*, 1441.
67. Graustein, W. C.; Cromack, K., Jr.; Sollins, P. *Science (Washington, DC)* **1977**, *198*, 1252–1254.
68. Lundstrom, V,; Ohman, L. O. *J. Soil Sci.* **1990**, *41*, 359.
69. Hiebert, F. K.; Bennett, P. C. *Science (Washington, DC)* **1992**, *258*, 278–281.
70. Mesuere, K.; Fish, W. *Langmuir,* in press.

RECEIVED for review October 23, 1992. ACCEPTED revised manuscript April 27, 1993.

5

Reaction Rates and Products of Manganese Oxidation at the Sediment–Water Interface

Bernhard Wehrli[1], Gabriela Friedl[1], and Alain Manceau[2]

[1]Swiss Federal Institute for Water Resources and Water Pollution Control (EAWAG), CH–6047 Kastanienbaum, Switzerland
[2]Environmental Geochemistry Group, LGIT–IRIGM, University Joseph Fourier and Centre National de la Recherche Scientifique (CNRS), BP 53X, F–38041 Grenoble, France

Manganese(II) oxidation rates in a eutrophic lake were calculated from a 4-year record of sediment-trap data, and the structure of the prevailing manganese oxides were determined by extended X-ray absorption fine structure (EXAFS) spectroscopy. The oxidation rate near the sediment surface showed a distinct seasonal pattern, with maxima of up to 2.8 mmol/m^2 per day during summer. The average half-life of Mn(II) during stagnation in summer was 1.4 days. A review of published oxidation rates showed that this half-life, which cannot be explained with available data of abiotic surface catalysis, is within the typical range of microbiological oxidation. EXAFS revealed that the oxidation product consists mainly of vernadite (δ-MnO$_2$), an X-ray-amorphous Mn(IV) oxide.

INTEREST IN THE AQUATIC REDOX CHEMISTRY of manganese is at least as old as Werner Stumm's scientific career. The first Ph.D. student in his laboratory at Harvard University worked on the chemistry of aqueous Mn(II) and Mn(IV) (1). Since then aquatic chemists have refined their analytical tools (2), their conceptual (3) and numerical (4) models of manganese cycling, and their understanding of heterogeneous redox reactions in general (5, 6). This chapter examines the biogeochemical and mineralogical aspects of the manganese redox cycle, with an emphasis on lakes. Davison (3) showed how the limnological view can be scaled to marine systems.

Aquatic Geochemistry of Manganese

The redox cycle of manganese in lakes (Figure 1) and oceans is triggered at the oxic–anoxic boundary, which is often located near the sediment–water interface. In this zone, settling manganese oxide particles undergo reductive dissolution. Burdige and Nealson (7) showed that, under conditions of microbial sulfate reduction, the rate of reductive dissolution of manganese oxides is limited by bacterial sulfide production. Sulfide and Fe(II) (8) rapidly reduce manganese oxides by direct chemical mechanisms. Ehrlich (9) reviewed the growing literature on microbial reduction of manganese oxides. Bacteria were found that reduce manganese oxides even in the presence of oxygen (10).

Transport of dissolved Mn(II) accumulated by reductive dissolution is governed by molecular diffusion in pore waters and by eddy diffusion in the stratified hypolimnia of lakes. Transport follows the concentration gradients,

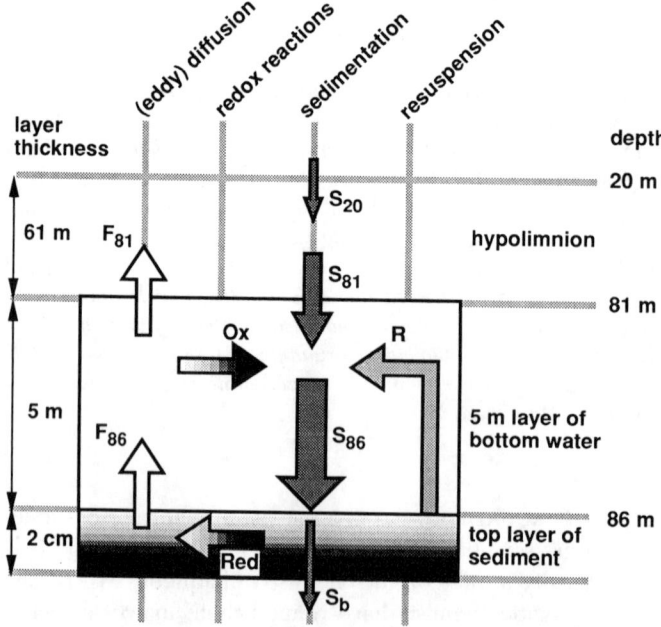

Figure 1. Box model for the calculation of Mn redox cycling near the sediment–water interface. Sedimentation rates are measured with sediment traps. The burial rate S_b is estimated from dated sediment cores. In situ sampling techniques (flux chambers and peepers) are used to quantify the diffusive flux across the sediment–water interface F_{86}. The resuspension rate R is estimated from the increase in the mass flux of settling material between the 81- and 86-m horizons.

which decrease toward the oxic zone. At the sediment–water interface such diffusive fluxes can be calculated from concentration profiles in pore waters (11) or by direct measurements with benthic flux chambers (12, 13). Murray (14) showed that both techniques may yield values comparable with fluxes calculated from hypolimnetic mass balances.

Oxidation of dissolved Mn(II) by molecular oxygen is very slow in the absence of microorganisms or mineral particles. Even after 8 years, Diem and Stumm (15) found no precipitated manganese oxides in their experiments with 10 μM Mn(II) at pH 8.5 under air-saturated conditions. However, Davies and Morgan (16) reported that Mn(II) is oxidized completely within a few hours under similar conditions in the presence of a 10 mM suspension of goethite (α-FeOOH). Because Mn(II) is only weakly adsorbed at a pH below the zero point of charge (ZPC) of iron hydroxides ($pH_{ZPC} \sim 8$), the oxygenation rate is rather slow at pH 7.5. This value is typical of the hypolimnia of productive lakes during stagnation.

In natural systems microbiological oxidation may offer a faster pathway, particularly at pH < 8 and low concentrations (<5 μM) of particulate oxides. Hastings and Emerson (17) showed that sporulated cultures of marine bacillus SG-1 at pH 7.5 accelerated the oxidation of Mn(II) by a factor of 10^4 with respect to the abiotic catalysis on a colloidal MnO_2 surface. A radiotracer study of microbial Mn oxidation in a marine fjord revealed half-lives as short as 2 days (18). Perhaps microorganisms can use the entire redox cycle of manganese. A study indicates that the vegetative cells of spores that mediate Mn(II) oxidation also reduce manganese oxides (19).

The mineralogy of manganese oxides formed in lakes and oceans by oxidative precipitation is still an active field of research with sometimes intense discussions [e.g., the todorokite–buserite problem (20–22)]. Laboratory studies of abiotic Mn(II) oxidation at pH 9 yielded Mn_3O_4 (hausmannite) as the initial product. The hausmannite oxidized within 8 months to γ-MnOOH (manganite) (23). In a more dilute system, Kessick and Morgan (24) obtained a Mn(III) phase by oxidizing a Mn(II) solution in ammonia buffer with O_2. Formation of such Mn(III) minerals in lakes has been suggested (25). However, in such chemical experiments the solubility product of $Mn(OH)_2$ is often exceeded during the preparation of the system, a condition that is rarely encountered in lakes and oceans. Therefore, the sequence of oxidation products in such experiments may not be characteristic of natural environments. The solid oxidation products found in ocean sediments (26, 27) are chiefly Mn(IV) oxides with an oxygen to manganese ratio in the range 1.9–2.0.

Two reviews (28, 29) covered the mineralogical literature on marine manganese oxides until the late 1970s. Both groups of authors concluded that birnessite, vernadite (δ-MnO_2), and todorokite are the dominant forms of manganese oxides in marine concretions and nodules. Vernadite is the most common mineral. Various obstacles have hindered the determination of the crystal structures of these minerals:

1. No large single crystals for X-ray diffraction are available, and the fine-grained material yields poor powder diffraction patterns.
2. Different amounts of counterions and substituting elements induce structural changes in the lattice.
3. The three minerals exhibit different forms of structural disorder.

Complementary structural techniques have recently helped to clarify the structure of birnessite and todorokite and to limit possible structural models for vernadite. High-resolution transmission electron microscopy (HRTEM) revealed the tunnel structure of todorokite in marine manganese deposits (30). Post and Bish (31) used these results to carry out a Rietveld refinement on powder diffraction data of todorokite, which confirmed the basic (3 × 3) tunnel structure of this mineral.

The structure of birnessite was refined on the basis of electron and X-ray diffraction data (32–36). It contains layers of edge-shared MnO_6 octahedra separated by about 7.2 Å. One out of six octahedral sites is unoccupied, and vacancies are balanced by counterions like Na^+, K^+, or Mg^{2+}. Extended X-ray absorption fine structure (EXAFS) spectroscopy (37) showed that vernadite can be viewed as a three-dimensional mosaic of single and multiple octahedral chains with a variable length of between 1 and n octahedra. These chains are joined by shared corners.

Our main objectives in this study were to determine manganese oxidation rates from different types of flux measurements at the sediment–water interface in a productive lake and to evaluate the oxidation states and structures of the oxidation products. We chose a eutrophic lake that is artificially oxygenated. The intense manganese cycle at this site is quantified with sediment-trap data, benthic chamber experiments, pore water measurements, and sediment-core analyses. From the detailed mass balance of the manganese cycling at the bottom of Lake Sempach, we determined a time series of manganese oxidation rates. The resulting average rate constant was used to discriminate between microbial and abiotic pathways of manganese oxidation. The structure and dominant oxidation state of the manganese oxide found in sediment-trap material were analyzed by EXAFS spectroscopy.

Study Site and Methods

Lake Sempach. Lake Sempach is a eutrophic, hard-water lake in central Switzerland with a surface area of 14.4 km^2 and a mean depth of 46 m.

The relatively small drainage basin of 61.4 km² is responsible for the long hydraulic residence time of about 17 years. The limnological P and N cycles of Lake Sempach (*38–40*) and the sedimentation rates (*41*) have been studied extensively in recent years. Before 1984 the deep hypolimnion was periodically anoxic (*42*). In 1984 an internal lake restoration program was started. Since then the deep hypolimnion has been aerated during summer. Pure oxygen in the form of small gas bubbles is introduced at a rate of 2.5–3 metric tons per day. As a consequence, the oxygen concentration has been kept above 3 mg/L, even at the deepest point of the lake (87 m). The whole lake is artificially mixed with a plume of compressed air between fall and spring.

Lake Sempach therefore offers a unique opportunity to study the diagenetic cycling of manganese in an oxic environment with rather high sedimentation rates of particulate organic carbon (POC). A time series of POC, particulate organic nitrogen (PON), and phosphorus sedimentation rates is available from biweekly sediment-trap analyses since 1984. The average POC sedimentation rates in the past 7 years have been 15.5 and 17.8 mmol/m² per day at depths of 20 and 81 m, respectively. Sedimentation rates of Mn, Fe, and Ca were monitored at these two levels and at the sediment–water interface at a depth of 86 m during 1988–1991.

Determination of Sedimentation Rates. Sediment traps with an inner diameter of 9 cm and a length of 70 cm were sampled biweekly. The design of a special trap to sample the particle flux at the sediment–water interface was described by Gächter and Meyer (*39*). No poisons were applied. The sediment-trap station was in the center of the lake near the deepest point. When the traps were recovered from the lake, small plugged holes at the side were unplugged to remove the overlying water. The collected material was freeze-dried with portions of supernatant water and weighed. For metal analyses the dried material was digested in boiling H_2SO_4–H_2O_2 and analyzed by graphite furnace atomic absorption spectrometry (GF–AAS).

Sediment cores were taken with a gravity corer. They were sliced with 0.5-cm resolution and processed the same way as sediment-trap material. One of the cores (code SE–8802) was dated by Wieland et al. (*41*), who used the ^{137}Cs record from bomb fallout and the Chernobyl accident. The authors determined the sediment accumulation rate at this site to be 1.84 g/m² per day. This value agrees with the average sedimentation rate of 2.17 g/m² per day gleaned from the sediment-trap record of the years 1984–1991 at a depth of 81 m.

Measurements of Diffusive Fluxes. Benthic chamber experiments were carried out by using an automatic flux chamber (sediment lander). Our lander followed closely the design of Devol (*43*). The two benthic flux chambers have an area of 400 cm² each. They penetrated the sediment to a depth of 30 cm. One hour after deployment, two lids at the top of the chambers were closed automatically. About 8 L of water above the sediment–water interface was enclosed. A tracer (RbBr) was injected into the stirred chambers to determine the exact volume and to verify that the chambers were tightly closed. Samples were taken automatically with syringes at preset intervals of 2 h. Springs on the syringes were released with burning wires controlled by an electronic timer. After 24 h the device was brought back to the surface. The subsamples from the syringes were stabilized with HNO_3 and analyzed by GF–AAS within a few days for total Mn. Four benthic chamber experiments were performed at the deepest site in Lake Sempach on July 12, October 11, and October 19, 1990, and on January 29, 1991.

Dialysis pore water samplers (peepers) were used to determine seasonal variations in the concentration profiles of dissolved manganese at the deepest site in Lake Sempach. The peeper method to collect pore water samples was reviewed by Adams (44). The design of our peepers followed suggestions by Brandl and Hanselmann (45). Depth resolution was 15 mm. The peeper cells were covered with a nondegradable polysulfone membrane (HT–200 Tuffryn from Gelman) with a pore size of 0.2 μm. Initially, the cells were filled with twice-distilled water and equilibrated for at least 24 h in twice-distilled water saturated with 1 atm of N_2 to remove dissolved oxygen. The peepers were mounted on a tripod, which was lowered to the bottom of the lake. Equilibration lasted about 3 weeks. The tripod allowed retrieval of the peepers from a depth of 87 m to the ship within a few minutes. Subsamples of 1–2 mL were placed in polycarbonate tubes, acidified with 20 μL of concentrated HNO_3, stored at 4 °C, and analyzed by GF–AAS for Mn within about 10 days.

EXAFS Spectroscopy of Mn Oxides.

Samples from the sediment trap at a depth of 86-m were processed on the ship in a glove box with an N_2 atmosphere. Particles were filtered and dried under N_2 in a desiccator. Dried material was covered with gas-tight Captone film (3M) and stored under nitrogen. EXAFS spectroscopy of sediment-trap material from Lake Sempach was performed at the synchrotron facility of the Laboratoire pour l'Utilisation du Rayonnement Electromagnétique (LURE) in Orsay, France, with running conditions of 1.85 GeV and 260–300 mA. [For information on EXAFS spectroscopy, see books by Teo (46) and Koningsberger and Prins (47)]. We used the EXAFS IV spectrometer at LURE. The incident beam was monochromatized with a Si(111) double crystal, and Mn K-edge spectra were recorded in fluorescence mode by using a plastic scintillator. The sediment samples were cooled to 77 K to maximize the signal-to-noise ratio. Structural parameters were determined from EXAFS spectra by using standard procedures (46). The mineral Na–birnessite was used as a reference standard for the analysis of the oxygen and manganese coordination shells.

Results

Oxidation Rate of Mn near the Sediment–Water Interface.

We used a box model approach to calculate rates of the Mn redox cycle from sediment-trap data and in situ sampling techniques (peeper and lander experiments). Figure 1 depicts an overview of the relevant processes and length scales. Sedimentation rates of particles were determined with sediment traps at 81 time intervals of about 2 weeks during 1988–1991 and at three depths (z = 20, 81, and 86 m).

Figure 2 compares the particle sedimentation rate at the two trap stations at depths of 81 and 86 m. The mass flux of particles P_z is always larger just above the sediment surface (86 m) than at the 81-m level. Large maxima during summer are due to algal blooms and biogenic calcite precipitation. To calculate the average composition of settling particles, stoichiometries of CH_2O and NH_3 were assumed for POC and PON, respectively. The results are given in Table I, with the average sedimentation rate of particles at the three depths.

Apart from some apparent dissolution of calcite during settling through the water column, only small changes occurred at depths between 20 and 81 m. However, in the two deep traps (81 and 86 m) the average Mn concentration increased by factors of 8 and 15, compared with the trap at 20 m. The trap at the sediment–water interface showed a significant increase in the overall sedimentation rate, which points toward an important contribution from

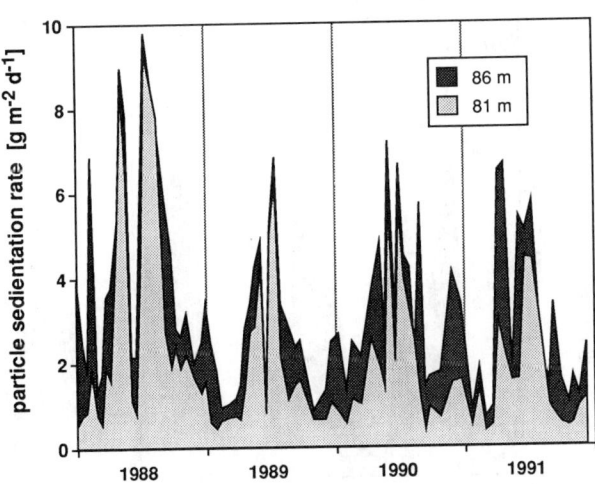

Figure 2. Time series of the particle sedimentation rate in Lake Sempach. Fluxes are higher near the sediment surface in 86-m depth because of resuspension. The large maxima during summer correspond to algal blooms and subsequent biogenic calcite precipitation.

Table I. Average Composition of Settling Particles in Lake Sempach, 1988–1990

Depth z (m)	P_z^a (g/m² per day)	Organic Matter[b] (wt %)	$CaCO_3$ (wt %)	FeOOH (wt %)	MnO_2 (wt %)	Allochthonous Material[c] (wt %)
20	1.96	23	51	1.0	0.3	25
81	2.17	26	42	2.1	2.3	27
86	3.10	25	39	2.6	4.5	28

[a] Average particle sedimentation rate for the period 1984–1991.
[b] Calculated from POC and PON measurements by using the stoichiometries CH_2O and NH_3 for organic carbon and nitrogen, respectively.
[c] Allochthonous material (clay, minerals, etc.), including SiO_2 from diatoms.

sediment resuspension. Manganese sedimentation rates (in millimoles per square meter per day) were calculated from the analytical data for each time interval of about 2 weeks:

$$S_z = c_z \cdot P_z \qquad (1)$$

where the subscript z refers to the depth of the sediment trap, S denotes the Mn sedimentation rate, c stands for the Mn concentration in the settling particles (mmol/g), and P represents the particle sedimentation rate (g/m² per day). The seasonal variations of $S_z(t)$ are shown in Figure 3.

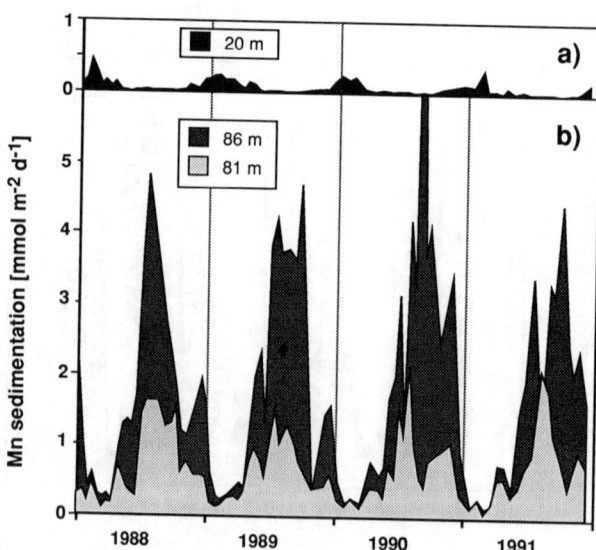

Figure 3. Mn sedimentation rates in Lake Sempach. Part a: Flux of particulate manganese from the epilimnion. Part b: Mn sedimentation at 81 m and at the sediment surface (86 m).

The particulate input of Mn from the epilimnion at a depth of 20 m was about an order of magnitude smaller than the sedimentation rates observed in the deep waters. Maxima in the Mn flux from the epilimnion occurred during winter. During this time the lake is well-mixed and the plume of compressed air in the center of the lake transports dissolved Mn(II) from the sediment–water interface directly to shallower depths. By contrast, the Mn cycle in the hypolimnion is accelerated during summer when the sedimentation rates of organic matter are at a maximum.

The large difference in Mn sedimentation rates at depths of 81 and 86 m (Figure 3b) points toward intense redox cycling in this 5-m layer of bottom water above the sediment. Two processes contribute to this sharp increase in Mn sedimentation toward the bottom of the lake: (1) oxidation of dissolved Mn(II) and (2) resuspension of solid sediment particles (Figure 1). We show later in this chapter that adsorption of Mn(II) does not contribute significantly to the increased sedimentation rate of manganese. The resuspension rate R of manganese in millimoles per square meter per day can be estimated from the difference in mass flux at the two depths and the concentration of manganese in the resuspended sediment.

$$R = c_{86}(P_{86} - P_{81}) \qquad (2)$$

Here c_{86} denotes the manganese concentrations in material from the trap at 86 m, and $(P_{86} - P_{81})$ refers to the difference in the mass fluxes measured at 86 and 81 m.

Several factors may account for large resuspension rates. The retrieval and deployment of the trap at the sediment surface may resuspend some particulate matter. Natural resuspension may result from storms and sediment-focusing mechanisms. Postdepositional remobilization may increase the sedimentation rate of ^{210}Pb at the deepest point of Lake Sempach (41). Because we cannot discriminate among different resuspension processes, we assumed that the Mn concentration in the resuspended material is equal to that in the sedimenting particles at a depth of 86 m. Particulate MnO_x is rapidly reduced at the sediment surface; therefore, this procedure tends to overestimate the resuspension term.

Given the correction for resuspension in equation 2, we can now calculate the rate of the Mn oxidation process (Ox, in millimoles per square meter per day) within a layer 5 m above the lake bottom:

$$Ox = S_{86} - S_{81} - R \qquad (3)$$

The resulting time series of the calculated Mn(II) oxidation rate is shown in Figure 4. During the first quarter of the year the oxidation rate is close to zero, which indicates that during this time Mn(II) is oxidized either within the sediment or at shallower depths.

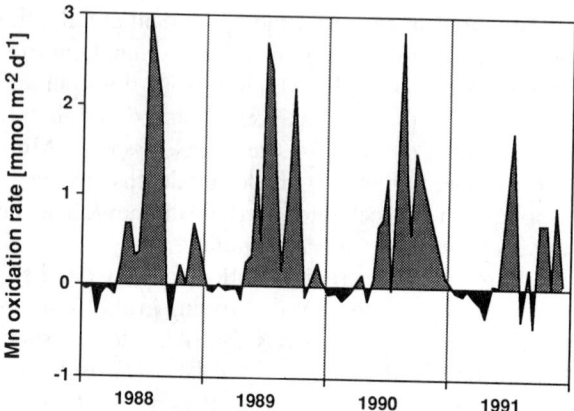

Figure 4. Calculated Mn oxidation rates in the layer of bottom water at depths between 81 and 86 m (see eq 2). The slightly negative values during the first quarter of the year are due to an overestimation of the resuspension flux.

Maximum oxidation rates occurred in June and July, reaching values as high as 2.8 mmol/m² per day. For the stagnation period from the beginning of April to the end of September, average values of S_{81}, S_{86}, R, and Ox are given in Table II. The 4-year average of the Mn redox cycle during stagnation in our 5-m layer above the sediment can be summarized as follows:

$$S_{86} = S_{81} + Ox + R \tag{4a}$$

$$S_{86} = 0.97 + 0.77 + 0.79 \text{ mmol/m}^2 \text{ per day} \tag{4b}$$

$$S_{86} = 2.53 \text{ mmol/m}^2 \text{ per day} \tag{4c}$$

Thus in the deepest 5 m of the water column, the Mn content of the settling particles increased by 80% through oxidation of Mn(II). Resuspension intensified the overall flux of settling Mn particles by another 80%.

Burial Rate of Mn. Analysis of sediment cores provided a closer look at the dynamics of Mn within the sediments. Figure 5 combines results from three sediment cores taken at the deepest site of Lake Sempach. Two cores were taken in January 1991 and one in May 1988. The arrows indicate the flux-averaged Mn concentrations of settling material at depths of 20 and 81 m, respectively. The time scale in Figure 5 was calculated with ^{137}Cs dating (41) by using $P_s = 1.84$ g/m² per day as the sediment accumulation rate.

Profiles of total Mn concentrations indicate that the high Mn content in the settling material must be reduced within months. The Mn concentration in the top 5 mm of the sediment was only about 20% of the average Mn content of settling particles at depths of 81 m. Within 2 years the Mn con-

Table II. Rates of Mn Cycling

Flux	Symbol	Method	Average	Max.	Min.	Time
Trap–core balance[a]						
Sedimentation at 20 m	S_{20}	trap	0.072	24.2	0.014	1988–1991
Burial rate	S_B	cores	0.067			1968–1988
Balance in 5-m layer[b]						
Sedimentation at 81 m	S_{81}	trap	0.97	2.16	0.27	1988–1991
Sedimentation at 86 m	S_{86}	trap	2.53	9.31	0.38	1988–1991
Resuspension	R	trap	0.79	7.5	(−0.58)	1988–1991
Oxidation of Mn	Ox	calc.	0.77	2.83	(−0.42)	1988–1991
Diffusive fluxes[c]						
Diffusion across	F_{86}	lander	5.5			July 1990
Sediment–water interface	F_{86}	lander	1.5	1.9	1.1	Oct. 1990
	F_{86}	lander	0.5	0.6	0.4	Jan. 1991
	F_{86}	peeper	0.46			June 1990
	F_{86}	peeper	0.28			July 1990
	F_{86}	peeper	0.18			Oct. 1990
	F_{86}	peeper	0.07			April 1991

[a] Values for the whole period 1988–1990.
[b] Values for the stagnation periods only (April–September).
[c] Data from in situ sampling experiments; the incubation time was 1 day for lander experiments and 3 weeks for peeper sampling.

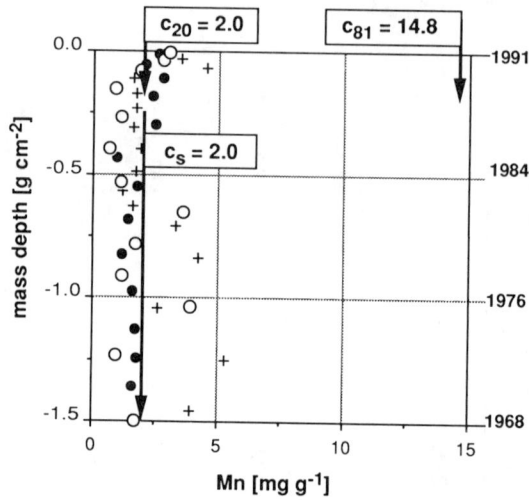

Figure 5. Total Mn concentration in three sediment cores from the deepest site of Lake Sempach (86-m depth). Arrows at the top indicate the average Mn concentration in settling material at depths of 20 and 81 m. The time scale is calculated from ^{137}Cs dating and is valid for the two cores from 1991 only. Key: ●, concentrations in a core from 1988; ○ and +, concentrations in two cores from 1991.

centrations decreased to a constant value of $c_s \sim 0.2$ wt %. By using this concentration and the accumulation rate P_s we obtained the burial rate S_b of manganese.

$$S_b = c_s P_s = 0.067 \text{ mmol/m}^2 \text{ per day} \tag{5}$$

This value comes very close to the 4-year average of the Mn sedimentation rate in the epilimnion trap ($S_{20} = 0.072$ mmol/m² per day). Thus, the "excess" Mn in settling particles at the 81- and 86-m stations is recycled locally.

Diffusive Fluxes of Mn(II). If Mn is not transported to the 5-m layer of bottom water by lateral turbulent diffusion or advection, we should observe maximum values of the diffusive fluxes similar to the oxidation rate across the sediment–water interface (up to 3 mmol/m² per day). Profiles of Mn concentrations in the pore water are shown in Figure 6a. The steep concentration gradients near the sediment–water interface are not at steady state; the gradients are at maximum in summer and decrease to a minimum in spring. Sharp peak profiles were observed in June and July 1990. The diffusive flux of Mn(II), F_{86} in millimoles per square meter per day, was estimated from pore water profiles by using Fick's law with a correction for the porosity, ϕ (48):

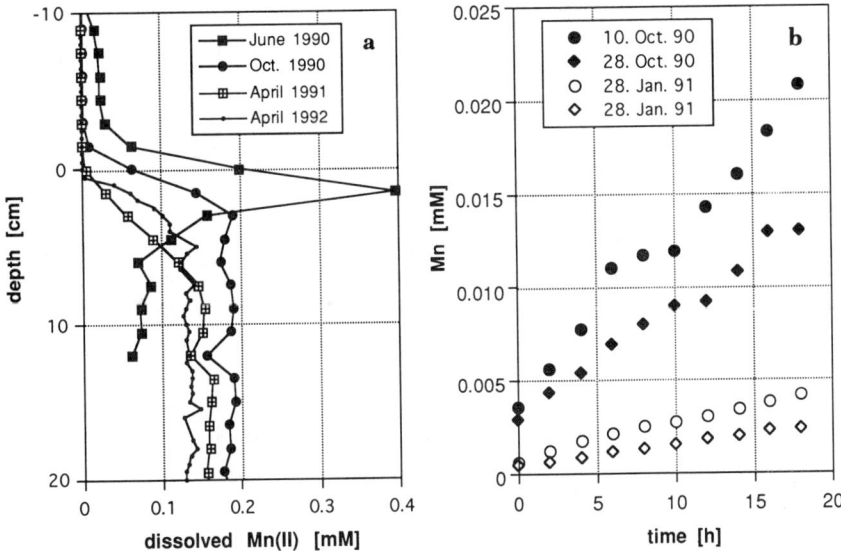

Figure 6. Part a: Seasonal variations of dissolved Mn(II) concentration gradient in pore water of Lake Sempach at the deepest site. Part b: Accumulation of dissolved Mn(II) in flux chambers of a sediment lander. Data from January are parallel determinations of two chambers at the same site.

$$F_{86} = D_{Mn}\phi^2 \frac{\partial [Mn(II)]}{\partial z} \qquad (6)$$

where D_{Mn} represents the molecular diffusion coefficient (48) of Mn^{2+} at 5 °C (0.32×10^{-5} m^2/day), $\partial[Mn(II)]/\partial z$ stands for the concentration gradient of Mn(II) at the sediment–water interface (mmol/m^4), and ϕ denotes the porosity ($\phi = 0.95$) at the sediment surface. The fluxes obtained from equation 6 are listed in Table II.

In April 1991 a flux of 0.07 mmol of Mn(II) per m^2 per day from the deeper sediment was obtained from a well-resolved gradient. A very similar profile was observed 1 year later. On the other hand, the diffusive flux of 0.46 mmol/m^2 per day (June 1990) was clearly too low to sustain Mn oxidation rates of up to 3 mmol/m^2 per day during summer. The 15-mm resolution of the dialysis samplers probably was too coarse to determine such gradients reliably.

In this case, flux chamber measurements may provide more accurate estimates of the diffusive fluxes (Figure 6b). Fluxes are calculated in millimoles per square meter per day according to

$$F_{86} = h \frac{\partial [Mn(II)]}{\partial t} \qquad (7)$$

where h is the height of the enclosed water column in the flux chamber (m) and $\partial[Mn]/\partial t$ represents the accumulation rate of Mn in the benthic chamber (mmol/m^3 per day).

A small correction of a few percent must be applied to fluxes calculated from equation 7, because every 60-mL sample is replaced with lake water from outside the chamber. The Mn concentration of this lake water has also been analyzed. The measured fluxes vary between 5.5 mmol/m^2 per day in July and a minimum of 0.5 mmol/m^2 per day in April. We therefore conclude that the high oxidation rates in summer are balanced by similar Mn reduction rates at the sediment–water interface and that pore water analyses on the millimeter scale would be required to quantify the steep Mn(II) gradients during summer.

EXAFS Spectroscopy of Mn Oxides from Sediment Traps. The coordinative environment of manganese in sediment-trap material was evaluated with EXAFS spectra recorded at the Mn K-edge. Figure 7 shows the radial distribution function (RDF) around Mn atoms of a sediment-trap sample. The RDFs of pyrochroite, manganite, Na–birnessite, todorokite, and vernadite are also shown. A comparison of EXAFS data with results from X-ray diffraction (XRD) is shown in Table III for all three oxidation states of manganese.

The radial distribution functions in Figure 7 represent the Fourier transform of EXAFS spectra. They display several peaks according to the nearest atomic shells surrounding the central Mn atoms. The first peak in the RDF allows the calculation of the Mn–O distance and the coordination number of the first shell. The second peak corresponds to the Mn–Mn distances.

In Mn(OH)$_2$, the Mn–O and the Mn–Mn peaks are shifted to longer distances compared with the Mn(IV) oxides. This shift reflects the increase of interatomic distances in going from the Mn(IV) to Mn(II) (hydr)oxides. The difference is large enough to allow a precise determination of the average oxidation state of Mn ions in the sediment-trap sample. The analysis of the first RDF peak in the Lake Sempach sample yields a Mn–O distance of 1.93 Å (Table III), a value to be compared with those of Mn(II) ($d_{Mn-O} = 2.23$ Å) and Mn(IV) oxides ($d_{Mn-O} = 1.90–1.94$ Å). These distances clearly indicate the prevalence of Mn(IV) in the sediment-trap sample. If present, Mn(II) does not exceed 20% of the total Mn.

As shown in Figure 8, Mn–Mn distances in these oxides depend on the way octahedra are linked to each other. As a consequence, Mn–Mn distances and numbers of nearest Mn neighbors derived from EXAFS spectra can be used to determine the short-range structure of X-ray amorphous hydrous Mn oxides (49, 50).

The sensitivity of EXAFS to changes in the local structure of MnO$_2$ polymorphs is illustrated in Figure 7, where the RDF of birnessite (a phyllomanganate) and todorokite (a tectomanganate) are compared. The RDF of phyl-

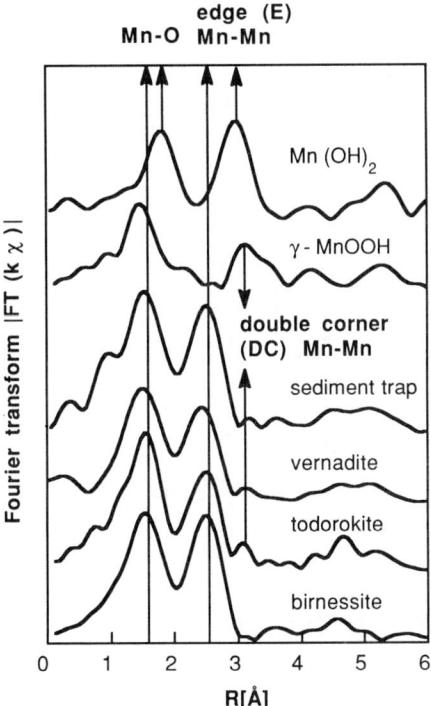

Figure 7. Radial distribution functions (RDF), not corrected for phase shift from EXAFS spectra, of sediment-trap material from Lake Sempach and from reference oxides. Pyrochroite, $Mn(OH)_2$, and birnessite [a Mn(IV) oxide] have the same layered structure with edge-sharing Mn octahedra. Todorokite is a Mn(IV) oxide with a 3 × 3 tunnel structure. A shift to longer distances occurs in going from the Mn(IV) oxide birnessite to the Mn(II) hydroxide pyrochroite. Contributions from double-corner Mn–Mn linkages are clearly seen in sediment-trap material and in todorokite and vernadite but not in the layered minerals birnessite and pyrochroite.

lomanganates, whose structure is built of sheets of edge-sharing Mn octahedra, displays only one Mn–Mn peak corresponding to a distance of 2.90 Å (see Table III). This interatomic distance reflects the average of d_{Mn-Mn} = 2.85 and 2.95 Å obtained from the refinement of powder diffraction data (Figure 8) (36). EXAFS spectroscopy is not sensitive enough to resolve these two contributions from edge-sharing octahedra with a Δd = 0.1 Å. However, tectomanganates show two Mn–Mn contributions from edge-sharing ($d_{Mn-Mn} \approx 2.90$ Å) and corner-sharing ($d_{Mn-Mn} \approx 3.50$ Å) octahedra.

The RDF of the sediment-trap material clearly shows two Mn(IV)–Mn(IV) contributions at distances characteristic of edge- and corner-sharing linkages. Therefore, we can exclude the mineral birnessite. This result is consistent with the XRD pattern obtained for sediment-trap material; it displays

Table III. Structural Parameters of Mn Oxides

Sample[a]	Method[b]	R_{Mn-O}[c] (Å)	N[d]	$\Delta\sigma$[e] (Å)	R_{Mn-O} (Å)	N	$\Delta\sigma$ (Å)
Pyrochroite $Mn(OH)_2$	XRD	2.23	6		3.34	6	
Manganite	XRD	1.85	2		2.76	1	
		1.92	2		2.94	1	
γ-MnOOH		2.30	2		3.87	8	
Na-birnessite	XRD	1.94	6		2.90	5	
$Na_4Mn_{14}O_{30} \cdot 10H_2O$	EXAFS	1.94	6.0	0.00	2.90	5	0.00
Todorokite	XRD	1.89	6		2.90	4.7	
$(Na,Ca,Mg)_1Mn_6O_{12} \cdot H_2O$	EXAFS	1.90	6.0	0.02	2.91	4.7	0.00
					3.48	2.5	0.00
Vernadite	EXAFS	1.89	6.0	0.01	2.87	3.7	0.00
δ-MnO_2					3.46	0.4	0.00
Sediment trap	EXAFS	1.93	6.0	0.01	2.88	4.2	0.00
					3.50	0.6	0.00

[a] Only approximate stoichiometries of birnessite and todorokite are given.
[b] XRD data were taken from the literature (31, 36, 52); the EXAFS fits were calculated by using Na–birnessite as a reference standard.
[c] Interatomic distance; the precision of determinations by EXAFS is ≈ 0.02 Å for R_{Mn-O} and ≈ 0.04 Å for R_{Mn-Mn}.
[d] Average numbers of nearest neighbors in the atomic shell; the precision of the EXAFS results is $\approx 20\%$.
[e] Difference of Debye–Waller factor between the reference (Na–birnessite) and the sample; this factor accounts for static and thermal disorder.

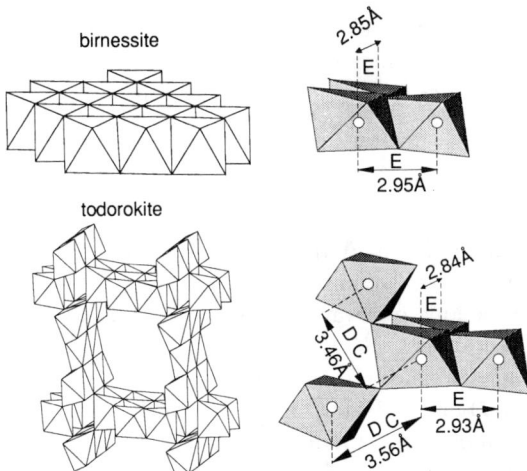

Figure 8. Local structures of birnessite and todorokite. The linkage of octahedra along edges and double corners is abbreviated by E and DC, respectively. Distances are from Rietveld refinements of powder diffraction data (31, 36). They can be compared with the average distances obtained from EXAFS in Table III.

two broad hk bands at 2.5 and 1.45 Å and no basal reflection near 7 Å (data not shown). Layered mineral particles such as birnessite always yield a basal (001) reflection, even in a highly disordered state (37).

The structural parameters of the Lake Sempach samples also disagree with those obtained from Mn(III) oxides such as γ-MnOOH (manganite). Its structure has been determined by X-ray and neutron diffraction (51, 52) and is related to that of the manganese dioxide pyrolusite. However, the Jahn–Teller effect distorts the Mn(III)O$_6$ octahedra with four oxygen neighbors at short distances and two at longer distances. Within the lattice only two Mn neighbors at d_{Mn-Mn} = 2.76 and 2.94 Å share a common edge, although eight corner-sharing positions are occupied at the fairly long distance of d_{Mn-Mn} ≈ 3.8–3.95 Å.

Because of these structural peculiarities, the RDF of manganite does not resemble any of the others (Figure 7). It can be characterized by the absence of the first Mn–Mn peak due to the large difference in distance of only two Mn–Mn neighbors across edges. Conversely, the presence of eight next-nearest manganese neighbors across corners leads to a marked Mn peak at a longer distance. In sediment-trap samples the presence of an intense Mn–Mn contribution at 2.88 Å indicates that (if present) manganite impurities account for less than 20% of the total Mn.

In summary, the EXAFS data of the sediment trap material are compatible with both todorokite and vernadite (δ-MnO$_2$). However, the fact that this Mn oxide is X-ray-amorphous suggests that vernadite is the prevalent Mn mineral that forms in the deep waters of Lake Sempach. The formation of this oxide was inferred from electron microscopic (EM) images of very finely crumpled sheets of Mn oxides in two English lakes (53) and from EM and electron diffraction work on particulate marine Mn samples (28).

Discussion

Oxidation Kinetics of Mn(II). This section addresses the question of whether the Mn(II) oxidation rates shown in Figure 4 can be explained by microbiological or abiotic pathways. Several incubation studies of Mn(II) with natural water, natural particulate matter, or pure cultures reported evidence for microbial catalysis of Mn(II) oxidation (4, 18, 54–58). In bottom waters of Lake Zurich (58) and in water samples from the marine fjord of Saanich Inlet (18) maximum Mn oxidation occurred at around 33 and 20 °C, respectively. These results strongly suggest microbial catalysis. In the case of abiotic catalysis, a steady increase in the oxidation rate with temperature is to be expected (16). Working with water samples from the bottom of Lake Zurich that were spiked with Mn(II) at 2 or 10 μM, Diem (58) found a Michaelis–Menten-type rate law for Mn(II) oxidation:

$$-\frac{d[\mathrm{Mn(II)}]}{dt} = V_{max} \frac{[\mathrm{Mn(II)}]}{[\mathrm{Mn(II)}] + K} \qquad (8)$$

where V_{max} and K are the maximal velocity and the Michaelis constant, respectively, with average values of V_{max} = 4.0 μM/day and K = 0.43 μM. A manganese half-life of about 1 day is obtained from these values. Table IV

Table IV. Half-Lives of Mn(II) in Microbiological Oxidation

$t_{1/2}$ (days)	T (°C)	Location	Reference
		Marine sites	
17	21	Tamar Estuary	54
1.8–69	9	Saanich Inlet	18
		Freshwater sites	
3.6	20	Lac de Bret	55
3.5	—[a]	Lake Greifen	4
2.5	—[a]	Oneida Lake	56
1.1	10	Lake Zurich	58
0.94	7.5	Rhone River	57
1.4	5	Lake Sempach	this study

[a] Information not available.
SOURCE: Modified from reference 4.

shows that different authors working in different systems found similar half-lives for microbiological Mn(II) oxidation.

Davies (59) predicted rates of surface-catalyzed oxidation of Mn(II) in the presence of iron oxyhydroxides for freshwater and estuarine environments on the basis of kinetic studies. His results are summarized in Table V. The following rate law was reported (60) for autocatalysis by MnO_2:

$$\frac{d[Mn(II)]}{dt} = k_{Mn}[Mn(II)][OH^-]^2[MnO_2][O_2] \quad (9)$$

with a rate constant $k_{Mn} = 5 \times 10^{18}$ M^4/day. Typical half-lives for lake water conditions are also given in Table V. During the summer stagnation (April–September) for the years 1988–1991, the following average concentrations (at pH 7.6) were found at depths between 80 and 85 m in Lake Sempach: $[O_2]$ = 0.24 mM, [Mn(II)] = 0.3 μM, $[MnO_2]$ = 2.1 μM, and [FeOOH] = 0.6 μM.

A comparison of these values with the conditions for the calculations in Table V yields estimated half-lives of Mn(II) in Lake Sempach of about 1 year for the surface-catalyzed pathways. The slow kinetics is mainly due to the low pH and the correspondingly small fraction of adsorbed Mn(II).

The expected half-lives on the order of 1 day for microbial catalysis and 1 year for the surface-controlled process can now be compared with the average oxidation rate in Lake Sempach, which is given in Table II. The average oxidation rate of 0.77 mmol/m^2 per day within a 5-m layer of deep water in Lake Sempach corresponds to a rate of R_{ox} = 0.15 μM/day. This value and a pseudo-first-order rate law with the average concentration of dissolved $[Mn(II)]_{av}$ = 0.3 μM

$$R_{ox} = k_{ox}[Mn(II)]_{av} \quad (10)$$

Table V. Half-Lives of Mn(II) in Surface-Catalyzed Oxidation

Oxidation Process	pH 7.5	pH 8.0	pH 8.5
Catalyzed by FeOOH[a]			
Freshwater	290	37	5
Estuarine water	470	94	57
Catalyzed by MnO_2[b]			
Freshwater	2800	280	28

NOTE: All values are half-lives in days.
[a] Calculations from Davies (59); conditions: 25 °C, [FeOOH]$_p$ = 2μM, and oxygen saturation.
[b] Calculated by using the rate law of Brewer (60) (eq 9); conditions: 25 °C, $[MnO_2]_p$ = 2 μM, and $[O_2]$ = 0.25 mM.

yield a rate constant of $k_{ox} = 0.5$ per day and a corresponding half-life of $t_{1/2} = 1.4$ days. This fast rate cannot be explained with available data for abiotic oxidation mechanisms (Table V), but it is well within the range of reported microbial oxidation rates (Table IV).

However, some caution is required when transferring rate constants for surface catalysis to the field situation. Most kinetic experiments (16, 61) on the surface-catalyzed oxidation of Mn(II) have been performed with well-crystallized minerals such as goethite or lepidocrocite (α- or γ-FeOOH). It has been shown (62) that the catalytic effect of amorphous hydrous ferric oxide on the oxidation of Fe(II) by O_2 is much larger than the promotion by α- or γ-FeOOH. The estimated abiotic half-lives in Table V should therefore be regarded as upper boundaries.

Mineralogy of Oxidation Products. Our EXAFS experiments suggest that X-ray amorphous vernadite is formed during microbiological oxidation of manganese. Manceau et al. (37) showed that the EXAFS spectra of vernadite are identical to those of synthetic hydrous manganese oxide (which is often called δ-MnO_2). The mineral in the sediment-trap samples from Lake Sempach is clearly different from birnessite, because a small contribution from corner-sharing octahedra was revealed by EXAFS spectroscopy. On the basis of data shown in Figure 7, we excluded the formation of significant amounts (>20%) of hausmannite (Mn_3O_4) and manganite (γ-MnOOH). The former would produce a shift in the Mn–O peak in the RDF, and the latter would be easily detected in the RDF by a Mn–Mn peak of distorted octahedra at a long distance of d_{Mn-Mn}. Although the formation of these minerals at redox gradients in the hydrosphere was suggested by several authors (17, 23, 25, 63), we found no spectroscopic evidence for their presence in the bottom waters of Lake Sempach.

The results of EXAFS spectroscopy have two major implications:

1. The large specific surface area of vernadite (typically >200 m^2/g) (64) and its low pH_{ZPC} (~2.5) provide an efficient scavenging mechanism for metal cations in the bottom waters of productive lakes such as Lake Sempach. The manganese cycle is therefore likely to interfere with the accumulation of metal ions and radiotracers such as ^{210}Pb. Anions are less strongly adsorbed to the negatively charged surface of δ-MnO_2. However, a study of phosphate adsorption to hydrous manganese oxide (65) indicates that in hard-water lakes adsorption of Ca^{2+} may significantly enhance the adsorption of HPO_4^{2-}. This possible link of P and Mn cycles at the sediment–water interface should be studied in more detail.

2. Vernadite on the surface of microorganisms may be part of a catalytic system in this oxidative biomineralization process. To clarify this point, a more detailed study of the oxidation kinetics of Mn(II) adsorbed to δ-MnO$_2$ should complement the available data (1).

3. The direct oxidation of Mn(II) to Mn(IV) provides an efficient transfer of oxidizing equivalents to the sediment surface. Recent peeper results from Lake Sempach showed that large mineralization rates during summer release enough dissolved ions to the overlying water so that a density stabilization occurs (66). The stagnation of water layers near the sediment surface inhibits the transport of dissolved oxidants such as O$_2$ and NO$_3^-$. However, the flux of particulate MnO$_2$ is at maximum during such periods.

Conclusions

Analysis of a 4-year time series of Mn sediment-trap data yielded Mn(II) oxidation rates with a strongly variable seasonal pattern. The average half-life of Mn(II) (1.4 days) cannot be explained by published abiotic oxidation rates. It is, however, in line with typical microbiological oxidation rates.

EXAFS spectroscopy revealed that the products of Mn(II) oxidation consist chiefly of vernadite, an X-ray-amorphous Mn(IV) oxide. No evidence was found for reduced forms of Mn oxides such as hausmannite (Mn$_3$O$_4$) or manganite (γ-MnOOH). In a subsequent study we used EXAFS to elucidate the structure of reduced Mn phases that control Mn(II) solubility within the sediment.

In situ measurements of Mn fluxes with benthic chambers and dialysis samplers confirmed the seasonal variability of the Mn redox cycle. They indicated that the reduction of particulate Mn oxides is a fast process that occurs close to the sediment surface within a time scale similar to that of Mn(II) oxidation. A maximum release rate of 5.5 mmol/m^2 per day was measured in July. This rate indicates a close coupling between the oxidation of Mn(II) in the bottom waters and the reduction of MnO$_2$ at the sediment surface.

Acknowledgments

The time series of sediment-trap analyses in Lake Sempach was initiated by René Gächter. We thank him for access to these data and Erwin Grieder, Antonin Mares, André Steffen, and Alois Zwyssig for the analytical work. Christian Dinkel performed the sediment lander and peeper experiments, and Judith Hunn analyzed the sediment cores. We thank the staff of the LURE synchrotron facility. This chapter benefited from valuable comments by James J. Morgan, Carola Annette Johnson, Dieter Diem, Noel Urban, and three anonymous reviewers. We acknowledge the support of the Schweizer Nationalfonds (NFP–24) and of the European Environmental Research Organisation (EERO). CNRS provided a visiting fellowship to B. Wehrli.

References

1. Morgan, J. J. Ph.D. Thesis, Harvard University, 1964.
2. De Vitre, R. R.; Buffle, J.; Perret, D.; Baudat, R. *Geochim. Cosmochim. Acta* **1988**, 52 1601–1613.
3. Davison, W. In *Chemical Processes in Lakes;* Stumm, W., Ed.; Wiley-Interscience: New York, 1985; pp 31–53.
4. Johnson, C. A.; Ulrich, M.; Sigg, L.; Imboden, D. *Limnol. Oceanogr.* **1991**, *36*, 1415–1426.
5. Wehrli, B. In *Aquatic Chemical Kinetics;* Stumm, W., Ed.; Wiley: New York, 1990; pp 311–336.
6. Stumm, W. *Chemistry of the Solid–Water Interface;* Wiley: New York, 1992; pp 309–335.
7. Burdige, D. J.; Nealson, K. H. *Geomicrobiol. J.* **1986**, *4*, 361–387.
8. Postma, D. *Geochim. Cosmochim. Acta* **1985**, *48*, 1023–1033.
9. Ehrlich, H. L. *Geomicrobiology*, 2nd ed.; Marcel Dekker: New York, 1990; pp 347–440.
10. Ehrlich, H. L. *Geomicrobiol. J.* **1986**, *5*, 423–431.
11. Robbins, J. A.; Callender, E. *Am. J. Sci.* **1975**, *275*, 512.
12. Sundby, B.; Andersson, L. G.; Hall, P. O. J.; Iverfeldt, A.; Rutgers van der Loeff, M. M.; Westerlund, S. F. G. *Geochim. Cosmochim. Acta* **1986**, *50*, 1281–1288.
13. Johnson, K. S.; Berelson, W. M.; Coale, K. H.; Coley, T. L.; Elrod, V. A.; Fairey, W. R.; Lams, H. D.; Kilgore, T. E.; Nowicki, J. L. *Science (Washington, DC)* **1992**, *257*, 1242–1244.
14. Murray, J. W. In *Sources and Fates of Aquatic Pollutants;* Hites, R. A.; Eisenreich, S. J., Eds.; Advances in Chemistry 216; American Chemical Society: Washington, DC, 1987; pp 153–183.
15. Diem, D.; Stumm, W. *Geochim. Cosmochim. Acta* **1984**, *48*, 1571–1573.
16. Davies, S. H. R.; Morgan, J. J. *J. Colloid Interface Sci.* **1989**, *129*, 63–77.
17. Hastings, D.; Emerson, S. *Geochim. Cosmochim. Acta* **1986**, *50*, 1819–1824.
18. Tebo, B. M.; Emerson, S. *Appl. Environ. Microbiol.* **1985**, *50*, 1268–1273.
19. de Vrind-de Jong, E. W.; de Vrind, J. P. M.; Boogerd, F. C.; Westbroek, P.; Rosson, R. A. In *Origin, Evolution and Modern Aspects of Biomineralization in Plants and Animals;* Crick, R. E., Ed.; Plenum: New York, 1989; pp 489–496.
20. Burns, R. G.; Burns, V. M.; Stockman, H. W. *Am. Miner.* **1983**, *68*, 972–980.
21. Burns, R. G.; Burns, V. M.; Stockman, H. W. *Am. Miner.* **1985**, *70*, 205–208.
22. Giovanoli, R. *Am. Miner.* **1985**, *70*, 202–204.
23. Murray, J. W.; Dillard, J. G.; Giovanoli, R.; Moers, H.; Stumm, W. *Geochim. Cosmochim. Acta* **1985**, *49*, 463–480.
24. Kessick, M. A.; Morgan, J. J. *Environ. Sci. Technol.* **1975**, *9*, 157–159.
25. Giovanoli, R. *Chimia* **1976**, *30*, 102–103.
26. Murray, J. W.; Balistrieri, L. S.; Paul, B. *Geochim. Cosmochim. Acta* **1984**, *48*, 1237–1247.
27. Dillard, J. D.; Crowther, D. L.; Murray, J. W. *Geochim. Cosmochim. Acta* **1982**, *46*, 755–759.

28. Chukhrov, F. V.; Gorshkov, A. I.; Beresovskaya, V. V.; Sivtsov, A. V. *Miner. Deposita* **1979**, *14*, 249–261.
29. Burns, R. G.; Burns, V. M. In *Marine Minerals;* Burns, R. G., Ed.; Mineralogical Society of America: Washington, DC, 1979; pp 1–46.
30. Turner, S.; Buseck, P. R. *Science (Washington, DC)* **1982**, *212*, 1024–1027.
31. Post, J. E.; Bish, D. L. *Am. Miner.* **1988**, *73*, 861–869.
32. Giovanoli, R.; Stähli, E.; Feitknecht, W. *Helv. Chim. Acta* **1970**, *53*, 209–220.
33. Giovanoli, R.; Stähli, E.; Feitknecht, W. *Helv. Chim. Acta* **1970**, *53*, 453–464.
34. Strobel, P.; Charenton, J. C.; Leglet, M. *Rev. Chim. Miner.* **1987**, *24*, 199–220.
35. Chukhrov, F. V.; Sakharov, B. A.; Gorshkov, A. I.; Drits, V. A. *Int. Geol. Rev.* **1985**, *27*, 1082–1088.
36. Post, J. E.; Veblen, D. R. *Am. Miner.* **1990**, *75*, 477–489.
37. Manceau, A.; Gorshkov, A. I.; Drits, V. A. *Am. Miner.* **1992**, *77*, 1144–1157.
38. Gächter, R. *Schweiz. Z. Hydrol.* **1987**, *49*, 170–185.
39. Gächter, R.; Meyer, S. I. In *Sediments: Chemistry and Toxicity of In-Place Pollutants;* Baudo, R.; Giesy, J. P.; Munatu, H., Eds.; Lewis Publishers: Chelsea, MI, 1990; pp 131–162.
40. Höhener, P. Ph.D. Thesis, Swiss Federal Institute of Technology (ETH), Zurich, Switzerland, 1990.
41. Wieland, E.; Santschi, P. H.; Höhener, P.; Sturm, M. *Geochim. Cosmochim. Acta* **1993**, *57*, 2959–2979.
42. Gächter, R.; Imboden, D.; Bührer, H.; Stadelmann, P. *Schweiz. Z. Hydrol.* **1983**, *45*, 246–266.
43. Devol, A. *Deep Sea Res.* **1987**, *34*, 1007–1026.
44. Adams, D. D. In *CRC Handbook of Techniques for Aquatic Sediments Sampling;* Mudroch, A.; MacKnight, S. D., Eds.; CRC: Boca Raton, FL, 1991; pp 171–202.
45. Brandl, H.; Hanselmann, K. W. *Aquat. Sci.* **1991**, *53*, 55–73.
46. Teo, B. K. *EXAFS: Basic Principles and Data Analysis;* Springer: New York, 1986; pp 1–349.
47. Koningsberger, D. C.; Prins, R. *X-ray Absorption: Principles, Applications, and Techniques of EXAFS, SEXAFS, and XANES;* Wiley: New York, 1988; pp 1–673.
48. Lerman, A. *Geochemical Pocesses, Water and Sediment Environments;* Wiley-Interscience: New York, 1979; p 92.
49. Manceau, A.; Combes, J. M. *Phys. Chem. Miner.* **1988**, *15*, 283–295.
50. Manceau, A.; Charlet, L. *J. Colloid Interface Sci.* **1992**, *148*, 425–442.
51. Buerger, M. J. *Z. Kristallogr.* **1936**, *95*, 163–174.
52. Dachs, H. *Z. Kristallogr.* **1963**, *118*, 303–326.
53. Tipping, E.; Thompson, D. W.; Davison, W. *Chem. Geol.* **1984**, *44*, 359–383.
54. Vojak, P. W.; Edwards, L. C.; Jones, M. V. *Estuarine Coastal Shelf Sci.* **1985**, *20*, 661–671.
55. De Vitre, R. Ph.D. Thesis, University of Geneva, Geneva, Switzerland, 1986.
56. Chapnick, S. D.; Moore, W. S.; Nealson, K. H. *Limnol. Oceanogr.* **1982**, *27*, 1004–1014.
57. Balikungeri, A.; Robin, D.; Haerdi, W. *Toxicol. Environ. Chem.* **1985**, *9*, 309–325.
58. Diem, D. Ph.D. Thesis, Swiss Federal Institute of Technology (ETH), Zurich, Switzerland, 1983.
59. Davies, S. H. R. In *Geochemical Processes at Mineral Surfaces;* Davis, J. A.; Hayes, K. F., Eds.; ACS Symposium Series 323, American Chemical Society: Washington, DC, 1986; pp 487–502.
60. Brewer, P. G. In *Chemical Oceanography;* Riley, J. P.; Skirrow, G., Eds.; Academic: Orlando, FL, 1975; Vol. 1, pp 415–496.
61. Sung, W.; Morgan, J. J. *Geochim. Cosmochim. Acta* **1981**, *45*, 2377–2383.

62. Tamura, H.; Kawamura, S.; Nagayama, M. *Corros. Sci.* **1980,** *20,* 963–971.
63. Hem, J. D.; Lind, C. J. *Geochim. Cosmochim. Acta* **1983,** *47,* 2037–2046.
64. Healy, T. W.; Herring, A. P.; Fuerstenau, D. W. *J. Colloid Interface Sci.* **1966,** *21,* 435–444.
65. Kawashima, M.; Tainaka, Y.; Hori, T.; Koyama, M.; Takamatsu, T. *Water Res.* **1986,** *20,* 471–475.
66. Wüest, A. *Limnologica* **1994,** *24,* 93–104.

RECEIVED for review October 23, 1992. ACCEPTED revised manuscript April 19, 1993.

Redox Chemistry of Iodine in Seawater

Frontier Molecular Orbital Theory Considerations

George W. Luther, III, Jingfeng Wu, and John B. Cullen

College of Marine Studies, University of Delaware, Lewes, DE 19958

> *The thermal and photochemical conversion of iodide to iodate in seawater is examined by means of thermodynamic, kinetic, and frontier molecular orbital theory considerations. The reaction is not significant because of lack of available oxidants in the ocean and their slow or negligible reactivity. Oxidation of iodide by triplet oxygen ($^3O_2$), O_3, and H_2O_2 leads to the reactive intermediates HOI and I_2, which are electron acceptors. These intermediates re-form iodide on reaction with reduced material and will react with natural organic matter faster than most other reduced species in seawater. The conversion of iodide to iodate may occur in seawater by bacterial processes and at the microlayer, where stronger oxidants may be present. Oxidation occurs in the atmosphere when volatile iodine species are released from the sea surface. The chemistry of iodine in the ocean is coupled to organic carbon production because of the difficulty of oxidizing iodine to iodate. The analysis reinforces Stumm's idea that the kinetics of redox processes governs the chemistry of bioactive elements in the ocean.*

THE pϵ OF THE SEA was the subject of a 1978 paper by Werner Stumm (1), in which he discussed several redox couples and showed "that the various redox components are not in equilibrium with each other and that the real system cannot be characterized by a unique pϵ." Among the couples he looked at were the I^-–IO_3^- and the O_2–H_2O couples. He concluded that "biological mediation of the four-electron reduction of O_2 justifies modeling of the equilibrium system in terms of

$$O_2(g) + 4H^+ + 4e^- \rightleftharpoons 2H_2O \qquad \log K = 83.1 \ (25 \ °C)$$

This equilibrium system acquires at pH 8, for pO_2 0.2 atm, a pϵ of 12.5. As we have seen, at this pϵ all biochemically important elements should exist virtually completely in their highest naturally occurring oxidation states."

We hope to reinforce his ideas by showing for the $I^--IO_3^-$ couple that a combination of one- or two-electron stepwise redox processes is not adequate for equilibrium modeling of iodine in oceanic waters and that the kinetics of redox processes governs oceanic iodine chemistry. In fact, a rich chemistry relating to carbon dynamics is the result.

This chapter considers the oxidation of iodide in seawater by natural oxidants (O_2, H_2O_2, and O_3). The oxidation of iodide to iodate is considered slow, yet the six-electron $I^--IO_3^-$ redox couple normally used to represent the process (or predict stability) is thermodynamically favorable (2). We will discuss both one- and two-electron-transfer processes with these oxidants, focusing on the first step of electron transfer and using the frontier molecular orbital theory approach in conjunction with available thermodynamic and kinetic data. The analysis shows that the chemical oxidation of I^- to IO_3^- is not a very important process in seawater, except perhaps at the surface microlayer.

Thermal or dark reactions do not appear to affect the conversion at seawater pH from both thermodynamic and kinetic considerations. In the photic zone of the ocean, some of these processes are thermodynamically unfavorable but can be activated by sunlight or biological processes (3). During the oxidation of I^-, the intermediates HOI and I_2 are produced. These reactive electron acceptors will reform I^- or organic iodine (RI) compounds when organic matter is present. The ultimate question we wish to address through an analysis of iodide oxidation in seawater is how iodine chemistry can be used as an indicator of carbon dynamics in the ocean.

Experimental Procedure

Water samples were obtained from the northwest Atlantic Ocean at 36° 20' N and 74° 44' W on June 30, 1990. Box cores were obtained from the northwest Atlantic Ocean at 38° 20' N and 73° 31' W on June 26, 1990 (station TI–15, the Wilmington Canyon) and at 36° 20' N and 74° 44' W on October 21, 1990 (station ML–10). The cores were subsampled with a whole core squeezer, as per Jahnke (4). The pore waters were obtained at selected depths by nitrogen pressurization, which expressed the water from the squeezer into polypropylene syringes that had been purged with nitrogen gas and were attached to Luer lock taps screwed into the side of the squeezer. Both pore water and water samples were filtered through 0.4-μm (Nuclepore) filters prior to analysis.

Iodide and iodate in seawater were measured by the methods of Luther et al. (5) and Herring and Liss (6), respectively, using an EG&G Princeton Applied Research (PAR) model 384B-4 with a Model 303A static dropping mercury electrode. Iodide was measured in rainwater by the method of Luther et al. (5) after 150 μL of 0.565 M NaCl was added as supporting electrolyte. After this measurement, 150 μL of 0.4 M sodium sulfite was added to the solution and iodide was determined again. The difference between these two measurements is operationally defined as inorganic oxidized iodine (e.g., HOI and I_2). The sodium sulfite and chloride solutions were iodide-free, as determined by analyzing blanks of these solutions.

Iodide in pore waters was measured by ion chromatography with electrochemical detection at a gold amalgam working electrode. The equipment consisted of a high-pressure liquid chromatography (HPLC) pump [Scientific Systems, Inc. (SSI) model 200] and SSI guardian (model 210), an injection valve (7125 Rheodyne), a strong anion column (Bio-Rad), and an electrochemical detector (EG&G PAR model 400).

The eluent was 0.1 M sodium acetate buffered to pH 6 in 70% methanol: 30% water. The procedure was run in the isocratic mode at 1.0 mL/min with a nominal pressure of 700–800 psi. The applied potential to the working electrode was 0 V versus the Ag–AgCl reference electrode. Chloride and bromide did not interfere with the determination of thiosulfate and iodide because they are weakly electroactive at the applied potential. Retention times in minutes are 2.0 (Cl^-), 3.9 (Br^-), 8.4 (I^-), and 12.5 ($S_2O_3^{2-}$).

Results and Discussion

Seawater Iodine Data.
Iodine is a bioactive element that exists predominantly as I^- and IO_3^- in seawater, with a total iodine concentration of about 470 nM at 35‰. Thermodynamically, in fully oxygenated seawater (pH 8.05, pε 12.5), iodine should exist entirely as IO_3^- (1, 2, 7). However, I^- concentrations in surface seawater reach 50–150 nM through biological processing of IO_3^- (2, 8) (Figure 1A).

The oceanic profiles of iodide are mirror images of those for iodate. Figure 1A shows our data for both iodine species over depth in the northwest Atlantic Ocean. These data are similar to previous reports in other oceanic systems. The profiles indicate that I^- is formed in the photic zone. I^- in these samples was typically stable for several months in our reanalysis of stored and refrigerated samples.

I^- solutions are stable to oxidation for long periods of time in the presence of O_2. The half-time for the iodide-to-iodate conversion is unknown (2). Conversely, the reaction of HS^- (which is isoelectronic with I^-) with O_2 has a half-

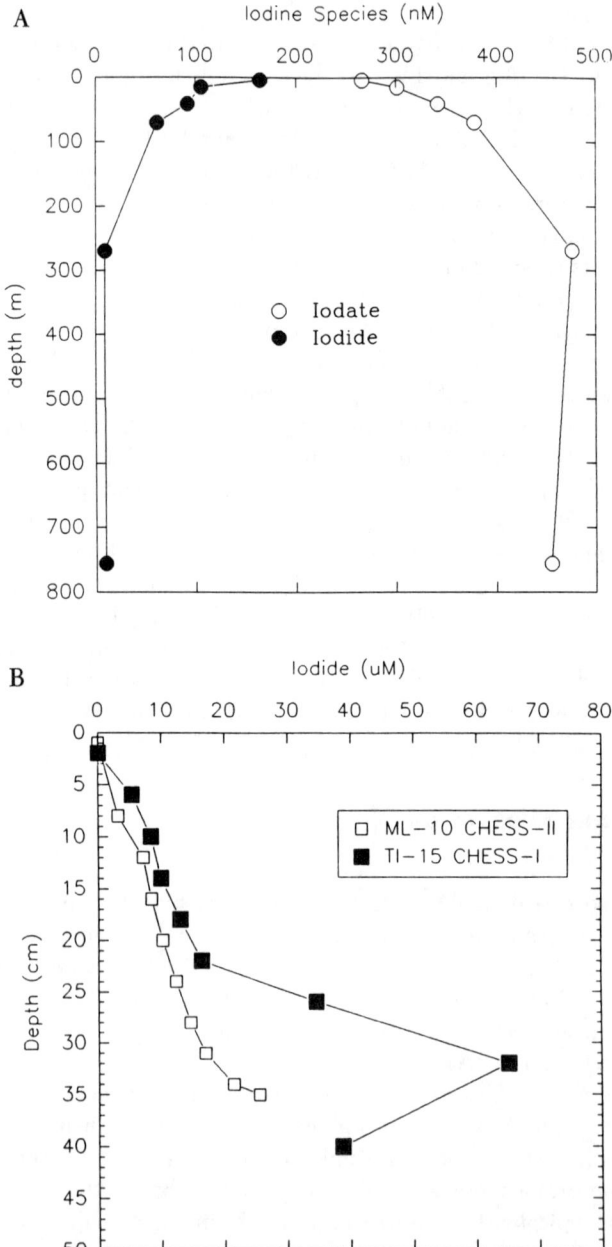

Figure 1. Part a: Typical iodide and iodate profiles obtained for the water column of the western Atlantic Ocean. Part b: Typical pore-water iodide profiles obtained from slope sediments of the western Atlantic Ocean.

life of 1 day (9). Iodide can be oxidized to IO_3^- by) O_2 with UV radiation, which forms peroxides and other strong oxidants (10), and by chlorine water or hypochlorite solutions (10, 11). The chlorine reaction is quantitative and has been used for total iodine analysis.

The mechanism for the iodide-to-iodate conversion in seawater is still not well understood. This conversion is not a facile process. Here we compare iodide with other isoelectronic ions, including SH^- and its congeners Cl^- (0.545 M in seawater) and Br^- (840 μM). The analysis indicates the conservative nature of Cl^- under photochemical and thermal conditions.

Available data in the literature (12) show that the primary oxidants in the photic zone of the ocean other than IO_3^- (<470 nM) are O_2 (about 250 μM) and H_2O_2 (100 nM). Ozone does not penetrate the surface microlayer, and its depletion has been estimated to occur within the first few micrometers (13). Nitrate, which is present in seawater (up to 50 μM in deep waters), is not considered to be an oxidant at seawater pH because as an anion it has Lewis base properties like I^-. In fact, the energies of the donor orbitals for NO_3^- and I^- are within 0.05 eV (14, 15).

The present discussion will focus first on one-electron- (free-radical) and two-electron- (nucleophilic) transfer reactions of I^- with these oxidants and then on singlet oxygen. The first intermediates, I_2 and HOI, are considered during these discussions. HOI reactions, including disproportionation, are also presented in further detail in a separate section.

Triplet Oxygen ($^3O_2$): An Outer-Sphere Process? The first step for the reaction of iodide with oxygen can be written as

$$I^- + O_2 \longrightarrow I + O_2^- \qquad E° = -1.49 \text{ V} \qquad (1)$$

Along with the orbital diagram for π electron transfer (Figure 2), equation 1 implies an outer-sphere redox (second-order) mechanism. The reduction potential ($E°$) is calculated from the recently tabulated free-radical reduction potentials for a variety of half-reactions (16); Table I presents reduction potentials of interest. Thus, the oxidation of I^- by O_2 is thermodynamically unfavorable by an abiotic, thermal, one-electron transfer process. Chloride, bromide, and bisulfide one-electron oxidations with O_2 are also thermodynamically unfavorable with $E°$ values of -2.57, -2.08, and -1.24 V, respectively.

If we assume that the second-order reaction given in equation 1 can be expressed as an equilibrium reaction, then $K = k_f/k_r$ (K is the equilibrium constant, k_f is the forward rate constant of the reaction, and k_r is the reverse rate constant of the reaction.) If $E° = -1.49$ V, then $K = 6.2 \times 10^{-26}$. For the reverse reaction, which is spontaneous, we can assume an upper diffusion-controlled limit for k_r of 1×10^{10} M^{-1} s^{-1} (17). Thus, $k_f = 6.2 \times 10^{-16}$, and we calculated a second-order half-life of about 200 billion years at O_2 concentrations of 250 μM in the photic zone of the ocean. These rates and the

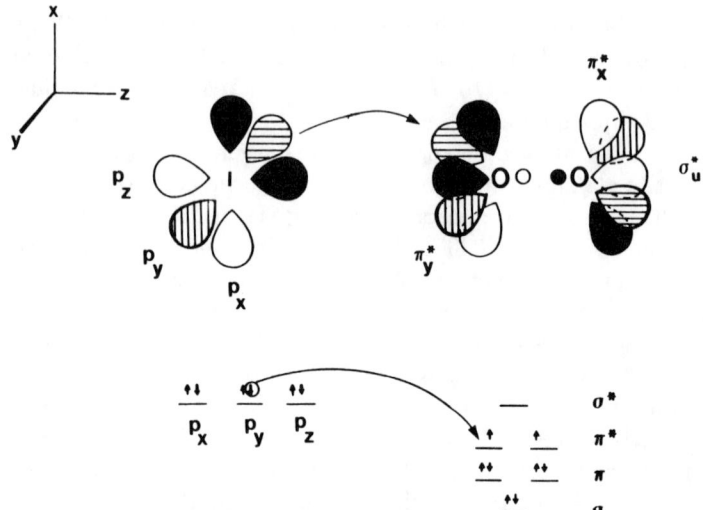

Figure 2. Molecular orbital diagram for the reaction of I^- and triplet O_2. The geometrical convention for this and the other figures is as follows. The y axis is perpendicular to the plane of the paper (xz). The dark and light lobes are in the xz plane; the orbital lobes with horizontal and vertical lines are in the yz plane. The different shading and lines represent different signs of the original wave function (Ψ). Overlap or electron transfer occurs when the signs (shading or lines) match.

Table I. Aqueous Reduction Potentials for Reductants and Oxidants

Reaction	$E°$ (V)
Reductants	
$Cl + e^- \rightarrow Cl^-$	2.41
$Br + e^- \rightarrow Br^-$	1.92
$I + e^- \rightarrow I^-$	1.33
$HS + e^- \rightarrow HS^-$	1.08
Oxidants	
$OH + e^- \rightarrow OH^-$	1.90
$O_3 + e^- \rightarrow O_3^-$	1.01
$^1O_2 + e^- \rightarrow {}^1O_2^-$	0.83
$H_2O_2 + e^- \rightarrow OH + OH^-$	−0.03
$^3O_2 + e^- \rightarrow {}^3O_2^-$	−0.16
Reactions of Triplet O_2 with Halides	
$Cl^- + {}^3O_2 \rightarrow Cl + {}^3O_2^-$	−2.57
$Br^- + {}^3O_2 \rightarrow Br + {}^3O_2^-$	−2.08
$I^- + {}^3O_2 \rightarrow I + {}^3O_2^-$	−1.49

SOURCE: Data are taken from reference 16.

conversion of IO_3^- to I^- by biological processes suggest that either a continuous buildup of I^- or a cycling process is likely. No continuous buildup of I^- occurs in the upper ocean, according to the data of several groups. However, Jickells et al. (8) reported seasonal changes in iodine speciation in the Sargasso Sea, surface-water iodate concentrations being highest in the nonproductive winter months when iodate-rich deep waters are mixed with surface waters.

The unfavorable thermodynamics for these outer-sphere, one-electron oxidations can be understood from the geometry and energy of the frontier molecular orbitals involved. Molecular oxygen exists in the triplet state, in which the lowest unoccupied molecular orbitals (LUMO) are singly occupied $\pi_x^1 \pi_y^1$ with an energy of -0.44 eV (Figure 2). The highest occupied molecular orbitals (HOMO), which are nonbonding orbitals for HS^- and I^-, are -2.32 and -3.06 eV, respectively.

These data provide two explanations for slow or no reactivity. First, the reactions with O_2 must occur by one-electron transfers (eq 1) rather than two-electron transfers because the anions cannot donate two electrons to partially filled O_2 orbitals (Figure 2). This scenario would be a violation of the Pauli exclusion principle.

The differences between the HOMO and LUMO of the reactants alone indicate that the donation of an electron from HS^- to O_2 costs 1.88 eV of energy, whereas for the I^- reaction the cost is 2.62 eV—a difference of 0.74 eV. If the energy difference between HOMO and LUMO is less than 6 eV, reaction can usually occur if symmetry and other energetic considerations are favorable - in these cases they are not. This cost in energy includes the energy required to unpair the electrons from the HOMO before reaction. The unpairing energy requirement is higher for I^- than for HS^-; the 5p orbitals of I^- are larger and more diffuse than the 3p orbitals of HS^-, so I^- can better accommodate two electrons in a more energetically favorable ground state. Thus, the reaction of I^- with O_2 must be slower than the corresponding reaction with HS^-.

However, these reactions may be photochemically accessible. In analogy to charge-transfer transitions, the difference between the HOMO and LUMO orbitals in an encounter complex can approximate the minimum amount of energy required for photochemical electron transfer (18). The reaction of HS^- with O_2 could occur with low-energy visible excitation because the 1.88-eV difference between the HOMO and LUMO orbitals of the reactants corresponds to 660 nm. Similarly, the reaction of I^- with O_2 could occur with UV or high-energy visible excitation (the 2.62-eV HOMO–LUMO gap corresponds to 473 nm). For the reactions of Cl^-, Br^-, I^-, and SH^- with O_2, the HOMO–LUMO gaps are 3.175 eV (390 nm), 2.924 eV (424 nm), 2.621 eV (473 nm), and 1.88 eV (660 nm), respectively. The general reaction is

$$X^- + {}^3O_2 \longrightarrow X + {}^2O_2^-$$

Photochemically, more excitation is required for these reductants with O_2 in the order $Cl^- > Br^- > I^- > HS^-$. However, these photochemical reactions may not be very important because the highest I^- concentrations occur at the surface (e.g., Figure 1A). Light penetration in seawater is greatest for wavelengths near 474 nm, but only 40% of the incident energy reaches 50 m. At 400 nm, 50% of the energy penetrates to 10 m (19). The minimum energy needed for reaction and its shallow penetration into seawater for photochemical excitation are a further indication of the conservative nature of Cl^- compared to the other halides.

In this photochemical case, a reactive I atom may be produced that can react with other I atoms to form I_2 or other anions (Cl^-, OH^-, etc.) to produce radical anion species (eqs 2a and 2b)

$$I + I \longrightarrow I_2 \quad (2a)$$

$$I + X^- \longrightarrow IX^- \quad (2b)$$

where X is halide or OH^-. These species would reform I^- and HOI readily because their unoccupied or singly occupied $\sigma°$ orbitals are excellent electron acceptors (Figures 3A and 3B). The unoccupied $\sigma°$ LUMO of I_2 has an energy of -2.55 eV, which is substantially lower than that given for O_2. The disproportionation reaction of I_2 with OH^- is fast in laboratory solutions (<1 s) and results in HOI formation (20). For this reaction there is no energy gap between the HOMO and LUMO orbitals (Table II).

The HOMO of OH^- is higher than the LUMO of I_2, which indicates that two-electron transfer from OH^- to I_2 should be facile. However, the reaction is reported to be slower in seawater, and I_2 appears to react with reduced organic matter (21–23). HOI in turn reacts with organic matter to produce I^- and RI (3, 24). Thus iodide oxidation leads to electron acceptors as intermediates that can eventually re-form iodide. In contrast, sulfide oxidation leads to strong electron donors such as the polysulfides, which oxidize back to SO_4^{2-}, as the intermediates (25). Thus, halides and sulfide have a significant difference in their environmental cycling, despite their isoelectronic nature.

However, these photochemical reactions may not be very important. Butler and Smith (10) noted that intense UV irradiation of seawater with excess peroxide added to the sample resulted in IO_3^- formation. However, the reaction was retarded in the absence of added peroxide, and iodate also reacted with natural organic matter. Richardson et al. (26) noted that BrO_3^- formation was retarded in natural estuarine waters during O_3 oxidation of Br^-.

Metal Catalysis with O_2 as Oxidant. Oxidation of sulfide can be enhanced by trace metals (27). However, iodide oxidation is not likely to be catalyzed by such a process. Luther (25) indicated that sulfide transfers electrons to oxygen through the metal. The main requirement appears to be a

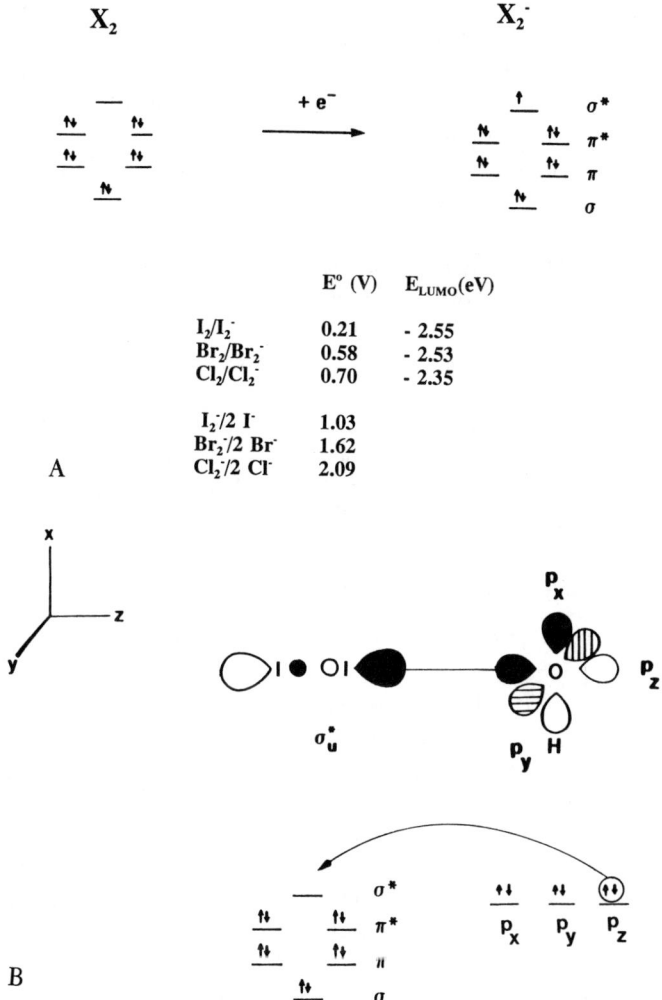

Figure 3. Part A: Molecular orbital diagrams for X_2 and X_2^- indicating the electron-acceptor properties of X_2. Part B: Molecular orbital diagram for the reaction of I_2 with OH^-.

metal sulfide complex with the oxygen binding to the metal trans to the sulfide. The metal acts as a pump to facilitate electron transfer through the π system of the reactants.

Metal–iodide complexes are not likely to be very important in seawater because there is a higher concentration of chloride (6 orders of magnitude higher) and bromide (3 orders of magnitude). Although the stability constants for metal–iodide complexes are stronger than for chloride and bromide complexes, the stronger complexation does not always negate the concentration effect. For example, the HgOHX complexes have log β values of 3.67 (Cl), 5.68 (Br), and 8.90 (I) at 25 °C and ionic strength = 0.5 (28). Thus, trace metals should not act as efficient catalysts for iodide oxidation in seawater. In support of this theory, experimental results from the Black Sea (29) show that I⁻ is present in zones with high metal concentrations and low sulfide concentrations, where anaerobic oxidation of sulfide occurs.

H_2O_2: An Inner-Sphere Process. An outer-sphere, one-electron-transfer process is not thermodynamically favorable for the reaction of H_2O_2 with I⁻ ($E° = -1.36$ V), Br⁻ ($E° = -1.95$ V), and Cl⁻ ($E° = -2.44$ V). However, a two-electron inner-sphere transfer to the vacant σ* LUMO of peroxide is possible and can result in the breaking of the O–O bond. The LUMO energy for H_2O_2, estimated to be -1.2 eV, gives an energy difference of 1.84 eV for the HOMO–LUMO gap for the reactants. Unlike equation 1 which is thermodynamically unfavorable, reaction 3a is favorable with the Gibbs free energy of reaction ($\Delta G_r°$) = -70.81 kJ/mol.

$$I^- + H_2O_2 \longrightarrow [I-H_2O_2]^- \longrightarrow HOI + OH^- \quad (3a)$$

$$I^- + H_2O_2 \longrightarrow [I-H_2O_2]^- \longrightarrow 2OH^- + I^+ \quad (3b)$$

In equations 3a and 3b an intermediate, $[I-H_2O_2]^-$ (analogous to I_3^-), is probably formed. This intermediate may break down in seawater either directly to HOI and OH⁻ as in equation 3a or to I⁺ as in equation 3b. In turn, I⁺ can react with any anion (halide, hydroxide, etc.) or organic compound, R, which may be nitrogenous or unsaturated (eqs 4a and 4b). HOI, IX, and I_2 are also reactive with organic matter (3; eqs 4c and 4d). I_2 will also react with chloride to form I_2Cl^- (28; eq 4e). The IX species are more reactive than the X_2 species (30) and rapidly disproportionate to HOI and X⁻.

$$I^+ + OH^- (X^-, I^-) \longrightarrow HOI (IX, I_2) \quad (4a)$$

$$I^+ + R-H \longrightarrow RI + H^+ \quad (4b)$$

$$I_2 + R \longrightarrow RI_2 \quad (4c)$$

$$HOI + R-H \longrightarrow R-I + H_2O \quad (4d)$$

$$I_2 + Cl^- \longrightarrow I_2Cl^- \quad (4e)$$

Equation 3a is important because it involves oxygen-atom transfer to the iodine, which is necessary for the eventual formation of IO_3^-. The reaction should be kinetically facile because donation of an electron pair from I⁻ to an empty orbital of H_2O_2 is favorable from energetic, electron spin, and symmetry considerations (*see* Figure 4). This reaction is analogous to the oxidation of HS⁻ by peroxide, as discussed by Luther (*25*).

A similar reaction process should occur for Br⁻, because $\Delta G_r° = -1.4$ kJ/mol, but such a reaction is thermodynamically unfavorable for Cl⁻ with a $\Delta G_r° = +28.2$ kJ/mol. These data further indicate the conservative nature of Cl⁻ in seawater and the incorporation of iodine and bromine into organic matter in the ocean (*3, 31*).

Butler and Smith (*10*) noted that intense UV irradiation of seawater resulted in IO_3^- formation only at high concentrations of added peroxide. Because peroxide concentration is low in seawater, this process may not be accessible to a high degree.

Because phytoplankton produce H_2O_2 (*32*) and have unsaturated carbon networks and nitrogenous organic compounds, soluble and particulate organic iodine compounds should form rapidly. A Redfield ratio was partially established for iodine by Elderfield and Truesdale (*33*). (The element of interest has a constant atomic ratio with carbon in phytoplankton. The ratio is constant usually throughout the world's oceans.) During bacterial remineralization in the water column, iodide is released back to the water column. Also, in surface waters, a small loss of iodine occurs in association with the loss of biological material. This loss is observed in a flux of particulate iodine to sediments in the deep sea (*34*) and in the formation of particulate iodine in surface waters in phase with primary production cycles (*8, 35*). The particulate iodine, which reaches the ocean sediments after death of the biological material, releases

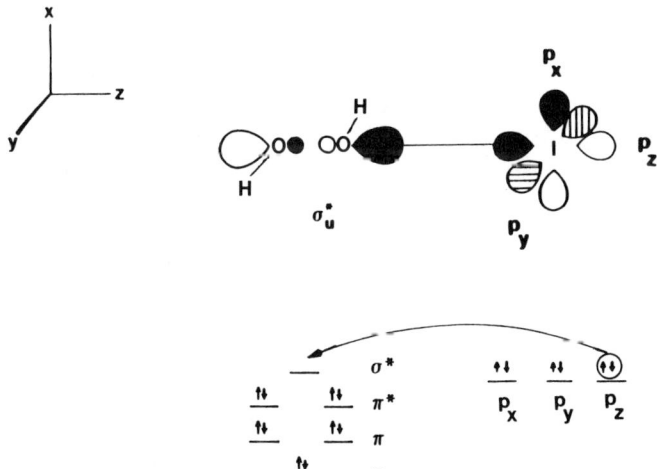

Figure 4. Molecular orbital diagram for the reaction of I⁻ and H_2O_2. It is similar to Figure 3B.

the iodine as I^- on diagenetic alteration of the organic matter (36–38; see Figure 1B). Thus, the highest I^- concentrations in marine waters (e.g., Figure 1A) occur in the photic zone, where primary productivity is high.

In pore waters the dissolved I^- concentrations, which are higher than the total iodine concentration in seawater, can be observed with decomposition of particulate organic matter (e.g., Figure 1B). In Figure 1B the I^- pore-water concentrations are in micromolar units, compared to the total iodine concentration in seawater of <0.5 μM. Thus, the pore waters contain several times the amount of iodine that is found in seawater. Increased I^- concentrations and significant amounts of organic iodine are found in anoxic basins such as the Black Sea (29, 39, 40).

O_3. Because O_3 does not penetrate the ocean microlayer from the atmosphere, its reactions are significant only in the microlayer. The reaction of iodide with ozone has been documented at the surface microlayer of the ocean (41), and I_2 was the only volatile iodine product detected. The reaction is believed to be an important trap for ozone impinging on the sea surface (42). A one-electron, outer-sphere, electron-transfer process (eq 5) is not thermodynamically favored by the reduction potentials in Table I.

$$I^- + O_3 \longrightarrow I + O_3^- \qquad E° = -0.32 \text{ V} \qquad (5)$$

Such a process is also unfavorable for Br^- ($E° = -0.91$) and Cl^- ($E° = -2.40$). O_3 has an empty orbital with good electron-acceptor properties (unlike triplet oxygen, but like H_2O_2). As a result, I^- and the other halides react better as nucleophiles than as free radicals in aqueous solution. Because of the energy difference between the HOMO and LUMO orbitals for all the halides and ozone, a photochemical process may be accessible in the visible region to near-IR region at the microlayer. However, these longer wavelengths are strongly absorbed by seawater (19).

Other alternatives with O_3 must be considered for I^- oxidation, including a possible thermal inner-sphere process. As with peroxide, O_3 has an empty antibonding orbital. This π-like orbital can accept electrons better than O_2 and peroxide (lower E_{LUMO} in Table II). Figure 5 shows the geometric arrangement of I^- with O_3.

The orbital accepting electrons ($2b_1$) from I^-, the Lewis base, is derived from the overlap of three parallel p orbitals from each oxygen atom. An intermediate of the form $[I-O_3]^-$ is probably analogous to the I_3^- and $[I-H_2O_2]^-$ species. Addition of electrons to this orbital lowers the π bonding in O_3 but not the σ bonding, so there is no apparent oxygen-atom transfer. However, the breakup of this intermediate may depend on the concentration of the reactants in solution. Figure 6 shows two possible pathways (eqs 6a and 6b) for the decomposition of the intermediate $[I-O_3]^-$.

Table II. HOMO and LUMO Energies for Reductants and Oxidants

Reductants	E_{HOMO} (eV)	Oxidants	E_{LUMO} (eV)
BrO^-	−1.5	O_2	−0.44
OH^-	−1.825	H_2O_2	−1.2
ClO^-	−2.0	OH	−1.82
HS^-	−2.32	O_3	−2.10
IO^-	−2.5	HOI	−2.5
I^-	−3.061		
Br^-	−3.364		
Cl^-	−3.615		
HOI	−8.5		

SOURCE: Data are taken from references 14 and 15.

$$I^- + O_3 \longrightarrow [I-O_3]^- \xrightarrow{H_2O, I^-(X^-)} I_2 \,(IX) + O_2 + 2OH^- \quad (6a)$$

$$I^- + O_3 \longrightarrow [I-O_3]^- \longrightarrow IO^- + O_2 \quad (6b)$$

Pathway 1 (eq 6a) shows no oxygen-atom transfer and the production of I^+, as discussed for the H_2O_2 reaction (eq 3). The I^+ can form a variety of products, in particular I_2 from excess I^- as discussed for the H_2O_2 reaction. The $[I-O_3]^-$ intermediate should break down to oxygen and hydroxide on H_2O and anion attack. The reaction of ozone with iodide to form I_2 is quantitative in the presence of excess I^- and can be used for the analysis of ozone (43). In seawater this process seems less likely to occur because I^- is a trace constituent.

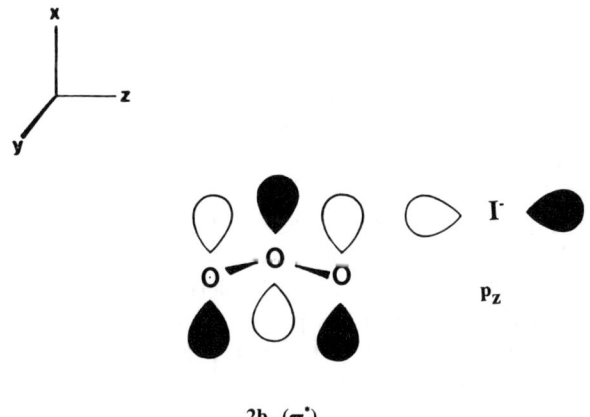

Figure 5. Molecular orbital diagram for the reaction of I^- and O_3.

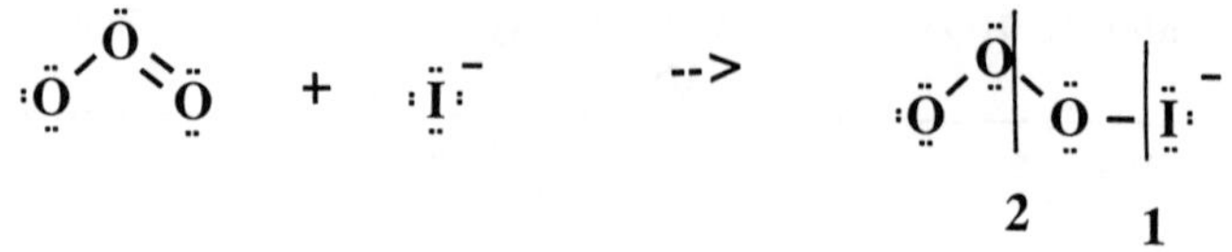

$$O_3 + I^- \longrightarrow [I-O_3]^-$$

$$\text{1)} \quad [I-O_3]^- \xrightarrow{H_2O,\ xs\ I^-} I_2 + O_2 + 2\ OH^-$$

$$\text{2)} \quad [I-O_3]^- \longrightarrow IO^- + O_2$$

Figure 6. Possible breakdown pathways for the intermediate of the reaction of I^- and O_3.

Although Garland and Curtis (41) found volatile I_2 evolution when O_3 was passed over seawater in batch reactors, they did not analyze the seawater for any change in iodine speciation. I_2 and other interhalogen compounds, IX, should disproportionate to HOI in water. The first step in this thermal reaction does not lead to oxygen-atom transfer from O_3 unless OH^- or H_2O hydrolyzes I_2 and IX.

The second pathway that leads to oxygen-atom transfer with excess O_3 seems more plausible in the seawater microlayer (eq 6b). Hoigné et al. (42) indicated, by analogy to bromide and chloride reactions with ozone in laboratory studies, that the first continuously generated product of iodide reaction with aqueous ozone was IO^-–HOI (44). They estimated the second-order rate constant to be greater than 10^6 M^{-1} s^{-1}. Aqueous ozone can also be an effective oxygen atom transfer reagent for iodide via photochemical initiation. Ozone decomposes primarily to peroxide in aqueous solution (45) under UV light. Peroxide reactions were already discussed.

However, the reaction of I^- with O_3 may lead to different end products from that reported by Haag and Hoigné (44) on Br^- with O_3. For the Br^- reaction, BrO_3^- is produced. The reaction goes toward completion at higher pH, where the species OBr^- can exist (pK_a of HOBr = 8.69). The reaction is slower or negligible when HOBr is present as an intermediate; that is, the reaction rate decreases at lower pH. Their sequence of reactions for BrO_3^- formation is given in equations 7a–7c.

$$O_3 + Br^- \longrightarrow O_2 + OBr^- \tag{7a}$$

$$O_3 + OBr^- \longrightarrow 2O_2 + Br^- \tag{7b}$$

$$2O_3 + OBr^- \longrightarrow 2O_2 + BrO_3^- \qquad (7c)$$

$$O_3 + HOBr \longrightarrow \text{no reaction} \qquad (7d)$$

The pK_a of HOI is near 11, so a mechanistic sequence similar to that in equations 7a–7c seems unlikely for ozone oxidation of I$^-$ at seawater pH. Table II shows that the nucleophile BrO$^-$, which can exist at seawater pH, has a higher (more positive) E_{HOMO} than O$_3$; this difference indicates a possible reaction. However, HOI with a very stable orbital (not OI$^-$) exists at seawater pH; thus, a large barrier to reaction that prevents IO$_3^-$ formation is evident. The rate of oxidation for the H$_2$S–HS$^-$ system shows a similar dependence on pH (9); that is, an increase in pH increases the HS$^-$ content and the rate of reaction.

Singlet O$_2$. As with O$_3$, I$^-$ reaction with ^{1}O$_2$ is not favorable thermodynamically as an outer-sphere, one-electron-transfer process ($E° = -0.50$ V). However, a reaction between these two reactants was reported by Haag and Hoigné (46) with a second-order rate constant similar to that of I$^-$ with O$_3$. A two-electron inner-sphere transfer is possible because the singly occupied π_x^* and π_y^* orbitals from triplet oxygen are now empty in one case and filled in the other (Figure 7). Thus, I$^-$ can attack the empty π^* as shown in Figure 7, and form an intermediate [I–O$_2$]$^-$. As with O$_3$, the accepting orbital is from the π system, which results in loss of π bonding but not σ bonding in O$_2$. The σ^* orbital remains empty. Thus, oxygen-atom transfer does not appear to be possible directly from the ^{1}O$_2$ (equation 8a and 8b).

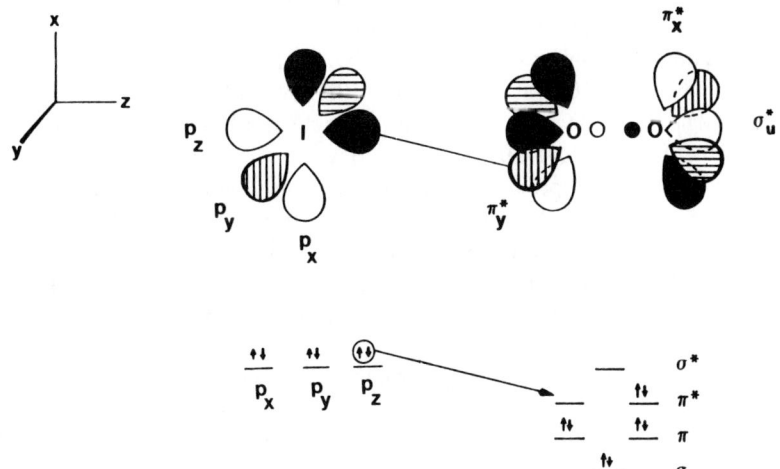

Figure 7. Molecular orbital diagram for the reaction of I$^-$ and singlet O$_2$.

$$I^- + {}^1O_2 \longrightarrow [I-O_2]^- \xrightarrow{2H_2O, I^-(X^-)} H_2O_2 + I_2 \text{ (IX)} + 2OH^- \quad (8a)$$

$$I^- + {}^1O_2 \longrightarrow [I-O_2]^- \xrightarrow{H_2O} HO_2^- + HOI \quad (8b)$$

Equation 8b indicates that the breakdown of the $[I-O_2]^-$ intermediate on hydrolysis leads to less favorable products than equation 8a, which involves hydrolysis and anion attack in solution. Thus, I^+ should be formed with the same result as discussed for H_2O_2 and O_3 (pathway 1, eq 6a).

OH Radical. Although OH radical production was documented in seawater by Mopper and Zhou (47), its steady-state concentration is near 10^{-18} M, significantly less than the singlet oxygen concentration of 10^{-13} M (46, 48). Mopper and Zhou (47) indicated that the OH radical appears to decompose rapidly by reaction with Br^- in seawater. Br^- concentration in seawater is 840 μM, whereas the I^- concentration is <100 nM.

Intermediate HOI. Reactions 2b, 3a, 4a, and 4d indicate that HOI is an important intermediate and is also isoelectronic with H_2O_2. At least three factors prevent it from reacting to form IO_3^- in seawater. First, HOI will not build up to significant levels in seawater with the oxidants available. Second, it is very reactive with organic matter (3). Third, HOI has electron-acceptor properties similar to H_2O_2 (Table II) and should not react readily with other electron acceptors, O_2 and H_2O_2. The pK_a of both HOI and H_2O_2 are both >11, indicating that the protonated compounds are the predominant species at seawater pH (8.1). The LUMO orbitals for these compounds are -1.2 eV (H_2O_2) and -2.5 eV (HOI). The HOMO orbitals are more stable by about 6 eV, a result indicating that a large barrier to reaction exists between these species. Thus, thermal reactions of HOI with H_2O_2 and with itself (the disproportionation of HOI) at seawater pH (8.1) are not likely.

The disproportionation of HOI at pH ≥ 13 has been reported (49). Disproportionation reactions are likely at pH values near the pK_a, where both HOI (an electron acceptor) and OI^- (an electron donor) species are present. The energies of the LUMO of HOI are close to the HOMO of IO^-. Thus, a two-electron transfer from IO^- to HOI is likely (Figure 8). Because the HOMO of IO^- and the LUMO of HOI are predominantly iodine in character (50), equation 9a is predicted to be more appropriate than equation 9b.

$$IO^- + HOI \longrightarrow OI_2 + OH^- \longrightarrow IO_2^- + H^+ + I^- \quad (9a)$$

$$IO^- + HOI \longrightarrow HOIO + I^- \longrightarrow IO_2^- + H^+ + I^- \quad (9b)$$

The formation of the intermediate OI_2 is expected over HOIO for several reasons. First, the reactivity of hypohalous acids is usually as OH^- and X^+ (51,

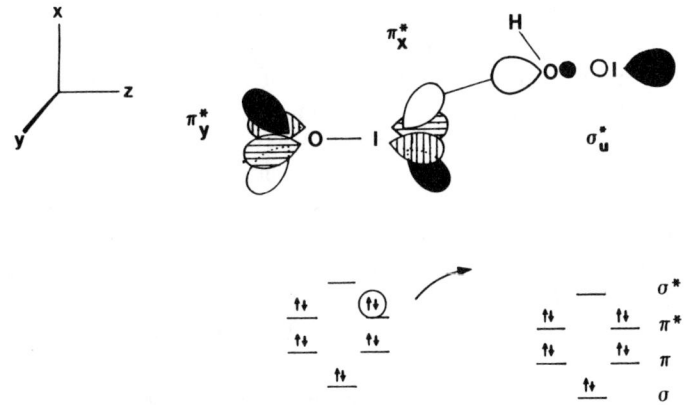

Figure 8. Molecular orbital diagram for the reaction of OI⁻ and HOI. It is similar to Figures 3B and 4.

52). Second, the iodine in IO⁻ should tend to attack the softer atom, I, rather than the harder atom, O. Third, because both HOIO and OI_2 have not been synthesized and are probably equally unstable, most likely the formation of the thermodynamically more stable OH⁻ ($\Delta G_r°$ = −157.3 kJ/mol) is favored over I⁻ ($\Delta G_r°$ = −51.59 kJ/mol) in the intermediate step. The hydrolysis of OI_2 should be similar to that for I_2; H_2O and OH⁻ lead to the formation of IO_2^-. IO_2^- is also unstable in solution and should react quickly to form IO_3^- with oxidants (e.g., O_3 and H_2O_2) that are effective oxygen-atom-transfer reagents or with HOI in a reaction sequence similar to equation 9a.

This sequence of reactions indicates that conversion of I⁻ to IO_3^- in seawater is not likely or is certainly not a significant chemical process, particularly in the presence of organic matter, which converts the intermediates (I_2 and HOI) into organic iodine or I⁻. Thermal or dark chemical reactions are not likely in deep waters indicating that any I⁻ detected in deep waters occurs primarily from mixing processes and/or the decomposition of particulate organic matter sinking to the sediments.

The oxidation of I⁻ may occur in seawater by algae (3) and by bacteria, as shown in sediments by Kennedy and Elderfield (36, 37). Also, volatile iodine forms released to the atmosphere may be oxidized to IO_3^- and returned to the ocean via dry or wet deposition (53). The seasonal data at one location (8) suggest net limited iodine cycling by iodide oxidation.

The Atmosphere's Role in Iodide Oxidation. The production of gaseous iodine compounds [such as CH_3I in a manner analogous to $(CH_3)_2S$ (54) or I_2 (41)] would permit the volatilization of iodine to the atmosphere. This iodine may be easier to oxidize to IO_3^- in the atmosphere by photolysis or by electrical storms, which produce stronger oxidants such as OH radical.

Ozone is also a better-oxygen-atom-transfer reagent in the atmosphere because UV light results in the formation of O_2 and O atoms, which react with halides. The role of metals in the catalysis of iodide oxidation is also more likely in the atmosphere because of the enrichment of iodide relative to the other halides in rainwater and aerosols (13).

We measured iodide by the method of Luther et al. (5) in subsamples of one rainwater event collected from the western Atlantic Ocean during June 1991. Iodide was 7.1 nM when sulfite was not added. When sulfite was added to the rainwater, which was then analyzed immediately, 10.1 nM iodide was measured. This value indicates that organic forms, which can react with sulfite to release I^- (55) or more oxidized forms (I_2 or HOI) of iodine, are present in the rainwater.

However, we believe that the inorganic species predominate because they react instantaneously with sulfite (56), whereas organic iodine compounds have significant reaction times (57). We did not add sulfurous acid, which converts IO_3^- to I^-. On the basis of this data, we suggest that oxidation of I^- to IO_3^- in the marine atmosphere may be more important than the corresponding reaction in seawater for the recycling of I^- to IO_3^-. Our data support the earlier radiochemical work of Luten et al. (53), who found similar results.

Conclusion

The oceanic iodine cycle can be represented by Figure 9. Iodate is reduced primarily by biotic processes to I^-. Iodine incorporated into phytoplankton and other organisms is also remineralized as I^- into the water column. Because I^- apparently does not build up in the deep waters studied to date, an unknown I^- oxidation process may occur. This process may be biological; unknown amounts of I^- may be oxidized back to IO_3^- by algae and bacteria. Chemical oxidation to IO_3^- appears unlikely, unless larger quantities of peroxide than have been reported previously are present in the ocean.

However, I_2 and HOI are the likely intermediates formed by both abiotic and biotic processes during I^- oxidation. These intermediates, if formed, can be reduced back to I^-; react with organic matter to form particulate or dissolved organic iodine (RI) compounds; or be volatilized to the atmosphere as I_2, HOI, or CH_3I. Iodate can also react directly with humic material during reduction to form RI (58), but this reaction is more important in sedimentary environments and has not been documented in the photic zone.

Particulate RI compounds sink to the sediments, where I^- is released during diagenesis. Some decomposition occurs in the deep waters, releasing low levels of I^- to the deep waters where it is not likely to be chemically reoxidized. Some of the volatile iodine compounds or sea salt spray containing iodide are reoxidized in the atmosphere, perhaps to iodate, which is redeposited to the ocean as wet or dry deposition.

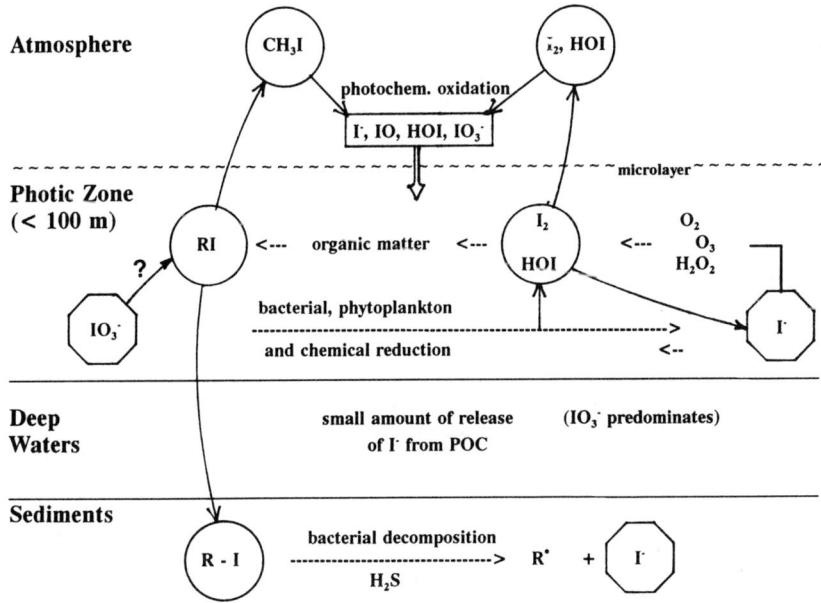

Figure 9. An oceanic cycle for iodine indicating the major pathways for transport and reaction of iodine species.

I⁻ does not appear to be oxidized to IO_3^- to any significant degree in the photic zone or the surface microlayer by chemical reactions. However, iodide oxidation may occur by biological mechanisms as found with macroalgae (3). Kennedy and Elderfield (36, 37) provide evidence for iodide oxidation (presumably bacterial) in sediments, but biologically mediated oxidation has not been shown yet as a significant process in the photic zone.

This cycle is consistent with the known reaction chemistry of iodine, and the profiles of iodine species in the ocean and in sedimentary pore waters. The cycle allows for cycling of iodine in the marine environment through biological fixation. Thus, iodine appears to be an excellent carbon tracer in oceanic waters because of the difficulty in oxidizing I⁻ to IO_3^-. This chemistry is in contrast to that of the sulfide system (25).

The preceding analysis of I⁻ oxidation to IO_3^- further demonstrates the concepts that Stumm (1) expounded on in his 1978 paper, "What is the pε of the Sea?" In particular, the kinetics of redox processes governs the chemistry of bioactive elements in the ocean. We, as a community of aquatic chemists from diverse environmental chemistry disciplines, are indebted to him for his application of the principles of physical and inorganic chemistry to understanding the environment.

Acknowledgments

This work was supported by grants from the National Science Foundation (OCE-8916804 and OCE-9217245) and the National Oceanic and Atmospheric Administration (NA16RG0162-02). We thank the crew of the R/V Cape Henlopen for their efforts in obtaining the cores. We thank T. Ferdelman and J. Kostka for processing the pore-water samples.

References

1. Stumm, W. *Thalassia Jugosl.* **1978**, *14*, 197–208.
2. Wong, G. T. F. *Rev. Aquat. Sci.* **1991**, *4*, 45–73.
3. Neidleman, S. L.; Geigert, J. *Biohalogenation: Principles, Basic Roles, and Applications;* Halsted (John Wiley & Sons): New York, 1986; p 203.
4. Jahnke, R. A. *Limnol. Oceanogr.* **1988**, *33*, 483–487.
5. Luther, G. W., III; Branson-Swartz, C.; Ullman, W. J. *Anal. Chem.* **1988**, *60*, 1721–1724.
6. Herring, J. R.; Liss, P. S. *Deep-Sea Res.* **1974**, *21*, 777–783.
7. Stumm, W.; Morgan, J. J. *Aquatic Chemistry;* John Wiley & Sons: New York, 1981; 780 pp.
8. Jickells, T. D.; Boyd, S. S.; Knap, A. H. *Mar. Chem.* **1988**, *24*, 61–82.
9. Millero, F. J.; Hubinger, S.; Fernandez, M.; Garnett, S. *Environ. Sci. Tech.* **1987**, *21*, 439–443.
10. Butler, E. C. V.; Smith, J. D. *Deep-Sea Res.* **1980**, *27A*, 489–493.
11. Takayanagi, K.; Wong, G. T. F. *Talanta* **1986**, *33*, 451–454.
12. Zika, R. G.; Moffett, J. W.; Petasne, R. G.; Cooper, W. J.; Saltzman, E. S. *Geochim. Cosmochim. Acta* **1985**, *49*, 1173–1184.
13. Thompson, A. M.; Zafiriou, O. C. *J. Geophys. Res.* **1983**, *88*, 6696–6708.
14. Drzaic, P. S.; J. Marks, Brauman, J. I. In *Gas Phase Ion Chemistry;* Academic: Orlando, FL, 1984; Vol. 3, pp 167–211.
15. Lias, S. G.; Bartmess, J. E.; Liebman, J. F.; Holmes, J. L.; Levin, R. D.; Mallard, W. G. *J. Phys. Chem. Ref. Data* **1988**, *17 (Suppl. 1)*.
16. Stanbury, D. M. In *Advances in Inorganic Chemistry;* Sykes, A. G., Ed.; Academic: Orlando, FL, 1989; Vol. 33, pp 69–138.
17. Atkins, P. W. *Physical Chemistry*, 2nd ed.; W. H. Freeman: San Francisco, CA, 1982; p 1095.
18. Pearson, R. G. *J. Am. Chem. Soc.* **1986**, *108*, 6109–6114.
19. Pickard, G. L.; Emery, W. J. *Descriptive Physical Oceanography*, 4th ed.; Pergamon: New York, 1982; p 249.
20. Eigen, M.; Kustin, K. *J. Am. Chem. Soc.* **1962**, *84*, 1355–1361.
21. Truesdale, V. W. *Deep-Sea Res.* **1974**, *21*, 761–766.
22. Truesdale, V. W.; Moore, R. M. *Mar. Chem.* **1992**, *40*, 199–213.
23. Truesdale, V. W. *Mar. Chem.* **1993**, *42* 147–166.
24. Wong, G. T. F. *Mar. Chem.* **1980**, *9*, 13–24.
25. Luther, G. W., III In *Aquatic Chemical Kinetics;* Stumm, W., Ed.; John Wiley & Sons: New York, 1990; pp 173–198.
26. Richardson, L. B.; Burton, D. T.; Helz, G. R.; Rhoderick, J. C. *Water Res.* **1981**, *15*, 1067–1074.
27. Vazquez, G. F.; Zhang, J. Millero, F. J. *Geophys. Res. Lett.* **1989**, *16*, 1363–1366.
28. Hogfeldt, E. *Stability Constants of Metal–Ion Complexes. Part A: Inorganic Ligands;* IUPAC Chemical Data Series 21; Pergamon: New York, 1982; p 310.

29. Luther, G. W., III; Campbell, T. *Deep-Sea Res.* **1991**, *38 (Suppl. 2A)*, S875–S882.
30. Troy, R. C.; Kelley, M. D.; Nagy, J. C.; Margerum, D. W. *Inorg. Chem.* **1991**, *30*, 4838–4845.
31. Gschwend, P. M.; MacFarlane, J. K.; Newman, K. A. *Science (Washington, DC)* **1985**, *227*, 1033–1035.
32. Palenik, B.; Zafiriou, O. C.; Morel, F. M. M. *Limnol. Oceanogr.* **1987**, *32*, 1365–1369.
33. Elderfield, H.; Truesdale, V. W. *Earth Planet. Sci. Lett.* **1980**, *50*, 105–114.
34. Deuser, W. G.; Ross, E. H.; Anderson, R. F. *Deep-Sea Res.* **1981**, *28*, 495–505.
35. Jickells, T. D.; Deuser, W. G.; Belastock, R. A. *Mar. Chem.* **1990**, *29*, 203–219.
36. Kennedy, H. A.; Elderfield, H. *Geochim. Cosmochim. Acta* **1987**, *51*, 2489–2504.
37. Kennedy, H. A.; Elderfield, H. *Geochim. Cosmochim. Acta* **1987**, *51*, 2505–2514.
38. Shimmield, G. B.; Pedersen, T. F. *Rev. Aquat. Sci.* **1990**, *3*, 255–279.
39. Luther, G. W., III; Ferdelman, T.; Culberson, C. H.; Kostka, J.; Wu, J. *Est. Coast. Shelf Sci.* **1991**, *32*, 267–279.
40. Ullman, W. J.; Luther, G. W., III; de Lange, G.; van der Sloot, H. *Mar. Chem.* **1990**, *31*, 153–170.
41. Garland, J. A.; Curtis, H. *J. Geophys. Res.* **1981**, *86*, 3183–3186.
42. Hoigné, J.; Bader, H.; Haag, W. R.; Staehelin, J. *Water Res.* **1985**, *19*, 993–1004.
43. Greenwood, N. N.; Earnshaw, A. *Chemistry of the Elements*; Pergamon: Oxford, England, 1984; p 1542.
44. Haag, W. R.; Hoigné, J. *Environ. Sci. Tech.* **1983**, *17*, 261–267.
45. Hoigné, J. In *Process Technologies for Water Treatment*; Stucki, S., Ed.; Plenum: New York, 1984; pp 121–143.
46. Haag, W. R.; Hoigné, J. In *Water Chlorination*; Jolley, R. L.; Bull, R. J.; Davis, W. P.; Katz, S.; Roberts, M. H.; Jacobs, V. A., Eds.; Lewis: Chelsea, MI, 1985; Vol. 5, pp 1011–1020.
47. Mopper, K.; Zhou, X. *Science (Washington, DC)* **1990**, *250*, 661–664.
48. Zepp, R. G.; Wolfe, N. L.; Baughman, G. L.; Hollis, R. C. *Nature (London)* **1977**, *267*, 421–423.
49. Li, C. H.; White, C. F. *J. Am. Chem. Soc.* **1943**, *65*, 335–339.
50. Gimarc, B. J. *Molecular Structure and Bonding: The Qualitative Molecular Orbital Approach*; Academic: Orlando, FL, 1979; p 224.
51. Johnson, D. W.; Margerum, D. W. *Inorg. Chem.* **1991**, *30*, 4845–4851.
52. Troy, R. C.; Margerum, D. W. *Inorg. Chem.* **1991**, *30*, 3538–3543.
53. Luten, J. B.; Woittiez, J. R. W.; Das, H. A.; De Ligny, C. L. *J. Radioanal. Chem.* **1978**, *43*, 175–185.
54. Lovelock, J. E.; Maggs, R. J.; Wade, R. J. *Nature (London)* **1973**, *241*, 194–196.
55. Bauman, L.; Stenstrom, M. K. *Environ. Sci. Tech.* **1989**, *23*, 232–236.
56. Kolthoff, J. M.; Sandell, E. B.; Meehan, E. J.; Bruckenstein, S. *Quantitative Chemical Analysis*; Macmillan: London, 1969; p 1199.
57. Barbash, J. E.; Reinhard, M. In *Biogenic Sulfur in the Environment*; ACS Symposium Series 393; American Chemical Society: Washington, DC; 1989; pp 101–138.
58. Francois, R. *Geochim. Cosmochim. Acta* **1987**, *51*, 2417–2427.

RECEIVED for review October 23, 1992. ACCEPTED revised manuscript April 13, 1993.

Oxidation–Reduction Environments

The Suboxic Zone in the Black Sea

James W. Murray[1], Louis A. Codispoti[2], and Gernot E. Friederich[2]

[1] School of Oceanography, University of Washington, Seattle, WA 98195
[2] Monterey Bay Aquarium Research Institute, 160 Central Avenue, Pacific Grove, CA 93950

> *A well-defined suboxic zone was observed at the oxic–anoxic interface in the Black Sea. The redox zones for many elements, well-separated within this suboxic zone, were sampled by using a high-resolution pump-profiling sampler. Although characteristic features in the profiles occur at distinctly different depths at different locations, they are associated with the same density layer. A single vertical density scale can therefore be used to describe the features of the oxic–anoxic interface. In terms of a simple one-dimensional vertical exchange model, the distributions suggest that the upward flux of sulfide is not oxidized by oxygen. Instead, it may be oxidized by settling particulate manganese and iron oxides, it may be oxidized anaerobically during phototrophic reduction of CO_2, or the vertical model may not be appropriate and sulfide may be oxidized by oxygen transported intermittently by horizontal ventilation events. The vertical models also suggest that the upward fluxes of ammonia, manganese(II), and possibly iron(II) are oxidized by nitrate.*

AQUATIC OXIDATION–REDUCTION (REDOX) PROCESSES control the distribution of many major and minor elements in natural environments (*1*). Equilibrium redox calculations can be used to indicate the boundary conditions toward which a natural system must be proceeding. Real systems are frequently far from equilibrium because photosynthesis traps the energy of the sun in the form of energy-rich chemical bonds and thus creates nonequilibrium chemical species. The return to equilibrium (even when mediated by bacteria)

0065–2393/95/0244–0157$08.00/0
© 1995 American Chemical Society

is frequently slow. Werner Stumm and his students lead the field in developing approaches for studying many of the redox reactions that are important in the environment.

The most common redox end-members in marine systems are oxic environments, where physical transport replaces oxygen faster than it can be consumed, and anoxic conditions, where oxygen consumption occurs at a rate faster than it can be replaced. Anoxic environments are characterized by elevated concentrations of hydrogen sulfide and methane. The boundaries between oxic and anoxic environments are fascinating study sites because they contain a rich population of oxidation–reduction reactions with a resulting change in the speciation of many elements. As a result, these interfaces are natural settings for process studies such as the scavenging reactions associated with particulate oxide and sulfide phases under in situ conditions (2) and the determination of how the particle flux through the water column and sediment preservation are modified by the transition from oxic to anoxic conditions (3, 4). The flux of electrons and concentrations of the different species combine to create a dynamic balance that determines the position of the oxic–anoxic interface (5).

A set of possible oxidation–reduction half-reactions that may occur in the oxic–anoxic interface region are given in Table I (6). These half-reactions can be used to calculate electron free energy levels of oxidized and reduced forms of oxygen, nitrogen, sulfur, carbon, manganese, and iron, as shown in Figure 1. The arrows indicate the pathways of the flow of electrons. Such diagrams, first introduced by Stumm and Morgan (7), illustrate the ideal sequence of chemical species that might be observed in an unperturbed system at equilibrium.

Table I. Some Oxidation–Reduction Half-Reactions for the Species Shown in Figures 4, 5, and 6

Reaction	$pe°$	$pe°_w$
$1/4\ O_2(g) + H^+ + e^- = 1/2\ H_2O$	20.75	13.00
$1/5\ NO_3^- + 6/5\ H^+ + e^- = 1/10\ N_2(g) + 3/5\ H_2O$	21.05	11.75
$1/2\ NO_3^- + H^+ + e^- = 1/2\ NO_2^- + 1/2\ H_2O$	14.15	6.40
$1/2\ MnO_2(s) + 2H^+ + e^- = 1/2\ Mn^{2+} + H_2O$	20.8	5.30
$1/8\ NO_3^- + 5/4\ H^+ + e^- = 1/8\ NH_4^+ + 3/8\ H_2O$	14.9	5.21
$1/6\ SO_4^{2-} + 4/3\ H^+ + e^- = 1/48\ S_8(col) + 2/3\ H_2O$	5.9	−4.43
$1/8\ SO_4^{2-} + 9/8\ H^+ + e^- = 1/8\ HS^- + 1/2\ H_2O$	4.25	−4.47
$1/16\ S_8(col) + 1/2\ H^+ + e^- = 1/2\ HS^-$	−0.8	−4.68
$1/6\ N_2 + 4/3\ H^+ + e^- = 1/3\ NH_4^+$	4.68	−5.65
$Fe(OH)_3(am) + 3H^+ + e^- = Fe^{2+} + 3\ H_2O$	16.0	−7.25
$1/4\ CO_2(g) + H^+ + e^- = 1/4\ CH_2O + 1/4\ H_2O$	−0.2	−7.95

NOTE: The equilibrium constants ($pe° = 1/n\ \log K$) are taken from Morel (6). The $pe_w°$ (pH 7.75) values are calculated for a pH value representative of the oxic–anoxic interface in the Black Sea (24).

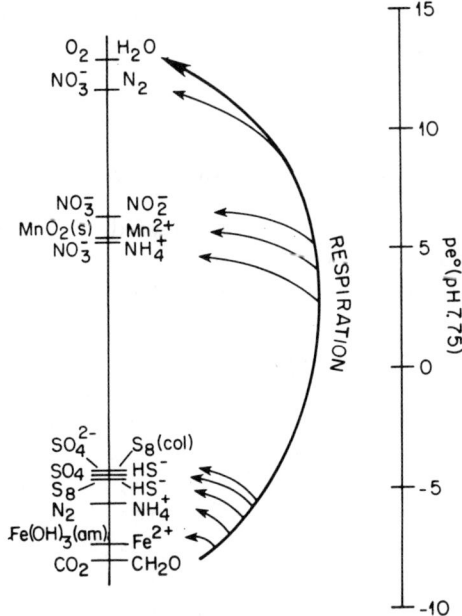

Figure 1. Electron free energy levels calculated for the approximate pH of the oxic–anoxic interface of the Black Sea (pH 7.75). Dissolved species other than H^+ are assumed to have unit activity. The strongest oxidants are at the top, and the strongest reductants are at the bottom. Such diagrams are a simple way to evaluate the feasibility of redox reactions. For example, ammonia and Mn^{2+} oxidation by nitrate may be feasible, but the actual free energy available will depend on the in situ concentrations at the site of reaction. All such reactions are, most likely, mediated by bacteria. The vertical separation of the different oxidants from organic matter (CH_2O) is proportional to the energy available from the different respiration reactions (1).

One difficulty in studying oxic–anoxic interfaces is that they are typically dynamic regimes where the reactions are compressed into small spatial scales that make detailed sampling resolution difficult. It is usually not possible to observe the ideal sequence of redox species as predicted in Figure 1. Suboxic transition zones in which both oxygen and sulfide concentrations are low and the intermediate redox reactions are well resolved have not been frequently observed or sampled with high resolution in the oceanographic water column. One environment where suboxic zones have been observed is in the centimeter-scale interstitial water of marine sediments with moderate amounts of organic carbon (8). Because of the small vertical scale, it is difficult to study these sediments with high resolution.

During the spring and summer of 1988, using a high-resolution pump-profiling system in the Black Sea, we made the unusual observation of a well-defined suboxic zone at the oxic–anoxic interface. A summary of those observations and their implications for the sequence of redox reactions is presented here.

Oxic–Anoxic Interface in the Black Sea

The Black Sea has long been a major site for studying anoxic oceanographic conditions (9, 10). It is the world's largest stable anoxic basin. The water column, from about 100 m to the bottom at depths greater than 2000 m, is characterized by the absence of oxygen and by elevated concentrations of hydrogen sulfide and methane. The water column becomes anoxic because of the oxidation of organic matter sinking from the euphotic zone. A shallow and sharp salinity-determined density gradient prevents exchange between the surface and deep water.

A comprehensive set of chemical and biological studies using modern oceanographic techniques was conducted as part of the U.S–Turkish Black Sea Expedition on the R/V Knorr in 1988 (11, 12) (Figure 2). The expedition generated many interesting and surprising results, some of which were published in volumes edited by Murray (13) and by Izdar and Murray (14). Two of the most intriguing observations were the detection of a suboxic zone at

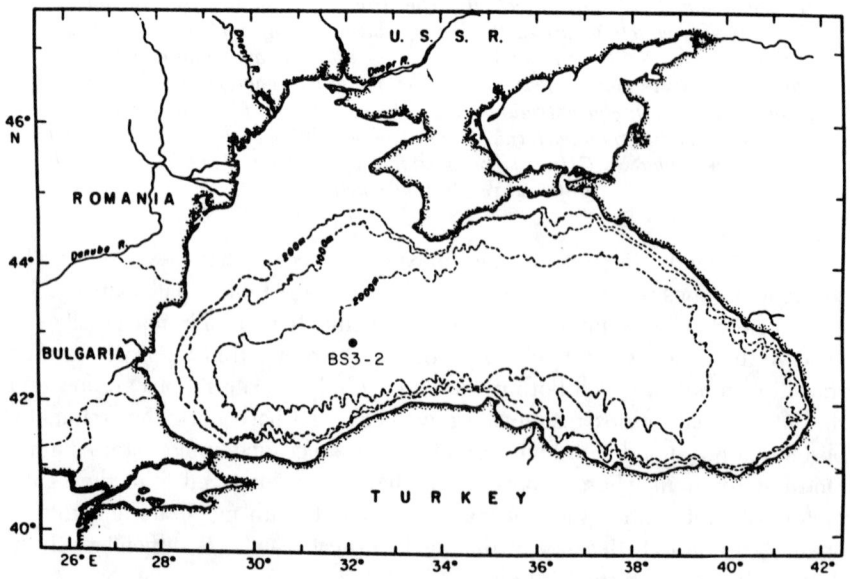

Figure 2. Chart of the Black Sea showing station BS3–2. The exclusive economic zones of the former U.S.S.R. and Turkey are indicated.

the oxic–anoxic interface and evidence that the depth of the interface had recently shoaled (15).

Suboxic Zone. Most previous profiles in the Black Sea suggested coexistence in the distributions of oxygen and sulfide at the oxic–anoxic interface, although at relatively low concentrations (e.g., 10, 16, 17). The Russian literature referred to this zone of overlap as the "S-layer" (18). The coexistence of dissolved oxygen and sulfide has been difficult to justify, considering the rapid kinetics of their reaction (19). During the 1988 expedition no overlap was observed between the oxygen and sulfide profiles (Figure 3a). Instead, a suboxic zone that ranged in thickness from 20 to 50 m was present. In this suboxic zone oxygen varied from less than 2 to 10 μM and sulfide was less than 5 nM (20). Neither oxygen nor sulfide exhibited any perceptible vertical gradients.

A nitrate maximum was approximately coincident with the upper boundary of the suboxic zone. Nitrate decreased to zero within the suboxic zone, before the depth where sulfide began to increase (Figure 3b). Nitrite maxima were common near the upper and lower boundaries of the nitrate maximum, probably corresponding to zones of nitrification (top) and denitrification (bottom). Dissolved manganese, iron, and ammonia all started to increase at the lower boundary of the nitrate maximum (Figures 3b–3d). The increases in manganese and ammonia were both steep at density anomaly values greater than $\sigma_t = 15.85$, through the rest of the suboxic zone, and into the anoxic deep water. Iron, however, increased through the suboxic zone by a factor of 4 and then increased more rapidly after the first appearance of sulfide (Figure 3d).

Sholkovitz (21) argued that it is difficult to separate the Mn and Fe redox cycles because dissolved Mn and Fe are well correlated in the suboxic zone. This theory is true to the extent that they both increase in the suboxic zone, but there is a clear difference in their distributions. The sharp increase in dissolved iron begins about 0.15 density units deeper than dissolved Mn (22, 23).

The observation of the suboxic zone immediately leads to three questions:

1. Is the suboxic zone a new feature, whose origin is due to changes in the circulation of the Black Sea, or is it a transient seasonal feature, perhaps the result of the spring bloom or a ventilation event?
2. Was its observation simply the result of better sampling resolution and analytical techniques?

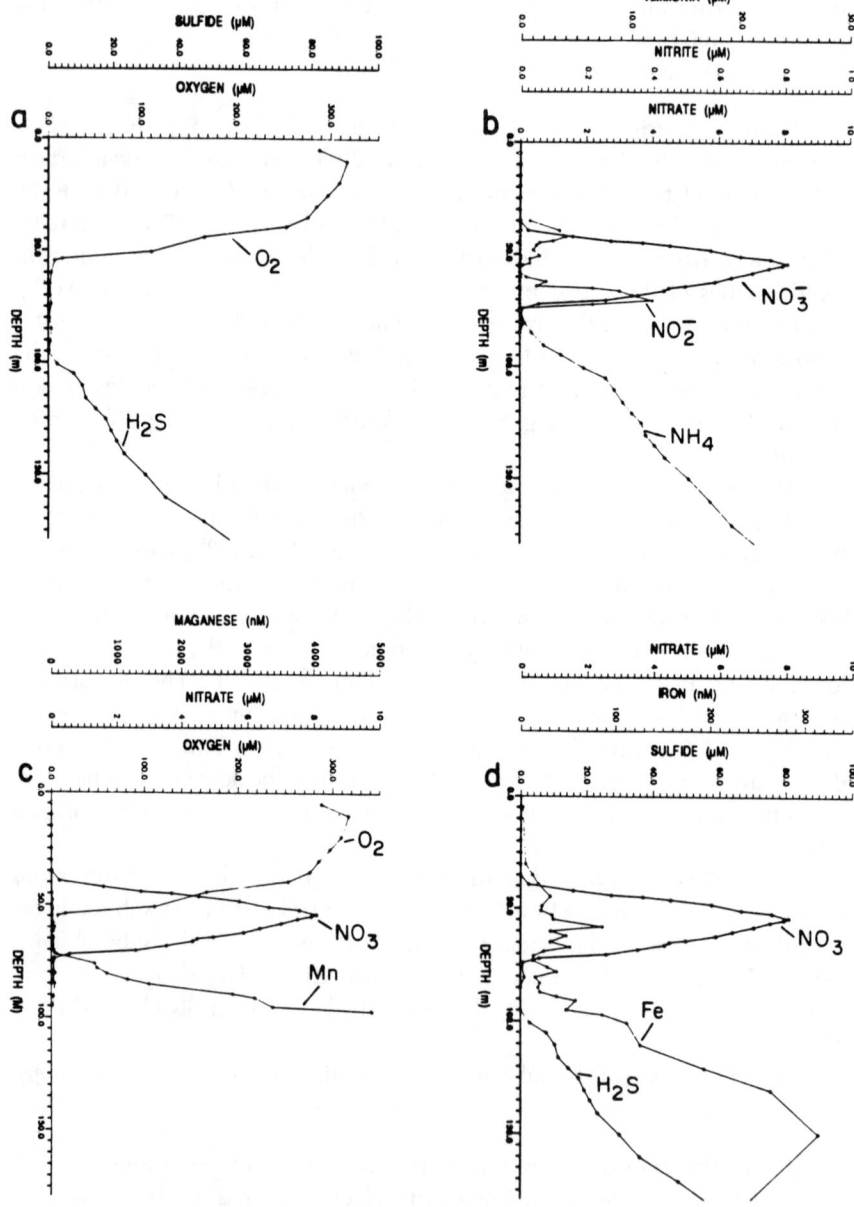

Figure 3. Distributions of species as a function of depth across the oxic–anoxic interface at station BS3-2 in the western basin of the Black Sea. Part a: O_2 and sulfide. Part b: NH_4^+, NO_3^-, and NO_2^-. Part c: Dissolved Mn, NO_3^-, and O_2. Part d: NO_3^-, dissolved Fe, and sulfide.

3. Given the observed distributions, what is the oxidant of hydrogen sulfide and other reduced substances like ammonia, manganese(II), and iron(II)?

Shoaled Interface. A second controversial observation was the suggestion that the oxic–anoxic interface had shoaled. When the 1988 data were compared with some earlier data [e.g., *Atlantis II*, 1969 (24); *Chain*, 1975 (16)] from the same location, it appeared that the depth of first appearance of sulfide had become shallower (15). It also appeared that the salinity in the upper pycnocline had increased since 1969.

These changes were thought to be either by natural climate-induced variability or by a decrease in river inflow from Europe (25). The new appearance of a suboxic zone and the increase in salinity were possibly related because the increase in salinity of the surface layer could change the rate and depth of ventilation of the pycnocline. Mixing events in other anoxic basins like Saanich Inlet, British Columbia, have been known to result in similar suboxic zones (26–28).

This interpretation has been controversial. Mesoscale features (e.g., eddies) and internal waves can result in a tremendous amount of local variability (29). Using depth as the vertical scale appears to result in widely varying and contradictory results (30–32). Even though some earlier studies had reported that the upper boundary of the sulfide zone was rising (33, 34), more recent analysis of a larger U.S.S.R. database (extending back as far as the 1920s) suggests that systematic changes may not, in fact, be occurring (35, 36).

One of the main problems with most of these analyses is that depth is a poor choice for comparison of data from different times and locations. Density is a more appropriate frame of reference for demonstrating whether or not changes have occurred (17). The parameter σ_t is used to describe the density anomaly of seawater relative to freshwater at temperature t where $\sigma_t = \rho_t - \rho_{00}$ and where $\rho_{00} = 1000$ kg m^{-3}. The parameter ρ_t (kilograms per cubic meter) is the true density of seawater at a salinity S and temperature t. A density of 1015.0 kg/m^3 has a $\sigma_t = 15.0$ kg m^{-3}.

Tugrul et al. (37) and Saydam et al. (38) analyzed a new data set collected on an R/V *Bilim* cruise in September 1991. They compared their new data with the 1969 *Atlantis II* and 1988 *Knorr* data sets, using density rather than depth for the vertical scale. They suggested that the density of the first increase in sulfide did not change from 1969 to 1988, but that the suboxic zone was enlarged toward the surface by about 0.3 σ_t units. They argued that the oxygen and nitrate data both supported this change. Buesseler et al. (18) also examined a more extensive data set (1965 to the present) and concluded that systematic changes in the density of the first appearance of sulfide have not occurred. Previous work showed that characteristic inflections in the water-column profiles of several elements are associated with specific density values, regardless of when and where they were obtained [e.g., Mn, Spencer and Brewer (*39*), Lewis and Landing (*40*); NO_3, Codispoti et al. (*41*); Cs isotopes, Buesseler et al. (*42*); physicochemical extrema and mesoplankton, Vinogradov and Nalbandov (*17*)].

A difficulty in applying a simple one-dimensional vertical explanation for the sequence of redox reactions, and for changes in the depth of the oxic–suboxic interface, is that horizontal transport appears to play a dominant role. Diffusion and mixing are much more rapid along, rather than across, equal density or isopycnal surfaces. Reactions occurring at the side boundaries of the Black Sea may have a strong influence on the distribution of properties observed in the interior (e.g., *40*). In addition, the chemocline of the Black Sea appears to be subject to rapid lateral ventilation (*15, 42*) with waters of different histories and pathways. Both dissolved and particulate components are affected in this way (*40*). Unfortunately, data are not yet available for developing one- or two-dimensional horizontal and vertical interpretations.

This chapter presents a detailed examination of the oxic–anoxic interface in the Black Sea, using data obtained with a high-resolution pump-profiling system during the 1988 U.S.–Turkish Black Sea Expedition. Density, rather than depth, will be used as the master variable to describe the features across the oxic–anoxic interface. This precise data set will be used to discuss the suboxic reaction zones across the oxic–anoxic interface.

Methods and Data

The data presented are for samples collected by using a pump-profiling system designed to minimize atmospheric contamination. This pump sampler was attached at the end of a 400-m electrical cable with a conductivity, temperature, and depth device (CTD, Seabird SBE-9/11). The nylon hose was interfaced to an autoanalyzer for detailed analyses of NO_3^-, NO_2^-, NH_4^+, H_4SiO_4, PO_4^{3-}, and H_2S. Discrete samples were taken for oxygen, manganese, and iron.

The flow rate of the system was about 4 L/min, and the time for passage of a sample from the inlet to the point of sampling was about 4 min. The pump was lowered at a rate of 6–10 m/min, and data were acquired every 3 s, and this system gave a sampling rate of two or three data points per meter. Features in the water column can be resolved on a scale better than 2 m. Nutrient analyses were made by using a computer-controlled rapid-flow analysis (RFA, Alpkem) system. Lag times were updated frequently by holding the pump head and CTD at a specific depth for several minutes.

High-level oxygen was analyzed by using the conventional Winkler titration (*43*). Oxygen concentrations below 25 μM were determined by using the colorimetric method of Broenkow and Cline (*44*). The details of the sampling and analytical procedures and data quality were given by Friederich et al. (*45*) and Codispoti et al. (*41*). The complete data set was given by Friederich et al. (*45*). The manganese and iron data were taken from Lewis and Landing

(40). The usual procedure for the pump profiles was to define the features in the water column with a down cast and conduct discrete sampling on the up cast. The data presented are from cruises 2, 3, and 4 of the 1988 Black Sea Expedition on the R/V Knorr (11). The complete data set was used in this analysis.

The vertical pump profiles of nitrate and sulfide as a function of depth are shown in Figure 4. Much variability occurs from site to site when using the depth scale. For example, the increase in sulfide begins at depths ranging from 90 to 170 m, depending on the location. The increase begins at shallower depths in the central gyres and deeper depths near the margins. The profile with several minima below 350 m is located near the Bosporus, and these inflections are associated with new water input from the Mediterranean Sea (41). The nitrate profiles are also variable, and the nitrate maxima is much deeper and broader near the margins than in the interior.

When the data are plotted versus density, all of the profiles from different locations fall together in a narrow range. The data for dissolved oxygen, sulfide, iron, and manganese are shown in Figure 5, and the data for dissolved nitrate, nitrite, ammonia, and phosphate are shown in Figure 6. Features in the water column occur at different depths at different locations, but they always occur close to the same density surface. The only exceptions appear to occur in the region close to the Bosporus, where the Black Sea inflow interleaves with ambient water.

The high quality of the data set allows for precise analysis of the characteristic density values of inflections and changes in the slope of the various

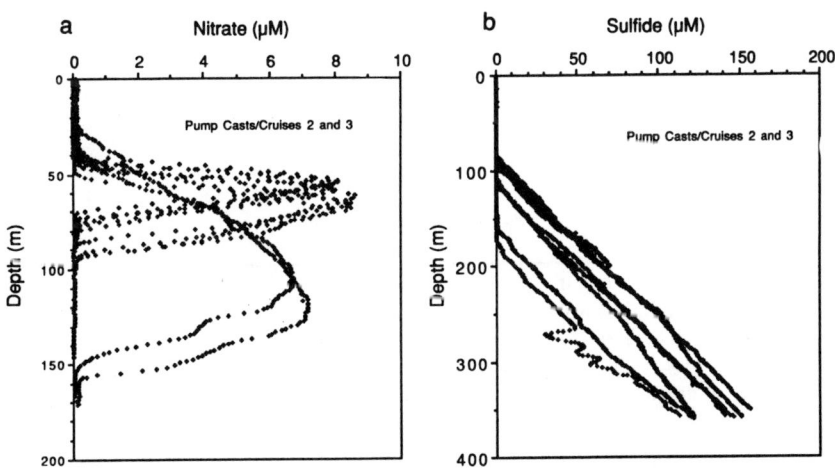

Figure 4. Pump cast profiles of (a) nitrate and (b) sulfide versus depth. The two casts with deep NO_3 maxima correspond to the deeper two sulfide profiles. Both come from the Rim Current region near the Bosporus.

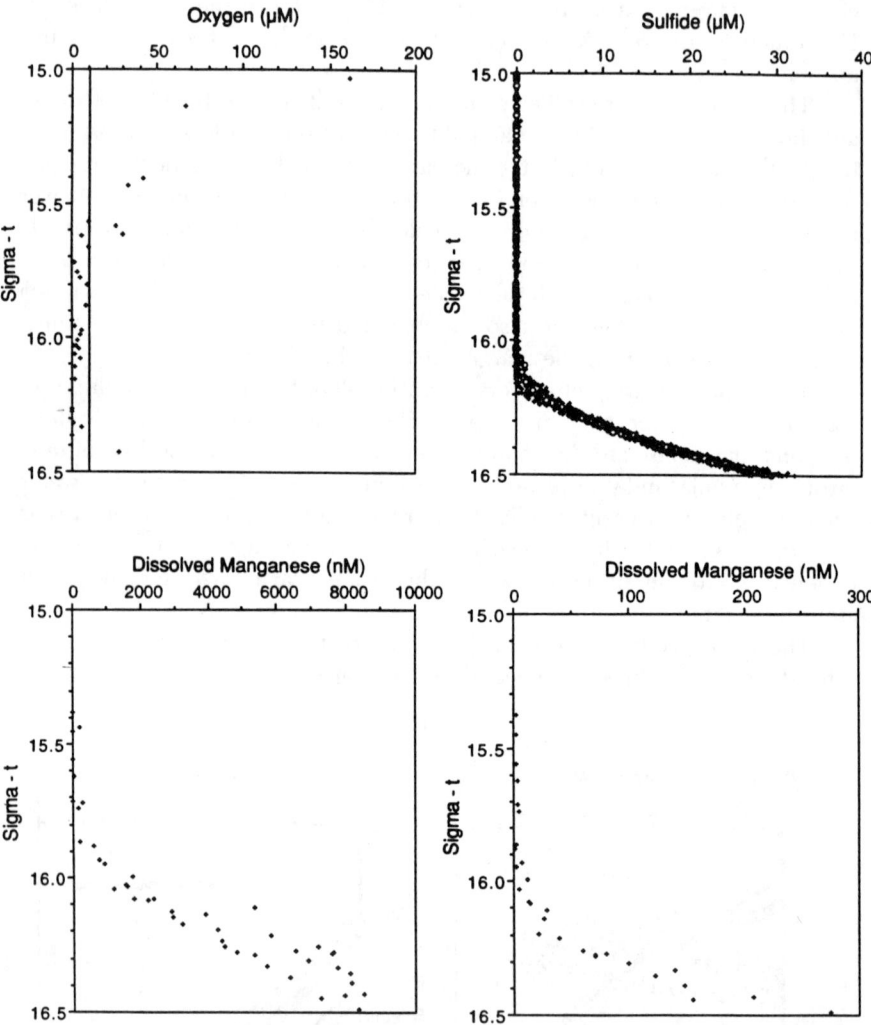

Figure 5. Pump cast profiles of oxygen, sulfide, iron, and manganese versus density expressed as σ_t. The oxygen values are from bottle casts on cruises 3 and 4. A marker line is shown for the 10 µM oxygen level. The iron and manganese are from bottle casts on cruise 3. The sulfide is from pump casts on cruises 2, 3, and 4.

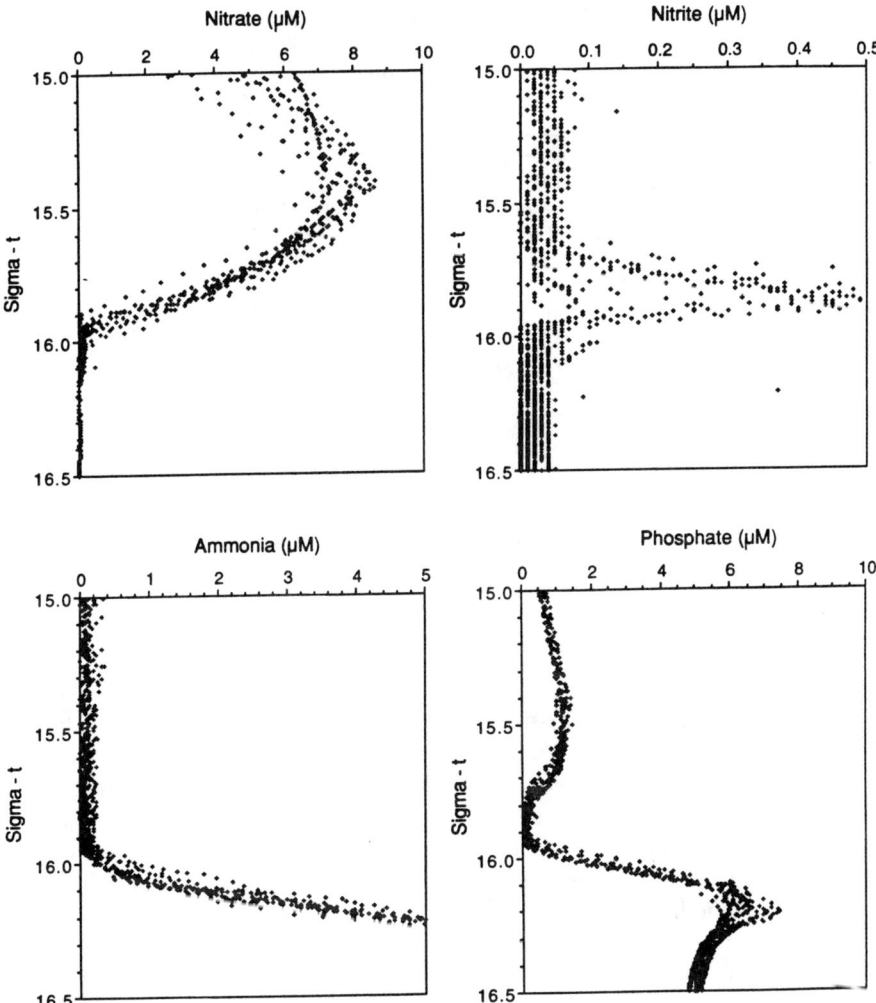

Figure 6. Pump cast profiles of nitrate, nitrite, ammonia, and phosphate versus density (σ_t). These profiles come from pump casts on cruises 2, 3, and 4.

property profiles. These values are listed in Table II. The uncertainties in those σ_t values is about 0.05. This data set should serve as a well-defined reference point for evaluating future changes. For example, the depth of the first appearance of sulfide may vary in the future, depending on the local hydrodynamics, ventilation, and the salt budget of the Black Sea. A change in the density of this feature would suggest additional changes in the biogeochemical dynamics.

Table II. Density Anomaly (σ_t) Values of Characteristic Features in the Water-Column Profiles

Feature	σ_t (kg/m^3)
PO_4^{3-} shallow maximum	15.50[a]
$O_2 < 10$ μM	15.65
NO_3^- maximum	15.40
$Mn_d < 200$ nM	15.85
Particulate Mn maximum	15.85
PO_4^{3-} minimum	15.85
NO_2^- maximum	15.85
$NO_3^- < 0.2$ μM	15.95
$NH_4^+ > 0.2$ μM	15.95
$Fe_d < 10$ nM	16.00
$H_2S > 1$ μM	16.15
PO_4 deep maximum	16.20

NOTE: The density values of these features, determined from the pump profile data, have a range of about 0.05 density units.
[a] Broad density range.

Discussion

Vertical Reaction Zones. The fact that the characteristic features in the vertical profiles from different locations (Figures 5 and 6) fall on the same density surfaces (Table II) means that the distributions and reactions across the oxic–anoxic interface can be discussed in terms of a composite profile with a vertical density scale. The chemical species in the 1988 Black Sea data set have good vertical resolution so that the reaction zones can be clearly identified.

Many of the questions about the origin of the suboxic zone and the redox reaction zones would be easier to answer if we could calculate vertical fluxes. Unfortunately, neither the mechanism nor the rate of vertical transport are well understood. Estimates of the vertical advection velocity (w) and eddy diffusion coefficients (K_z) are available in the literature (e.g., 5, 32, 39, 40), but they are probably not realistic, considering the importance of horizontal ventilation discussed earlier.

We can calculate the vertical (e.g., cross isopycnal) molar- and electron-equivalent gradients as a function of depth and use them as a constraint for proposed reactions. The approach used was first to calculate the vertical molar depth gradients (moles per liter per meter equals moles per meter to the fourth power) and then to multiply those gradients by the number of electrons required for the appropriate redox reactions (see Table I). These gradients were calculated against depth rather than density because depth is the traditional unit for gradients. If the gradients are divided by the density, their magnitude would be about 1.7% lower and their units would be moles per kilogram per meter.

The concentration profiles are nearly linear for at least 10–20 m above the interface and for 50 m or more below the interface. The resulting vertical electron-equivalent gradients for a station representative of the central western gyre (BS3-2) (for location see Figure 2 and for the profiles see Figure 3) are shown in Figure 7. The calculations in this figure assume that O_2 and NO_3^- are reduced to H_2O and N_2 and that HS^-, Mn^{2+}, Fe^{2+}, and NH_4^+ are oxidized to SO_4^{2-}, $MnO_2(s)$, $FeOOH(s)$, and N_2, respectively. The oxidation of NH_4 to N_2 is hypothesized because the water-column distributions (Figure 3) show that it is not being oxidized to NO_3.

Oxygen–Sulfide–Metal Oxides. The oxygen and sulfide distributions are of particular interest because they are used to define the oxic–anoxic interface (46). Dissolved oxygen decreases to concentrations less than 10 μM by $\sigma_t = 15.65$, and sulfide does not begin to increase rapidly until $\sigma_t = 16.15$. The region between these two density values is the suboxic zone. The density separation between the decreasing oxygen and increasing sulfide concentrations is $\Delta\sigma_t = 0.50$. Murray et al. (15) defined the upper boundary of the suboxic zone as less than 5 μM. The value of 10 μM is used here instead because a few randomly distributed data points are in the 5–10 μM range. The downward O_2 electron gradient is 52.8×10^{-3} mol e^-/m^4, and the upward sulfide electron gradient is 5.1×10^{-3} mol e^-/m^4 (Figure 7). Most of the downward oxygen flux is consumed by heterotrophic aerobic respiration (46). The oxidant of the upward sulfide flux is less certain because of the barrier of the density separation from oxygen.

Previously, oxygen and sulfide were observed to coexist over some finite depth (or density) region, and it was assumed that oxygen oxidized the upward flux of sulfide at the oxic–anoxic interface. For example, Vinogradov and Nalbandov (17) summarized data from several cruises (1984–1989) and observed that the first appearance of sulfide was at $\sigma_t = 16.18$ (range of 15.95–16.30) (Table II). This value is in excellent agreement with our density of $\sigma_t = 16.15$

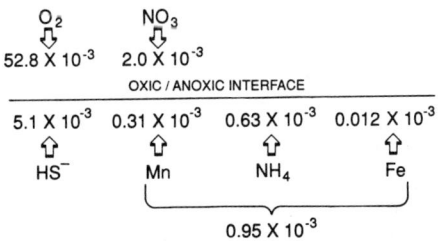

Figure 7. Electron-equivalent gradients (in moles of electrons per meter to the fourth power) into the oxic–anoxic interface from above and below.

for 1988. On the other hand, they observed that O_2 did not disappear until 16.59 (range of 16.40–16.70), which is much deeper than our value of 15.65 for the horizon at which oxygen decreases to less than 10 μM. As a result, there is a significant overlap in their oxygen and sulfide data. The sulfide data sets are in good agreement, and the oxygen profiles differ because of the difficulty in sampling and analyzing oxygen at low concentrations.

Recent oxygen and sulfide analyses by Turkish oceanographers on several research cruises on the R/V *Bilim* confirmed the existence of a suboxic zone (37, 38), although their data may contain a 5-μM blank due to air contamination. The density of the features for the 1988 *Knorr* and the *Bilim* data sets agree very well.

The newest oxygen and sulfide data from the R/V *M. Lomonosov*, Cruise 55 in October 1992, is of excellent quality (A. S. Romanov and S. K. Konovalov, Marine Hydrophysical Institute, Sevastopol, personal communication). Special precautions were taken to prevent contamination and to remove oxygen contamination of the reagents. A well-defined suboxic zone was observed, and oxygen decreased to less than 10 μM at $\sigma_t = 15.83 \pm 0.05$. Sulfide increased to 1 μM at $\sigma_t = 16.20 \pm 0.05$. These results are significant because the suboxic zone has been observed consistently since 1988 by American, Turkish, and Ukrainian research groups when precautions are taken to avoid contamination. Most likely the suboxic zone was always present but was not observed because of the difficulty in avoiding oxygen contamination at low concentrations.

In our data set the suboxic zone represents a density barrier that separates the O_2 and sulfide gradients. Because oxygen and sulfide have such different concentration ranges, it is possible to argue that sufficient O_2 to oxidize the sulfide somehow leaks through the suboxic zone. The O_2 gradient that would be required to oxidize the observed sulfide gradient is about 1.3×10^{-3} mol/m^4. This amount is equivalent to an increase in O_2 of about 13 μM every 10 m. This gradient would be easily detectable by the microoxygen techniques used in this study, and it was not observed (Figure 5A). The absence of sufficient oxygen gradient implies, in the context of a simple vertical model, that sulfide is not oxidized by oxygen. Three main alternatives have been proposed to explain this dilemma, but none of them is easy to quantify and all have significant uncertainties.

Sulfide Oxidation by Metal Oxides. Millero (47) proposed that sulfide is oxidized by settling particulate MnO_2 (s) and possibly FeOOH(s) formed by bacteria in shallow oxic waters. Luther et al. (20) proposed that dissolved Mn(III) was responsible. Lewis and Landing (40) did not observe a significant particulate Fe oxide maximum above the sulfide interface. They did observe, however, two maxima in particulate manganese in the central basin. A shallow maximum, located at 76 m, may not contain oxidized Mn (48). The deeper maximum, at 105 m, does contain oxidized Mn. Both maxima have similarly high specific rates of Mn(II) oxidation potential and specific oxidation rates (48) but Mn(III,IV) reduction may occur simultaneously in the shallower zone. There is no agreement about the relative importance of in situ versus horizontal transport for the creation of either of these maxima (30, 40, 42, 48).

The deep maximum in particulate Mn, which is probably Mn(III,IV) manganate material, appears to be much larger at the margins than in the central parts of the Black Sea. Particulate Mn and Fe oxides must have a very short residence time in the suboxic zone before settling back into the anoxic waters. If they control sulfide oxidation, the reaction is probably most important at the margins of the Black Sea, with the effects propagating into the interior (40). Tebo (48) argued that the Mn cycle may be the key process for maintaining the broad suboxic zone. Oxidized Mn may oxidize reduced sulfur, whereas reduced Mn(II) may contribute to oxygen consumption.

Another difficulty with the metal oxide hypothesis is that the observed sum of the vertical electron equivalent gradients of Mn(II) and Fe(II) is much less than that of sulfide (Figure 7). In a simple vertical, steady-state system where the upward flux of Mn(II) and Fe(II) results in oxidized particulate metal oxides, which in turn settle to oxidize sulfide, the electron gradients of Mn(II) + Fe(II) would equal that for sulfide. The fact that they do not equal it suggests that the vertical flux of Mn(II) + Fe(II) would not produce sufficient particulate metal oxides. This problem would be solved if the particulate oxides were produced primarily at the boundaries and transported into the interior (40).

Anaerobic Sulfide Oxidation. An alternative explanation is that sulfide is oxidized anaerobically in association with phototrophic reduction of CO_2 to organic carbon (46, 49). This hypothesis is supported by the discovery of considerable quantities of bacteriochlorophyll pigments within and below the suboxic zone (50). The integrated quantities of these pigments appear to exceed that of the chlorophyll a in the overlying oxygenated portion of the euphotic zone. The light levels at the depth of the bacteriochlorophyll maximum, however, are very low ($<<0.1\%\ I_o$, where I_o is the incident radiation) and the carbon assimilation rates necessary to verify the hypothesis are difficult to calculate or measure.

Ventilation as Source of Suboxic Zone. Ventilation processes may be the ultimate explanation for the origin of the suboxic zone. Murray et al. (32), Buesseler et al. (42), and Ozsoy et al. (51) suggested that significant ventilation occurs, especially in the upper 500 m of the Black Sea. Oguz et al. (29) showed that horizontal variability in salinity and currents associated with eddies and filaments can be seen to almost 500 m.

One end-member view is that the pycnocline originates due to horizontal ventilation of water produced by entrainment of the fresh, cold, intermediate layer water by the inflowing warm, salty Mediterranean water. Variations in the mixing regime result in a continuous variation in the degree of entrainment, so that the upper pycnocline reflects more entrainment and the deeper pycnocline reflects less. Dissolved oxygen in these ventilating injections would react with sulfide to oxidize it to sulfate. Lewis and Landing (40) showed that horizontal mixing of oxygen, at levels below the Winkler detection limits, could provide the necessary oxygen to account for Mn oxidation but probably not that of sulfide and other reduced substances. Such ventilation could create a suboxic zone and account for the imbalance in the gradients of oxygen and sulfide.

Nitrogen Transformations. Nitrogen transformations are always complicated at oxic–anoxic interfaces because of the large number of oxidation states available for N. The most interesting aspect about N distributions in the Black Sea is that the concentration of NH_4^+ decreases to below the detection limit at $\sigma_t = 15.95$ (Figure 6), which is about 0.35 density units deeper than the isopycnal where O_2 decreases to less than 10 μM (Table II). Nitrate is below detection in the euphotic zone and increases below this depth region because of nitrification. The NO_3^- maximum is at $\sigma_t = 15.40$ (Figure 6), which is about the density at which O_2 decreases to less than 10 μM. Nitrate then decreases with depth, most likely because of denitrification.

A NO_2^- maximum ($\sigma_t = 15.85$) (Figure 6) almost always coincides with the zone of denitrification. Occasionally, but not always, a NO_2^- maximum corresponds to the zone of nitrification. The simplest explanation for the decrease in NO_3^- with depth is heterotrophic denitrification according to

$$5\{CH_2O\} + 4NO_3^- = 2N_2 + 5HCO_3^- + H^+ + 2H_2O \tag{1}$$

Ammonia decreases to background values at $\sigma_t = 15.95$. This loss presents a problem because ammonia needs an oxidant. A similar problem exists for Mn(II) and Fe(II), both of which decrease to low concentrations within the suboxic zone, well below where O_2 decreases to less than 10 μM. The electron balance (Figure 7) indicates that there are sufficient equivalents of NO_3^- to account for the oxidation of ammonia, Mn(II), and Fe(II) according to reactions 2–4.

$$3NO_3^- + 5NH_4^+ = 4N_2 + 9H_2O + 2H^+ \tag{2}$$

$$2NO_3^- + 5Mn^{2+} + 4H_2O = N_2 + 5MnO_2(s) + 8H^+ \tag{3}$$

$$2NO_3^- + 10Fe^{2+} + 24H_2O = N_2 + 10Fe(OH)_3(s) + 18H^+ \tag{4}$$

The downward gradient of NO_3^- equivalents equals 2.0×10^{-3} mol/m^4, whereas the sum of the upward gradients of Mn(II) + Fe(II) + NH_4^+ equals

0.95×10^{-3} mol/m^4. Calculations of the energetics of all three reactions indicate that in each case the free energy of reaction is favorable at in situ concentrations. The following observations suggest that the reactions are possible.

1. NO_3^-, NH_4^+, Mn(II), and Fe(II) all decrease to low concentrations at about the same density levels (15.95, 15.95, 15.85, and 16.00, respectively).
2. Their electron equivalent gradients match within a factor of 2.
3. Their energetics are favorable.

All of these reactions are problematical. For example, reaction 2 was proposed by Richards (52) and explored experimentally (53), but the organism responsible grows very slowly and has not yet been isolated. Tebo (48) stated that particulate Mn formation has never been observed in the absence of oxygen. The water-column distributions suggest that nitrate is the oxidant for NH_4^+, Mn(II), and Fe(II) in the suboxic zone of the Black Sea, and future studies should address these reactions.

An interesting ramification of this proposed nitrogen cycle is that, under the steady-state conditions considered here, no new nitrogen would reach the euphotic zone from deeper water. This means that the only new production, as defined by Dugdale and Goering (54), and the export production, or the sinking flux of particulate N, must be the result of riverine and atmospheric inputs of N. The sinking flux of particulate N through the suboxic zone, measured by using drifting sediment traps at 80 m, is on the order of 3 mg of N/m^2 per day, which is only about 3% of the autotrophic N fixation (55). The particulate export production to the anoxic zone, therefore, appears to be extremely small.

Phosphate. The phosphate profile (Figure 6d) is the most complicated of those shown here. A broad PO_3^- maximum is centered at approximately $\sigma_t = 15.50$. This is about the density at which O_2 decreases to less than 10 µM and agrees with the location of the NO_3^- maximum. A well-defined PO_4^{3-} minimum in the suboxic zone is centered at $\sigma_t = 15.85$. This is the same density at which dissolved Mn increases sharply to concentrations greater than 200 nM. The similarity suggests that Mn cycling and the formation of MnO_2 (s) has a significant influence over the distribution of PO_4^{3-}.

Iron oxide has frequently been proposed as a scavenging agent for PO_4^{3-}. Shaffer (56) proposed that iron oxides were more important than manganese oxides. Kawashima et al. (57) proposed that manganese oxides have very low affinities for phosphate, but that binding can be significantly enhanced by the presence of Mg^{2+}. In our data set dissolved iron increases by a factor of 4 through this region and then increases sharply to greater than 10 nM at a density surface of $\sigma_t = 16.00$. This value is below the PO_4^{3-} minimum and gives the impression that Fe may not be as tightly coupled as Mn with PO_4^{3-}. Nevertheless, because both Mn and Fe increase across the PO_4^{3-} minimum, it is not possible to distinguish between MnO and FeOOH as carrier phases (21). Finally, a pronounced PO_4^{3-} maximum at $\sigma_t = 16.20$ is at or just below the density at which sulfide increases to greater than 1 µM.

The most likely cycle proposes particulate $MnO_2(s)$ and FeOOH(s) formation in the suboxic zone through oxidation of Mn(II) and Fe(II) by nitrate. The newly formed $MnO_2(s)$ and FeOOH(s) scavenge dissolved PO_4^{3-}. The particulate oxides settle until they reach the sulfide, at which point they are reduced and release their adsorbed PO_4^{3-}.

The close relationship of the PO_4^{3-} profile to other features in the water column led Tugrul et al. (37) to propose using PO_4^{3-} (which is relatively easy to measure) as a tracer for the first appearance of sulfide (which is a more difficult measurement) and the presence of a suboxic zone. Buesseler et al. (18) took this idea one step further and suggested that the PO_4^{3-} minima in historical data sets suggest that a suboxic zone existed at that time. This hypothesis is difficult to evaluate, because it is most likely that formation of oxidized particulate forms of Fe and Mn do not give an indication of whether or not oxygen concentrations were within the suboxic range.

Acknowledgments

This research was supported by NSF Grant No. OCE8614400. This manuscript benefited from comments by Ken Buesseler, George Luther, and the ACS reviewers. This publication is University of Washington contribution No. 1964.

References

1. Stumm, W.; Morgan, J. J. *Aquatic Chemistry*; John Wiley: New York, 1981; p 780.
2. Wei, C.-L.; Murray, J. W. *Deep-Sea Res.* **1992**, *38*, S855–S873.
3. Hay, B. J.; Honjo, S.; Kempe, S.; Ittekkot, V. A.; Degens, E. T.; Konuk, T.; Izdar, E. *Deep-Sea Res.* **1990**, *37*, 911–928.
4. Calvert, S. E.; Vogel, J. S.; Southon J. R. *Geology* **1987**, *15*, 918–921.
5. Brewer, P. G.; Murray. J. W. *Deep-Sea Res.* **1973**, *20*, 803–818.
6. Morel, F. M. M. *Principles of Aquatic Chemistry*; John Wiley: New York, 1983; p 446.
7. Stumm, W.; Morgan, J. J. *Aquatic Chemistry*; John Wiley: New York, 1971; p 583.
8. Froelich, P. N.; Klinkhammer, G. P.; Bender, M. L.; Luedtke, N. A.; Heath, G. R.; Cullew, D.; Dauphin, P.; Hammond, D.; Hartman, B.; Maynard, V. *Geochim. Cosmochim. Acta* **1979**, *43*, 1075–1090.
9. Degens, E. T.; Ross D. A. *The Black Sea: Geology, Chemistry, and Biology*; American Association of Petroleum Geologists: Tulsa, OK, 1974; p 635.
10. Sorokin, Yu. I. In *Estuaries and Enclosed Seas. Ecosystems of the World*; Ketchum, B. H., Ed.; Elsevier: Amsterdam, The Netherlands, 1983; Vol. 26, pp 253–292.

11. Murray, J. W.; Izdar, E. *Oceanography* **1989**, *2*, 15–21.
12. Murray, J. W. *Deep-Sea Res.* **1991**, *38*, S655–S661.
13. Murray, J. W. (Ed.) *Deep-Sea Res.* **1991**, *38*, S655–S1266.
14. *Black Sea Oceanography*; Izdar, E.; Murray, J. W., Eds.; Kluwer: Dordrecht, The Netherlands, 1991; p 487.
15. Murray, J. W.; Jannasch, H. W.; Honjo, S.; Anderson, R. F.; Reeburgh, W. S.; Top, Z.; Friederich, G. E.; Codispoti, L. A.; Izdar, E. *Nature (London)* **1989**, *338*, 411–413.
16. Karl, D. M. *Limnol. Oceanogr.* **1975**, *23*, 936–949.
17. Vinogradov, M. Ye.; Nalbandov, Yu. R. *Oceanology* **1990**, *30*, 567–573.
18. Buesseler, K. O.; Livingston, H. D.; Ivanov, L.; Romanov, A. *Deep-Sea Res.* **1994**, *41*, 283–296.
19. Millero, F. J.; Hubinger, S.; Fernandez, M.; Garnett, S. *Environ. Sci. Technol.* **1987**, *21*, 439–443.
20. Luther, G. W.; Church, T. M.; Powell, D. *Deep-Sea Res.* **1991**, *38*, S1121–S1138.
21. Sholkovitz, E. R. *Geochim. Cosmochim. Acta* **1992**, *56*, 4305–4307.
22. German, C. R.; Holliday, B. P.; Elderfield, H. *Geochim. Cosmochim. Acta* **1991**, *55*, 3553–3558.
23. German, C. R.; Holliday, B. P.; Elderfield, H. *Geochim. Cosmochim. Acta* **1992**, *56*, 4309–4313.
24. Brewer, P. G. *Hydrographic and Chemical Data from the Black Sea*; Woods Hole Oceanographic Institution Technical Note 71–65; Woods Hole Oceanographic Institute: Woods Hole, MA, 1971.
25. Tolmazin, D. *Prog. Oceanogr.* **1985**, *15*, 217–276.
26. Emerson, S.; Kalhorn, S.; Jacobs, L.; Tebo, B. M.; Nealson, K. H.; Rosson, R. A. *Geochim. Cosmochim. Acta* **1982**, *46*, 1073–1080.
27. Tebo, B. M.; Nealson, K. H.; Emerson, S.; Jacobs, L. *Limnol. Oceanogr.* **1984**, *29*, 1247–1258.
28. Anderson, J. J.; Devol, A. H. *Deep-Sea Res.* **1987**, *34*, 927–944.
29. Oguz, T.; Latun, V. S.; Latif, M. A.; Vladimirov, V. V.; Sur, H. I.; Markov, A. A.; Ozsoy, E.; Kotovshchikov, B. B.; Eremeev, V. V.; Unluata, U. *Deep-Sea Res.* **1993**, *40*, 1597–1612.
30. Kempe, S.; Liebezeit, G.; Diercks, A.-R.; Asper, V. *Nature (London)* **1990**, *346*, 419.
31. Murray, J. W. In *Black Sea Oceanography*; Izdar, E.; Murray, J. W., Eds.; Kluwer: Dordrecht, The Netherlands, 1991; pp 1–16.
32. Murray, J. W.; Top, Z.; Ozsoy, E. *Deep-Sea Res.* **1991**, *38*, S663–S689.
33. Fashchuk, D. Ya.; Ayzatullin, T. A. *Oceanology* **1986**, *26*, 171–178.
34. Leonov, A. V.; Ayzatullin, T. A. *Oceanology* **1987**, *27*, 174–178.
35. Bezborodov, A. A.; Eremeev, V. N. *Sov. J. Phys. Oceanogr.* **1991**, *2*, 407–410.
36. Bezborodov, A. A.; Eremeev, V. N. *Sov. J. Phys. Oceanogr.* **1992**, *3*, 61–66.
37. Tugrul, S.; Basturk, O.; Saydam, C.; Yilmaz, A. *Nature (London)* **1992**, *359*, 137–139.
38. Saydam, C.; Tugrul, S.; Basturk, O.; Oguz, T. *Deep-Sea Res.* **1993**, *40*, 1405–1412.
39. Spencer, D. W.; Brewer, P. G. *J. Geophys. Res.* **1971**, *76*, 5877–5892.
40. Lewis, B.; Landing, W. M. *Deep-Sea Res.* **1991**, *38*, S773–S804.
41. Codispoti, L. A.; Friederich, G. E.; Murray, J. W.; Sakamoto, C. M. *Deep-Sea Res.* **1991**, *38*, S691–S710.
42. Buesseler, K. O.; Livingston, H. D.; Casso, S. A. *Deep-Sea Res.* **1991**, *38*, S725–S746.
43. Carpenter, J. H. *Limnol. Oceanogr.* **1965**, *10*, 141–143.
44. Broenkow, W. W.; Cline, J. D. *Limnol. Oceanogr.* **1969**, *14*, 450–454.

45. Friederich, G. E.; Codispoti, L. A.; Sakamoto, C. M. *Bottle and Pumpcast Data from the 1988 Black Sea Expedition;* Monterey Bay Aquarium Research Institution Technical Report 90–3; Monterey Bay Aquarium Research Institution: Monterey Bay, CA, 1990; p 224.
46. Jannasch, H. W. In *Black Sea Oceanography;* Izdar, E.; Murray, J. W., Eds.; Kluwer: Dordrecht, The Netherlands, 1991; pp 271–286.
47. Millero, F. J. *Deep-Sea Res.* **1991,** *38,* S1139–S1150.
48. Tebo, B. M. *Deep-Sea Res.* **1991,** *38,* S883–S905.
49. Jorgensen, H. W.; Fossing, H.; Wirsen, C. O.; Jannasch, H. W. *Deep-Sea Res.* **1991,** *38,* S1083–S1104.
50. Repeta, D. J.; Simpson, D. J.; Jorgenson, B. B.; Jannasch, H. W. *Nature (London)* **1989,** *342,* 69–72.
51. Ozsoy, E.; Top, Z.; White, G.; Murray, J. In *Black Sea Oceanography;* Izdar, E.; Murray, J. W., Eds.; Kluwer: Dordrecht, The Netherlands, 1991; pp 17–42.
52. Richards, F. A. In *Chemical Oceanography;* Riley, J. P.; Skirrow, G., Eds.; Academic: London, 1965; pp 611–646.
53. Hamm, R. E.; Thompson, T. G. *J. Mar. Res.* **1941,** *4,* 11–27.
54. Dugdale, R. C.; Goering, J. J. *Limnol. Oceanogr.* **1967,** *12,* 685–695.
55. Karl, D. M.; Knauer, G. A. *Deep-Sea Res.* **1991,** *38,* S921–S942.
56. Shaffer, G. *Nature (London)* **1986,** *321,* 515–517.
57. Kawashima M.; Tainaka, Y.; Hori, T.; Koyama, M.; Takamatsu, T. *Water Res.* **1986,** *20,* 471–476.

RECEIVED for review October 23, 1992. ACCEPTED revised manuscript June 2, 1993.

8

Cycles of Trace Elements (Copper and Zinc) in a Eutrophic Lake

Role of Speciation and Sedimentation

Laura Sigg, Annette Kuhn, Hanbin Xue, Elke Kiefer, and David Kistler

Federal Institute for Water Resources and Water Pollution Control (EAWAG), Swiss Federal Institute of Technology, Zürich (ETHZ), CH–8600 Dübendorf, Switzerland

> *Zinc and copper in a eutrophic lake (Lake Greifen) are affected in distinct ways by interactions with settling particles and ligands in solution, with respect to their speciation and to their removal rates to the sediments. In the water column of Lake Greifen, Cu is strongly complexed by organic ligands, whereas a large fraction of Zn is present in weak complexes and free aquo ions. The sedimentation rate of Zn exhibits a maximum during summer, and this behavior indicates removal from the water column together with biological material; depletion of Zn from the epilimnion results from this removal process. The removal of Cu by sedimentation is less efficient during summer. Zn and Cu also sediment together with manganese oxides, which are precipitated during lake overturn. The role of algae in this system with regard to the binding of Cu and Zn in the particulate phase and to the production of ligands in solution is discussed.*

THE FATE OF TRACE ELEMENTS in a eutrophic lake is strongly linked to biological processes, as has been demonstrated for the oceans (1–3). Photosynthetic production of algae and the subsequent sedimentation of algal material affect the removal of metal ions to the sediments (4). High sedimentation rates in eutrophic lakes indicate efficient mechanisms for elimination of metal ions. As a result, low concentrations of trace metals are found in the water column.

0065–2393/95/0244–0177$08.00/0
© 1995 American Chemical Society

Seasonal variations of primary productivity and of thermal stratification generate variable physicochemical conditions, such as concentrations of various components, pH, and the occurrence of anoxic conditions, in the water column of eutrophic lakes. These variations influence the binding of trace elements either to the particulate phase (algae and inorganic sediment components) or to dissolved ligands and also affect the concentrations in the water column. Zinc and copper in a eutrophic lake (Lake Greifen) are affected in distinct ways by interactions with settling particles and ligands in solution, with respect to their speciation and to their removal rates to the sediments.

Trace Elements and Algae

Interactions of trace elements with algae in the marine environment are being extensively studied (1–3, 5–12). Copper and zinc, both essential micronutrients required by phytoplankton, may be toxic at elevated concentrations (11–13). The biological effects of copper and zinc are strongly dependent on their speciation; the activity of the free metal ion has been shown to be a key parameter (13). Toxic effects of Cu on marine algae have been observed in the range of pCu ≈ 10–12 (11–13).

Various species of marine algae grew optimally at free zinc ion concentrations up to $[Zn^{2+}] \approx 10^{-8}$ M, but they were limited at $[Zn^{2+}] < 10^{-11}$ M (6, 11). Algae take up essential elements such as copper and zinc and incorporate them into their cell material. Binding to various functional groups on the surfaces of algae is a prerequisite for this uptake and may also be important as an adsorption mechanism (1).

The elementary composition of algae may be understood in terms of Redfield ratios of the major elements (typical, nearly constant ratios, e.g., of C:N:P) (14); the extension of Redfield ratios to trace elements is widely discussed (2, 3, 15, 16). Essential elements, such as Cu and Zn, are expected to have characteristic Zn:P and Cu:P ratios in algae. However, these ratios may vary over a certain range, depending on the available concentrations in water and on the species of algae (6). The limitation of algal productivity in the oceans by various trace elements and the influence of trace-metal concentrations on the occurrence of various algal species are currently being discussed (2, 3, 6).

Ligands, which bind metal ions and decrease the free metal ion concentration, may be released from the algae (1). The release of strong copper chelators was demonstrated in several studies of algal cultures (17–20). These strong chelators may be used by algae in maintaining an optimum concentration range of free metal ions. The speciation of copper and zinc has feedback effects on their adsorption and uptake by the algae, as well as on their toxic effects (1, 2). Adsorption to inorganic particulate phases such as iron and manganese oxides is also affected by the presence of ligands in solution.

The inputs and concentrations of trace elements in lakes are elevated in comparison to the oceans; thus, limiting concentration ranges of trace elements are less likely to occur in lakes. The mutual interactions of trace elements and algae have received much less attention in eutrophic lakes than in the oceans. However, low concentrations of trace elements are observed in the water column of productive lakes (4, 16), in spite of large inputs of these elements. The speciation of trace elements in lakes and the interactions between trace elements, dissolved ligands, algae, and settling particles are poorly known.

We will illustrate some of the processes involved in a eutrophic lake. Points of interest are the factors that determine the speciation of copper and zinc, the role of biologically produced ligands, how copper and zinc are bound in the settling particles, and whether the settling particles reflect the composition of algae with respect to trace elements.

Experimental Procedure

Sampling Site. Lake Greifen (Switzerland) is a highly eutrophic lake with a surface area of 8.5 km^2 and a volume of 150×10^6 m^3. Its average depth is 17.7 m, with a maximum of 32.2 m. Water samples were collected from the water column at the deepest point of the lake. The tributaries of the lake are strongly loaded with nutrient inputs and pollutants from sewage and agriculture. The hypolimnion (below about 10-m depth) is seasonally anoxic because of high algal productivity from about June to December. A very active manganese cycle takes place in this system (21, 22).

Collection of Settling Particles in Sediment Traps. Sediment traps (Plexiglas tubes with a height-to-diameter ratio of 10:1) were exposed in the deepest part of the lake at 15- and 28-m depths; two identical tubes were exposed at each depth (23). The line carrying the traps was moored by using an acoustic release device at the bottom (24). The particulate material in the traps was collected approximately every 3 weeks for 15 months and was subsequently freeze-dried until analysis. Sedimentation rates were quantified by weighing the dry material. The material from both tubes at the same depth was mixed for analysis after weight determination.

Analysis of Sediment-Trap Material. The freeze-dried material from the sediment traps was digested with HCl–HNO$_3$ in a microwave digestion device (MLS–1200). Teflon beakers, cleaned with NHO$_3$, were used. Fe, Mn, Ca, Zn, Cu, and Cr were determined by inductively coupled plasma emission spectrometry, P by the molybdate spectrophotometric method (25); and organic C and N on a C, H, N analyzer (Heraeus).

Determination and Speciation of Cu and Zn. Water samples were collected with a sampler (Go-Flo, General Oceanics), which had been carefully cleaned with HNO$_3$. Samples were filtered in the laboratory (0.45

μm) within a few hours of sampling; samples for determination of total dissolved Zn and Cu were acidified to 0.01 M HNO_3. Samples for speciation determinations were kept in polyethylene bottles in the dark at 4 °C until analysis, which was performed within 1–3 days. All bottles and filters used were carefully cleaned with 0.01 M HNO_3.

Zinc was determined in filtered samples (<0.45 μm) by flame atomic absorption spectrometry after preconcentration. For the preconcentration, 8-hydroxyquinoline was added to the acidified samples as a chelating agent, and the samples were subsequently buffered to pH 8. The hydroxyquinoline complexes were collected on C_{18} columns (Baker) and were eluted with 0.6 M HCl (modified after reference 26). The preconcentration factor was 40. Blank values of this method were ≤2–3 nM Zn; reproducibility was ±1.5 nM. Recovery of Zn spikes added to lake water was 90–100%.

Zinc complexation was evaluated by ligand exchange with ethylenediaminetetraacetic acid (EDTA) and anodic stripping voltammetry of the labile zinc (27). Labile and nonlabile Zn species were distinguished by voltammetry. The labile Zn complexes were evaluated by competition with added EDTA. The Zn–EDTA complex is voltammetrically nonlabile. The decrease in labile Zn concentration was thus measured as a function of added EDTA, and the concentration of [Zn^{2+}] was calculated from the [Zn–EDTA] formed and the equilibrium with free EDTA.

Total dissolved copper was determined in filtered samples by graphite furnace atomic absorption spectrometry by direct injection of the acidified samples. The limit of determination was about 3 nM.

Copper complexation was evaluated by ligand exchange with catechol and cathodic stripping voltammetry of the copper–catechol complexes (28, 29). Titration curves with Cu were obtained by spiking aliquots of lake-water samples with different Cu concentrations. The free [Cu^{2+}] concentration was calculated from the concentration of copper–catechol complexes formed in equilibrium with free catechol.

Dissolved organic carbon (determined by combustion at 680 °C on a Shimadzu TOC–500 instrument) was about 3.5 mg/L and varied little in samples from different depths and times. Alkalinity (determined by Gran plot titrations) was 3–4 mmol/L, and pH varied with depth and time in the range 7.5–8.5 (21). Dissolved phosphate and silicate were measured in filtered samples by standard automated (Auto-Analyzer) methods (25).

Results and Discussion

Zn and Cu in the Water Column. Total concentrations of dissolved Zn (<0.45 μm) in the water column of Lake Greifen were 10–40 nM; total concentrations of dissolved Cu (<0.45 μm) were 5–20 nM. Systematic variations of total dissolved Zn were observed over the seasonal cycle, with a depletion of Zn from the epilimnion during summer stagnation (Figures 1 and 2). During lake overturn, homogeneous concentrations of Zn were measured throughout the water column (30).

The depth profiles of Zn during summer may be compared to those of major nutrients such as phosphate and silicate. The depletion from the epilimnion indicates a removal mechanism similar to that for nutrients. Zn, however, was never completely depleted from the epilimnion. Under anoxic con-

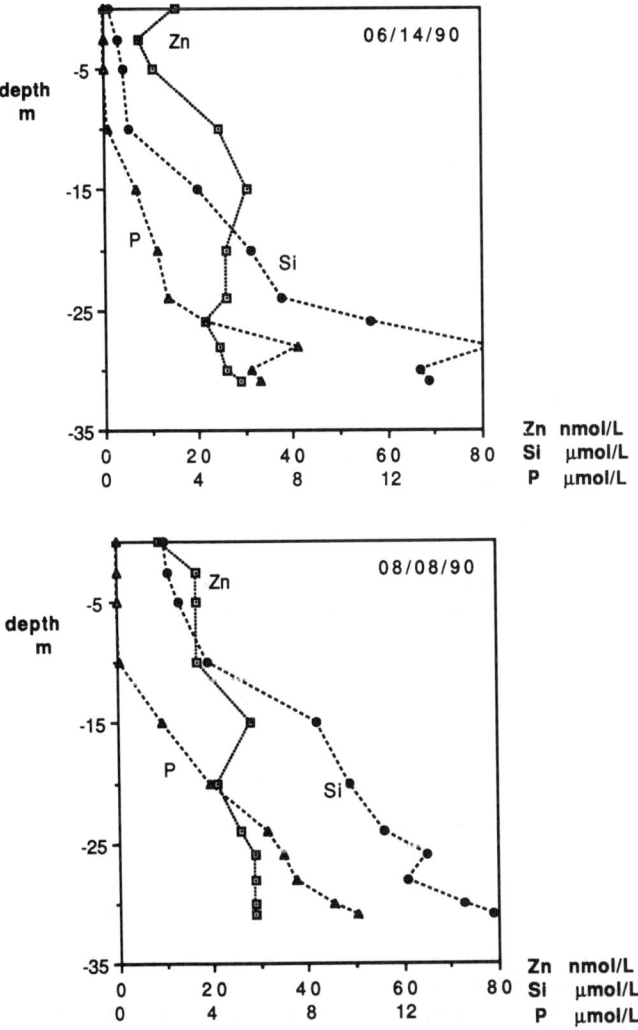

Figure 1. Concentration versus depth profiles of dissolved Zn (<0.45 μm) in the water column of Lake Greifen on two days during summer stagnation, in comparison to depth profiles of the nutrients P and Si.

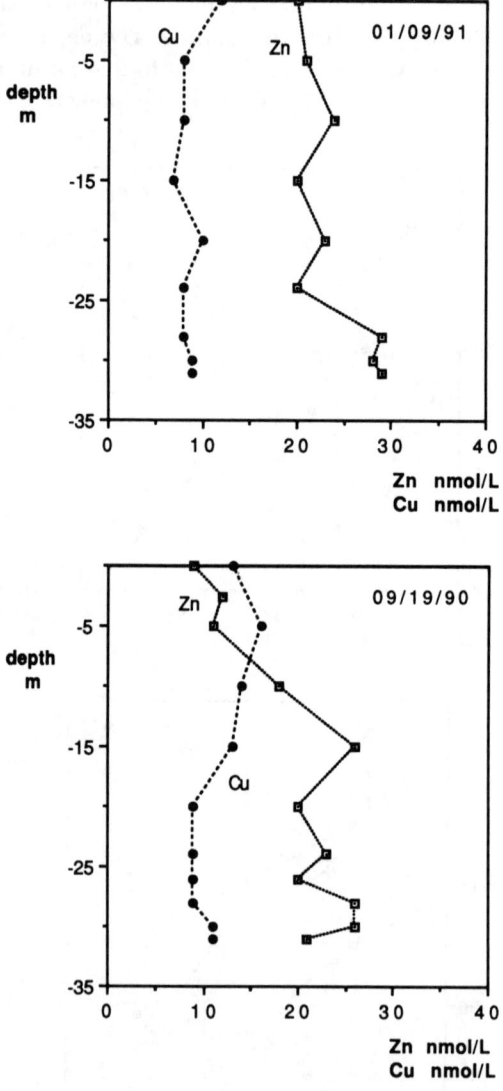

Figure 2. Concentration versus depth profiles of dissolved Zn and dissolved Cu (<0.45 μm) in the water column of Lake Greifen in the mixed water column (January 9, 1991) and at the end of summer stagnation (September 19, 1990).

ditions phosphate and silicate concentrations increase at the bottom of the lake, surpassing their lake-overturn concentrations (Figure 1) because of the release of these elements from the sediments. This release effect does not seem to be significant for Zn.

Results on total dissolved copper were less systematic than for Zn. Some analytical difficulties were encountered because of the concentration range, which is close to the detection limit of the method used. No depletion from the epilimnion was observed during summer. On the contrary, higher Cu concentrations were observed in the epilimnion than in the hypolimnion during summer stagnation.

The ratio of total dissolved Zn to Cu in the water column was thus Zn:Cu ≈ 3–4 (mol/mol) in the mixed lake-water column. This ratio remained rather constant in the hypolimnion, although it decreased to Zn:Cu ≈ 0.5–1 (mol/mol) in the epilimnion during summer (Figure 2).

The speciation of Cu obtained by the ligand-exchange method indicated very strong complexation of Cu; pCu was in the 14–16 range (Figure 3) (28). These results indicated the presence of very strong ligands for copper, at concentrations in excess of the total dissolved Cu. The free [Cu^{2+}] concentrations were 6–7 orders of magnitude lower than the concentrations of total dissolved copper. More than 99% of the total dissolved copper must thus be present in the form of organic complexes.

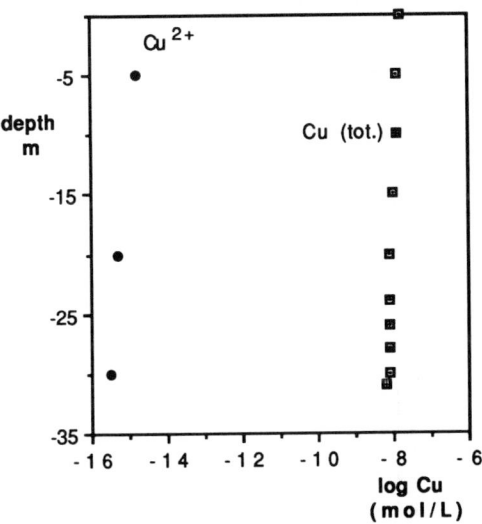

Figure 3. *Concentration versus depth profiles of dissolved Cu (<0.45 μm) and of Cu^{2+} in the water column of Lake Greifen [as determined by ligand exchange with catechol (28)] on August 8, 1990. Concentrations are given on a logarithmic scale, because [Cu^{2+}] is 7 orders of magnitude lower than Cu (total).*

In contrast to Cu, a substantial fraction of Zn was present as free zinc ions and in weak organic complexes (Figure 4) (27). Labile Zn includes inorganic Zn species and weak organic complexes. From 5 to 10% of the total dissolved Zn was evaluated by ligand exchange with EDTA to be free Zn^{2+}; pZn was thus in the 8.5–9 range. About 30–40% of the total Zn appeared to be present in weak organic complexes, whereas about 50% was present in forms that are not labile to the anodic stripping voltammetry determinations and thus are bound in stronger organic ligands or possibly in colloidal particles. Strong organic ligands for Zn, if really present, may be available only at lower concentrations than total dissolved Zn.

The ratios of the free aquo ions $[Zn^{2+}]:[Cu^{2+}]$ are thus in the 2×10^5–10^6 range. These ratios reflect the different tendencies of Cu and Zn to bind to organic ligands.

Settling Particles and Sedimentation Rates. The composition of the settling particles (Table I) and the sedimentation rates as a function of time (Figure 5) give some insights into the mechanisms of the binding of Zn and Cu to the particulate phase and of removal to the sediments. The overall composition of the settling particles reflects the seasonal variations in productivity and the redox cycles in this lake (30). Organic material represents a substantial fraction of the dry weight, with organic C ≈ 5–16%.

During the summer stagnation with high productivity from June to September, a significant fraction of the settling material consists of calcium car-

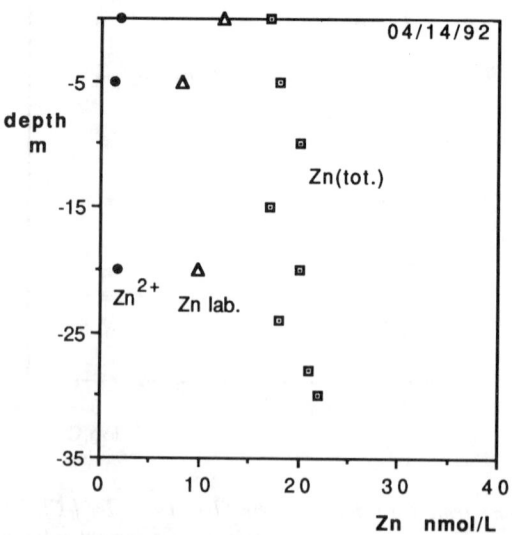

Figure 4. Concentration versus depth profiles of total dissolved Zn [Zn(tot.)], Zn^{2+}, and voltammetrically labile Zn (Zn lab.) in the water column of Lake Greifen on April 14, 1992. $[Zn^{2+}]$ amounts to 5–10% of total dissolved Zn.

Table I. Composition of the Settling Particles Collected in Lake Greifen

Date of Sampling	Depth (m)	Org. C (mmol/g)	P (mmol/g)	N (mmol/g)	Ca (mmol/g)	Fe (mmol/g)	Mn (mmol/g)	Zn (µmol/g)	Cu (µmol/g)
11/01/89– 11/21/89	15	9.12	0.129	1.11	4.38	0.10	0.013	2.29	0.55
11/01/89– 11/21/89	28	7.81	0.104	1.03	5.03	0.13	0.020	2.29	0.57
11/21/89– 12/12/89	15	13.35	0.205	1.93	2.73	0.15	0.204	3.36	0.68
11/21/89– 12/12/89	28	12.41	0.191	1.82	3.00	0.18	0.207	2.60	0.71
12/12/89– 01/15/89	15	8.35	0.135	1.03	3.83	0.18	0.677	4.74	1.31
12/12/89– 01/15/89	28	8.68	0.131	1.04	3.83	0.19	0.664	4.74	1.29
01/15/89– 02/26/90	15	5.41	0.075	0.60	4.25	0.25	0.080	3.52	0.71
01/15/89– 02/26/90	28	5.40	0.078	0.58	4.25	0.28	0.098	2.45	0.71
04/30/90– 05/21/90	15	9.65	0.107	1.47	3.00	0.08	0.017	1.68	0.31
04/30/90– 05/21/90	28	9.17	0.145	1.81	2.25	0.11	0.160	2.91	0.50
05/21/90– 06/12/90	15	3.40	0.038	0.40	8.23	0.02	0.006	1.07	0.13
05/21/90– 06/12/90	28	3.75	0.052	0.42	8.05	0.04	0.051	1.38	0.16
06/12/90– 07/12/90	15	4.45	0.049	0.48	7.73	0.04	0.005	1.07	0.17
06/12/90– 07/12/90	28	4.13	0.050	0.48	7.90	0.04	0.016	1.07	0.16
07/12/90– 08/02/90	15	4.20	0.040	0.54	8.03	0.02	0.002	0.92	0.13
07/12/90– 08/02/90	28	4.21	0.041	0.56	8.05	0.02	0.004	0.92	0.14
08/02/90– 08/30/90	15	7.15	0.083	0.87	7.00	0.04	0.004	0.76	0.20
08/02/90– 08/30/90	28	6.53	0.073	0.84	7.13	0.04	0.007	0.61	0.25
08/30/90– 09/27/90	15	6.78	0.066	0.82	6.45	0.05	0.005	0.92	0.20
08/30/90– 09/27/90	28	7.52	0.065	0.93	6.83	0.08	0.006	1.83	0.35
09/27/90– 10/22/90	15	10.31	0.079	1.14	4.40	0.11	0.005	1.68	0.44
09/27/90– 10/22/90	28	10.41	0.083	1.21	4.33	0.16	0.007	1.53	0.47
10/22/90– 11/08/90	15	6.87	0.080	0.93	4.63	0.15	0.025	1.53	0.49
10/22/90– 11/08/90	28	9.15	0.104	1.12	4.40	0.16	0.018	1.53	0.57
11/08/90– 11/27/90	15	8.00	0.095	1.13	4.38	0.16	0.058	2.29	0.63
11/08/90– 11/27/90	28	7.46	0.087	0.94	4.75	0.20	0.031	2.14	0.61
11/27/90– 12/18/90	15	8.32	0.110	1.22	4.05	0.18	0.415	5.20	1.17
11/27/90– 12/18/90	28	7.59	0.132	1.00	4.30	0.22	0.218	4.28	1.10
12/18/90– 01/09/91	15	6.79	0.114	0.92	4.20	0.14	0.553	5.05	1.09
12/18/90– 01/09/91	28	6.56	0.114	0.93	4.18	0.16	0.693	5.96	1.35

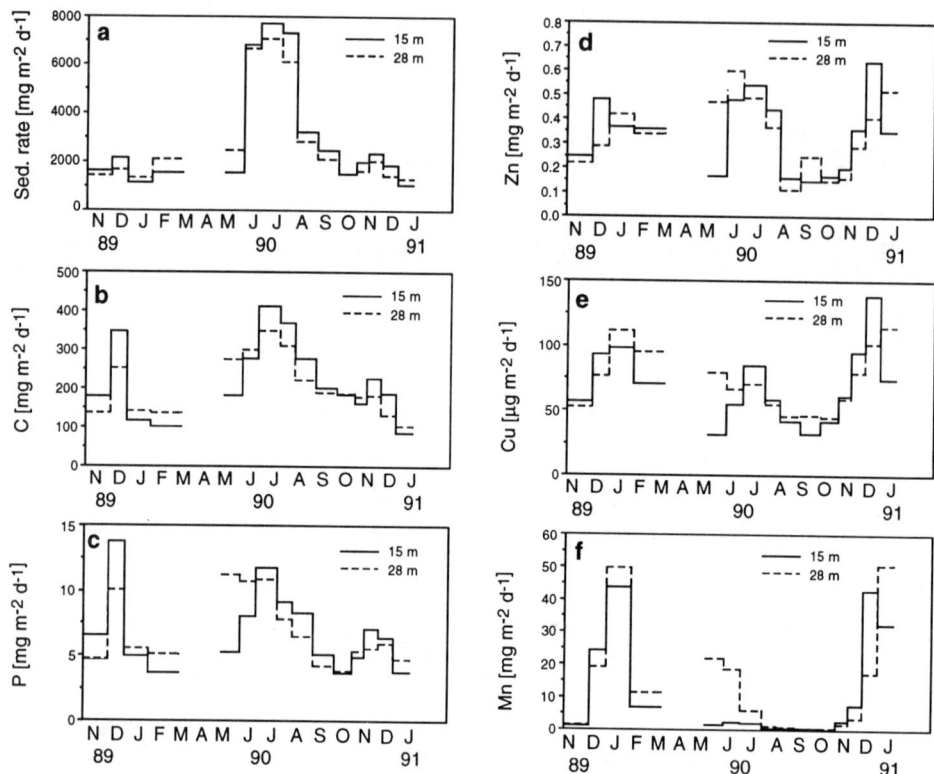

Figure 5. Overall sedimentation rates (a) and sedimentation rates of organic C, P, Zn, Cu, and Mn (b–f, respectively) versus time (at 15- and 28-m depth). Lake Greifen overturn occurs in December–January.

bonate (up to 80% of dry weight, measured as Ca, Table I), which is precipitated in the epilimnion as a consequence of the pH increase during photosynthesis. The manganese oxide content is high during lake overturn (December–January), because Mn(II) is oxidized within days to weeks of contact with oxygen when the lake mixes (22). This Mn(II) had previously been released from the sediments into the anoxic hypolimnion. Large amounts of manganese oxide are precipitated and sedimented within a few weeks in Lake Greifen.

Both sediment traps (at 15 and 28 m) are below the productive layers. From July to November 1990, oxygen was absent from the hypolimnion below 10-m depth, so that both traps were within the anoxic hypolimnion. The differences in composition between the traps are relatively small (Table I). The organic C and P content, as well as the concentrations of the other elements, are generally very similar at both depths. Differences in the manganese content appear during certain time periods, such as during May–June 1990, when anoxic conditions were already established in the deepest few meters of the lake, but not in the whole hypolimnion.

The overall sedimentation rate (Figure 5a) exhibits a clear maximum during summer. This maximum corresponds to a maximum sedimentation rate of calcium carbonate as well as of organic material, as indicated by the sedimentation rates of organic C and of P (Figures 5b and 5c). The sedimentation rate of manganese, in contrast, shows a high maximum during lake overturn and a minimum during summer (Figure 5f).

The Zn and Cu contents in the settling particles and the sedimentation rates indicate a combination between the sedimentation with organic material and with manganese oxides. The Zn and Cu contents in the settling particles during summer stagnation are negatively correlated with Ca and weakly correlated with organic C and P (with positive coefficients) (Table II). The Zn sedimentation rate (Figure 5d) has a maximum during summer, which corresponds to the maximum of the sedimentation rate of organic C and P. This summer maximum is less marked for Cu (Figure 5e).

The high sedimentation rates of Zn and Cu during summer may be interpreted as an association with organic material rather than with calcium carbonate. Zn and Cu concentrations in the particles decrease with increasing $CaCO_3$ content (Tables I and II). Zn and Cu are remarkably high in the samples with high Mn content. Another maximum of the Zn sedimentation rate occurs in December, together with the maximum Cu sedimentation rate. Thus the precipitation of large amounts of manganese oxides over a short time affects Zn and Cu.

Zn and Cu contents in these particles are well correlated with each other ($r = 0.956$ over all samples), with an average Zn:Cu ratio ≈ 3.8; in the summer samples this ratio is up to 8. These relationships are similar to previous observations of correlation in settling particles from Lake Zurich (4). However, the correlations of Zn and Cu with organic C and P are less clear for Lake Greifen.

Table II. Correlation Matrix for the Concentrations in Settling Particles during Summer Stagnation

Element	C	P	N	Ca	Fe	Mn	Zn	Cu
C	1							
P	0.803	1						
N	0.891	0.95	1					
Ca	−0.885	−0.916	−0.955	1				
Fe	0.79	0.663	0.677	−0.796	1			
Mn	0.204	0.661	0.561	−0.510	0.238	1		
Zn	0.565	0.733	0.758	−0.763	0.597	0.78	1	
Cu	0.832	0.774	0.781	−0.851	0.962	0.368	0.692	1

NOTE: The measurement period was April 30, 1990, through November 8, 1990. The number of samples was 16.

The removal rates from the epilimnion by sedimentation can be calculated on the basis of the sedimentation rates measured at 15-m depth (Figure 5) and of the total amounts of Zn and Cu in the water column in the epilimnion. Expressed as a first-order process, the removal rate constants of Zn during the summer sedimentation maximum (June–August 1990) are about 0.1–0.15 day^{-1}, whereas for Cu they are about 0.01 day^{-1}. During the other periods, removal rate constants of Zn are about 0.01–0.03 day^{-1} and of Cu 0.005–0.008 day^{-1}. During overturn (December 1990) the removal rate constant of Zn is 0.03 day^{-1} and of Cu 0.01 day^{-1}. In comparison, the water renewal rate constant in Lake Greifen is 0.0025 day^{-1}. The difference in removal rates of Zn and Cu during summer means that Zn is much more efficiently removed than Cu from the epilimnion during this time, although the differences are smaller at other times of the year.

Zn:P and Cu:P Ratios in Settling Particles. If zinc and copper are mostly bound in algal material in the settling particles, their ratios to P should be indicative of Zn:P and Cu:P ratios in algae. This may be the case in samples that contain only a small fraction of manganese and iron oxides. P is expected to be mostly bound in organic material in these particles. The molar organic C:P ratios are in the 60–130 range (80–110 during summer) with an average of 85 over the whole year. This range is typical of algal material. Organic C, P, and N are well correlated in these settling particles, a result indicating a common source in biological material (Table II). The Zn:P and Cu:P ratios are represented as a function of time in Figure 6.

The samples from June to September are probably the most representative ones for algal material, because they contain mostly organic material and calcium carbonate. In these samples, the Zn:P ratios are 0.01–0.03 (Zn:C = 1–2 × 10^{-4}) and the Cu:P ratios are 0.002–0.004 (Cu:C = 3–4 × 10^{-5}). The average (over the whole year) Zn:P ratio is 0.03, and the average Cu:P ratio is 0.006. The variations of these ratios over the year (Figure 6) correspond to the varying composition of the particles. Higher ratios of Zn and Cu

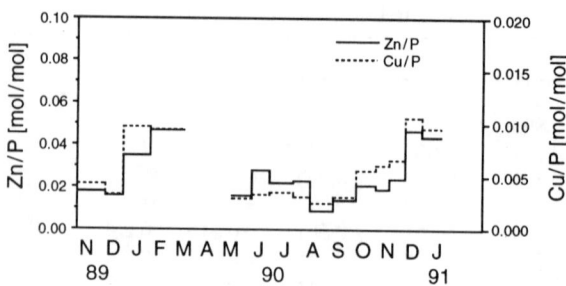

Figure 6. *Ratios of Zn:P and of Cu:P in the settling particles (trap at 15-m depth) from Lake Greifen as a function of time.*

to P are observed in the samples with a high fraction of manganese oxides, in which a substantial fraction of the Zn and Cu may be bound to Mn oxides.

These Zn:P and Cu:P ratios (or the corresponding Zn:C and Cu:C ratios) may be compared to those observed in other systems and in algae (Table III). The composition of the settling particles in Lake Zurich is quite similar to that in Lake Greifen; this similarity is reflected in these ratios.

In Lake Michigan (*31*), diatoms appear to play an important role in the sedimentation of Cu and Zn. In Lake Windermere (*32*), the influence of allochthonous material may be more important. A recent study of Lake Windermere (*33*) indicated a Zn:P ratio of 0.03 from the regression analysis of dissolved concentrations during a diatom bloom. Mountain Lake (*34*) is a small acidic lake.

Few data on such ratios in algae are available; it is especially difficult to find data on freshwater algae. The work on marine algae by Sunda and Huntsman (*6*) indicates that the Zn:C ratios depend on the concentration of available Zn in solution. For the higher range of pZn $\approx$ (Ap 8.5–9, which is similar to our data for Lake Greifen, the Zn:C ratios are in a range similar to those observed in our data. For Cu:P, the ratios given in reference 35 for marine plankton appear to be in a lower range than the values measured in the settling particles of lakes. Cu and Zn were measured in cultures of *Cyclotella cryptica*, a freshwater diatom, and of the green algae *Chlamydomonas reinhardtii* (*36*). The elementary composition of the algae was measured after growing them in a medium with pCu $\approx$ 12 and pZn $\approx$ 7.7. The ratios Zn:P and Cu:P are low in comparison with the settling particles and even with the examples of marine algae.

Comparison of these different data does not yet allow definite conclusions on the significance of algae for the composition of settling particles. However, the ratios of Cu and Zn to P are relatively similar in settling particles from different lakes; the presence of allochthonous particles and of manganese oxides tends to increase these ratios. Appropriate data on the trace-element content of freshwater algae are still lacking for comparison. Work is presently in progress in our laboratory to determine the relative importance of the binding of Cu to algae surfaces in comparison to copper uptake inside algae and to determine the dependence of the copper content of algae on various factors.

Discussion

With regard to speciation, these results indicate interesting differences between Cu and Zn, which are in line with the generally much higher tendency of Cu to form complexes with organic ligands as derived, for example, from the Irving–Williams series (*37*). The complexation of Cu, however, implies the presence of very strong complexing ligands at nanomole concentration levels in the water column of Lake Greifen. It seems likely that these ligands are of

Table III. Ratios of Zn and Cu to P and Organic C in Settling Particles from Lakes and Algae

Source	Zn:P	Zn:C	Cu:P	Cu:C	Ref.
Lakes					
Lake Greifen	0.01–0.06	1–6×10^{-4}	2–12×10^{-3}	3–15×10^{-5}	this study
Lake Zurich	0.02–0.07	2–7×10^{-4}	4–20×10^{-3}	4–20×10^{-5}	4
Lake Michigan		5×10^{-4}		15×10^{-5}	31
Lake Windermere		5×10^{-4}		43×10^{-5}	32
Mountain Lake		2×10^{-4}		8×10^{-5}	34
Algae					
Marine algae	2–5×10^{-3}	0.3–1×10^{-4}			6
Marine plankton	1–2×10^{-3}		5–6×10^{-4}		35
Cyclotella cryptica		2–3×10^{-5}	1–5×10^{-4}	1–7×10^{-6}	36
Chlamydomonas reinhardtii	2–4×10^{-3}	0.3–1×10^{-4}	0.3–0.8×10^{-4}	0.5–2×10^{-6}	36

NOTE: All ratios are given on a mole per mole basis. Blanks indicate data are not available.

biological origin; they may be released during algae blooms, as indicated by seasonal variations of pCu in Lake Greifen (28). The presence of these ligands in Lake Greifen may be linked to the high algal productivity in this lake. However, the nature of these ligands has not been further elucidated.

One may ask how selective these ligands are for Cu in comparison to Zn. Only a fraction of dissolved Zn is bound to strong ligands, and this behavior implies either that the selectivity of these ligands for Cu is very high or possibly that the concentration of available strong ligands is exceeded by the Zn concentration. These questions should be further investigated to get more insight into the sources of the strong ligands for Cu and their possible significance in maintaining low levels of [Cu^{2+}] in productive lakes.

In this regard, it would be of interest to know whether the levels of free [Cu^{2+}] and [Zn^{2+}] in this lake correspond to an optimum level for freshwater algae and to what extent the ligands responsible for the complexation of Cu are released by growing algae. Both pCu and pZn in Lake Greifen would be below the toxic levels observed for marine algae (6, 11–13).

Possible mechanisms of binding to the settling particles should be examined, taking into account the speciation in solution of Cu and Zn. An intriguing relationship exists between total dissolved copper, free copper ion, and copper bound in particles. The ratio of [Cu^{2+}] to total dissolved Cu is about 10^{-6}–10^{-7}. As determined from the concentration in settling particles and in the water column, the distribution coefficients for Cu are $K_d \approx 1$–5×10^4 L/kg, with variations over the year.

Considering a simple competition model for binding to ligands in solution and to surfaces, the Cu content in the settling particles seems to be too high. For example, the surface complexation constants given in reference 38 can be used to evaluate the binding of Cu to an iron hydroxide surface. With [Cu^{2+}] = 10^{-14}–10^{-16}, about 5.10^{-9}–5.10^{-11} mol/g of Cu bound to the solid phase would be obtained at pH 8, which is much lower than the concentrations measured in the settling particles.

In a similar way, it seems difficult to explain how Cu is bound to the manganese oxides that are formed in the water column and would thus have to scavenge Cu from solution at these low [Cu^{2+}] levels. Straightforward competition between complexation in solution and on surfaces may be too simplistic. The formation of ternary complexes of metal ions with surfaces and organic ligands is a possible binding mechanism. It was demonstrated in some model systems (39), but is poorly known for natural systems.

In the samples dominated by biological materials, more specific binding mechanisms may be involved, such as binding to strong ligands at the surfaces of algae and uptake into algae. Uptake into algae is possible at very low free [Cu^{2+}] levels. However, the Cu:P ratios in the settling particles seem to be high in comparison to algae. Algal debris, including specific ligands, could also play a role. A lack of equilibration between bound Cu (e.g., in allochthonous solid phases such as iron oxides) and Cu in solution may account for the apparently too high Cu content of the solid phase.

In comparison, it is easier to account for the binding of Zn to the solid phase, because its free aquo ion concentration is much higher. The Zn:P ratios in summer samples appear to be close to what may be expected in algae. Both binding to the surfaces of algae and uptake into algae may in this case explain the concentrations of zinc observed in the settling particles. Organic material represents a substantial part of the settling particles during summer, so it seems likely that a significant fraction of Zn is bound in algal material. Another significant fraction of Zn in solution is present as free Zn ions or weak complexes. Thus Zn^{2+} possibly becomes adsorbed to inorganic surfaces such as manganese oxides and iron hydroxides. Binding to the active surfaces of freshly precipitated manganese oxides is a likely mechanism for Zn.

The Zn:Cu ratio in the particles is highest in summer, whereas the ratio of total dissolved Zn:Cu is low at that time. This effect may be related to the more efficient complexation of Cu in solution during periods of high productivity.

A study of Lake Windermere (33) on the cycling of Zn and Cu during a diatom bloom also showed stronger removal of Zn than of Cu by the algae.

Algae may thus influence both the total dissolved and the free aquo ion concentrations of Cu and Zn in the water column of the lake by binding in the settling material or by releasing complexing ligands into solution. These two effects are linked to each other because binding in the particulate phase, either by uptake or by adsorption, also depends on the speciation in solution.

Conclusions

Copper in the water column of the eutrophic Lake Greifen is complexed by very strong ligands, which are probably biologically produced. In contrast, a large fraction of zinc is present in weak complexes and free aquo ions. During summer stagnation, Zn is efficiently removed by sedimentation from the epilimnion, in connection with the sedimentation of algal material, and is depleted from the epilimnion. In comparison, the extent of removal of Cu with settling particles is less significant during summer. The precipitation of manganese oxides during lake overturn plays a significant role in the sedimentation of Zn and Cu.

Considering the strong complexation of Cu in solution, the binding mechanisms of Cu to particulate matter are unclear. The role of algae in this system, although not yet fully elucidated, may include uptake of Zn and Cu, their subsequent transport to the sediments, and the release of complexing ligands into solution.

Acknowledgments

We thank Werner Stumm for his continuous interest and enthusiasm for this subject, especially for the Redfield ratios, better known to many students as "Butterbrotstöchiometrie".

References

1. Sunda, W. G. *Biol. Oceanogr.* **1988/1989**, *6*, 411–442.
2. Bruland, K. W.; Donat, J. R.; Hutchins, D. A. *Limnol. Oceanogr.* **1991**, *36*, 1555–1577.
3. Morel, F. M. M.; Hudson, R. J. M.; Price, N. M. *Limnol. Oceanogr.* **1991**, *36*, 1742–1755.
4. Sigg, L.; Sturm, M.; Kistler, D. *Limnol. Oceanogr.* **1987**, *32*, 112–130.
5. Sunda, W. G.; Huntsman, S. A. *Limnol. Oceanogr.* **1983**, *28*, 924–934.
6. Sunda, W. G.; Huntsman, S. A. *Limnol. Oceanogr.* **1992**, *37*, 25–40.
7. Sunda, W. G.; Swift, D. G.; Huntsman, S. A. *Nature (London)* **1991**, *351*, 55–57.
8. Harrison, G. I.; Morel, F. M. M. *J. Phycol.* **1983**, *19*, 495–507.
9. Harrison, G. I.; Morel, F. M. M. *Limnol. Oceanogr.* **1986**, *31*, 989–997.
10. Price, N. M.; Morel, F. M. M. *Nature (London)* **1990**, *344*, 658–660.
11. Brand, L. E.; Sunda, W. G.; Guillard, R. R. L. *Limnol. Oceanogr.* **1983**, *28*, 1182–1198.
12. Brand, L. E.; Sunda, W. G.; Guillard, R. R. L. *J. Exp. Mar. Biol. Ecol.* **1986**, *96*, 225–250.
13. Sunda, W. G.; Guillard R. R. L. *J. Mar. Res.* **1976**, *34*, 511–529.
14. Redfield, A. C.; Ketchum, B. H.; Richards, F. A. In *The Sea;* Hill, M. N., Ed.; Wiley: New York, 1963; Vol. 2.
15. Morel, F. M. M.; Hudson, R. J. M. In *Chemical Processes in Lakes;* Stumm, W., Ed.; Wiley: New York, 1985; Chapter 12.
16. Sigg, L. In *Chemical Processes in Lakes;* Stumm, W., Ed.; Wiley: New York, 1985; Chapter 13.
17. McKnight, D. M.; Morel F. M. M. *Limnol. Oceanogr.* **1979**, *24*, 823–837.
18. Zhou, X.; Wangersky, P. J. *Mar. Chem.* **1989**, *26*, 239–259.
19. Van den Berg, C. M. G.; Wong, P. T. S.; Chau, Y. K. *J. Fish. Res. Board. Can.* **1979**, *36*, 901–905.
20. Xue, H.; Sigg, L. *Water Res.* **1990**, *24*, 1129–1136.
21. Sigg, L.; Johnson, C. A.; Kuhn, A. *Mar. Chem.* **1991**, *36*, 9–26.
22. Johnson, C. A.; Ulrich, M.; Imboden, D.; Sigg, L. *Limnol. Oceanogr.* **1991**, *36*, 1415–1426.
23. Kuhn, A. Ph.D. Thesis No. 9783, Swiss Federal Institute of Technology, Zurich, Switzerland, 1992.
24. Sturm, M.; Zeh, U.; Müller, J.; Sigg, L.; Stabel, H. H. *Eclogae Geol. Helv.* **1982**, *75*, 579–588.
25. *Standard Methods for the Examination of Water and Wastewater*, 17th ed.; Chesceri, L. S.; Greenberg, A. E.; Trussell, R. R., Eds.; American Public Health Association: Washington, DC, 1989.
26. Watanabe, H.; Goto, K.; Taguchi, S.; McLaren, J. W.; Berman, S. S.; Russell, D. S. *Anal. Chem.* **1981**, *53*, 738–739.
27. Xue, H.; Sigg, L. *Anal. Chim. Acta* **1994**, *284*, 505–515.
28. Xue, H.; Sigg, L. *Limnol. Oceanogr.* **1993**, *38*, 1200–1213.

29. Van den Berg, C. M. G. *Anal. Chim. Acta* **1984**, *164*, 195–207.
30. Kuhn, A.; Johnson, C. A.; Sigg, L. In *Environmental Chemistry of Lakes and Reservoirs;* Baker, L. A., Ed.; Advances in Chemistry 237; American Chemical Society: Washington, DC, 1993; pp 473–497.
31. Shafer, M. M.; Armstrong, D. E. In *Organic Substances and Sediments in Water;* Baker, R. A., Ed.; Lewis: Chelsea, MI, 1991; Vol. 2, pp 15–47.
32. Hamilton-Taylor, J.; Willis, M.; Reynolds, C. S. *Limnol. Oceanogr.* **1984**, *24*, 695–710.
33. Reynolds, G. L.; Hamilton-Taylor, J. *Limnol. Oceanogr.* **1992**, *37*, 1759–1769.
34. Nriagu, J. O.; Wong, H. K. T. *Sci. Tot. Environ.* **1989**, *87/88*, 315–328.
35. Collier, R.; Edmond, J. *Progr. Oceanogr.* **1984**, *13*, 113.
36. Kiefer, E. Ph.D. Thesis No. 10786, Swiss Federal Institute of Technology, Zurich, Switzerland, 1994.
37. Stumm, W.; Morgan, J. J. *Aquatic Chemistry;* Wiley: New York, 1981.
38. Dzombak, D. A.; Morel, F. M. M. *Surface Complexation Modeling: Hydrous Ferric Oxide;* Wiley: New York, 1990.
39. Schindler, P. W. In *Mineral–Water Interface Geochemistry;* Hochella, M. F.; White, A. F., Eds.; Reviews in Mineralogy 23; Mineralogical Society of America: Washington, DC, 1990; pp 281–307.

RECEIVED for review October 23, 1992. ACCEPTED revised manuscript May 3, 1993.

9

Metals and Microbiology

The Influence of Copper on Methane Oxidation

Mary E. Lidstrom and Jeremy D. Semrau

Department of Environmental Engineering Science, 138–78, California Institute of Technology, Pasadena, CA 91125

> *Methane is oxidized under aerobic conditions by a group of bacteria called methanotrophs. These widespread bacteria play an important role in the global cycling of methane. Two types of methane oxidation systems are known, a ubiquitous particulate methane monooxygenase (pMMO) and a cytoplasmic soluble methane monooxygenase (sMMO) found in only a few strains. These enzymes have different catalytic characteristics, and so it is important to know the conditions under which each is expressed. In those strains containing both sMMO and pMMO, the available copper concentration controls which enzyme is expressed. However, the activity of the pMMO is also affected by copper. Data on methane oxidation in natural samples suggest that methanotrophs are not copper-limited in nature and express the pMMO predominantly.*

THE AEROBIC OXIDATION OF METHANE is carried out by bacteria called methanotrophs (1). These bacteria grow on methane as their sole carbon and energy source, oxidizing a portion of the methane to CO_2 and fixing a portion into cell material. They are obligate aerobes because the methane oxidation reaction requires molecular oxygen.

Two major classes of methanotrophs are known, called Type I and Type II strains (2). These two groups are phylogenetically and physiologically distinct, and are easily distinguished by the characteristic pattern of a complex internal membrane system. Type I strains (*Methylococcus*, *Methylomonas*, and *Methylobacter*) contain stacked bundles of disc-shaped membranes through the center of the cell, whereas Type II strains (*Methylosinus* and *Meth-*

ylocystis) contain peripheral rings of membranes (2). A third class of methanotrophs, called Type X, consists of a few *Methylococcus* strains. They have Type I membranes but also have some physiological characteristics of Type II strains and are phylogenetically distinct (2).

Methanotrophs, which are widespread in aquatic and terrestrial environments, carry out methane oxidation in most habitats where methane and oxygen coexist (3). On a global scale, these bacteria are responsible for a major portion of the biological methane consumption that occurs on the earth's surface. Therefore, a great deal of interest exists concerning their role in global methane cycling (4).

Two Methane Oxidation Systems

To understand the role of these bacteria in methane cycling, the methane oxidation system must be studied. In methanotrophs, methane is oxidized to methanol by an enzyme called the methane monooxygenase (MMO) (1), which uses methane, molecular oxygen, and reducing equivalents to produce methanol and water. All known methanotrophs contain a membrane-bound MMO, called the particulate methane monooxygenase (pMMO). The presence of this enzyme system is correlated with the complex internal membrane system found in all known methanotrophs.

However, a few species of methanotrophs also have the ability to produce a second cytoplasmic enzyme, called the soluble methane monooxygenase (sMMO) (1, 2) (Table I). When the sMMO is present, the complex internal membrane systems are absent. The sMMO has been intensively studied in the past few years, and much is known about this enzyme (5, 6) (Table II). It was purified and characterized from three strains, and in all three cases it consists of three components containing a total of five polypeptides. The genes for these polypeptides, were cloned and sequenced from two strains and show

Table I. Methane Monooxygenase Enzymes in Characterized Methanotrophic Isolates

Strain	pMMO	sMMO
Type I		
Methylomonas	+	−
Methylobacter	+	−
Type II		
Methylocystis	+	−
Methylosinus	+	+
Type X		
Methylococcus capsulatus Bath	+	+

Table II. Properties of the pMMO and sMMO

Property	pMMO	sMMO
Components	unknown[a]	A (hydroxylase): 20, 45, and 60 kDa B (electron transfer): 16 kDA C (NADH reductase): 39 kDA
Metals	copper	iron
K_s (CH_4)	1–5 μM	20–100 mM
Substrates	C_1–C_4 alkanes C_2–C_4 alkenes	Alkanes, alkenes, cyclic and aromatic hydrocarbons

[a] Polypeptides of 45, 36, and 27 kDa have been implicated.

high similarity (5). This enzyme has a remarkably broad substrate range, oxidizing a variety of straight-chain, cyclic, and aromatic hydrocarbons. The active site was shown to contain a binuclear iron cluster, and work is in progress to delineate the catalytic mechanism (5).

In contrast to the sMMO, the pMMO has been only poorly characterized. This enzyme is extremely unstable and has never been purified. However, induction experiments suggest that three major membrane polypeptides are involved in this enzyme and that at least one contains copper (6) (Table II). The pMMO has a more restricted substrate range, and oxidizes only straight-chain alkanes and alkenes. This enzyme has been reported to have a lower apparent K_m (half-saturation constant or Michaelis constant) for methane than the sMMO (7, 8), and so the two enzymes differ with respect to components, kinetics, and, presumably, catalytic mechanism.

Both enzymes are capable of oxidizing small halogenated solvents such as trichloroethylene (TCE) by a cooxidation reaction, and a great deal of interest exists concerning the possible use of methanotrophs in detoxifying these ubiquitous contaminants (9). Cells expressing the sMMO cooxidize TCE at rates that are approximately 2 orders of magnitude higher than cells expressing the pMMO (9). However, evidence exists that the pMMO can reduce TCE to lower levels than the sMMO (9), and so the preferred enzyme for detoxification will depend on the application.

Regulation of the sMMO and pMMO by Copper

In strains that have the ability to produce both enzymes, the major controlling parameter for the expression of the two MMOs appears to be copper availability. Under conditions of copper excess, the pMMO predominates, whereas the sMMO predominates during copper limitation (10). These results have caused researchers to speculate that the pMMO is a copper enzyme, and that the synthesis of the pMMO is controlled at the level of transcription by the availability of copper.

The speciation and concentration of copper in the growth medium for these experiments have not been well-documented, and therefore the concentration of copper that limits growth is not well-defined. Current information from the literature suggests that when copper is added as copper sulfate in the presence of ethylenediaminetetraacetic acid (EDTA), copper limitation occurs at a level of approximately 10^6 atoms of copper per cell (10). No information is available concerning the details of copper utilization by methanotrophs, and therefore the copper species used and its biologically relevant concentration are unknown.

Studies of the pMMO

In collaboration with Sunney Chan (Caltech), we initiated laboratory studies of the pMMO in attempts to characterize it biochemically and catalytically. Using electron paramagnetic resonance (EPR) spectroscopy as a tool, Chan and his co-workers obtained evidence that the pMMO contains copper in the catalytic site (11). Therefore, most likely the response of the pMMO to available copper reflects the need for copper in the active enzyme.

In my laboratory we studied the effect of copper on whole cells of methanotrophs. These experiments involved growing different pure cultures of methanotrophs in medium containing different initial amounts of copper, added as copper sulfate, as described previously (9), and analyzing both growth rate and methane oxidation kinetics. We did not measure free copper in these experiments, so the effects noted have not yet been correlated to copper speciation but rather to total copper added at the start of growth.

Methane oxidation kinetics was assessed as follows. In these experiments, cells were inoculated into medium containing different initial amounts of copper sulfate and grown to an optical density at 600 nm of 0.5–0.7. Aliquots were then placed in closed vials at different initial dissolved methane concentrations. These aliquots were incubated under optimal conditions, and headspace samples were taken at four different time points (1–4 h) for determination of methane concentrations by gas chromatography. From these data, initial methane consumption rates determined for different methane concentrations were used to obtain the Michaelis–Menten parameters K_s (half-saturation constant) and V_{max} (rate at substrate saturation). Under these conditions sMMO was not expressed, as described previously (9).

Growth rates in cells growing in flasks did not vary when copper sulfate was added at initial concentrations of between 2 and 20 μM. In addition, under similar conditions the whole-cell V_{max} values for methane oxidation did not vary significantly. However, the K_s values in whole cells were strongly altered by the copper concentration in which the cells were grown. This effect was strain-specific. In strains that contain both the sMMO and pMMO, a small but reproducible decrease in the K_s was observed as the copper concentration in the growth medium increased. In strains containing only the pMMO, the effect was dramatic. The apparent K_s decreased an order of magnitude as the copper concentration in the growth medium was increased. In cells grown with 2 μM total copper in the medium, K_s values were 40–80 μM (depending upon the strain), whereas at 20 μM copper they were 4–8 μM. These data from whole cells have been confirmed by measuring apparent K_m values in cell-free extracts.

We showed that the high K_s values correlate with a low in vitro pMMO activity and the loss of a characteristic EPR copper signal. These data strongly suggest that the pMMO can exist in two states, a copper-deficient form with high K_m and a copper-sufficient form with low K_m. Final proof of this hypothesis will require obtaining purified enzyme from cells grown at high and low copper. However, these data show that when cells expressing pMMO are grown at low copper concentrations, they are not as effective in oxidizing methane at low concentrations as cells grown at higher copper levels.

Environmental Significance

These data show that the copper concentration available to growing cells not only determines whether the pMMO will be expressed, but strongly influences the kinetics of methane oxidation by whole cells containing the pMMO (Table III). This information has potentially important environmental implications. If methanotrophs were copper-limited in nature, we would expect to observe high whole-cell K_s values because of expression of sMMO and the copper-deficient pMMO. This situation should result in a poor ability of natural populations to survive at low (submicromolar) methane concentrations. Alternatively, if they are not copper-limited, the K_s values should be lower, improving the ability of these cells to grow at submicromolar methane concentrations.

Field studies suggest that methanotrophs are methane-limited in natural habitats (1, 3). The concentrations of methane in the zones of highest consumption are always at least an order of magnitude below the measured K_s values of the natural populations, which are usually in the 2–10 μM range. The low K_s values for methane consumption measured for natural samples suggest that the sMMO is not the predominant methane consumption system

Table III. Predicted Methane Oxidation Systems and K_s Values: Environmental Implications of Copper Availability

Condition	Methylosinus, Methylococcus	All Other Methanotrophs
Copper limitation	sMMO, K_s 25–100 μM	pMMO*[a], K_s 40–80 μM
Copper sufficiency	pMMO, K_s 1–5 μM	pMMO, K_s 1–5 μM

[a] pMMO* is Cu-deficient pMMO.

in these natural populations, and also that the cells are in a state of copper sufficiency. Because the copper concentrations in most natural samples are quite low, these values suggest that the methanotrophs have efficient copper-scavenging systems. We showed with pure cultures in the laboratory that these bacteria are capable of taking up high levels of copper and that this copper is predominantly located in the membranes of these cells (11).

In summary, the information available to date indicate that copper is a major regulatory parameter for methane oxidation in methanotrophs. Not only does it regulate the expression of sMMO and pMMO in those strains that contain both, it also controls the kinetics of methane oxidation by the pMMO.

These findings have important implications for methane oxidation in natural samples. First, they suggest that the pMMO is the predominant enzyme system for methane oxidation in natural populations and thus provide more impetus for understanding this enzyme system. Second, the response of natural populations to changes in methane concentrations will most likely depend on a complex set of parameters, of which available copper concentration may be the key. It is now important to study how methanotrophs utilize copper and how they respond to changes in copper and methane concentrations and to copper speciation, in order to predict how natural populations will respond to environmental perturbations.

Acknowledgment

This work was supported by grants from Advanced Research Projects Administration (ARPA) (N0001492J1901) and the Office of Naval Research (ONR) (N00014-91-J-1899).

References

1. Lidstrom, M. E. In *The Prokaryotes*, 2nd ed.; Balows, A.; Truper, H. S.; Dworkin, M.; Harder, W.; Schleifer, K. H., Eds.; Springer-Verlag: New York, 1991; pp 431–445.
2. Hanson, R. S.; Netrusov, A. I.; Tsuji, K. In *The Prokaryotes*, 2nd ed.; Balows, A.; Truper, H. S.; Dworkin, M.; Harder, W.; Schleifer, K. H., Eds.; Springer-Verlag: New York, 1991; pp 2350–2364.
3. Rudd, J. W. M; Taylor, C. D. *Adv. Aquat. Microbiol.* **1980**, *2*, 77–102.
4. Crutzen, P. *Nature (London)* **1991**, *350*, 380–381.
5. Murrell, J. C. *FEMS Microbiol. Rev.* **1992**, *88*, 233–248.
6. Anthony, C. *Adv. Microbiol. Physiol.* **1986**, *27*, 113–152.
7. Joergensen, L.; Degn, H. *FEMS Microbiol. Lett.* **1983**, *30*, 331–335.
8. Fox, B. G.; Borneman, J. G.; Wackett, L.; Lipscomb, J. D. *Biochemistry* **1990**, *29*, 6419–6427.
9. DiSpirito, A. A.; Gulledge, J.; Shiemke, A. K; Murrell, J. C.; Lidstrom, M. E.; Krema, C. L. *Biodegradation* **1992**, *2*, 151–164.

10. Stanley, S. H.; Prior, S. D.; Leak, D. J.; Dalton, H. *Biotechnol. Lett.* **1983**, 5, 487–492.
11. Chan, S. I.; Nguyen, H. T.; Shiemke, A. K.; Lidstrom, M. E. *Microbial Growth on C1 Compounds;* Intercept: Andover, England, 1993; pp 93–107.

RECEIVED for review October 23, 1992. ACCEPTED revised manuscript May 4, 1993.

Coagulation of Marine Algae

George A. Jackson

Department of Oceanography, Texas A&M University, College Station, TX 77843

> *Coagulation is recognized as an important mechanism for the movement of organic matter from the ocean's euphotic zone to deeper regions. Algal blooms, with their high concentrations of particles, appear to be particularly prone to coagulation and rapid removal from surface waters. Models of algal blooms incorporating coagulation are elucidating significant properties of marine algae. Among the potentially important qualities is the ability of the algae to change their surface properties and thereby affect the stickiness coefficient.*

THE ECOLOGY OF AQUATIC ORGANISMS, especially the smaller ones, is intrinsically about particles and their interactions. Large particles (predators) must collide with and consume smaller ones (bacteria and algae); as a result, the large particles get larger and then divide to form smaller particles or release (excrete) smaller ones. For this reason, coagulation theory has a natural place in describing the flow of material through aquatic ecosystems. Any such description must include the aspects of biology that distinguish living organisms from nonliving particles.

Among the questions that coagulation theory can help answer are how organisms feed on each other and how biological production leaves the euphotic zone. Coagulation theory and the related filtration theory have, in fact, been used to understand the nature of feeding in both planktonic and benthic animals (1 4). This chapter addresses the transport of organic matter downward, out of the euphotic zone.

Phytoplankton algal size depends on the species and the physiological state of the individual. Algal diameters generally range from 0.5 to 100 μm (5). Small organisms, such as algal cells, typically settle at rates of 1–10 m/day (5). At these rates, algal cells could fall for as long as 10 years before they reach the ocean floor 3500 m below. Not only should little of the production arrive at the bottom undegraded, but also any temporal relationship with the seasonal growth pattern in the overlying surface waters should be lost. Organic matter transport by rapidly settling fecal pellets can explain some of the rapid transport of organic matter from the surface, but it cannot explain the rapid accumulation of algal flocs on the North Atlantic seafloor after the spring algal bloom (6–9). Coagulation has been invoked as the process responsible for the collection of small organisms into larger, faster settling particles.

Marine Snow

Aggregates, also known as marine snow, have been observed and studied in surface waters by divers (10–14). Marine snow particles range in size from 2 to more than 70 mm (15) and are variable in composition. They can include diatoms, bacteria, protozoa, fecal pellets, discarded zooplankton feeding structures, and amorphous organic detritus (16). The lower size range is determined by a diver's ability to observe a particle rather than by a cutoff in the aggregate size distribution. Aggregates in the 2.4–75-mm size range settle at 74 ± 39 m/day, more rapidly than solitary algal cells (15). Settling data have been interpreted to show that aggregates have fractal dimensions of 1.4–1.5 (17). This value is significantly lower than the 3 expected for a solid structure and is in a range characteristic of aggregates produced by coagulation processes (18). Marine snow aggregates may be important for more than just their ability to move small algal cells rapidly to the seafloor. They offer localized habitats in the ocean in which the concentration of different organisms is large enough to produce a greater range of ecological interactions than in ocean water. Possibly, they could provide microenvironments sufficiently isolated from solution that nutrient recycling would be efficient enough to allow rapid algal growth rates in otherwise impoverished oceanic waters (19). Conversely, they might be so efficient at extracting nutrients from the marine environment that it would be advantageous for algal cells to be within them (20). Measurements show that oxygen concentrations within aggregates can be substantially different than those in surrounding waters. This fact raises the possibility that aggregates could offer specialized microchemical environments conducive to chemical reactions that might not be favorable in open waters (21).

Formation. Evidence that at least some of the marine snow particles are formed by coagulation has been accumulating. Sedimenting diatoms in a coastal Canadian bay have been observed to change the particle size spectra with depth in a manner consistent with coagulation of the settling particles (22). As already noted, the relatively small fractal dimensions of marine snow particles are consistent with their formation from coagulation processes.

Aggregates can be produced in the laboratory by using material from algal cultures (23) or material collected in the field in locations where marine snow particles exist (24). This material is placed in a reaction vessel designed to enhance particle contact rates. A typical reactor is a rotating cylinder whose axis is parallel to the floor (24).

Algal aggregates have been observed in regions of high primary productivity, including off the coast of California (10, 15), among the ice algae in the Antarctic (25), in the North Sea (26, 27), and in Bedford Basin, Canada (22). Their formation and subsequent sedimentation have also been invoked to explain the disappearance of high-latitude phytoplankton blooms (28–30).

Chemical Nature of Particles. The chemical nature of particle surfaces is an important aspect of the coagulation rate because it determines whether two particles will join to form a new, larger particle after a collision. The tendency of two particles to stick is expressed by using an experimentally determined parameter known as the collision efficiency factor, the sticking efficiency, or the stickiness, and denoted by the symbol α. It is usually thought of as the probability that two colliding particles will stick (31).

Measurements on algal cultures have shown that different species have varying values of α that depend on the species and its nutritional state (32). One algal species, Skeletonema costatum, has an essentially constant value for α of about 0.1 over a range of nutritional conditions. In contrast, another species, Thalassiosira pseudonana, has a small value of α (10^{-4}) when growing rapidly under nutrient-rich conditions and an increased α of about 0.1 when growing slowly under nutrient-depleted conditions. An α value of 0.1 is fairly high and should facilitate coagulation. Skeletonema's range of stickiness values could have a large effect on its coagulation tendencies and possibly lead to biological control over coagulation rates.

Direct observation of the probability of two colliding aggregates sticking has yielded α values for aggregates of between 0.60 and 0.88 (33). Although they indicate that the aggregates have a high "stickiness" quality, these values are not directly comparable to those for the algal cultures because the algal values were determined by the traditional technique of using α to fit data to the disappearance of particles within a shear reactor (32). Any discrepancies between predicted and actual contact rates that might exist because of inadequate theory are compensated for by changes in the calculated value of α.

Model System. Aggregates formed from diatoms after an algal bloom could provide a nice system to study the role of coagulation in marine snow formation. Because an algal bloom is an inherently unbalanced system, zooplankton grazing is an unimportant competitor of coagulation as a source of algal mortality. The high particle concentrations lead to frequent particle collisions and relatively rapid coagulation rates.

This relatively well-defined system facilitates laboratory experiments and theoretical simulations. The presence of marine snow particles composed of diatoms after diatom blooms implies that coagulation can occur rapidly in this system. The relationship between algal sedimentation on the ocean floor and the algal blooms at the surface suggests that this system is important to study if the vertical flux of organic material in the ocean is to be understood.

Kinetic Coagulation Theory

The rate at which two particles with masses m_i and m_j and concentrations n_i and n_j collide is given by $n_i n_j \beta_{ij}$, where β_{ij} is the coagulation kernel (34, 35). New particles of mass $(m_i + m_j)$ are formed at a rate of $\alpha n_i n_j \beta_{ij}$, where α is the stickiness coefficient. If all aggregates are composed of unit particles of the same size, then $m_i = i\, m_1$, and $(m_i + m_j) = m_{i+j} = (i + j)m_1$, where m_1 is the mass of the unit particle. If no new unit particles are produced and there is no nonaggregation process making particles, the change in concentration of particles of size i is the difference between the rate at which the particles are formed by collision of smaller particles and the rate at which they are lost to formation of larger particles.

$$\frac{dn_i}{dt} = 0.5\alpha \sum_{j=1}^{i} \beta_{j,i-j} n_j n_{i-j} - \alpha n_i \sum_{j=1}^{\infty} \beta_{i,j} n_j (1 + \delta_{ij}) \qquad (1)$$

where δ_{ij} is 1 if $i = j$ and is 0 otherwise. To simulate an algal bloom, one can add additional terms to represent algal increase by division and to represent algal loss to sinking.

$$\frac{dn_i}{dt} = \mu_i n_i + 0.5\alpha \sum_{j=1}^{i} \beta_{j,i-j} n_j n_{i-j} - \alpha n_i \sum_{j=1}^{\infty} \beta_{i,j} n_j (i + \delta_{ij}) - \frac{w_i}{z} n_i \qquad (2)$$

where μ_i is the specific growth rate of algae in aggregates with i cells, w_i is the settling rate of an aggregate of size i, and z is the thickness of the mixed layer (36–38). Equation 2 can be modified to incorporate situations in which the source particles are not all of the same size, and the aggregate sizes are continuously distributed rather than having masses that are integer multiples of the unit particles (34, 37).

With this formulation, chemical effects on coagulation are included in α and physical effects in β_{ij}. Particle contacts are usually considered to be caused by three mechanisms: differential sedimentation, shear (laminar and turbulent), and Brownian motion. Differential sedimentation contact occurs when two particles fall through the water at different rates and the faster particle overtakes the slower one. Shear contact occurs when different parts of the fluid environment move at different speeds relative to each other, and thus a particle that is moving with one fluid patch overtakes and collides with a particle in a slower fluid patch. Brownian motion contact occurs when two particles move randomly through their fluid in Brownian motion and collide with each other. Rates for each of these collision mechanisms depend on the sizes of the two colliding particles. The total collision kernel β_{ij} is calculated as the sum of the collision kernels describing each of these processes:

$$\beta_{ij} = \beta_{ds,ij} + \beta_{sh,ij} + \beta_{Br,ij} \tag{3}$$

where the subscripts ds, sh, and Br denote the terms for differential sedimentation, shear, and Brownian motion. The values of $\beta_{Br,ij}$ are usually small enough relative to the other two terms for particles greater than 1 μm in aquatic systems that $\beta_{Br,ij}$ can be neglected (39) and will be neglected here.

Hydrodynamic Models. The coagulation kernels are usually calculated for solid spheres with hydrodynamic models of different sophistication. The simplest calculation uses fluid flow in the absence of any effect of either particle on the flow. This flow level is known as rectilinear flow. The next level of sophistication involves calculating the flow around one particle, usually the larger of the two interacting particles. This level of calculation is known as curvilinear flow. Further levels of sophistication can be obtained by considering the particle trajectories as affected by the interacting flow fields of the particles, as well as any attractive or repulsive forces between them.

The coagulation kernel for the rectilinear case of differential sedimentation is

$$\beta_{ds,ij} = \pi (r_i + r_j)^2 |w_j - w_i| \tag{4}$$

where r_i and r_j are the radii of aggregates of particles of size i and j. The coagulation kernel for the curvilinear case is smaller

$$\beta_{ds,ij} = 0.5 \, \pi r_i^2 |w_i - w_j| \tag{5}$$

where $r_j \gg r_i$ (33).

The coagulation kernel for the rectilinear case of turbulent shear is

$$\beta_{sh,ij} = 1.3 \left(\frac{\epsilon}{\nu}\right)^{0.5} (r_i + r_j)^3 \tag{6}$$

where ϵ is the energy dissipation rate and ν is the kinematic viscosity (40). More complete descriptions of the interaction between two spheres are given elsewhere (41).

Hill (42) derived a curvilinear version of the turbulent shear coagulation kernel.

$$\beta_{sh,ij} = 9.8 \frac{p^2}{(1 + 2p)^2} \left(\frac{\epsilon}{\nu}\right)^{0.5} (r_i + r_j)^3 \tag{7}$$

where $p = r_i/r_j$ and $r_j > r_i$. Again, the effect of the hydrodynamic correction is to reduce the coagulation kernel, although the effect is relatively small for similar-sized particles.

The fractal nature of aggregates has two effects that are important for coagulation modeling: It makes them bigger for a given mass, and it alters the fluid flow through them. Because the particles are bigger, the coagulation kernels describing their interactions change, as do their settling velocities. The effect of this fractal nature was rigorously analyzed for the Brownian coagulation kernel (43, 44) and for the particle fall velocity (45). Adler (46) analyzed the flow fields in and around particles of uniform porosity for shear and differential sedimentation. Valioulis and List (47) used a relationship between average porosity and aggregate size developed by Sutherland and Tan (48) to apply Adler's porosity relationships. It is unclear how different the coagulation kernels are for aggregates with nonuniform porosity.

Biological Model. We developed an algal bloom model that includes nutrient and light limitation on the algal growth (37). Algal specific growth rate, μ, is usually related to nutrient concentration and average light irradiance. We used the relationships relating μ to nitrate concentration and light limitation. Light distribution was influenced by phytoplankton concentration and mixed-layer depth, as well as by solar radiation (49, 50). We incorporated decreases in nitrate concentration caused by algal uptake. Unless noted otherwise, the following calculations use the curvilinear flow model for the differential sedimentation kernel and the rectilinear flow model for the shear coagulation kernel. The aggregates have a fractal dimension of 2.17.

We further developed this model to include zooplankton interactions (38) by including algal losses to zooplankton grazing, zooplankton production of fecal pellets that can interact with other particles, and zooplankton population growth with the model of Fasham et al. (49).

Representative Results for Coagulation of Algal Bloom Models

In the absence of coagulation or grazing, an algal population seeded into an environment with a high nutrient concentration grows exponentially until it depletes the available nutrients or otherwise consumes all of a necessary resource, such as light. When the limiting resource is nitrate, the algae will slow and ultimately stop their growth (Figure 1). In the absence of any phytoplankton losses, the total nitrogen concentration—nitrate plus algal nitrogen—stays constant. The slow loss of algal cells by sinking, however, decreases both the total and the algal nitrogen concentrations.

When algal coagulation is included in this simulation, the formation of aggregates represents a loss of solitary algae that decreases the maximum algal

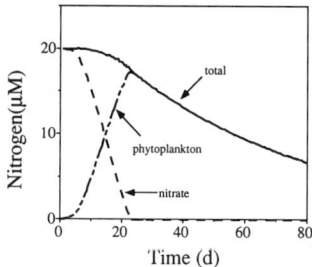

Figure 1. Concentrations of different nitrogen pools for the case in which there is no algal coagulation. Shown are the concentrations of nitrate, phytoplankton (particulate) nitrogen, and total nitrogen (sum of nitrate and particulate concentrations). Decreases in total nitrogen concentration are caused by sedimentation of solitary algal cells. The algae had radii of 10 μm. The mixed-layer depth was 50 m. The shear was 0.1/s. (Reproduced with permission from reference 37. Copyright 1992 American Society of Limnology and Oceanography.)

concentration (Figures 2 and 3). The formation of large, rapidly sinking aggregates increases the vertical flux of material (Figures 4 and 5), increasing the rate of algal disappearance. Most of this enhanced flux occurs in intermediate-sized particles (Figure 4).

Because of the nonlinear relationship between the coagulation rate and particle concentration (e.g., eq 2), coagulation rates increase rapidly as algal populations grow. As a result, aggregate formation occurs over a relatively short time interval (Figure 3). The interaction of the second-order coagulation rate and linear algal growth rate constrain the maximum concentration that an alga can have for a given set of conditions.

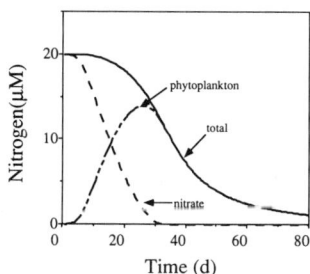

Figure 2. Concentrations of different nitrogen pools for the case in which there is algal coagulation. Shown are the concentrations of nitrate, phytoplankton (particulate) nitrogen, and total nitrogen (sum of nitrate and particulate concentrations). Decrease in total nitrogen concentration is caused by sedimentation of all particles. The peak algal concentration is lower and the total nitrogen loss rate is faster with coagulation than without. (Reproduced with permission from reference 37. Copyright 1992 American Society of Limnology and Oceanography.)

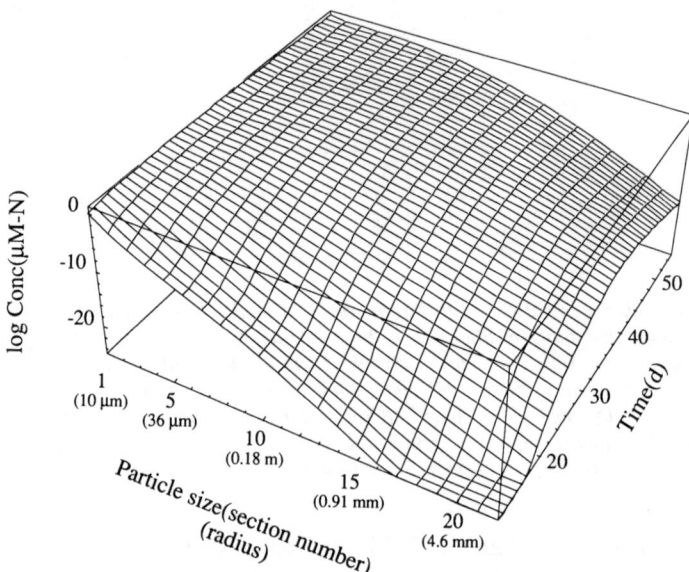

Figure 3. Concentration of nitrogen in the different size classes as a function of time. The maximum particle mass in each size interval is twice that of the smallest particle in the interval. Because of the fractal dimension used to relate mass and length, the maximum radius within a particle size class is 1.38 times the minimum radius. There are 22 size classes in all. Dividing algae compose the smallest size class, with minimum and maximum particle radii of 10 and 13.8 µm. The largest size class has particles with radii ranging from 0.8 to 1.11 cm. The rapidity with which coagulation changes the concentration of the largest particles is evident.

Large numbers of simulations have shown that the dominant effects of coagulation are the limitation that aggregations put on algal populations and the enhanced rates of export of particulate matter that result from the formation of the large aggregates (36, 37).

The importance of coagulation in determining the fate of algal production depends on the relative importance of other factors, including grazer consumption. For a spring bloom, where both the algae and their grazers are initially at low concentrations, whether the grazers or coagulation ultimately controls algal biomass depends on the initial concentration of the herbivores (Figure 6). Similarly, the vertical flux of particles can be controlled by large aggregates produced by aggregation or by fecal pellets produced by zooplankton (Figure 7). Clearly either factor can control the fate of algal cells under appropriate conditions.

Algae are not usually spheres. Among other traits, many of the diatoms have spines that may protrude from the cell for distances much greater than the size of the central cell. Such protrusions can make a cell effectively much larger and increase its coagulation kernels more than the equivalent spherical

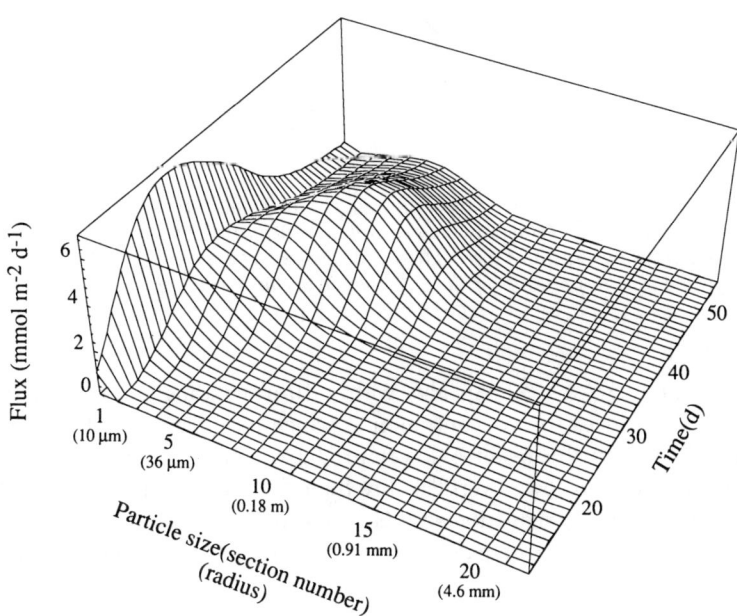

Figure 4. The particle flux spectrum. The vertical axis shows the rate of export of nitrogen by particles in the different sections; the other axes are as in Figure 3. The formation of high concentrations of aggregates causes the major loss to be in the midrange particles.

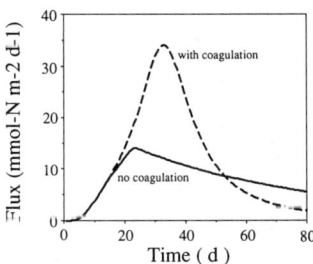

Figure 5. Total flux of particulate nitrogen out of the euphotic zone in the presence and absence of coagulation. Settling of individual cells causes the flux with noncoagulation; settling of large aggregates causes the vertical flux to be greater with coagulation. This figure is for the same cases as Figures 1–4. (Reproduced with permission from reference 37. Copyright 1992 American Society of Limnology and Oceanography.)

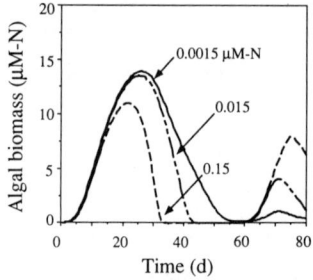

Figure 6. Particulate matter (algae plus aggregates) concentrations through time in the presence of herbivores, whose initial concentrations are different. The initial algal concentrations were 0.015 μM. Peak algal concentration is relatively unaffected by the presence of the two lowest initial herbivore concentrations— 0.0015 and 0.015 μM.

radius would suggest (38). An alga with a 10-μm radius and 50-μm spines dramatically reduces the concentration at which coagulation controls population numbers, and the result is lower production rates (Figure 8). The low production causes a lower vertical flux than for an equivalent alga without spines. In this case, coagulation results in a lower vertical flux. Effects on the fractal structure of an aggregate containing diatoms have not been explored.

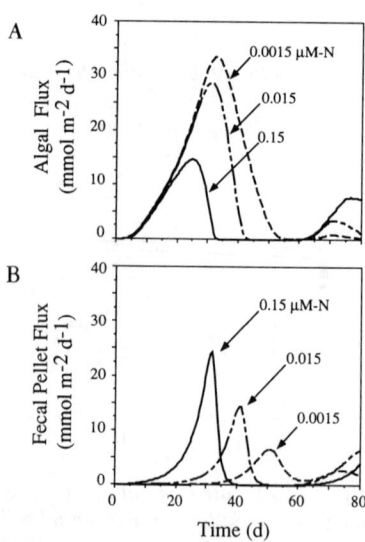

Figure 7. Vertical flux of particulate matter as algal biomass (A) and as fecal pellets (B) through time for different initial concentrations of herbivores. In this case, fecal pellets can add to vertical transport. This flux is recorded for the set of cases shown in Figure 6.

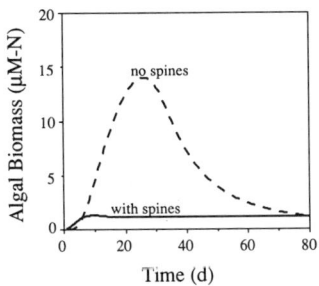

Figure 8. Effect of spines on algal concentrations. Spines increase the effective radius of a particle. Coagulation kernels have been modified by assuming that they make the effective capture radius greater by the length of the spine. There are no grazers. (Reproduced with permission from reference 38. Copyright 1993 Lewis Publishers.)

By changing the value of the stickiness coefficient, α, an alga can change its population dynamics and the rate at which it sediments out of the euphotic zone (Figure 9). The effect is most pronounced if the algal population achieves a much larger population size with a small α than it can with the larger α (Figure 10). Again, the nonlinear nature of the coagulation rate causes aggregation to proceed much faster for the higher initial population.

All of these calculations have been made with the curvilinear model for the differential sedimentation kernel. Flow through an aggregate should increase the kernel. Using the rectilinear kernel is a way to test the effect that such enhanced flow might have (38, 46). As should be expected, it causes a substantial decrease in the maximum algal population. The effect of increased

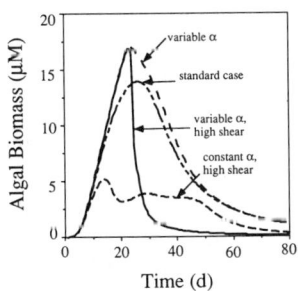

Figure 9. Effect of a change in stickiness coefficient. The standard case used a shear of 0.1/s and a constant stickiness of 0.25. For the variable stickiness case, the value of α increased from its initial value of 10^{-3} to 0.25 when nutrient depletion caused the specific growth rate to drop below 0.05/day. The final value of α was the same as that used in all of the other simulations presented here, 0.25. The high-shear cases used a shear of 1/s. (Reproduced with permission from reference 38. Copyright 1993 Lewis Publishers.)

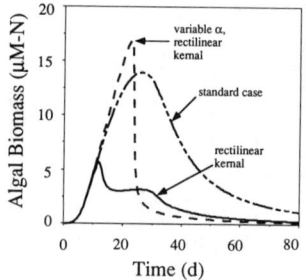

Figure 10. Effect of using a rectilinear coagulation kernel for the differential sedimentation instead of a curvilinear kernel. The standard case used a curvilinear differential sedimentation coagulation kernel (eq 5), a rectilinear shear kernel (eq 6), and a constant stickiness. The rectilinear cases differed by using a rectilinear differential sedimentation flow kernel (eq 4). The variable α case was the same as in Figure 9. Flow through an aggregate would enhance the efficiency of capture from the situation represented by the curvilinear kernel. The rectilinear kernel represents a possible upper bound on the coagulation kernel. The effect of the variable stickiness shown in Figure 9 and the rectilinear kernel is to dramatically increase the rate of coagulation and resulting loss to sedimentation. The rectilinear kernel causes coagulation to dominate at much lower algal concentrations. (Reproduced with permission from reference 38. Copyright 1993 Lewis Publishers.)

coagulation rates is particularly dramatic if it occurs with an alga having a variable α (Figure 10).

Discussion

There is a long history of interest in using coagulation theory to explain the nature of particulate distributions and interactions in the ocean (51–53). Most of this work focused on nonliving particles.

McCave (54) reviewed the expected coagulation rates for various particle-size spectra and concluded that coagulation could be important for determining the particle distribution near the benthic boundary layer, where particle concentrations were high and water motions fairly energetic. His paper was an important introduction for biologists to coagulation concepts, even though he concluded that biological processes would render coagulation processes unimportant for all particles except colloidal ones in the upper ocean layers.

O'Melia and his collaborators studied the role of coagulation in controlling the fate of nutrients such as phosphorus incorporated into algal biomass in freshwater lakes and reservoirs (35, 55, 56). Their models described algal production as being constant rather than variable, as is true in the models discussed here. They concluded that coagulation is rapid enough to determine the fate of particulate phosphorus and that coagulation is an important process in nutrient cycling in lakes.

The evidence is accumulating that coagulation can be an important process in determining the fate of primary production. The following facts argue for the importance of coagulation of algae in marine ecosystems:

1. Aggregates mimicking natural ones can be produced in the laboratory by coagulation with particles in natural seawater (24).
2. They form under the high particle concentrations associated with an algal bloom (26, 27).
3. They are needed to explain the disappearance of algae from surface waters (28, 29) and their rapid appearance in the benthos (6).

These examples of coagulation are marked by high concentrations of algae after a bloom. In many situations, algae in high concentrations do not appear to form large aggregates or cells sinking rapidly at rates characteristic of aggregates (57). Such situations may occur when there is insufficient shear to speed coagulation rates. Far more likely, the algal species may be too small, may have too low a stickiness, or may lack the spines to enhance collision rates. The importance of such factors, which can be species-specific, reinforces the importance of species in determining the properties of planktonic ecosystems. The role that variable stickiness can have is similar to the importance of growth stage for bioflocculation of bacteria in water-treatment systems (58).

The modeling results displayed here emphasized algal–algal interactions. However, other particles in the sea can also participate in the coagulation process. Indeed, Hill (42) argued that they may be necessary to initiate rapid coagulation. Whether or not this is the case, the interaction of aggregates composed predominantly of algae with other particles could enhance their removal as well. Such particles could include colloids (53, 59, 60) and other organisms. Hill (42) also argued that coagulation could account for significant amounts of the particle flux in oligotrophic regions, where particle concentrations are low but particle sedimentation rates are also low.

The modeling of algal coagulation demonstrated that such coagulation would be consistent with oceanic conditions. Effects of such coagulation would include the facts that aggregation could constrain the maximum phytoplankton concentration, that it would enhance transport of material out of the euphotic zone, and that the transition to an aggregation-dominated state could be very rapid.

The modeling also showed the importance of some of the underdeveloped aspects of coagulation theory, particularly with regard to the interaction of the hydrodynamics and the fractal structure of the aggregates on the coagulation kernels. Traditional coagulation theory emphasized understanding the interactions of two solid spheres. Some work has been done on the flow in and around a permeable sphere of uniform porosity and how that would affect the coagulation kernels (46). More such work is needed if we are to understand the formation of marine aggregates adequately.

Acknowledgments

This work was done in conjunction with Steve Lochmann, with the help of Jeff Haney. It was supported by Office of Naval Research Contract N0001487–K0005 and U.S. Department of Energy Grant DE–FG05–85–ER60341. The work was inspired by the work of Werner Stumm, Charles O'Melia, and James Morgan.

References

1. Koehl, M. A. R. *J. Theor. Biol.* **1983**, *105*, 1–11.
2. Fenchel, T. In *Flows of Energy and Materials in Marine Ecosystems;* Fasham, M. J. R., Ed.; Plenum: New York, 1984; pp 301–315.
3. Monger, B. C.; Landry, M. R. *Mar. Ecol. Prog. Ser.* **1990**, *65*, 123–140.
4. Shimeta, J.; Jumars, P. A. *Oceanogr. Mar. Biol.* **1991**, *29*, 191–257.
5. Smayda, T. R. *Oceanogr. Mar. Biol. Ann. Rev.* **1970**, *8*, 353–414.
6. Billet, D. S. M.; Lampitt, R. S.; Rice, A. L.; Mantoura, R. F. C. *Nature (London)* **1983**, *302*, 520–522.
7. Lampitt, R. S. *Deep-Sea Res.* **1985**, *32*, 885–897.
8. Lochte K.; Turley, C. M. *Nature (London)* **1988**, *333*, 67–69.
9. Thiel, H.; Pfannkuche, O.; Schriever, G.; Lochte, K.; Gooday, A. J.; Hemleben, V.; Mantoura, R. F. G.; Turley, C. M.; Patching, J. W.; Riemann, F. *Biol. Oceanogr.* **1990**, *6*, 203–239.
10. Silver, M. W.; Shanks, A. L.; Trent, J. D. *Science (Washington, DC)* **1978**, *201*, 371–373.
11. Trent, J. D.; Shanks, A. L.; Silver, M. W. *Limnol. Oceanogr.* **1978**, *23*, 626–635.
12. Shanks, A. L.; Trent, J. D. *Limnol. Oceanogr.* **1979**, *24*, 850–854.
13. Shanks, A. L.; Trent, J. D. *Deep-Sea Res.* **1980**, *27*, 137–144.
14. Alldredge, A. L. *Limnol. Oceanogr.* **1979**, *31*, 68–78.
15. Alldredge, A. L.; Gotschalk, C. C. *Deep-Sea Res.* **1988**, *36*, 159–171.
16. Alldredge, A. L.; Silver, M. W. *Prog. Oceanogr.* **1988**, *20*, 41–82.
17. Logan, B. E.; Wilkinson, D. B. *Limnol. Oceanogr.* **1990**, *35*, 130–136.
18. Jiang, Q.; Logan, B. E. *Environ. Sci. Technol.* **1991**, *25*, 2031–2038.
19. Goldman, J. C. *Bull. Mar. Sci.* **1984**, *35*, 462–476.
20. Logan, B. E.; Hunt, J. R. *Limnol. Oceanogr.* **1987**, *32*, 1034–1048.
21. Alldredge, A. L.; Cohen, Y. *Science (Washington, DC)* **1987**, *235*, 689–691.
22. Kranck, K.; Milligan, T. *Mar. Ecol. Prog. Ser.* **1988**, *44*, 183–189.
23. Kranck, K.; Milligan, T. *Mar. Ecol. Prog. Ser.* **1980**, *3*, 19–24.
24. Shanks, A. L.; Edmondson, E. W. *Mar. Biol.* **1989**, *101*, 463–470.
25. Riebesell, U.; Schloss, I.; Smetacek, V. *Polar Biol.* **1991**, *11*, 239–248.
26. Riebesell, U. *Mar. Ecol. Prog. Ser. a* **1991**, *69*, 273–280.
27. Riebesell, U. *Mar. Ecol. Prog. Ser. b* **1991**, *69*, 281–291.
28. Wassmann, P.; Vernet, M.; Mitchell, B. G.; Rey, F. *Mar. Ecol. Prog. Ser.* **1990**, *66*, 183–195.

29. Smith, W. O.; Codispoti, L. A.; Nelson, D. M.; Manley, T.; Buskey, E. J.; Niebauer, H. J.; Cota, G. F. *Nature (Washington, DC)* **1991**, *352*, 514–516.
30. Cadée, G. C. *Mar. Ecol. Prog. Ser.* **1985**, *24*, 193–196.
31. O'Melia, C. R. In *Aquatic Surface Chemistry;* Stumm, W., Ed.; Wiley Interscience: New York, 1987; pp 385–403.
32. Kiørboe, T.; Andersen, K. P.; Dam, H. G. *Mar. Biol.* **1990**, *107*, 235–245.
33. Alldredge, A. L.; McGillivary, P. *Deep-Sea Res.* **1991**, *38*, 431–443.
34. Pruppacher, H. R.; Klett, J. D. *Microphysics of Clouds and Precipitation;* D. Reidel: Dordrecht, The Netherlands, 1980.
35. O'Melia, C. R. *Schweiz. Z. Hydrol* **1972**, *34*, 1–33.
36. Jackson, G. A. *Deep-Sea Res.* **1990**, *37*, 1197–1211.
37. Jackson, G. A.; Lochmann, S. *Limnol. Oceanogr.* **1992**, *37*, 77–89.
38. Jackson, G. A.; Lochmann, S. E. In *Environmental Particles;* van Leeuwen, H. P.; Buffle, J., Eds.; Lewis Publishers: Chelsea, MI, 1993; Vol. 2, pp 387–414.
39. McCave, I. N. *Deep-Sea Res.* **1984**, *31*, 329–352.
40. Han, M.; Lawler, D. *J. Hydraul. Eng.* **1991**, *117*, 1269–1289.
41. Saffman, P. G.; Turner, J. S. *J. Fluid Mech.* **1956**, *1*, 16–30.
42. Hill, P. S. *J. Geophys. Res.* **1992**, *97*, 2295–2308.
43. Sutherland, D. N. *J. Colloid Interface Sci.* **1967**, *25*, 373–380.
44. Feder, J.; Jøssang, T. In *Scaling Phenomena in Disordered Systems;* Pynn, R.; Skjeltorp, A., Eds.; Plenum: New York, 1985; pp 99–149.
45. Rogak, S. N.; Flagan, R. C. *J. Colloid Interface Sci.* **1990**, *134*, 206–218.
46. Adler, P. M. *J. Colloid Interface Sci.* **1981**, *81*, 531–535.
47. Valioulis, I. A.; List, E. J. *Environ. Sci. Technol.* **1984**, *18*, 242–247.
48. Sutherland, D. N.; Tan, C. T. *Chem. Eng. Sci.* **1970**, *25*, 1948–1950.
49. Fasham, M. J. R.; Ducklow, H. W.; McKelvie, S. M. *J. Mar. Res.* **1990**, *48*, 591–639.
50. Evans, G. T.; Parslow, J. S. *Biol. Oceanogr.* **1985**, *3*, 327–347.
51. Hahn, H. N.; Stumm, W. *Am. J. Sci.* **1970**, *268*, 354–368.
52. Hunt, J. R. *Environ. Sci. Technol.* **1982**, *16*, 303–309.
53. Honeyman, B. D.; Santschi, P. H. *J. Mar. Res.* **1989**, *47*, 951–992.
54. McCave, I. N. *Deep-Sea Res.* **1984**, *31*, 329–352.
55. O'Melia, C. R.; Bowman, K. S. *Schweiz. Z. Hydrol.* **1984**, *46*, 64–85.
56. Weilenmann, U.; O'Melia, C. R.; Stumm, W. *Limnol. Oceanogr.* **1989**, *34*, 1–18.
57. Laws, E. A.; Bienfang, P. K.; Ziemann, D. A.; Conquest, L. D. *Limnol. Oceanogr.* **1988**, *33*, 57–65.
58. Busch, P. L.; Stumm, W. *Environ. Sci. Technol.* **1968**, *2*, 49–53.
59. Wells, M. L.; Goldberg, E. D. *Mar. Chem.* **1993**, *41*, 353–358.
60. Wells, M. L.; Goldberg, E. D. *Limnol. Oceanogr.* **1994**, *39*, 286–302.

RECEIVED for review October 23, 1992. ACCEPTED revised manuscript March 30, 1993.

11

Diversity of Anaerobes and Their Biodegradative Capacities

L. Y. Young[1,2] and M. M. Häggblom[1]

[1]Center for Agricultural Molecular Biology and [2]Department of Environmental Sciences, Cook College, Rutgers, The State University of New Jersey, New Brunswick, NJ 08903

> *Anaerobic biodegradation processes are becoming attractive alternatives for remediation of toxic contaminants in subsurface environments where oxygen may be limiting and aeration may be expensive. This chapter discusses biodegradation of chloroaromatic compounds such as chlorophenols and chlorobenzoates with enrichment cultures of anaerobes derived from Hudson River and East River sediment. On the basis of the stoichiometric equation describing the complete conversion of the chloroaromatic compound under methanogenic or sulfidogenic conditions, measured values for substrate utilization, electron-acceptor depletion, and product formation are compared to calculated values. Close agreement between measured and calculated values indicates that the substrate is completely utilized. Similar data are shown for the degradation of toluene under denitrifying conditions. Because loss of parent compound may indicate only a biotransformation and not necessarily biodegradation of the parent structure, this approach provides evidence for complete catabolism of the contaminant.*

T HE *EXXON VALDEZ* RAN AGROUND in Prince William Sound, Alaska, in March 1989 and released 10 million gallons of oil into the environment. The bioremediation efforts used as part of the cleanup generated interest in and acceptance of biological processes as a means of cleaning hazardous waste from the environment. Microbially mediated degradation is one of the few means by which toxic organic compounds can be rendered into their harmless components. Most physical and chemical methods change only the compartment in which the compound exists.

Biodegradation and bioremediation are not new concepts. Processes for waste treatment and composting have long relied on microorganisms for removal of unwanted organic matter. Now there is renewed attention to using microbial processes for removal of toxic organic compounds in environmental cleanup. The advantages of using biodegradative methods include reduced cost compared to other processes, conversion of toxic organic substances to nontoxic end products, compatibility with the environment, and low technology input. The disadvantages include, in some cases, slower rates of removal, incomplete conversion, the requirement for threshold concentrations, and incomplete removal of contaminants (also common in other processes).

Although aerobic microbial processes have been well investigated and widely used in biological treatment and remediation, anaerobic processes are becoming more attractive. Unlike most aerobic systems, anaerobes can mediate the dehalogenation of toxic alkanes, alkenes, and aromatics, including highly chlorinated polychlorinated biphenyls (PCBs). In addition, if oxygen is not required, mass-transfer limitations of oxygen are not a restriction. This aspect can be an advantage for unit processes as well as for treatment of contaminated sediment or soil environments where anaerobic habitats may predominate and aeration would be either difficult or expensive.

Investigations over the past few years illustrate the distribution of detoxifying and biodegradative abilities among widely differing groups of anaerobic microorganisms such as methanogenic consortia, sulfate reducers, denitrifiers, and iron and manganese reducers (1–3). The extent of these capabilities is under active investigation by a number of laboratories, yet thus far there has been limited application in treatment.

In this chapter we discuss some of our recent studies of several different groups of anaerobes and their biodegradative capabilities. The approach we take helps to distinguish between biodegradation and biotransformation of a compound. This distinction is important because biodegradation indicates complete mineralization and, hence, detoxification of a toxic compound. In contrast, biotransformation is only a partial metabolism of a compound in which one or a few biochemical reactions have taken place with the major structure of the original compound still intact. In some cases, biotransformation can produce metabolites with similar or even greater toxicity.

Background

In the absence of oxygen, microorganisms are the only group of organisms with the capability of using other inorganic electron acceptors for respiration and oxidation–reduction reactions. This capability is distributed among a number of diverse genera. These organisms are physiologically and phylogenetically distinct, and relatively little is known about their biodegradative abilities.

The environmentally significant and most extensively studied electron acceptors used, along with the reactions that can be mediated by microbes, are illustrated in equations 1–8.

$$O_2 + 4H^+ + 4e^- \longrightarrow 2H_2O \qquad (1)$$

$$2NO_3^- + 12H^+ + 10e^- \longrightarrow N_2 + 6H_2O \qquad (2)$$

$$NO_3^- + 10H^+ + 8e^- \longrightarrow NH_4^+ + 3H_2O \qquad (3)$$

$$MnO_2 + 4H^+ + 2e^- \longrightarrow Mn^{2+} + 2H_2O \qquad (4)$$

$$Fe^{3+} + e^- \longrightarrow Fe^{2+} \qquad (5)$$

$$SO_4^{2-} + 10H^! + 8e^- \longrightarrow H_2S + 4H_2O \qquad (6)$$

$$HCO_3^- + 9H^+ + 8e^- \longrightarrow CH_4 + 3H_2O \qquad (7)$$

$$2HCO_3^- + 10H^+ + 8e^- \longrightarrow CH_3COOH + 4H_2O \qquad (8)$$

These equations are listed in order, from those providing the most to the least energy. However, energy yield is not indicative of microbial preference because the biochemistry of electron transport and energy generation is genetically and not thermodynamically dictated. Many genera of anaerobes are strictly limited to a single electron acceptor, though some can use more than one.

Thus, depending on the environmental conditions and the electron acceptors available, the microbial oxidation of acetate, for example,

$$C_2H_4O_2 + 2H_2O \longrightarrow 2CO_2 + 8H^+ + 8e^- \qquad (9)$$

can be coupled to any of reduction reactions 1–8. The ability to use the different inorganic electron acceptors, however, is species-specific. For example, methanogens that mediate reaction 5 are strict anaerobes that do not use any of the other listed electron acceptors. Yet denitrification (reaction 2) and dissimilatory nitrate reduction (reaction 3) are mediated by facultative organisms that prefer oxygen if it is available. Sulfidogens (reaction 4) are also strict anaerobes, although some species are able to reduce nitrate to ammonium in the absence of sulfate (4). In the environment, in fact, carbon turnover mediated by sulfidogenesis was reported (5) to be 12 times higher than that mediated by oxygen in a coastal marsh. Methanogenesis can account for up to one-third of the annual primary productivity in a freshwater lake (6). It also has been estimated that up to two-thirds of the CO_2 production in rice paddy soils depends on Fe(III) reduction to Fe(II) (7). Hence, the anaerobic contribution to degradation of organic material and carbon cycling in the environment is not insignificant.

Although it is not yet certain whether anaerobes from these different groups can be used efficaciously for treatment purposes, work by various investigators in recent years reveals a range of novel anaerobic biodegradative processes. These reports serve to stimulate further work to understand these novel processes from a fundamental standpoint (biochemistry and genetics), as well as with an eye on their potential usefulness in waste treatment.

Chloroaromatic Compounds

Anaerobes are no longer considered less versatile than aerobes with regard to their biodegradative abilities. It is now widely accepted that fission and metabolism of the aromatic ring takes place in the absence of oxygen (8, 9) and is observed in methanogenic consortia (10, 11), sulfidogenic mixed and pure cultures (12–14), denitrifying mixed and pure cultures (15–17), and pure cultures of photosynthetic bacteria (18, 19). Furthermore reductive dehalogenation of aromatic compounds such as chlorobenzoates, chlorophenols, chlorobenzenes, and PCBs can occur under methanogenic conditions (1, 20–32). Dechlorination for some compounds precedes ring metabolism (e.g., chlorophenol), whereas for others the metabolism goes no further than the dechlorination step (e.g., PCBs). The degradation of chloroaromatic compounds under reducing conditions other than methanogenic has not been as extensively explored.

Anaerobic Metabolism. To examine the extent of anaerobic chloroaromatic metabolism, we undertook a study in which sediments from the upper Hudson River, the lower Hudson River, and the East River were used as inoculum (33, 34). Each monochlorophenol isomer (2-, 3-, and 4-chlorophenol, CP) and each monochlorobenzoate isomer (2-, 3-, and 4-chlorobenzoate, CB) was used as substrate. Duplicate or triplicate cultures were established under three anaerobic conditions: denitrifying, sulfidogenic, and methanogenic. The initial concentration of each of the chloroaromatic compounds was 0.1 mM; incubation was at 30 °C in the dark. Substrates were quantified by high-pressure liquid chromatography; N_2 and CH_4 were analyzed by gas chromatography; nitrate and sulfate were determined by colorimetric methods or by ion chromatography (33, 34).

Biodegradability. The biodegradability of the chloroaromatic compounds varied with the compound and the reducing conditions. Complete utilization of the parent compound took less than 30 days (e.g., 2-CP, methanogenic conditions, upper Hudson River, Figure 1) to more than 200 days (e.g., 2-CP, sulfidogenic conditions, East River, Figure 2). When the same substrate was re-amended to the culture, loss of parent compound occurred in less than 10 days for the methanogenic cultures and in less than 15 days for the sulfidogenic cultures, a result indicating acclimation of the microbial community to the substrate. No loss of chlorophenols or chlorobenzoates occurred in the autoclaved sterile controls.

Although suggestive, loss of the added parent compound does not necessarily indicate that biodegradation has taken place. Incomplete metabolism

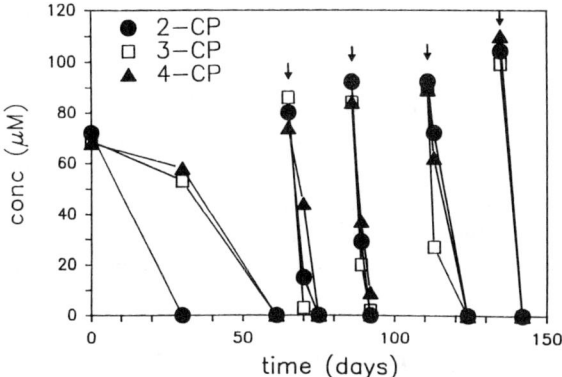

Figure 1. Depletion of 2-, 3-, and 4-chlorophenol isomers established in separate enrichment cultures under methanogenic conditions. Inoculum was obtained from the upper Hudson River, near Albany, NY. Data are average of duplicate cultures. Re-addition of substrate is indicated by arrows. (Adapted with permission from reference 34. Copyright 1993 American Society for Microbiology.)

and formation of metabolites may occur; in some cases, the metabolite may be more problematic than the original compound. For example, when pentachlorophenol (PCP) is metabolized aerobically by *Phanerochetes*, an O-methylation of the hydroxyl group takes place, producing pentachloroanisol, a compound that is recalcitrant and prone to bioaccumulate (35). A different but well-recognized anaerobic example is the dehalogenation of trichloroeth-

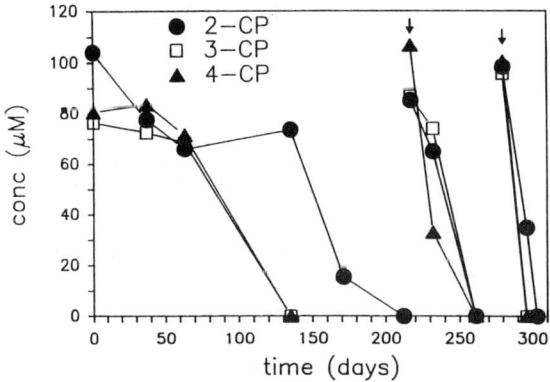

Figure 2. Depletion of 2-, 3-, and 4-chlorophenol isomers under sulfate-reducing conditions. Separate cultures in duplicate were established for each isomer. Inoculum was obtained from the East River bordering Manhattan Island, New York, NY. Re-addition of substrate is indicated by arrows. (Adapted with permission from reference 33. Copyright 1990 American Society for Microbiology.)

ylene to vinyl chloride, whereby a toxic solvent is biotransformed to a documented carcinogen (36).

Degradation to Inorganic End Products. Generally, we are able to ascertain if complete degradation to inorganic end products coupled to methanogenesis, sulfidogenesis, or denitrification has taken place by determining the stoichiometry of the complete degradation and then comparing the calculated with the measured values. Chlorophenol degradation and ring fission to methane and carbon dioxide under methanogenic conditions can be described with equation 10.

$$C_6H_5OCl + 4.5H_2O \longrightarrow 3.25CH_4 + 2.75CO_2 + H^+ + Cl^- \qquad (10)$$

Table I summarizes the results obtained for 2-, 3-, and 4-CP degradation in the methanogenic cultures from the upper Hudson River based upon the stoichiometry in equation 10. The methane production in the background controls resulting from endogenous carbon in the sediment inoculum has been subtracted from these values. The results show that the measured values of methane produced agree very well with those calculated from the equation. Thus, they support the observation that each of the monochlorophenol isomers was mineralized to CH_4 and CO_2.

Degradation with Sulfate Reduction. Equation 11 describes CP degradation coupled to sulfate reduction.

$$C_6H_5OCl + 3.25SO_4^{2-} + 4H_2O \longrightarrow$$
$$6HCO_3^- + 3.25H_2S + 0.5H^+ + Cl^- \qquad (11)$$

Table I. Methane Formation from Monochlorophenol Degradation by Methanogenic Hudson River Enrichments

CP	CP Metabolized (μmol)	CH_4 Formation (μmol)		Percent of Expected
		Calculated	Measured	
2-CP	68.0	221	234	106
3-CP	65.8	215	256	120
4-CP	65.6	214	208	98

NOTE: Chlorophenol and methane concentrations are the cumulative levels attained after repeated additions of substrate were made to, then utilized by, the culture. Methane levels in background cultures with no added chlorophenol (16 μmol) were subtracted from the values reported.
SOURCE: Reproduced with permission from reference 34. Copyright 1993 American Society for Microbiology.

Table II summarizes the results for degradation of the CP isomers in East River cultures under sulfate-reducing conditions based on the stoichiometry in equation 11. Sulfate loss in the background controls were subtracted from the cultures to which CPs were added. As noted in Table II, the measured sulfate depletion corresponded to that calculated and provided evidence that CP metabolism was coupled to sulfate reduction. In these studies sulfate reduction is supported by two additional experimental observations. First, molybdate, which is a specific inhibitor of microbial sulfate reduction, was documented to stop the CP degradation. Active controls that did not receive molybdate continued to degrade CP. Second, radiolabeled $^{35}SO_4^{2-}$ formed $^{35}S^{2-}$ in active cultures and not in control cultures (33).

Altered Aromatic Metabolites. Loss of parent compound is insufficient evidence to support the occurrence of biodegradation. Frequently, the disappearance of the parent aromatic compound results in the formation of an altered aromatic metabolite. This product indicates that, although biotransformation has taken place, biodegradation has not. An example is the anaerobic O-demethylation of chlorinated guaiacols to chlorocatechols mediated by the acetogenic bacteria, *Acetobacterium woodii* and *Eubacterium limosum* (37).

A different example is shown in Figure 3, which illustrates data from a methanogenic enrichment originating from a pond sediment (38). If only the 2,6-dichlorophenol (DCP) parent compound were being monitored, then one would conclude that loss occurs in less than 1 week. Its metabolism, however, occurs through a series of sequential dechlorination steps with formation of stoichiometric amounts of phenol. At this point, degradation of the organic carbon structure has not taken place. Metabolism of the phenol metabolite subsequently occurred, though its initial degradation rate was significantly slower than the dechlorination and initial biotransformation rate of the parent compound, 2,6-DCP. After the second addition of 2,6 DCP, the transient ap-

Table II. Sulfate Depletion from Monochlorophenol Degradation by Sulfidogenic East River Enrichments

CP	CP Metabolized (mM)	SO_4^{2-} Depletion (mM)		Percent of Expected
		Calculated	Measured	
2-CP	0.86	2.8	2.5	89
3-CP	0.82	2.7	2.4	89
4-CP	0.85	2.8	3.6	129

NOTE: Depletion of sulfate was measured after repeated additions of chlorophenol had been utilized. Sulfate loss in background cultures not receiving chlorophenols (1.5 mM) were subtracted from the values reported.
SOURCE: Reproduced with permission from reference 33. Copyright 1990 American Society for Microbiology.

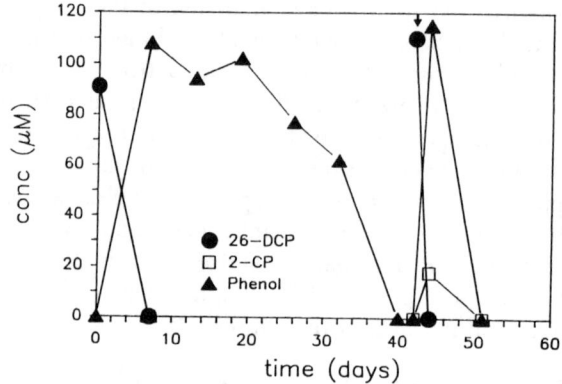

Figure 3. Metabolism of 2,6-dichlorophenol under methanogenic conditions. Duplicate cultures were established under methanogenic conditions with pond water. Re-addition of substrate is indicated by the arrow. (Adapted from 38).

pearance of a low level of 2-CP can be seen, as well as the stoichiometric formation of phenol again. With acclimation, however, phenol was degraded much more rapidly.

2,3,6-Trichlorophenol was also used as a substrate in this study. It was sequentially dechlorinated to 3-CP, which was not further degraded and accumulated in the cultures. It provides another example of biotransformation but not biodegradation of a compound.

Alkylbenzenes

The aromatic hydrocarbons—benzene, toluene, o-, m-, p-xylene, and ethylxylene (BTXs)—are relatively water-soluble gasoline components. These components can leak from underground gasoline storage tanks and, hence, are a hazard to groundwater supplies. Until recently, they were considered to be recalcitrant to biodegradation under anaerobic conditions. Aerobically, on the contrary, a great deal is known about the biochemistry and the molecular biology of the biodegradative pathways (39).

Given the low concentrations of oxygen in the subsurface environment, along with its poor solubility and poor mass-transfer efficiency, anaerobic treatment is an attractive alternative, provided that it is biologically possible. Studies by other researchers and in our own laboratory indicate that nitrate is an electron acceptor that can satisfactorily substitute for oxygen in microbially mediated processes and allow the degradation of a number of components of BTX (40–43).

Anaerobic BTX Biodegradation. Our investigation on anaerobic BTX biodegradation was initiated under denitrifying conditions with seven sediment and soil samples from a variety of contaminated and uncontaminated sources. Benzene, toluene, and the three isomers of xylene were initially added together at concentrations of 0.1 mM each as the sole carbon and energy source. Oxygen was removed from the headspace and replaced with argon, nitrate was the only available electron acceptor provided, and the vessels were sealed with butyl rubber stoppers lined with Teflon. N_2, N_2O, and CO_2 were measured by gas chromatography. Nitrate and nitrite were determined colorimetrically, and BTXs were measured by gas chromatography (44).

Over the initial incubation period, for all the sources of sediment and soil, toluene was the only alkylbenzene in the mixture that was completely metabolized. Some abiotic loss of benzene and xylenes occurred gradually over the 120-day incubation period. Utilization and degradation, however, was noted only for toluene. No loss of toluene was noted in sterile control cultures. Partial loss, but not biodegradation, of o-xylene was observed, whereas no activity was noted on any of the other compounds (Figure 4).

Whether toluene metabolism was coupled to nitrate reduction was determined in several ways. Activity on toluene by the enrichment culture could be transferred and sustained in the transferred cultures. As shown in Table III, active enrichment cultures were provided with substrate toluene, with nitrate or with both. In the absence of the electron-acceptor nitrate with

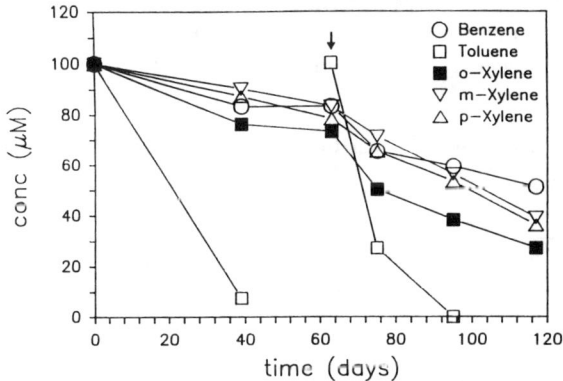

Figure 4. Depletion of benzene, toluene, and p-, m-, and o-xylenes under denitrifying conditions. All five alkylbenzenes were added together to triplicate enrichment cultures established under denitrifying conditions. Re-addition of toluene is indicated by the arrow. Seven different inocula were used. These results, typical of what was observed, were obtained with an inoculum from a digester at a domestic sewage treatment plant that receives industrial waste. (Adapted with permission from reference 44. Copyright 1993 American Society for Microbiology.)

Table III. Dependence of Toluene Loss and Growth on Nitrate by Mixed Enrichment Cultures

Initial Toluene (μM)	Initial KNO_3 (mM)	Growth (Turbidity)	Final Toluene (μM)	N_2 (μM)
300	0	−	320	12
0	2	−	n.a.[a]	12
330	2	+	0	54

NOTE: Toluene, dinitrogen gas, and turbidity were measured after 10 days of incubation. No activity was observed in sterile autoclaved controls.
[a] n.a. means not applicable.
SOURCE: Reproduced with permission from reference 44. Copyright 1991 American Society for Microbiology.

toluene present, no growth occurred, no denitrification took place, and only background levels of nitrogen were observed. Similarly, in the absence of toluene but with nitrate present, neither growth nor denitrification occurred. Only in the presence of both nitrate and toluene was activity observed. Furthermore, the mass balances were determined for carbon, nitrogen, and the electrons transferred in the mixed-culture enrichments (44). Using these mixed enrichments as a source, subsequent work yielded the isolation of a pure culture, designated strain T1, that uses toluene under denitrifying conditions (45).

Mass-Balance Determinations. Table IV illustrates the mass-balance determinations for carbon, nitrogen, and electrons for strain T1. Degradation is evidenced by the carbon transformation to CO_2 and cells. In ad-

Table IV. Mass Balances for Carbon, Nitrogen, and Electrons for Strain T1

Item	Source	Sink	Amount	Difference (%)
Carbon	toluene (25 mg C)	CO_2 Cells	12.80 (mg C) 7.25 20.10 total	−20
Nitrogen	NO_3^- (1.54 mmol N)	NO_2^- N_2O N_2	0.326 (mmol N) 0.363 0.917 1.610 total	4.5
Electron	toluene → CO_2 + cells (6.64 mmol e^-)	$NO_3^- \rightarrow NO_2^-$ $NO_3^- \rightarrow N_2O$ $NO_3^- \rightarrow N_2$	0.653 (mmol e^-) 1.450 4.580 6.680 total	0.60

SOURCE: Reproduced with permission from reference 43. Copyright 1992 American Society for Microbiology.

dition, later studies determined that the remainder of the carbon can be accounted for as a transformation product that did not undergo further biodegradation (43), thus closing the carbon balance. Nitrate reduction and denitrification is further confirmed by detection of the reduced nitrogen species, nitrite, nitrous oxide, and dinitrogen gas; these products agree closely with the nitrogen added as nitrate. Strain T1 is unable to use any other alkylbenzene as a carbon or energy source, although it can use other nonalkylbenzene aromatic substrates for growth. Although o-xylene does not support growth, it is transformed to a dead-end product. This transformation did not occur for any of the other alkylbenzenes.

The specific rate of toluene degradation at 200 μM was determined to be 56 nmol/min per milligram of protein. The isolated strain, designated T1, can grow in up to 3 mM toluene, which represents approximately half of its saturation concentration. Its physiological characteristics appear to be different from the other toluene-degrading strains reported in the literature (40, 42, 46, 47). A pathway for anaerobic toluene degradation proposed for strain T1 (43, 48) proceeds through phenylpropionyl–coenzyme A (CoA) as an intermediate. This route is different from other proposed pathways in which methyl group oxidation and hydroxylation in the *para* position has been suggested (42, 46, 47, 49).

The initial mixed-culture enrichments also yielded subcultures that degraded *m*-xylene. Determination of the mass balances for *m*-xylene, as carried out for toluene, confirmed that *m*-xylene is also completely degraded to CO_2 and cells under denitrifying conditions (44).

Summary and Conclusion

The ability to dehalogenate and to degrade aromatic and aliphatic compounds has been reported (31, 32) for sulfidogenic mixed cultures, and was also observed in denitrifying enrichments. From our recent studies reported herein, the different substrate specificities for degradation of monochlorophenol and monochlorobenzoate isomers under each reducing condition suggest that distinct populations are enriched for and responsible for the metabolic patterns.

Although it is generally believed that under anaerobic conditions dechlorination, either reductive or hydrolytic, precedes ring fission, this pathway has been clearly documented only for the methanogenic studies. The physical proof and transient appearance of each dechlorinated metabolite (e.g., 2,3,6-trichlorophenol → 2,3-dichlorophenol → 3-monochlorophenol), which is seen in methanogenic cultures, has not been observed for the sulfidogenic or denitrifying cultures. It is still unclear whether this indicates that there is indeed a different mechanistic pathway or whether the rates of metabolite transformation are so rapid that they preclude their detection.

The ability to metabolize and degrade the ring structure of the alkylbenzenes in the absence of oxygen has been documented under methanogenic and denitrifying conditions. BTX degradation under sulfidogenic conditions was also reported (50–52), and a toluene-utilizing sulfate-reducing bacterium was isolated (53).

Use of nitrate as an alternative electron acceptor for in situ treatment in an anaerobic biorestoration field demonstration project was reported by researchers from the U.S. Environmental Protection Agency (54, 55). A shallow aquifer contaminated with jet fuel spill was the study site in which BTX removal by nitrate addition, along with other essential nutrients, was examined. Significant depletion of benzene, toluene, and m- and p-xylenes (but not o-xylene) were observed. Benzene loss was attributed to oxygen, whereas toluene and xylene losses appeared to be related to the nitrate and nutrient additions (54, 55).

Toxic organic compounds are susceptible to biodegradation under a variety of environmental conditions. When assessing the environmental fate of these compounds, their effects on the environment and ecosystem, their risk to human exposure, and the effectiveness of bioremediation efforts, it is important to determine the extent of metabolism and whether biodegradation is complete. Clearly, the environment contains a wide variety of physiologically and phylogenetically diverse organisms with desirable degradative abilities. Whether they can be harnessed for treatment or cleanup purposes is under active investigation and may provide some attractive possibilities in the near future.

Acknowledgments

This work was supported in part by Grant ES04895 from the National Institute of Environmental Health Sciences, by Grants CR–818694 and R–816383 from the Environmental Protection Agency, by the Hudson River Foundation Grant 014/89A056, and by the Center for Agricultural Molecular Biology, Rutgers University.

References

1. Colberg, P. J. S. *Geomicrobiol. J.* **1990**, *8*, 147–164.
2. Grbic-Galic, D. *Geomicrobiol. J.* **1990**, *8*, 167–200.
3. Lonergan, D. J.; Lovley, D. R. *Microbial Oxidation of Natural and Anthropogenic Aromatic Compounds Coupled to Fe(III) Reduction in Humics and Soils;* Baker, R. A., Ed.; Lewis Publishers: Chelsea, MI, 1991; Vol. 1, pp 327–338.
4. Widdel, F. In *Biology of Anaerobic Microorganisms;* Zehnder, A. J. B., Ed.; John Wiley & Sons: New York, 1988; pp 469–585.
5. Howarth, R. W.; Teal, J. M. *Limnol. Oceanogr.* **1979**, *24*, 999–1013.
6. Strayer, R. F.; Tiedje, J. M. *Appl. Environ. Microbiol.* **1978**, *36*, 330–340.
7. Lovley, D. R. *Geomicrobiol. J.* **1987**, *5*, 375–399.
8. Young, L. Y. In *Microbial Degradation of Organic Compounds;* Gibson, D. T., Ed.; Marcel Dekker: New York, 1984; pp 487–523.
9. Evans, W. C.; Fuchs, G. *Annu. Rev. Microbiol.* **1988**, *42*, 289–317.

10. Ferry, J. G.; Wolfe, R. S. *Arch. Microbiol.* **1976**, *107*, 33–40.
11. Healy, J. B., Jr.; Young, L. Y. *Appl. Environ. Microbiol.* **1979**, *38*, 84–89.
12. Bak, F.; Widdel, F. *Arch. Microbiol.* **1986**, *146*, 177–180.
13. Szewzyk, R.; Pfennig, N. *Arch. Microbiol.* **1987**, *147*, 163–168.
14. Schnell, S.; Bak, F.; Pfennig, N. *Arch. Microbiol.* **1989**, *152*, 556–563.
15. Bakker, G. *FEMS Lett.* **1977**, *1*, 103–108.
16. Bossert, I. D.; Young, L. Y. *Appl. Environ. Microbiol.* **1986**, *52*, 1117–1122.
17. Kuhn, E. P.; Zeyer, J.; Eicher, P.; Schwarzenbach, R. P. *Appl. Environ. Microbiol.* **1988**, *54*, 490-496.
18. Dutton, P. L.; Evans, W. C. *Biochem. J.* **1969**, *113*, 525–536.
19. Harwood, C. S.; Gibson, J. *Appl. Environ. Microbiol.* **1988**, *54*, 712–717.
20. Suflita, J. M.; Horowitz, A.; Shelton, D. R.; Tiedje, J. M. *Science (Washington, DC)* **1982**, *218*, 1115–1117.
21. Boyd, S. A.; Shelton, D. R. *Appl. Environ. Microbiol.* **1984**, *47*, 272–277.
22. Fathepure, B. Z.; Tiedje, J. M.; Boyd, S. A. *Appl. Environ. Microbiol.* **1988**, *54*, 327–330.
23. Kohring, G.-W.; Rogers, J. E.; Wiegel, J. *Appl. Environ. Microbiol.* **1989**, *55*, 348–353.
24. Quensen, J. F., III; Tiedje, J. M.; Boyd, S. A. *Science (Washington, DC)* **1988**, *242*, 752–754.
25. Quensen, J. F., III; Boyd, S. A.; Tiedje, J. M. *Appl. Environ. Microbiol.* **1990**, *56*, 2360–2369.
26. Genthner, B. R. S.; Price, W. A., II; Pritchard, P. H. *Appl. Environ. Microbiol.* **1989**, *55*, 1472–1476.
27. Nies, L.; Vogel, T. M. *Appl. Environ. Microbiol.* **1990**, *56*, 2612–2617.
28. Hale, D. D.; Rogers, J. E.; Wiegel, J. *Microbiol. Ecol.* **1990**, *20*, 185–196.
29. Van Dort, H. M.; Bedard, D. L. *Appl. Environ. Microbiol.* **1991**, *57*, 1576–1578.
30. Alder, A.; Häggblom, M. M.; Oppenheimer, S. R.; Young, L. Y. *Environ. Sci. Technol.* **1993**, *27*, 530–538.
31. Häggblom, M. M. *FEMS Microbiol. Rev.* **1992**, *103*, 29–72.
32. Mohn, W. W.; Tiedje, J. M. *Microbiol. Rev.* **1992**, *56*, 482–507.
33. Häggblom, M. M.; Young, L. Y. *Appl. Environ. Microbiol.* **1990**, *56*, 3255–3260.
34. Häggblom, M. M.; Rivera, M. D.; Young, L. Y. *Appl. Environ. Microbiol.* **1993**, *59*, 1162–1167.
35. Lamar, R. T.; Larsen, M. J.; Kirk, T. K. *Appl. Environ. Microbiol.* **1990**, *56*, 3519–3526.
36. Vogel, T. M.; Criddle, C. S.; McCarty, P. L. *Environ. Sci. Technol.* **1987**, *21*, 722–736.
37. Häggblom, M. M.; Berman, M.; Frazer, A. C.; Young, L. Y. *Biodegradation* **1993**, *4*, 107–114.
38. Häggblom, M. M.; Rivera, M. D.; Young, L. Y. *Environ. Toxicol. Chem.* **1993**, *12*, 1395–1403.
39. *Microbial Degradation of Organic Compounds;* Gibson, D. T., Ed.; Marcel Dekker: New York, 1984.
40. Dolfing, J.; Zeyer, J.; Binder-Eicher, P.; Schwarzenbach, R. P. *Arch. Microbiol.* **1990**, *154*, 336–341.
41. Hutchins, S. R. *Environ. Toxicol. Chem.* **1991**, *10*, 1437–1448.
42. Schocher, R. J.; Seyfried, B.; Vazquez, F.; Zeyer, J. *Arch. Microbiol.* **1991**, *157*, 7–12.
43. Evans, P. J.; Ling, W.; Goldschmidt, B.; Ritter, E. R.; Young, L. Y. *Appl. Environ. Microbiol.* **1992**, *58*, 496–501.

44. Evans, P. J.; Mang, D. T.; Young, L. Y. *Appl. Environ. Microbiol.* **1991**, *57*, 450–454.
45. Evans, P. J.; Mang, D. T.; Kim, K. S.; Young, L. Y. *Appl. Environ. Microbiol.* **1991**, *57*, 1139–1145.
46. Lovley, D. R.; Lonergan, D. J. *Appl. Environ. Microbiol.* **1990**, *56*, 1858–1864.
47. Altenschmidt, U.; Fuchs, G. *Arch. Microbiol.* **1991**, *156*, 152–158.
48. Frazer, A. C.; Ling, W.; Young, L. Y. *Appl. Environ. Microbiol.* **1993**, *59*, 3157–3160.
49. Grbic-Galic, D.; Vogel, T. M. *Appl. Environ. Microbiol.* **1987**, *53*, 254–260.
50. Haag, F.; Reinhard, M.; McCarty, P. L. *Environ. Toxicol. Chem.* **1991**, *10*, 1379–1389.
51. Edwards, E. A.; Grbic-Galic, D. *Appl. Environ. Microbiol.* **1992**, *58*, 2663–2666.
52. Edwards, E. A.; Wills, L. E.; Reinhard, M.; Grbic-Galic, D. *Appl. Environ. Microbiol.* **1992**, *58*, 794–800.
53. Rabus, R.; Nordhous, R.; Ludwig, W.; Widdel, F. *Appl. Environ. Microbiol.* **1993**, *59*, 1444–1451.
54. Hutchins, S. R.; Wilson, J. T. In *In Situ Bioreclamation: Applications and Investigations for Hydrocarbon and Contaminated Site Remediation;* Hinchee, R. E.; Olfenbuttel, R. F., Eds.; Butterworth-Heinemann: Boston, MA, 1991; pp 157–172.
55. Hutchins, S. R.; Sewell, G. W.; Kovacs, D. A.; Smith, G. A. *Environ. Sci. Technol.* **1991**, *25*, 68–76.

RECEIVED for review April 13, 1993. ACCEPTED revised manuscript December 8, 1993.

12

The Chemical Effects of Collapsing Cavitation Bubbles

Mathematical Modeling

Anatassia Kotronarou and Michael R. Hoffmann*

W. M. Keck Laboratories, California Institute of Technology, Pasadena, CA 91125

A comprehensive mechanism is developed for aqueous-phase oxidation of S(–II), where [S(–II)] = [H_2S] + [HS^-] + [S^{2-}], by ·OH radical in the presence of oxygen. The oxidation of S(–II) is initiated by reaction with ·OH, but it is further propagated by a free-radical chain sequence involving O_2. This mechanism can adequately model the observed oxidation of S(–II) in air-saturated aqueous solutions sonicated at 20 kHz and 75 W/cm^2 at pH ≥ 10, assuming a continuous and uniform ·OH input into solution from the imploding cavitation bubbles. At this pH range, practically all S(–II) is present in the form of HS^- and cannot undergo thermal decomposition. Our work suggests that the use of simplified approaches for modeling the liquid-phase sonochemistry of a well-mixed solution may be justified when ·OH radical reactions predominate. For the immersion probe at 20 kHz and 75 W/cm^2, the effective ·OH uniform release into the bulk solution was estimated to be 3.5 µM/min with a corresponding steady-state ·OH concentration of ≤0.1 µM.

THE ACTION OF ULTRASONIC WAVES IN LIQUIDS can induce or accelerate a wide variety of chemical reactions (1, 2) The chemical effects of ultrasound have been explained in terms of reactions occurring inside, at the interface, or at some distance away from cavitating gas bubbles. In the interior of a collapsing cavitation bubble, extreme but transient conditions exist. Temper-

*Corresponding author

atures approaching 5000 K have been determined, and pressures of several hundred atmospheres have been calculated. Temperatures on the order of 2000 K have been determined for the interfacial region surrounding a collapsing bubble.

Sonochemical reactions are characterized by the simultaneous occurrence of pyrolysis and radical reactions, especially at high solute concentrations. Volatile solutes will undergo direct pyrolysis reactions within the gas phase of the collapsing bubbles or within the hot interfacial region. In these interfacial regions, both combustion and free-radical reactions (e.g., involving ˙OH derived from the decomposition of H_2O) are possible. Pyrolysis (i.e., combustion) in the interfacial region is predominant at high solute concentrations, whereas at low solute concentrations free-radical reactions are likely to predominate. In the bulk solution, the chemical reaction pathways are similar to those observed in aqueous radiation chemistry [as induced by aquated electrons (e_{aq}^-), gamma rays (γ), or X-rays]. However, combustion-like reactions may occur at low solute concentrations with nonvolatile surfactants and polymers.

The chemical effects of ultrasonic irradiation are a direct result of acoustic cavitation. Sound waves traveling through water with frequencies greater than 15 kHz force the growth and subsequent collapse of small bubbles of gas in response to the passage of expansion and compression waves. The greatest coupling occurs when the natural resonance frequency of the bubble equals the ultrasonic frequency (e.g., 20 kHz = a bubble diameter of 130 μm). The chemical effects are realized during and immediately after collapse of a vapor-filled cavitation bubble. During bubble collapse, which occurs within 100 ns, H_2O undergoes thermal dissociation to give H˙ atoms and ˙OH radicals.

$$H_2O \xrightarrow{v \geq 15 \text{ kHz}} H^\cdot + {}^\cdot OH \qquad (1)$$

We are motivated, in part, by the potential economical application of electrohydraulic cavitation for the rapid degradation of a wide range of chemical contaminants. Current approaches to the treatment of hazardous chemical wastes include high-temperature incineration; chemical oxidation with O_3, H_2O_2, and UV light; membrane separation; activated carbon adsorption; substrate-specific biodegradation; electron-beam bombardment; supercritical fluid extraction and oxidation; fixed-bed, high-temperature catalytic reactors; and steam gasification. These techniques are not totally effective, they are very often cost-intensive, they are often inconvenient to apply to mixed solid–liquid wastes, and they are not readily adapted to a wide variety of conditions.

The application of electrohydraulic cavitation in its various forms for the pyrolytic and oxidative control of hazardous chemicals in water has the potential to become economically competitive with existing technologies. In terms of convenience and simplicity of operation, electrohydraulic cavitation could prove to be far superior to many of the methods listed. For example, the relative efficiency of ultrasound in terms of the total power consumed per mole of p-nitrophenol degraded per liter of water is far superior to UV–photolytic degradation.

We recently reported (3) that ultrasonic irradiation of alkaline oxic aqueous solutions of bivalent sulfur, S(–II), at 20 kHz resulted in the rapid oxidation of S(–II). The observed distribution of the oxidation products was similar to that reported for γ-radiolysis of S(–II). The ultrasound-induced oxidation of S(–II) in alkaline solutions was attributed to the reaction of HS⁻ with 'OH radicals (3). These radicals form during ultrasonic irradiation of water as a result of the high-temperature decomposition of water vapor inside the hot cavitation bubbles (1, 2). Although the experimental results were qualitatively consistent with our proposed mechanism, some questions remained unanswered concerning the amount of 'OH released into the aqueous phase and the existence and relative importance of additional oxidants.

To address these questions, a comprehensive aqueous-phase mechanism that describes the free-radical chemistry of the S(–II) + 'OH + O_2 system was developed. This mechanism was subsequently used to model the ultrasonic oxidation of S(–II) in alkaline pH.

Experimental Approach

The free-radical chemistry of the sulfur system is complicated and includes a variety of species and reactions. We developed a mechanism that can accurately model the overall behavior of the S(–II) + 'OH + O_2 system with the minimum number of intermediates. Furthermore, we limited our interest to the neutral-to-alkaline pH region where the ultrasonic irradiation of S(–II) experiments were performed.

The chemical species included in the mechanism are as follows:

- Nonradical group species:
  S(–II) = H_2S + HS^-
  S(–I) = H_2S_2 + HS_2^- + S_2^{2-}
  S(IV) = HSO_3^- + SO_3^{2-}
- Nonradical single species: O_2, $S_2O_3^{2-}$, SO_4^{2-}, H_2O_2, NO_2^-, NO_3^-
- Radicals: 'OH, HS', $H_2S_2^{\cdot -}$, HSOH'⁻, $SO_2^{\cdot -}$, '$S_2O_3OH^{2-}$, '$S_4O_6^{3-}$, $O_2^{\cdot -}$, $SO_3^{\cdot -}$, $SO_4^{\cdot -}$, $HS_2^{\cdot}$, $H_2S_2O^{\cdot}$, $NO_2^{\cdot}$

S(–II), S(–I), and S(IV) are "group species", representing the sum of reactant species in rapid acid–base equilibrium. Equilibrium reactions relevant to our system are shown in Table I. Only the first four reactions are actually used in the mathematical model. The pK values for the reactions of the remaining

Table I. Equilibrium Reactions Relevant to the S(–II) System

Equilibrium Reaction	pK	ΔH	Ref.
$H_2O(l) \rightleftharpoons H^+ + OH^-$	14.00	13.34	4
$H_2S \rightleftharpoons H^+ + HS^-$	7.02	−5.30	4
$HSO_3^- \rightleftharpoons H^+ + SO_3^{2-}$	7.18	3.00	4
$HS_2^- \rightleftharpoons H^+ + S_2^{2-}$	9.70		5
$H_2S_2 \rightleftharpoons H^+ + HS_2^-$	5.00		5
$HS^- \rightleftharpoons H^+ + S^{2-}$	17.1	−12.00	4, 6
$SO_2 \cdot H_2O \rightleftharpoons H^+ + HSO_3^-$	1.91	4.00	4
$HSO_4^- \rightleftharpoons H^+ + SO_4^{2-}$	1.99	5.40	4
$HS_2O_3^- \rightleftharpoons H^+ + S_2O_3^{2-}$	1.60		4
$H_2O_2 \rightleftharpoons H^+ + HO_2^-$	11.65	−7.40	4
$HNO_2 \rightleftharpoons H^+ + NO_2^-$	3.15	−2.00	4
$HO_2 \rightleftharpoons H^+ + O_2^{\cdot -}$	4.46		7
$HSO_2^{\cdot} \rightleftharpoons H^+ + SO_2^{\cdot -}$	≤2		7
$HSO_3^{\cdot} \rightleftharpoons H^+ + SO_3^{\cdot -}$	≤2		8
$HSO_4^{\cdot} \rightleftharpoons H^+ + SO_4^{\cdot -}$	≤2		8

part of the table are considerably outside the pH region of interest, thus allowing us to neglect one of the two species involved in these equilibria (e.g., S^{2-} or $SO_2 \cdot H_2O$).

The 47 reactions included in the mathematical modeling of the system are shown in Table II. Broken arrows indicate multistep (i.e., nonelementary) reactions. In the few cases for which no data were available, estimates of the rate constants based on kinetic–thermodynamic considerations were used.

By analogy with the reactions of $^{\cdot}OH$ with H_2S and HS^-, the rate constant for the reaction of $HS^{\cdot}$ with H_2S (R5) is expected to be of the same order of

Table II. Reactions Included in Mathematical Modeling

ID No.	Reaction	k ($M^{-1} s^{-1}$)	Ref.
R1	$^{\cdot}OH + ^{\cdot}OH \rightarrow H_2O_2$	5.5×10^9	9
R2	$H_2S + ^{\cdot}OH \rightarrow HS^{\cdot} + H_2O$	1.5×10^{10}	9
R3	$HS^- + ^{\cdot}OH \rightarrow HSOH^{\cdot -}$	9.0×10^9	9
R4	$HS^{\cdot} + HS^{\cdot} \rightarrow H_2S_2$	6.5×10^9	10
R5	$HS^{\cdot} + H_2S \rightarrow H_2S_2^{\cdot -} + H^+$	5.5×10^9	EST[a]
R6	$HS^{\cdot} + HS^- \rightarrow H_2S_2^{\cdot -}$	5.4×10^9	10
R7	$HS^{\cdot} + O_2 \rightarrow HSO_2^{\cdot}$	7.5×10^9	10
R8	$HS^{\cdot} + H_2S_2^{\cdot -} \rightarrow H_2S_2 + HS^-$	9.0×10^9	10
R9	$HSOH^{\cdot -} + HS^- \rightarrow H_2S_2^{\cdot -} + OH^-$	2.0×10^9	10
R10	$HSOH^{\cdot -} + H_2S \rightarrow H_2S_2^{\cdot -} + H_2O$	3.0×10^9	10
R11	$HSOH^{\cdot -} + O_2 \rightarrow SO_2^{\cdot -} + H_2O$	1.0×10^9	EST

Table II. Continued

ID No.	Reaction	k ($M^{-1}\ s^{-1}$)	Ref.
R12	$H_2S_2^{\cdot-} + H_2S_2^{\cdot-} \rightarrow HS_2^- + 2HS^- + H^+$	9.5×10^8	10
R13	$H_2S_2^{\cdot-} + O_2 \rightarrow HS_2^- + O_2^{\cdot-} + H^+$	4.0×10^8	10
R14	$HS_2^- + SO_3^{2-} \rightarrow S_2O_3^{2-} + HS^-$	1.0×10^7	EST
R15	$HS_2^- + O_2 \rightarrow SO_2^{\cdot-} + HS^-$	5.0×10^7	EST
R16	$SO_2^{\cdot-} + O_2 \rightarrow SO_2 + O_2^{\cdot-}$	1.0×10^8	9
R17	$SO_3^{2-} + {}^{\cdot}OH \rightarrow SO_3^{\cdot-} + OH^-$	5.5×10^9	9
R18	$HSO_3^- + {}^{\cdot}OH \rightarrow SO_3^{\cdot-} + H_2O$	4.5×10^9	9
R19	$S_2O_3^{2-} + {}^{\cdot}OH \rightarrow S_2O_3OH^{2-}$	7.8×10^9	9
R20	$S_2O_3OH^{2-} + S_2O_3^{2-} \rightarrow S_4O_6^{3-} + OH^-$	6.0×10^8	11
R21	$S_4O_6^{3-} + {}^{3}\!/_4\, H_2O \relbar\!\!/\!\!\relbar\!\!\rightarrow {}^{1}\!/_2\, SO_3^{2-} + {}^{7}\!/_4\, S_2O_3^{2-} + {}^{3}\!/_2\, H^+$	$2.5 \times 10^{6\,b}$	11
R22	$O_2^{\cdot-} + {}^{\cdot}OH \rightarrow O_2 + OH^-$	1.0×10^{10}	9
R23	$SO_3^{\cdot-} + O_2 \rightarrow SO_5^{\cdot-} \relbar\!\!/\!\!\relbar\!\!\rightarrow {}^{1}\!/_2\, SO_4^{\cdot-} + {}^{1}\!/_2(SO_4^{2-} + O_2)$	1.0×10^8	12
R24	$SO_4^{\cdot-} + HSO_3^- \rightarrow SO_4^{2-} + SO_3^{\cdot-} + H^+$	1.3×10^9	13
R25	$SO_4^{\cdot-} + SO_3^{2-} \rightarrow SO_4^{2-} + SO_3^{\cdot-}$	2.0×10^9	8
R26	$SO_4^{\cdot-} + O_2^{\cdot-} \rightarrow SO_4^{2-} + O_2$	5.0×10^9	13
R27	$SO_4^{\cdot-} + OH^- \rightarrow SO_4^{2-} + {}^{\cdot}OH$	7.0×10^7	8
R28	$SO_4^{\cdot-} + H_2O_2 \rightarrow SO_4^{2-} + O_2^{\cdot-} + 2H^+$	1.2×10^7	8
R29	$S(IV) + O_2^{\cdot-} + H_2O \rightarrow SO_4^{2-} + {}^{\cdot}OH + OH^-$	1.0×10^5	7
R30	$S(IV) + H_2O_2 \rightarrow SO_4^{2-} + H_2O$	0.2	14
R31	$S(-II) + H_2O_2 \relbar\!\!/\!\!\relbar\!\!\rightarrow SO_4^{2-} + 4H_2O + H^+$	0.483	5
R32	$S_2O_3^{2-} + 4H_2O_2 \relbar\!\!/\!\!\relbar\!\!\rightarrow 2SO_4^{2-} + 3H_2O + 2H^+$	0.025	15
R33	$S(-II) + 3O_2^{\cdot-} \relbar\!\!/\!\!\relbar\!\!\rightarrow SO_4^{2-} + 2OH^-$	1.5×10^6	9
R34	$2SO_3^{\cdot-} \rightarrow S_2O_6^{2-} \rightarrow SO_4^{2-} + SO_3^{2-} + 2H^+$	5.3×10^8	8
R35	$OH + H_2O_2 \rightarrow O_2^{\cdot-} + H_2O$	2.7×10^7	9
R36	$NO_2^- + {}^{\cdot}OH \rightarrow NO_2 + OH^-$	1.0×10^{10}	7
R37	$NO_2 + {}^{\cdot}OH \rightarrow NO_3^- + H^+$	1.3×10^9	7
R38	$2NO_2 + S(IV) + H_2O \rightarrow 2NO_2^- + SO_4^{2-} + 3H^+$	1.7×10^7	16
R39	$SO_4^{\cdot-} + S(-II) \rightarrow SO_4^{2-} + HS^{\cdot}$	1.0×10^9	EST
R40	$H_2S_2^{\cdot-} \xrightarrow{O_2} HS^{\cdot} + HS^-$	$5.3 \times 10^{5\,b}$	10
R41	$SO_2^{\cdot-} + HS^- \rightarrow SO_4^{2-} + HS$	1.0×10^8	EST
R42	$SO_3^{\cdot-} + HS^- \rightarrow SO_3^{2-} + HS$	1.0×10^8	EST
R43	$S_2O_3^{2-} + 2O_2^{\cdot-} \relbar\!\!/\!\!\relbar\!\!\rightarrow SO_4^{2-} + SO_3^{2-}$	1.0×10^5	EST
R44	$HS_2^{\cdot-} + {}^{\cdot}OH \rightarrow HS_2 + OH^-$	1.0×10^9	EST
R45	$HS_2 + OH \rightarrow H_2S_2O$	1.0×10^9	EST
R46	$HS_2 + HSOH^- \rightarrow H_2S_2O + HS^{\cdot}$	1.0×10^9	EST
R47	$H_2S_2O + 4\,{}^{\cdot}OH \relbar\!\!/\!\!\relbar\!\!\rightarrow S_2O_3^{2-} + 2OH^- + 4H^+$	1.0×10^9	EST

[a] EST: estimated (see text).
[b] Rate constant in reciprocal seconds.

magnitude as the rate constant for HS$^{\cdot}$ + HS$^-$ (R6); a value of 5.5×10^9 M^{-1} s^{-1} is therefore used. The choice of k_{r5} is not critical for our application because H$_2$S is not the main S(–II) species at alkaline pH.

The rate constant for the reaction of HSOH$^{\cdot-}$ with O$_2$ (R11) is expected to be of the same order of magnitude as for the reaction of HSOH$^{\cdot-}$ with HS$^-$ (R9) and lower than the rate constants for the reaction of HS$^{\cdot}$ with those species (R7, R6); the value of 10^9 M^{-1} s^{-1} is therefore used for R11. The overall rate of S(–II) oxidation does not seem to be sensitive to k_{r11}. At pH 10 and [S(–II)]$_0$ = 200 μM, the kinetics and product distribution remained the same even when k_{r11} was set to 5×10^9 M^{-1} s^{-1}.

Although S$_2$O$_3^{2-}$ is a known product of the radiolysis of S(–II), no pathways that lead to its formation have been proposed. We included two possible pathways: the reaction of HS$_2^-$ with SO$_3^{2-}$ (R14) and oxidation of HS$_2^-$ by $^{\cdot}$OH (R44) followed by successive addition and abstraction reactions (R45 and multistep R47). The rate constant of R14 is not expected to be higher than the rate constant for the reaction of SO$_3^{2-}$ with NO$_2^{\cdot}$ (R38) and was therefore set at 10^7 M^{-1} s^{-1}. The rate constants for reactions R44–R47 are expected to be near the diffusion-controlled limit and were set at 10^9 M^{-1} s^{-1}. Even if $k_{r14} = 10^8$ M^{-1} s^{-1} and $k_{r44-47} = 5 \times 10^9$ M^{-1} s^{-1}, these reactions cannot account for the observed formation of S$_2$O$_3^{2-}$. However, the values of these rate constants are not critical for the overall rate of S(–II) oxidation.

The rate constant for the reaction of HS$_2^-$ with O$_2$ (R15) is expected to be higher than the rate constant for the reaction of HS$_2^-$ with SO$_3^{2-}$ (R14); a value of 5×10^7 M^{-1} s^{-1} was therefore used for k_{r15}. Sulfate radical is a very strong oxidant and is expected to react with S(–II) near the diffusion-controlled limit; the value of k_{r39} was set at 10^9 M^{-1} s^{-1}. Sulfur dioxide and sulfite radicals are expected to react with S(–II) slower than SO$_4^{\cdot-}$, and a value of 10^8 M^{-1} s^{-1} was used for k_{r41} and k_{r42}. Reaction R41 should be seen as multistep where the nucleophilic adduct SO$_2$HS^{2-} formed by the reaction of SO$_2^{\cdot-}$ and HS$^-$ reacts further with O$_2$.

The model parameters are shown in Table III. The model accepts a constant and continuous input of $^{\cdot}$OH, H$_2$O$_2$, and NO$_2^-$ (OH$_{input}$, H$_2$O$_{input}$, and NO$_{2\ input}$, respectively) so that it can simulate the continuous release of those species from the collapsing bubbles during sonication. Provision has been made for reaeration of the solution; an overall oxygen-transfer coefficient [$K_l(O_2)$] is used, and the rate of O$_2$ addition to the system is $K_l(O_2) \cdot$ ([O$_2$]$_{sat}$–[O$_2$]), where [O$_2$]$_{sat}$ is the oxygen saturation concentration at the given temperature and pressure and [O$_2$] is the actual oxygen concentration in solution.

The values of the model parameters shown in Table III represent the conditions of our sonication experiments. Their selection is discussed in the next section of this chapter.

Table III. Conditions of Sonication Experiments

Fixed Model Parameter	Value
OH_{input}	3.5 μM/min
$H_2O_{2\ input}$	2.0 μM/min
$NO_{2\ input}$	1.0 μM/min
$K_1(O_2)$	$3.7 \times 10^{-4}\ s^{-1}$
T	298 K
P	1 atm (101.3 kPa)
pH	10.0

Because activation energies for most of the reactions of Table II were not available, T and P are fixed at standard conditions. The pH is also treated as a fixed variable to simulate our buffered sonication experiments. The computer code can be modified to correct the rate constants for different T or P (assuming that the necessary thermodynamic data for the reactions of Table II become available), or to treat pH as a variable species (by including H^+ as a 23rd chemical species).

Given the initial conditions (concentrations of the 22 chemical species at $t = 0$), the concentrations of the chemical species with time are found by numerically solving the set of the 22 stiff ordinary differential equations (ODE). An ordinary differential equation system solver, EPISODE (17) is used. The method chosen for the numerical solution of the system includes variable step size, variable-order backward differentiation, and a chord or semistationary Newton method with an internally computed finite difference approximation to the Jacobian equation.

Model Results and Discussion

During ultrasonic irradiation of aqueous solutions, ·OH radicals are produced from dissociation of water vapor upon collapse of cavitation bubbles. A fraction of these radicals that are initially formed in the gas phase diffuse into solution. Cavitation is a dynamic phenomenon, and the number and location of bursting bubbles at any time cannot be predicted a priori. Nevertheless, the time scale for bubble collapse and rebound is orders of magnitude smaller than the time scale for the macroscopic effects of sonication on chemicals (2) (i.e., nanoseconds to microseconds versus minutes to hours). Therefore, a simplified approach for modeling the liquid-phase chemistry resulting from sonication of a well-mixed solution is to view the ·OH input into the aqueous phase as continuous and uniform. The implicit assumption in this approach is that the kinetics of the aqueous-phase chemistry are not controlled by diffusion limitations of the substrates reacting with ·OH.

The mathematical model presented here was first used to simulate the oxidation of S(–II) at pH 10, $[S(-II)]_0 = 196$ μM, and $[O_2]_0 = 240$ μM (air saturation). These initial conditions correspond to the conditions of one of the sonication experiments conducted in a stainless steel cell at 20 kHz and ≈75 W/cm² with a Branson 200 sonifier and a ½-inch direct-immersion horn (3). The rates of H_2O_2 and NO_2 input to the system (i.e., H_2O_{2input} and NO_{2input}, respectively) were set at the experimentally observed zero-order formation rates for those species in deionized water buffered at pH 10 and sonicated under the same conditions. Various runs were performed with different values of OH_{input}.

Figure 1 shows the model-calculated evolution of total sulfide with time for six different ˙OH input values. As expected, the initial S(–II) oxidation rate increases with increasing OH_{input}. But that initial linear decrease of [S(–II)] is "halted" at [S(–II)] ≃ 55 μM in all cases. Figure 2 explains why; the main oxidant in our system turns out to be molecular oxygen present in the solution. The ratio of $[O_2]_{depleted}:[S(-II)]_{oxidized}$ is about 1.7 in all cases, a result showing that most of the oxygen contained in the oxidized forms of sulfur that are the final products (i.e., SO_4^{2-}, SO_3^{2-}, and $S_2O_3^{2-}$) comes from O_2 and not from ˙OH. Figure 3 shows schematically the reactions and species that proved to be the most significant. The oxidation of S(–II) is initiated by reaction with ˙OH, but is further propagated by a free-radical chain-reaction sequence involving O_2.

The value of ˙OH_{input} that results in an initial rate of S(–II) oxidation equal to the experimentally observed oxidation rate at the stated conditions (i.e., 7.5

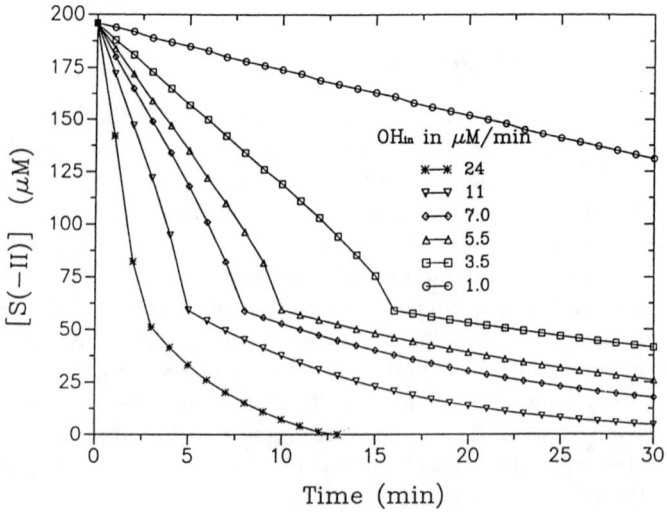

Figure 1. [S(–II)] profiles for various OH_{input} values; pH 10, $[S(-II)]_0 = 196$ μM, and $[O_2]_0 = 240$ μM.

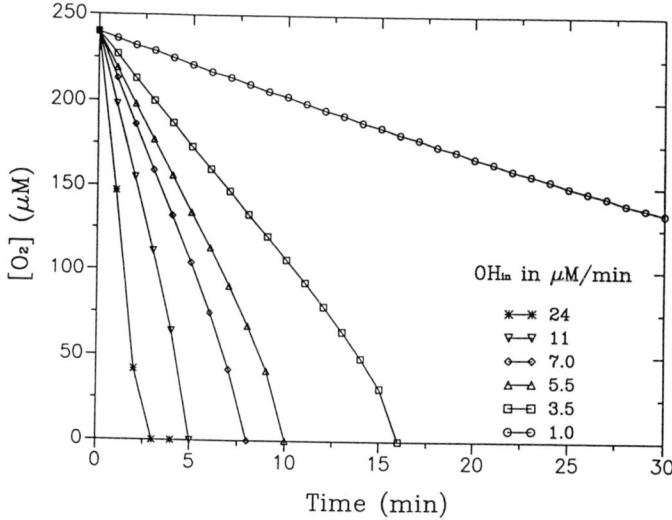

Figure 2. $[O_2]$ profiles for various OH_{input} values; pH 10, $[S(-II)]_0 = 196$ μM, and $[O_2]_0 = 240$ μM.

μM/min at $[S(-II)]_0 = 196$ μM and pH 10) is 3.5 μM/min. To explain the experimentally observed linear S(–II) decrease even after all of the initial O_2 should have been depleted, it is necessary to assume oxygen transfer from headspace to solution during sonication. Figure 4 shows the calculated [S(–II)] and $[O_2]$ profiles for $·OH_{input} = 3.5$ μM/min and six different values of the overall oxygen-transfer coefficient $K_1(O_2)$. The value of $K_1(O_2)$ does not affect the initial oxidation rate, but it determines the subsequent form of the [S(–II)] curve. If not enough oxygen is added into solution, oxygen is depleted

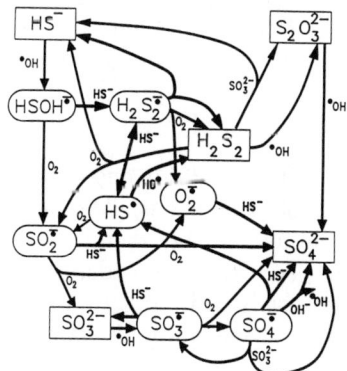

Figure 3. Main pathways for the S(–II) + OH system in the presence of O_2 at alkaline pH.

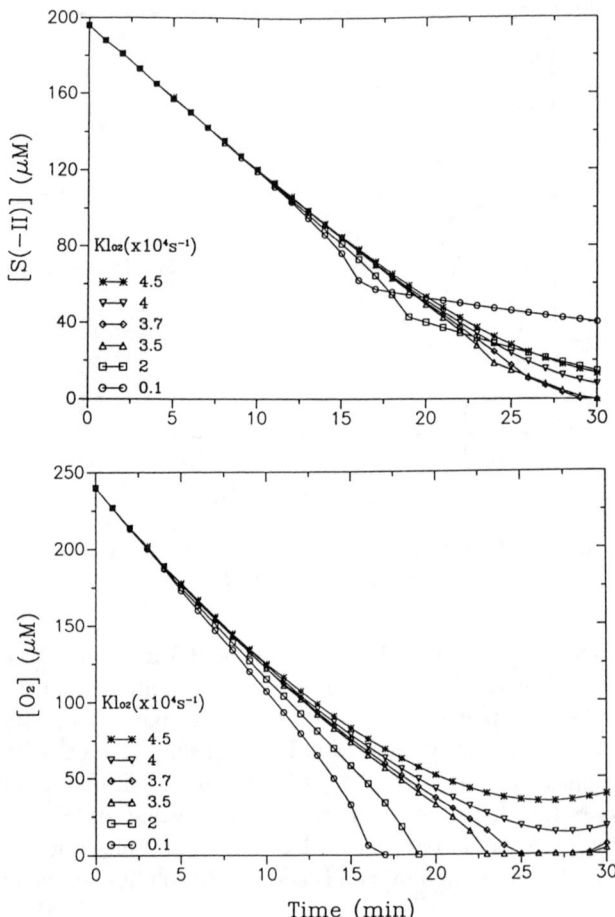

Figure 4. Top: [S(–II)] profiles for different oxygen-transfer coefficients; pH 10, OH_{input} = 3.5 µM/min, $[S(-II)]_0$ = 196 µM, and $[O_2]_0$ = 240 µM. Bottom: Corresponding $[O_2]$ profiles.

before all of the initial sulfide is oxidized, and the result is a sudden decrease in the oxidation rate. If a lot of oxygen is added to the system, $[O_2]$ near the end of the reaction is constantly higher than [S(–II)] and a slowdown of the rate of [S(–II)] oxidation occurs [the intermediate species involved in the chain mechanism will react preferably with O_2 and less with S(–II)]. The value of $K_1(O_2)$ that results in a continuous linear [S(–II)] decrease is 3.7 × 10^{-4} per second = 32 per day.

Experiments using the same direct-immersion horn that was used in S(–II) sonication, but in a completely different reaction vessel under different

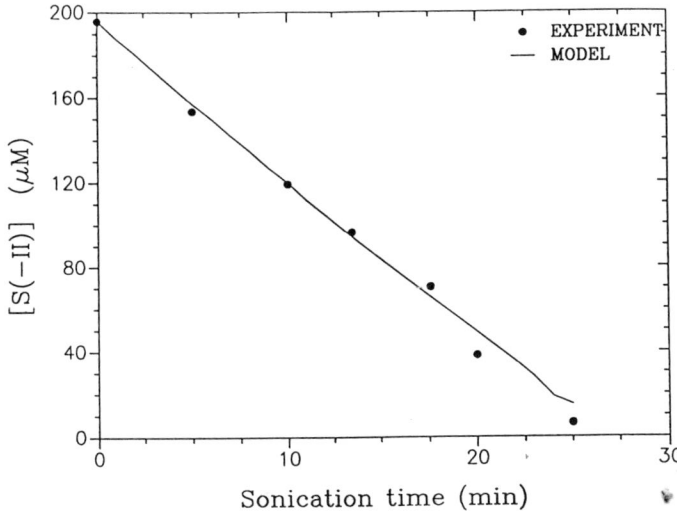

Figure 5. Comparison between [S(–II)] decrease predicted by the free-radical chemistry model and that observed upon ultrasonic irradiation of S(–II) at pH 10, $[S(-II)]_0$ = 196 µM, and $[O_2]_0$ = 240 µM.

conditions (e.g., higher T and open to the atmosphere), gave an oxygen-transfer coefficient that was greater than 32 per day. Therefore, the value of $K_1(O_2)$ was accepted as reasonable and was used in all subsequent modeling work. The value of $K_1(O_2)$ is specific for a particular reactor, because reaeration will depend on the mixing pattern and the ratio of the surface area to the total volume of the solution.

Figure 5 compares the experimentally observed [S(–II)] profile (Figure 4 top) with the model results by using the parameter values shown in Table III. In agreement with the experimental results, the values of $H_2O_{2\ input}$ and $NO_{2\ input}$ are not critical for the rate of S(–II) oxidation. Therefore, the initially chosen values (i.e., 2.5 and 1.0 µM/min, respectively) were not changed.

The parameters of Table III were then used to model the oxidation of S(–II) at pH 10 and different initial sulfide concentrations, $[S(-II)]_0$. The agreement with the experimental data was very good at low $[S(-II)]_0$; the calculated S(–II) profiles for $[S(-II)]_0$ = 7 and 45 µM are shown in Figure 6, together with the experimental S(–II) profiles. As $[S(-II)]_0$ increases, oxygen gets depleted before complete oxidation of the initial sulfide is achieved. Oxygen depletion results in a sudden decrease in the overall oxidation rate. Figure 7 shows the calculated [S(–II)] and [O_2] profiles for $[S(-II)]_0$ = 300 µM and two different values of the oxygen-transfer coefficient; the solid line represents the case in which [O_2] is fixed at the initial air-saturation value [by providing a very high $K_1(O_2)$ value], and the broken line is expected to represent our experimental conditions. For that value of $[S(-II)]_0$, continuous aeration of the solution does not improve the S(–II) oxidation rate significantly.

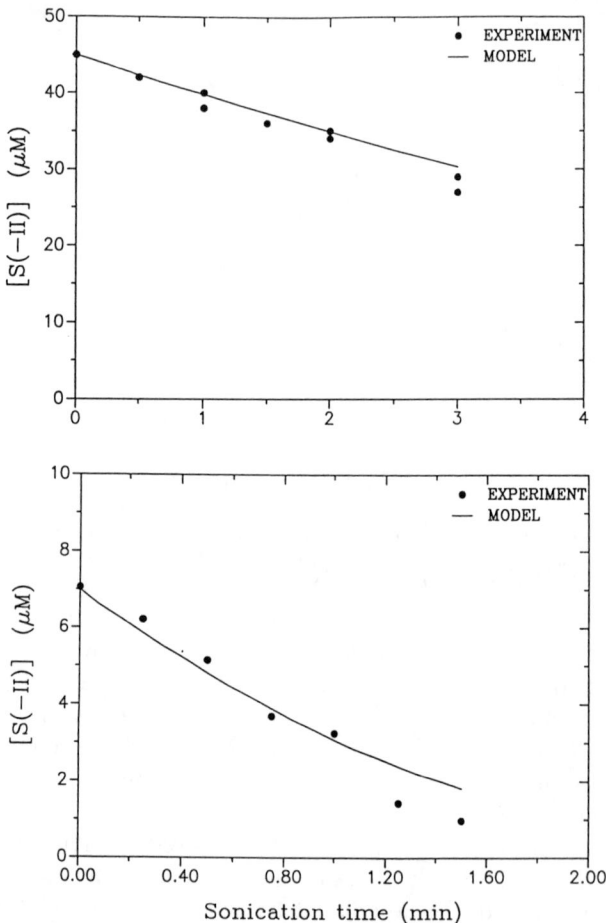

Figure 6. Comparison between [S(–II)] decrease predicted by the free-radical chemistry model and that observed upon ultrasonic irradiation of S(–II) at pH 10, $[O_2]_0$ = 240 μM, and two different $[S(–II)]_0$ low values: 45 (top) and 7 (bottom) μM.

The same is shown to be true up to $[S(–II)]_0 \simeq 450$ μM (Figure 8). At higher $[S(–II)]_0$, the observed S(–II) oxidation rate is lower than the rate at $[O_2]$ = constant = 240 μM; Figure 9 illustrates that point for $[S(–II)]_0$ = 955 μM.

Figure 10 shows the agreement between the experimentally determined effect of $[S(–II)]_0$ on the initial sulfide oxidation rate, k_0, for $[S(–II)]_0 \leq \sim 450$ μM. In both cases, k_0 increases linearly with $[S(–II)]_0$. The following linear relationship was found from the model: $k_0 = 4.4 + 0.016\ [S(–II)]_0$, where k_0 is in micromolars per minute and $[S(–II)]_0$ is in micromolars. The value of the

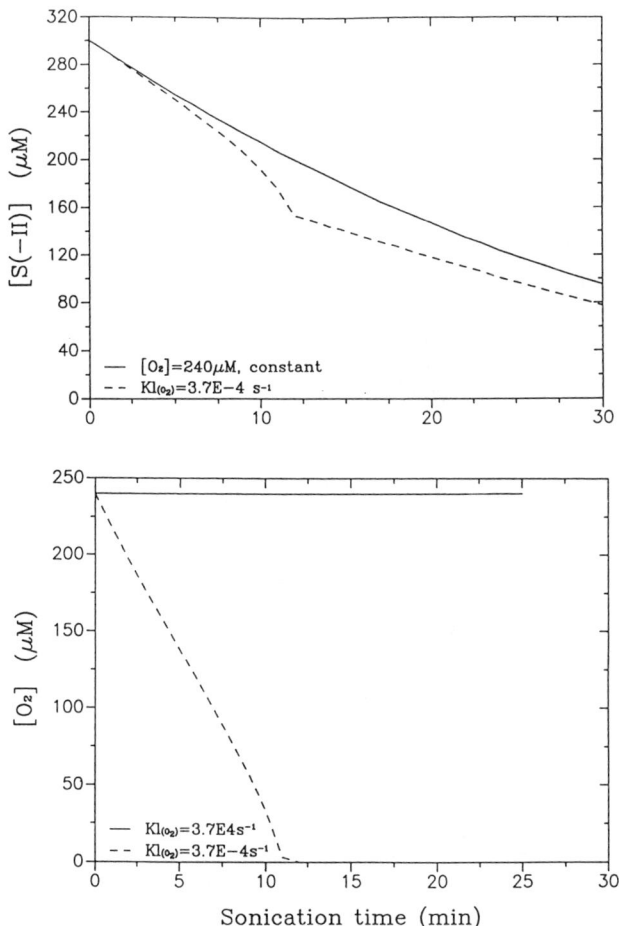

Figure 7. Top: Comparison of calculated [S(-II)] profiles with [O_2] kept constant at air saturation (solid line) and with the oxygen-transfer coefficient of Table III (broken line) at pH 10, OH_{input} = 3.5 μM/min, [S(-II)]$_0$ = 300 μM, and [O_2]$_0$ = 240 μM. Bottom: Corresponding [O_2] profiles.

intercept represents the contribution of (S)OH, whereas the slope reflects the free-radical chain sequence (3). For [S(-II)]$_0 \geq$ ≈300 μM, free-radical chain sequence exceeds the contribution of •OH (i.e., O_2 becomes the principal oxidant).

The two pathways that are used to model $S_2O_3^{2-}$ formation cannot outcompete the reaction of H_2S_2 with O_2. This fact results in an underprediction of [$S_2O_3^{2-}$], as can be seen in Figure 11. Sulfite, on the other hand, is modeled rather well. This situation can also be seen in Figure 12, where the profiles of [S(-II)] and its oxidation products are presented for pH 10.6. Nevertheless,

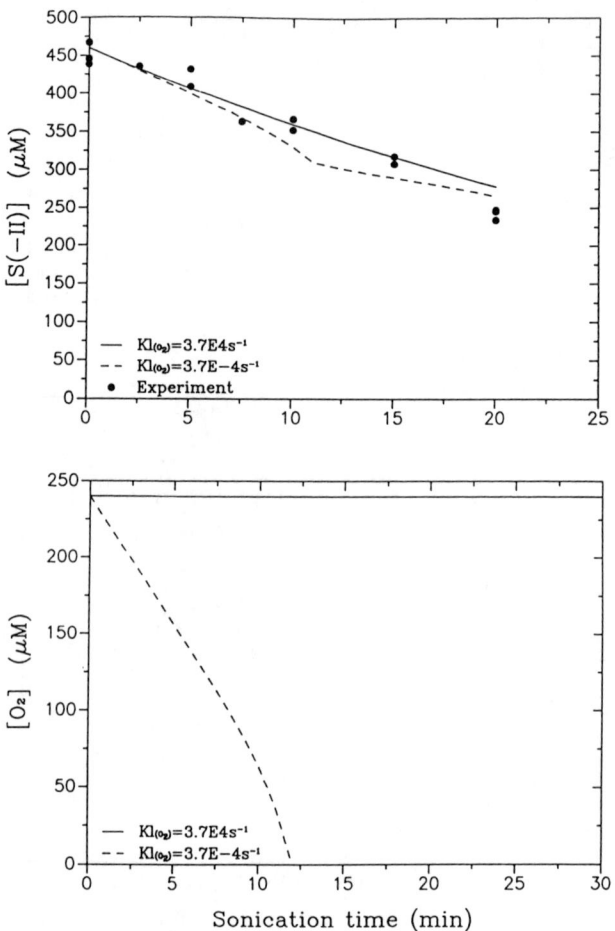

Figure 8. (Top): Comparison of sonolysis data with calculated [S(–II)] profiles, with [O_2] kept constant at air saturation (solid line) and with the oxygen-transfer coefficient of Table III (broken line) at pH 10, OH_{input} = 3.5 μM/min, $[S(–II)]_0$ ≃ 450 μM, and $[O_2]_0$ = 240 μM. Bottom: Corresponding calculated [O_2] profiles.

the free-radical chemistry mechanism adequately describes the overall oxidation of S(–II) upon sonication at alkaline pH.

Figures 13, 14, and 15 present the S(–II) and S_{ox}, where S_{ox} = $[SO_4^{2-}]$ + $[SO_3^{2-}]$ + $2[S_2O_3^{2-}]$, profiles observed upon sonication of S(–II) aqueous solutions at pH 9.0, 8.5, and 7.4, respectively. The broken lines represent the corresponding concentrations of those species predicted by the free-radical chemistry mechanism. The free-radical mechanism underpredicts the rate of S(–II) disappearance at pH ≤ 8.5. Nevertheless, the total amount of the oxidized S(–II) that was found in the form of the three species that form S_{ox} is

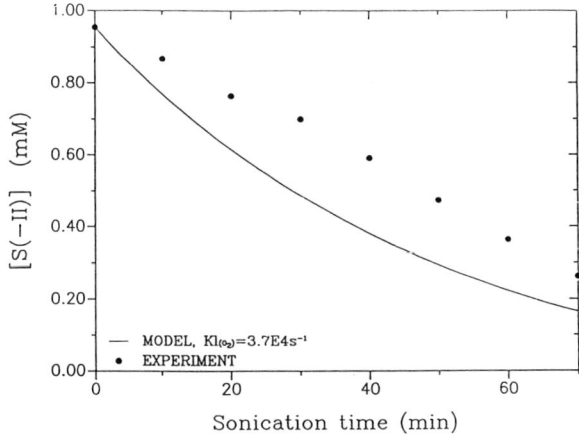

Figure 9. Comparison of sonolysis data with calculated [S(–II)] profiles with [O_2] kept constant at air saturation (solid line) at pH 10, OH_{input} = 3.5 µM/min, $[S(-II)]_0 \simeq 955$ µM, and $[O_2]_0 = 240$ µM.

not much higher than the calculated S_{ox}. In the chemical model, S_{ox} represents the total amount of sulfide that has been oxidized because the three species included in S_{ox} are the only final oxidation products), whereas in the case of the experimental data S_{ox} represents only a part of the total oxidation products (3).

These results provide further evidence that an important pathway for S(–II) sonolysis at pH ≤ 8.5 is thermal decomposition of H_2S within the

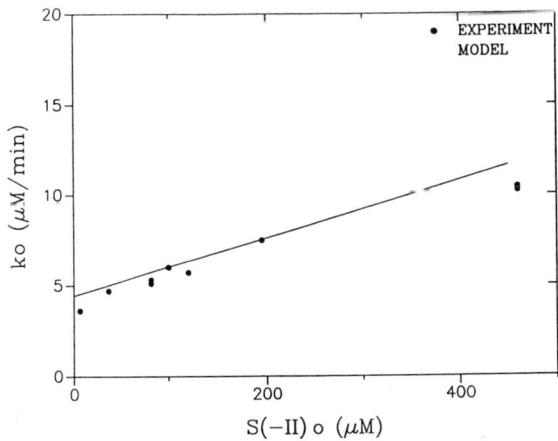

Figure 10. Effect of $[S(-II)]_0$ on initial zero-order S(–II) oxidation rate, k_0.

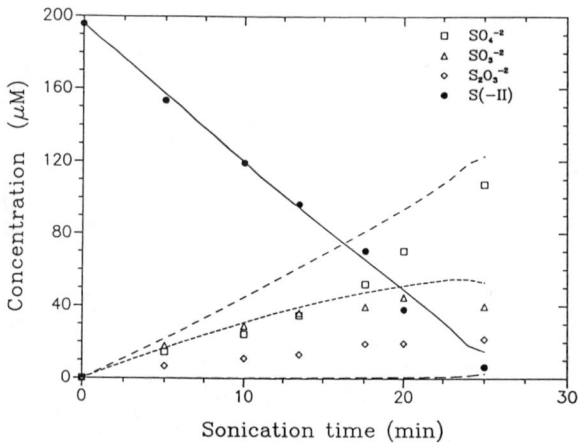

Figure 11. [S(–II)] profile and oxidation product distribution at pH 10, and [S(–II)]$_0$ = 196 µM/min. Model results (solid and broken lines) and experimental data (symbols) are shown.

cavitation bubbles or within the gas–liquid interface. Furthermore, they seem to suggest that elemental sulfur is the main product of that alternative sonolysis pathway.

In conclusion, the sonolysis of S(–II) in the pH range where that species is primarily in the form of HS⁻ and is not expected to undergo thermal decomposition can be modeled with an aqueous free-radical chemical mecha-

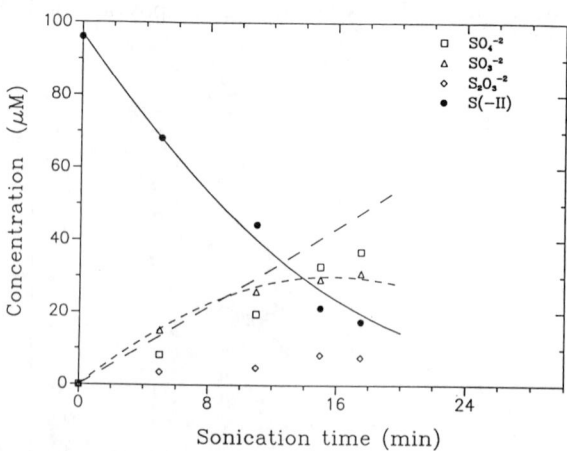

Figure 12. [S(–II)] profile and oxidation product distribution at pH 10.6 and [S(–II)]$_0$ = 96 µM/min. Model results (solid and broken lines) and experimental data (symbols) are shown.

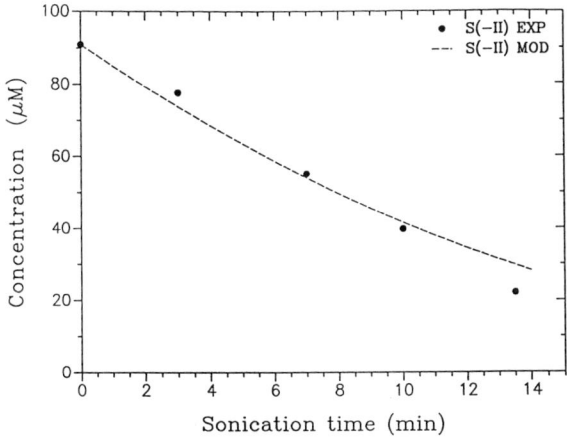

Figure 13. Comparison between [S(–II)] decrease predicted by the free-radical chemistry model and that observed upon ultrasonic irradiation of S(–II) at pH 9.0, [S(–II)]$_0$ = 91 μM, and [O$_2$]$_0$ = 240 μM.

nism and a continuous constant release of •OH into solution. This simplified approach is not valid at the pH range ≤8.5 in which a significant part of the total sulfide is in the form of H$_2$S, which can participate in the gas-phase, high-temperature chemistry that takes place inside and near collapsing cavitation bubbles. An extended chemical mechanism that would include both gas- and liquid-phase chemistry is needed to model S(–II) sonolysis in the

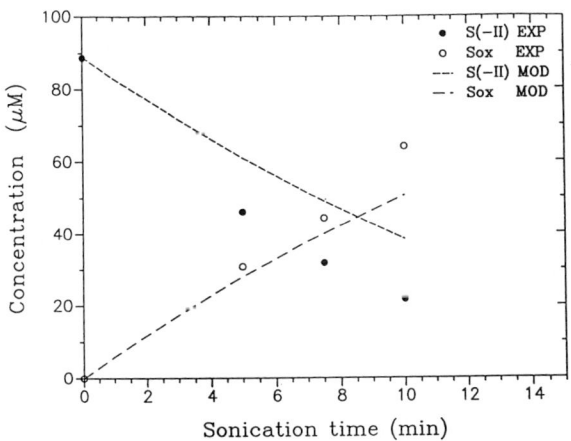

Figure 14. Comparison between [S(–II)] decrease predicted by the free-radical chemistry model and that observed upon ultrasonic irradiation of S(–II) at pH 8.5, [S(–II)]$_0$ = 88 μM, and [O$_2$]$_0$ = 240 μM.

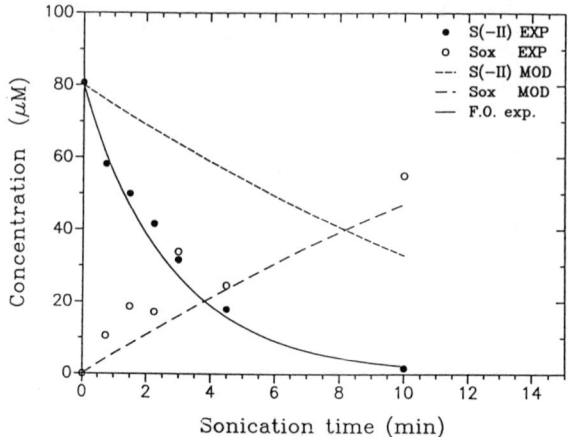

Figure 15. Comparison between [S(–II)] decrease predicted by the free-radical chemistry model and that observed upon ultrasonic irradiation of S(–II) at pH 7.4, $[S(-II)]_0 \simeq 80$ μM, and $[O_2]_0 = 240$ μM.

general case. However, the development of the next-level model is not feasible at this stage because not enough is known about the number and the size distribution of cavitation bubbles. In addition, little is known about the physical conditions inside the bubble and its surrounding region upon collapse.

Molecular oxygen is important for the sonolysis of S(–II) at alkaline pH because it propagates a free-radical chain reaction that is initiated by ·OH. Furthermore, the enhancement of oxygen transfer upon sonication with a direct-immersion horn is considerable. These results may have important implications for the application of ultrasonic irradiation for the destruction of chemical contaminants in water systems.

Acknowledgments

Many are grateful to German Mills for useful and stimulating discussions. This work was funded in part by U.S. Environmental Protection Agency (Exploratory Research Office Grant R815041–01–0) and the Advanced Research Projects Agency (DoD–ONR Grant N0014–92–J–1901).

References

1. Riesz, P.; Kondo, T. *Free Rad. Biol. Med.* **1992**, *13*, 247.
2. Suslick, K. S. In *Ultrasound: Its Chemical, Physical, and Biological Effects*; Suslick, K. S., Ed.; VCH: New York, 1988.
3. Kotronarou, A.; Mills, G.; Hoffmann, M. R. *Environ. Sci. Technol.* **1992**, *25*, 2940.
4. Smith, R. M.; Martell, A. E. *Critical Stability Constants; Volume 4: Inorganic Complexes;* Plenum, New York, 1976.

5. Hoffmann, M. R. *Environ. Sci. Technol.* **1977**, *11*, 61.
6. Giggenbach, W. *Inorg. Chem.* **1971**, *10*, 1333.
7. Pandis, S. N.; Seinfeld, J. H. *J. Geophys. Res.* **1989**, *94*, 1105.
8. Neta, P.; Huie, R. E.; Ross, A. B. *J. Phys. Chem. Ref. Data* **1988**, *17*, 1027.
9. Buxton, G. V.; Greestock, C. L.; Helman, W. P.; Ross, A. B. *J. Phys. Chem. Ref. Data* **1988**, *17*, 513.
10. Mills, G.; Schmidt, K. H.; Matheson, M. S.; Meisel, D. *J. Am. Chem. Soc.* **1987**, *91*, 1590.
11. Mehnert, R.; Brede, O. *Radiat. Phys. Chem.* **1984**, *23*, 463.
12. Chameides, W. L.; Davis, D. D. *J. Geophys. Res.* **1982**, *87*, 4863.
13. Jacob, D. J. *J. Geophys. Res.* **1986**, *91*, 9807.
14. Mader, P. M. *J. Am. Chem. Soc.* **1958**, *80*, 2634.
15. Hoffmann, M. R.; Edwards, J. O. *Inorg. Chem.* **1977**, *16*, 3333.
16. Clifton, C. L.; Altsein, N.; Huie, R. E. *Environ. Sci. Technol.* **1988**, *22*, 586.
17. Byrne, G. D.; Hindmarsh, A. C. *ACM Trans. Math. Software* **1975**, 71.

RECEIVED for review October 23, 1992. ACCEPTED revised manuscript June 11, 1993.

13

Photoreactions Providing Sinks and Sources of Halocarbons in Aquatic Environments

Richard G. Zepp and Leroy F. Ritmiller

Environmental Research Laboratory, U.S. Environmental Protection Agency, Athens, GA 30613

> *This chapter discusses laboratory and field studies used to test concepts and develop mathematical relationships that describe photochemical reactions providing sinks or sources for halocarbons in aquatic environments. Photochemical sinks involve direct photoreactions of aromatic halocarbons and photoredox reactions between halocarbons and natural substances. Direct photoreactions of halogenated aromatic compounds involve dehalogenation through photohydrolysis and other reactions. Reductive dehalogenation of certain halocarbons is enhanced through their sorption to natural organic matter (NOM). Possible mechanisms for the enhancement include reduction mediated by photoejected electrons, NOM excited states, or complexes. Sources of halocarbons in the sea may include haloperoxidase-catalyzed reactions between photochemically produced H_2O_2 and NOM or reactions involving photochemically produced halide radicals.*

ORGANOHALOGENATED COMPOUNDS, referred to here as halocarbons, are widely distributed in the environment (1–4). Some industrially produced compounds, such as polychlorinated biphenyls and DDT [1,1'-(2,2,2-trichloroethylidene)bis[4-chlorobenzene]] are biologically refractory and toxic pollutants of water and land. Volatile synthetic halocarbons such as chlorofluorocarbons (CFCs), a significant component of the greenhouse gases in the atmosphere, cause depletion of the ozone layer (5). Halogenated compounds also are produced naturally. The widespread occurrence of natural

This chapter not subject to U.S. copyright
Published 1995 American Chemical Society

halocarbons in terrestrial and aquatic environments was reviewed by Gribble (1) and by Asplund and Grimvall (2).

Some halocarbons are produced by both natural and industrial sources. Unfortunately, the few data concerning the magnitude of the natural sources are conflicting. For example, based on measured concentrations of methyl bromide in the sea and troposphere, Singh et al. (6) estimated that the net annual global flux from ocean to air is about 0.3 teragrams (1 Tg is 1×10^{12} g). This net emission is much larger than that derived from human activities (0.05–0.08 Tg per year) (7, 8). On the other hand, Penkett and co-workers (7) concluded that human activities were the main source of methyl bromide.

Photochemical reductions and oxidations in aquatic environments provide sinks or sources for halocarbons. Such photoreactions are an important process in the dissipation of low-volatility halocarbons, such as halogenated agrochemicals, in aquatic environments (reference 9 contains lead references). For example, field studies of Crossland and Wolff (10) demonstrated the rapid dissipation of pentachlorophenol residues by its photoreaction in English ponds. Evidence emerged that volatile halocarbons such as 1,1,1-trichloroethane (methylchloroform) may have significant sinks in the aquatic environment (11). Reactions of halocarbons with photochemically produced reactive transients help provide these sinks.

Although halocarbon sinks are present in the sea, some evidence suggests that the ocean is a net source of the single-carbon halocarbons methyl chloride, methyl bromide, methyl iodide, and bromoform (8). Production of halocarbons in natural waters is very poorly understood. Reactive oxygen species produced photochemically in surface waters may participate in the production of halocarbons. Halide radicals, hypobromite and hypoiodite, react with the dissolved organic matter in natural waters to produce halocarbons. These species can be produced through reactions between halide ions and reactive oxygen species that are produced photochemically in the upper layers of the sea. Moreover, photoreduction of oxygen produces hydrogen peroxide. Through interactions with haloperoxidases, hydrogen peroxide may be involved in the natural production of halocarbons.

This chapter reviews past studies of other investigators and presents previously unpublished laboratory studies relating to photochemical reactions of halocarbons in aquatic environments. Kinetics considerations relevant to the modeling of halocarbons in natural waters are briefly considered. Then direct and indirect photoreactions that provide sinks for halocarbons are examined with consideration of the effects of sorption onto dissolved NOM and suspended sediments. Photoreductive dehalogenation is emphasized. Finally, the role of photochemically produced reactive oxygen species in the production of halocarbons in the aquatic environment is discussed. The main emphasis is on field and laboratory studies that provide predictive capability and a mechanistic understanding of the processes.

Experimental Section

Materials. Except as noted, all halocarbons used in the experiments were reagent grade and were used as received. 2,2-Bis(4-chlorophenyl)-1,1-dichloroethylene (DDE, obtained from Aldrich, Milwaukee, WI) was purified by recrystallization from 95% ethanol. 2-Chloroethanol (Aldrich) was purified as described by Zepp and co-workers (*12*). Other halocarbons and reagents were the purest grade that could be obtained commercially and were used as received. The water used was first deionized and then distilled from alkaline permanganate.

The Contech fulvic acid was obtained commercially from Contech ETC, Ottawa, Canada. Characteristics of this humic substance were previously described (*13*). Natural organic matter (NOM) isolated from the Suwannee River at the point where it drains the Okefenokee Swamp, Georgia, was fractionated by Leenheer and Noyes (*14*) according to hydrophobicity and acid–base properties. The symbols used for these fractions are defined as follows: SHPO-A, strong hydrophobic acids; WHPO-A, weak hydrophobic acids; HPO-N, neutral hydrophobic fraction; and HPI-A, hydrophilic acids. NOM also was obtained from the Greifensee, a highly eutrophic lake near Zürich, Switzerland (*15*). Finally, a soil-derived humic acid was obtained commercially from Fluka AG. Natural water samples were obtained from the Aucilla River near Lamont, Florida, and the Oglethorpe Pond, a small eutrophic water body near Lexington, Georgia.

Procedures. Continuous irradiations of halocarbon solutions were conducted with monochromatic radiation in a merry-go-round apparatus (*12*) or in a Schoeffel reaction chemistry system. Reactions were followed through analysis for remaining halocarbon or analysis of chloride ions produced by the photoreactions. Dark controls were used in all cases to correct for thermal production of chloride ions. Ferrioxalate actinometers were used to determine the irradiance (*16*). The irradiance at the photoreaction cell surface was typically about 10 nanoeinsteins/cm^2·s. The Fe(II) concentrations were determined by using a modified version of the ferrozine procedure described by Stookey (*17*). Electronic absorption spectra were obtained by using a Shimadzu model 265 spectrophotometer.

Analysis for DDE, γ-1,2,3,4,5,6-hexachlorocyclohexane (lindane), and p,p'-dichlorobenzophenone in natural water samples, solutions of humic substances, or suspensions of soils and sediments was conducted by first adding acetonitrile (25% by volume) to the irradiated solution or suspension, followed by the addition of an equal volume of isooctane. The resulting mixture was briefly agitated and then sonicated for 15 min. The isooctane layer was analyzed by gas chromatography with a Hewlett–Packard model 5890 gas chromatograph equipped with an electron-capture detector. Chloride ion concentrations were determined by ion-exchange chromatography with conductivity detection, as described by Zepp and co-workers (12).

Results and Discussion

Kinetics Considerations. Kinetics concepts and data concerning halocarbon sources and sinks can be used for a variety of purposes. For example, such information is required in mathematical models to evaluate the fate and exposure concentrations of low-volatility toxic organohalogens in water (18, 19). Moreover, kinetics relationships and data concerning physical, chemical, and biological processes are needed to predictively model aquatic sinks of volatile halocarbons (11).

The rate of any type of halocarbon photoreaction at a certain wavelength λ, r_λ, depends upon the rate of light absorption by the photoreactive chromophore, $I_{a,\lambda}$, and the quantum efficiency of the reaction, ϕ_r (eq 1).

$$r_\lambda = I_{a,\lambda} \phi_r \tag{1}$$

where

$$I_{a,\lambda} = 2.303 E_{0,\lambda} \epsilon_\lambda l [P] \tag{2}$$

$E_{0,\lambda}$ represents the scalar irradiance, ϵ_λ is the molar absorptivity (or molar extinction coefficient) of the chromophore, l is the light path length, and $[P]$ is the concentration of the chromophore that initiates the photoreaction (e.g., the halocarbon itself, a natural substance, or a complex of the halocarbon with a natural substance). The rate of light absorption depends on the spectral overlap between the light source and the spectrum of the chromophore that initiates the photoreaction.

The quantum yield is the fraction of absorbed radiation that results in photoreaction. According to the Kasha–Vavilov law, the quantum yield for a photoreaction in solution in which only a single substance is the chromophore (e.g., direct photoreactions) is generally wavelength-independent, although there are some exceptions. With indirect photoreactions in natural waters, a mixture of chromophores is involved. Thus, the apparent quantum yield for an indirect photoreaction in a natural water sample usually is significantly wavelength-dependent.

The light absorption rate of a given chromophore in a water body depends both on the solar spectral irradiance reaching the water surface and on changes in the spectral irradiance as it penetrates down into the water (20, 21). Light absorption and scattering in the stratosphere and troposphere affect the solar spectral irradiance, especially in the ultraviolet region, a part of the irradiance that has particularly pronounced photochemical and photobiological effects in natural waters. For example, changes in total ozone strongly affect the irradiance in the ultraviolet-B (UV-B) region (280–315 nm). As solar radiation penetrates down into natural water bodies, dissolved and particulate substances in the photic zone influence the irradiance and thus affect photoreaction rates through both light absorption and scattering (20–22). The solar spectral irradiance at a given wavelength decreases in an approximately exponential fashion with increasing depth z. This depth dependence can be expressed in terms of a diffuse attenuation coefficient, $K_{z,\lambda}$, which is the slope of a natural log plot of the irradiance versus depth.

$K_{z,\lambda}$ is usually expressed per meter and depends on wavelength and water composition. Typically, ultraviolet radiation penetrates less deeply than visible radiation. Attenuation coefficients vary over many orders of magnitude in natural waters, with the highest values (least light penetration) in inland water bodies and the lowest values (highest penetration) in open seawater. The photic zone for solar ultraviolet radiation, which is very important for halocarbon photoreactions, ranges from tens of meters in the open ocean and clear lakes to only a few centimeters in some inland wetlands. The spectral properties of water bodies are linked to water composition. Baker and Smith (20) developed algorithms that relate $K_{z,\lambda}$ to certain parameters such as chlorophyll a concentrations.

Assuming that the water column is mixed more rapidly than photoreaction occurs, it can be shown (22, 23) that $v_{z,\lambda}$, the depth-averaged rate for a photoreaction at depth z, is expressed approximately by:

$$v_{z,\lambda} = \frac{v_{0,\lambda}(1 - e^{-K\lambda z})}{zK_\lambda} \tag{3}$$

where $v_{0,\lambda}$ is the near-surface rate and K_λ is the diffuse attenuation coefficient for the mixed water column. At sufficiently great depths that all the active radiation is absorbed, $v_{z,\lambda}$ approximately equals $v_{0,\lambda}/[zK_\lambda]$. For example, in open seawater the average rate of a photoreaction involving UV-B radiation in the top 100 m is about 10% of the near-surface rate.

Halocarbon concentrations in natural waters are typically very low. At such low concentrations and with constant irradiance, the rate is usually directly proportional to halocarbon concentration; that is, a pseudo-first-order rate expression applies. The kinetic data sometimes are expressed as half-lives that equal $0.693/k_p$ (where k_p is the pseudo-first-order rate constant). More references and a detailed discussion of the modeling of aquatic photochemical and photobiological processes, including the effects of vertical mixing, were given by Plane et al. (23), Smith (24), Zepp and Cline (25), and Sikorski and Zika (26).

Other recent papers provide useful background discussions of this general area. Hoigné (27) gave an excellent discussion of aquatic photochemical kinetics with stress on factors affecting sources and sinks of reactive transients that mediate indirect photoreactions. Recent chapters by Sulzberger (28) and Faust (29) are highly recommended for more in-depth discussion of heterogeneous photoredox kinetics involving metal oxide surfaces.

Types of Halocarbon Photoreactions. In the following sections, we discuss several types of halocarbon photoreactions that provide sinks in aquatic environments. Emphasis in this chapter is placed on photoreactions that result in dehalogenation, in which halide ions (X^-) are produced from halocarbons (RX). Here the term "direct photoreactions" indicates reactions that involve direct light absorption by the halocarbon itself (eq 4).

$$RX \xrightarrow{h\nu} ROH + R^{\cdot} + X^- + \text{other products} \quad (4)$$

Direct photoreactions are mediated by halocarbon excited state(s) that react to form products. Dehalogenations can involve either photohydrolysis (i.e., photonucleophilic replacement of halogen by OH) or homolysis of the carbon–halogen bond to form free radicals. Photohydrolysis is the most important dehalogenation pathway for aromatic halocarbons.

Indirect photoreactions of halocarbons involve reactions with reactive transients that are produced on absorption of light by natural substances, with NOM playing a key role as the source of the transients. These transients include reductants such as solvated electrons (e_{aq}^-) (eq 5).

$$RX + e_{aq}^- \longrightarrow R^{\cdot} + X^- \quad (5)$$

Excited states of NOM (S^*) can transfer electronic energy to or possibly can participate in redox reactions with halocarbons (eq 6).

$$RX + S^* \longrightarrow R^{\cdot} + X^- + S^{\cdot +} \quad (6)$$

Also, reactive oxygen species such as hydroxyl radicals ($^{\cdot}OH$), organoperoxyl radicals ($RO_2^{\cdot}$), hydroperoxyl–superoxide radicals, and singlet molecular oxygen ($^1O_2$) can oxidize halocarbons. Dehalogenation often is not a major pathway in indirect photooxidations, however.

Complexation of halocarbons with natural substances can enhance the rates of photoreactions that provide sinks. Ionizable halocarbons, such as halogenated organic carboxylic acids, potentially could form complexes with photoreactive transition metals, such as iron. In addition, dissolved NOM and sediments are known to sorb or "bind" ionic and nonionic halocarbons, and sorbed halocarbons may photoreact more efficiently (eq 7).

$$RX + NOM \rightleftharpoons RX\text{--}NOM \xrightarrow{h\nu} R^{\cdot} + X^- + NOM^{\cdot+} \quad (7)$$

Evidence is presented here that sorption to NOM enhances the photoreaction rates of halocarbons. Possible enhancement mechanisms are photoinduced electron transfer involving photoejected electrons or NOM excited states and/or formation of photoreactive complexes with NOM-associated electron donors, such as nitrogen bases.

Direct and indirect photodehalogenations can involve free radicals ($R^{\cdot}$) that react mainly with NOM to produce the reduced hydrocarbon (eq 8) or with oxygen to produce various oxygenated products (eq 9).

$$R^{\cdot} + NOM \longrightarrow RH \quad (8)$$

$$R^{\cdot} + O_2 \longrightarrow \text{oxygenated products} \quad (9)$$

The overall photoreaction rate of a given halocarbon in a certain aquatic environment is the sum of the rates of the direct photoreactions of the uncomplexed halocarbon, indirect photoreactions involving reactive transients that are produced by natural substances, and photoreactions of halocarbon complexes. After first discussing the effects of sorption on photoreaction kinetics, we then discuss these various reaction pathways in more detail.

Sorption and Complexation Effects on Photoreactions. Sorption of hydrophobic halocarbons onto suspended sediments, biota, or NOM can have complex effects on photoreaction rates and quantum efficiencies. Hydrophobic or ionic halocarbons, with their great tendency to sorb on sediments or NOM, are most likely to be affected by heterogeneous photoreactions. A flurry of publications (e.g., 30–34 and references cited therein) provided abundant experimental evidence that extremely hydrophobic pollutants (e.g., polycyclic aromatic hydrocarbons, DDT, and mirex) have a strong tendency to associate with the particulate and dissolved organic matter in water bodies.

The various photochemical processes that occur with natural sorbents present are conceptualized in Scheme I.

Two general diffusional processes are important in such heterogeneous reactions. The first involves movement of the reactant molecules to and from the sorbent, and the second involves mass transport into and back out of unreactive sorbent components. If these transport processes are more rapid than the photoreactions and if reactive site limitations on the sorbent are not present, then equilibrium considerations apply and the overall reaction rate is described by equation 10.

$$v = C_T \sum F_i k_i \quad (10)$$

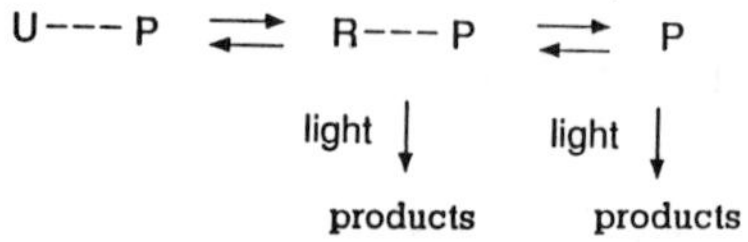

$$U\text{---}P \rightleftarrows R\text{---}P \rightleftarrows P$$
$$\qquad\qquad\quad \text{light} \downarrow \quad\ \text{light} \downarrow$$
$$\qquad\qquad\quad \text{products}\ \ \text{products}$$

Scheme I. Conceptual model for direct and indirect photoreactions in heterogeneous systems. Symbols: P represents the photoreactive halocarbon in solution, R–P represents halocarbon sorbed in reactive components of the sorbent, and U–P represents the halocarbon sorbed in unreactive components of the sorbent. Sorbents are sediments, natural organic matter, or biota. The types of reaction are direct with light absorption by P; and indirect with light absorption by sorbent.

where C_T is the total concentration of halocarbon, Σ denotes summation over all phases, k_i is the first-order rate constant for reaction in phase i in the system, and F_i denotes the fraction of halocarbon present in phase i (35). Either acceleration or retardation of photoreaction can occur, depending on the relative magnitudes of the rate constants in the various phases and the extent of sorption. If reactive site limitations on the sorbent are present, then the reaction kinetics can be described by Langmuir–Hinshelwood rate expressions of the form (35):

$$\text{rate} = \frac{kC_T}{1 + KC_T} \qquad (11)$$

where k and K are constants. Such expressions predict that plots of rate (and quantum yield) versus halocarbon concentration C_T are concave downward, with the rate becoming zero order (independent of C_T) with increasing C_T and first order (directly proportional to C_T) at low C_T.

Direct Photoreactions. *Halocarbons in the Aqueous Phase.* Direct photoreaction (eq 4) is important only for halocarbons (e.g., aromatic compounds) that significantly absorb radiation at wavelengths >295 nm, the cutoff for solar spectral irradiance at the earth's surface. Because saturated chlorinated and fluorinated organic compounds, including methylchloroform and chlorofluorocarbons, absorb solar radiation very weakly, their direct photoreaction is very slow in the sea and in fresh waters. As discussed in a later section, photoreactions of these compounds may be accelerated by sorption and indirect photoreactions in natural waters. Saturated and olefinic polybrominated and iodinated organic compounds have long absorption tails that extend beyond 295 nm. Direct photoreaction of such compounds in aquatic environments may be significant.

Although Zika and co-workers (36) investigated the direct photoreaction of methyl iodide, there are few other environmentally relevant studies of direct photoreactions of saturated halocarbons in water. Kropp (37), however, reviewed the photochemistry of saturated and olefinic halocarbons in organic solvents. These halocarbons photoreact by carbon–halogen bond homolysis to produce free radicals. In polar reaction media, electron transfer between radical pairs within the solvent cage produces ionic intermediates with certain halocarbons, such as iodinated compounds. Carbenes also are formed in these photoreactions.

Direct photoreactions of halogenated aromatic compounds in water have received considerably more attention. The products of direct photoreactions include photonucleophilic replacement of halogen by OH, reductive dehalogenation (i.e., replacement of halogen by hydrogen), photooxidation, and photoisomerization. Several such reactions are exemplified by the photoreactions of DDE in natural waters (eq 12) (38).

$$\text{(12)}$$

As shown in Table I, photoreactions of certain halogenated aromatic compounds and conjugated dienes are very rapid with full exposure to solar radiation. Field studies of Crossland and Wolff (10) showed that direct photoreaction of pentachlorophenol was its dominant fate in selected ponds in southern England (Figure 1).

Direct photoreaction is the only photoreaction pathway in distilled water. Quantum yields observed in air-saturated, distilled water have generally been

Table I. Direct Photoreaction Rates of Selected Halocarbons in Sunlight

Substance	Solvent	Quantum Yield	Half-Life (h)	Ref.
Pentachlorophenol	Water	0.014	1.0	10
2,3,7,8-Tetrachloro-	Water–CH_3CN	0.0022	52	76
dibenzodioxin	Hexane	0.049	2	
2,4,5-Trichlorophenol	Water	0.080	0.5	75
3,4-Dichloroaniline	Water	0.052	3	77
Picloram	Water	0.066	16	78
Trifluralin	Water	0.0020	0.5	25
Hexachloro-				
cyclopentadiene	Water	0.26	0.03	79
DDE	Water	0.24	12	38
	Hexane	0.30	9	

NOTE: Measurements were made at midday in midsummer at a latitude of 40° N.

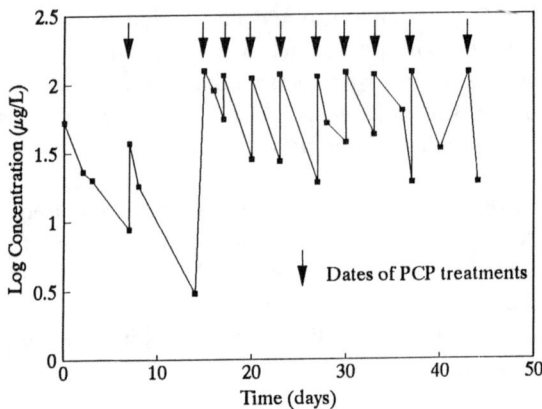

Figure 1. Mean concentration–time profile for pentachlorophenol in three outdoor ponds near Headcorn, Kent, United Kingdom, during April–May. Rapid loss between treatments was attributed to direct photoreaction. (Reproduced with permission from reference 10. Copyright 1985 Pergamon.)

observed to be nearly the same as those observed in the aqueous phase of natural fresh waters. Concentrations of natural substances such as nucleophiles that could undergo direct photoreactions with or quench photoreactions of halocarbons are relatively low in comparison with oxygen (0.3 mM), an excellent excited-state quencher. Quenching by oxygen reduces excited-state lifetimes to no more than about 1 μs and thereby limits the efficiency of diffusional excited-state reactions with constituents of natural waters. In marine waters, concentrations of bromide (about 0.8 mM) and chloride (about 0.55 M) are sufficiently high to affect direct photoreaction quantum yields in some cases.

Peijnenburg and co-workers (39) provided interesting empirical kinetics relationships for the direct photohydrolysis of a series of monocyclic aromatic halides. Although quantum yields for these reactions must be affected by a variety of competing excited-state primary processes (e.g., radiationless decay to ground state, intersystem crossing, and reaction), it was possible to estimate the reaction quantum yields by using quantitative structure–activity relationships (QSARs) based on carbon–halogen bond strength and steric factor descriptors.

Halocarbons Sorbed on Natural Organic Matter. Photoreactions of aromatic halocarbons that strongly absorb solar radiation can be greatly accelerated in natural water samples or in aqueous solutions of NOM or humic substances (Table II). Although these effects can be due in part to indirect photoreactions or formation of photoreactive complexes, the results in Table II can be most simply explained in terms of increases in direct photoreaction rates of sorbed halocarbon in comparison to halocarbon in aqueous solution.

Table II. Photoreaction Kinetics (313 nm) of p,p'-Dichlorobenzophenone and DDE in Selected Natural Water Samples and Humic Substance Solution

Reaction Medium	DOC (mg/L)	k_p (h^{-1})	S_λ	k_p/S	Rel. k_p/S^c
p,p'-Dichlorobenzophenone					
Aucilla River water[a]	27	0.11	0.53	0.20	>4
Contech fulvic acid[b]	35	0.13	0.53	0.25	>5
Distilled water	<1	<0.05	1.00	<0.05	1.0
DDE					
Aucilla River water[a]	27	1.2 ± 0.1	0.53	2.2	1.7
Santa Fe River water[a]	10	1.4 ± 0.2	0.78	1.8	1.4
Suwannee River water[a]	27	1.0 ± 0.1	0.53	1.9	1.5
Contech fulvic acid[b]	35	1.0 ± 0.1	0.53	1.9	1.5
Distilled water	<1	1.3 ± 0.1	1.00	1.3	1.0

[a] Natural water sample.
[b] Aqueous solution of isolated fulvic acid.
[c] Ratio of k_p/S to k_p for distilled water.

In Table II, k_p is the pseudo-first-order photoreaction rate constant, DOC is the dissolved organic carbon concentration, and S_λ is the light attenuation factor. S_λ, which represents the fractional reduction in photoreaction rate of a very dilute halocarbon in a well-mixed system, was calculated by

$$S_\lambda = \frac{1 - e^{-A_\lambda}}{A_\lambda} \qquad (13)$$

where A_λ was the absorbance of the solution at 313 nm, the wavelength of the monochromatic radiation used in the experiments. The ratio, k_p/S_λ, as shown in Table II, represents rate constants corrected for light attenuation in the system. Such corrected rate constants can be compared with rate constants in distilled water to assess the photochemical effects of the NOM or humic substance on halocarbon reaction rates. Quantum yields were not computed in this case, because we had no data on the effects of sorption on the molar absorptivities (and thus light absorption rates) of the halocarbons.

Negligible photoreaction was observed for p,p'-dichlorobenzophenone (DCB), a DDT oxidation product, in air-saturated, distilled water (half-life >15 h at 313 nm). Nevertheless, this halocarbon photoreacted (313 nm) with half-lives corrected for light attenuation of about 3 h in a filtered natural-water sample and a solution of Contech fulvic acid (Table II). The greater than four- to fivefold enhancement in photoreaction rate in this case probably results from hydrogen atom abstraction from the natural organic matter by the DCB in its excited triplet state (eq 14).

$$\underset{R}{\overset{O}{\underset{\|}{C}}}\underset{R}{} \longrightarrow \left[\underset{R}{\overset{O}{\underset{\|}{C}}}\underset{R}{} \right]^{*} \xrightarrow{\text{Humic Substances}} \underset{R}{\overset{OH}{\underset{|}{C}}}\underset{R}{}\cdot \quad (14)$$

(where R is $p - \text{ClC}_6\text{H}_5 -$)

Triplet ketones are excellent H-atom abstractors (*40*). Water molecules with their high-energy H–O bonds do not readily transfer H atoms to triplet ketones, but the C–H bonds or phenolic O–H bonds in NOM are less energetic and more readily attacked. No experiments were conducted to determine the extent to which DCB was sorbed to the NOM in these studies. The photoreaction with the NOM may in part have occurred via diffusion in bulk aqueous solution. However, photoreaction probably occurred in the sorbed phase for two reasons. First, DCB is quite hydrophobic and thus has a high tendency to partition into nonaqueous phases. Second, the excited-state lifetime of triplet DCB is greatly shortened by oxygen quenching in bulk solution; thus the efficiency of diffusional reaction with the NOM is greatly reduced.

Kinetic data for the hydrophobic halocarbon, DDE, provide another example of a photoreaction rate enhancement attributable to sorption on NOM. The direct photoreaction rates of DDE, corrected for light attenuation effects, are increased in filtered natural water samples (0.2 μm) or a solution of a soil fulvic acid (Contech) with high DOC (Table II). As shown in Table II, the degree of rate enhancement is approximately equal to the enhancement in direct photoreaction rates in going from an aqueous to hydrocarbon solution (Table I) (*38*).

Given other evidence that hydrophobic organic compounds such as DDE sorbed to the NOM aggregates are in a hydrocarbon microenvironment (*30–34*), the results in Table II may be attributable to a solvent effect on direct photoreaction. This hypothesis is supported by the finding (eq 12) that only direct photoreaction products were observed in organic extracts of the irradiated natural water samples.

The enhancement in photoreaction rates of DDE and dichlorobenzophenone on sorption to dissolved organic matter (DOM) seem to be inconsistent with other reports that the fluorescence of aromatic hydrocarbons is strongly quenched by sorption to DOM (*30, 33, 34*). After all, fluorescence, like photoreaction, is an excited-state process, and fluorescence quenching indicates that locally high concentrations of excited-state quenchers are present in the NOM. However, most NOM is fluorescent itself, a condition indicating that the quenchers in NOM cannot quench all excited states efficiently (*41*). Possible explanations for the difference in the effects of sorption on halocarbon photoreactions (enhancement) and PAH fluorescence (quenching) are

- Specific solute–sorbent interactions related to steric effects. Sorption of solutes to NOM has been explained in terms of association or "trapping" of the solutes in hydrophobic cavities embedded within NOM aggregates [discussed by Schlautman and Morgan (34)]. The NOM is a mixture of chromophores that are capable of quenching excited states or photosensitizing reactions. The net effect may depend on the position of the sorbed solute within hydrophobic cavities of this mixture. The position may, in turn, depend on the size and shape of the solute.

- Nature of the excited states. Quenching and photosensitization can occur by differing mechanisms that depend on the nature of the excited states that participate. For example, quenching or sensitization by energy transfer depend on the relative excited-state energies of the donor and acceptor molecules. The excited-state energy of the donor should exceed that of the acceptor for efficient energy transfer to occur. Moreover, quenching can occur by mechanisms that depend on the nature of the excited states. Heavy atoms, for example, quench π–π* excited states much more efficiently than n–π* excited states (42). Thus, the excited states mediating the DCB and DDE photoreactions (n–π* and/or n–σ* excited states) may be much less quenchable than the π–π* excited states of aromatic hydrocarbons that fluoresce.

Halocarbons Sorbed on Sediments. If a significant fraction of halocarbon is sorbed in an unreactive microenvironment, then the kinetics can become limited by exchange between the unreactive (U–P) and photoreactive (R–P) parts of the system. Intrasorbent transport limitations have been observed for extremely hydrophobic halocarbons sorbed on soils and sediments suspended in water. The photoreactions of DDE in sediment suspensions provide a good example of such transport limitation (Figure 2) (43). Plots of log concentration versus time were linear for DDE photoreaction in water, but nonlinear in the sediment suspensions (Figure 2). The degree of nonlinearity depended upon the equilibration time of the suspensions prior to irradiation.

The kinetic model shown in Scheme I was used to interpret the results, with P representing aqueous phase DDE, and R–P and U–P representing DDE sorbed in reactive and unreactive phases, respectively. The results indicated that U–P was accessed much more slowly than R–P, thus accounting

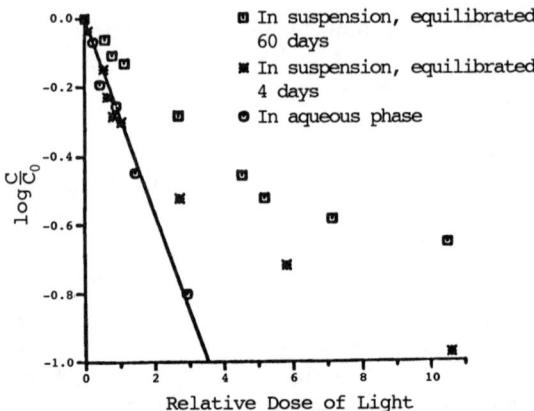

Figure 2. Effects of preequilibration on kinetics data for photoreaction of DDE in a suspension of Ohio River sediments. The term "Relative Dose of Light" refers to the irradiation time ratioed to the half-life observed in the aqueous phase. (Reproduced from reference 43).

for the dependence of the photoreaction kinetics on equilibration time prior to irradiation of the suspension. The aqueous and reactive phases were assumed to be in a rapid equilibrium. This type of intrasorbent diffusive limitation appears to be confined to extremely hydrophobic organic compounds sorbed on sediments or soils (35, 43). Given the data discussed, which indicate that DDE sorbed to DOM exhibits enhanced reactivity, most likely the unreactive phase(s) observed on the sediments are inner regions of the particles that are shielded from light.

Indirect Photoreactions. Enhancements of halocarbon photoreaction rates are not limited to compounds such as DDE and DCB, which directly absorb solar radiation. Such effects have also been observed with halocarbons that have little or no absorption of solar radiation. These reactions most likely occur through indirect mechanisms, in which the radiation is absorbed by natural chromophores. In this section, we provide evidence for such indirect photoreactions. Then, in subsequent sections, we discuss possible mechanisms for indirect photoreactions of halocarbons.

The photoreaction of lindane was markedly enhanced in solutions of a commercial soil fulvic acid (Contech) and in natural water samples with high DOM concentrations (DOC 10–27 mg of C/L) with estimated half-lives of 1–10 days versus a half-life in excess of 100 days in air-saturated distilled water (Table III). Moreover, Mudambi and Hassett (31) reported that the highly chlorinated pollutant, mirex, is reductively dehalogenated to photomirex with a half-life of 6–8 days under sunlight in Lake Ontario water (eq 15).

Table III. Photoreaction Kinetics (313 nm) of Lindane in Various Natural Water Samples and Humic Substance Solutions

Source of Water	DOC (mg/L)	k_p (h^{-1})	S_λ	k_p/S	Rel. k_p/S
Aucilla River	27	0.006	0.53	0.012	>6
Oglethorpe Pond	5	0.12	0.91	0.13	>65
Contech fulvic acid	27	0.069	0.63	0.11	>55
Fluka humic acid	35	<0.018	0.45	<0.04	—
Distilled water	<1	<0.002	1.00	<0.002	1.0

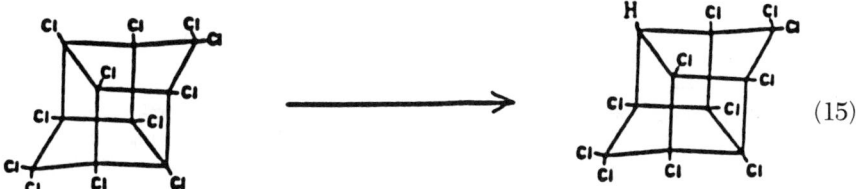

(15)

Mirex photoreduction in air-saturated distilled water was considerably slower (20–400 days). Finally, Zepp et al. (44) reported that the photoinduced dehydrochlorination of 1,1,1-trichloro-2,2-bis(p-methoxyphenyl)ethane (methoxychlor) is greatly accelerated in natural waters with high DOM concentrations.

Photoreductions in Aqueous Solution. Solvated electrons form on the irradiation of natural water samples or of aqueous solutions of natural organic matter isolated from surface waters (12, 45, 46). The solvated electron is a powerful reductant that reacts rapidly with electronegative substances such as chlorinated, brominated, and iodinated compounds. The interaction of photoejected electrons and a halocarbon is demonstrated by the laser flash photolysis data shown in Figure 3.

Trichloroacetate rapidly reacts with the solvated electrons produced by laser flash photolysis of natural organic matter isolated from the Suwannee River, and thus quenches the absorption of the electrons at 720 nm. The absorption is also quenched by the addition of other good electron acceptors, including oxygen, protons, or nitrous oxide. In natural waters, halocarbon concentrations are typically very low, and the dominant scavenger of solvated electrons is oxygen.

The reaction with halocarbons results in loss of halide, with concurrent formation of halocarbon-derived free radicals, R· (eq 5) (47). Zepp and coworkers (12) used the photoproduction of chloride ions from the halocarbon, 2-chloroethanol, a known solvated electron scavenger, in solutions of NOM

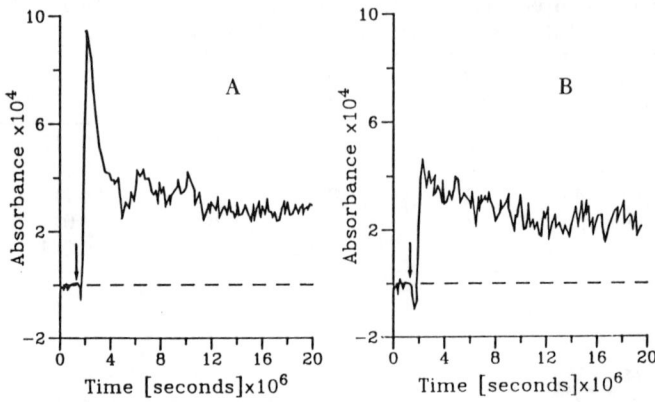

Figure 3. Effects of trichloroacetate (0.0050 M) on the absorption (720 nm) of solvated electrons produced by laser flash photolysis (355 nm) of argon-saturated solutions (pH 6.2) of natural organic matter isolated from the Suwannee River near Fargo, Georgia (12). A, no added trichloroacetate; B, with added trichloroacetate.

and humic substances to determine rates and quantum yields for electron production under continuous irradiation. The concentration of chloroethanol was sufficiently high that, based on its known rate constant for scavenging of solvated electrons (48), it quantitatively reacted with the electrons in bulk solution. As the results in Table IV indicate, the quantum yields (355 nm) varied considerably, depending on the source of the NOM.

Table IV. Quantum Yields (355 nm) for the Production of Solvated Electrons

Source of Natural Organic Solute	DOC (mg/L)	10^5 × Quantum Yield	
		Laser Flash	Continuous
Greifensee, Switzerland	32	760	12
Suwannee River, Okefenokee Swamp, Georgia[a]			
SHPO–A fraction	12	560	2.4
WHPO–A fraction	16	620	
HPO–N fraction	32	520	
HPI–A fraction	21	460	
Contech fulvic acid	4.4	400	2.3
Fluka humic acid	4.6	170	<0.8

[a] The fractions are explained in the Experimental section.
SOURCE: Reproduced from reference 12. Copyright 1987 American Chemical Society.

The quantum yields for the continuous irradiations were nearly 2 orders of magnitude lower than quantum yields observed in the laser flash photolysis studies (Table IV). Zepp et al. (12) attributed this difference either to biphotonic processes caused by high laser pulse intensity or, alternatively, to more efficient cage escape of electrons from the NOM aggregates caused by localized saturation of reducible functional groups within the aggregate in the laser experiments. The chloroethanol scavenging technique also was used to measure solvated electron photoproduction rates under April sunlight at Athens, Georgia, for dissolved organic matter from the Greifensee and Suwannee River.

These studies indicated that the photoproduction rates in solutions of varying composition were approximately proportional to the dissolved organic carbon (DOC) content. Assuming that the lifetime of the solvated electron in air-saturated water is 0.2 μs and that halocarbon concentrations are much lower than that of oxygen (and thus have little effect on the electron lifetime), the photoproduction rate observed in the Greifensee (DOC = 4 mg of C/L) corresponded to estimated near-surface pseudo-first-order photoreduction rate constants of about 10^{-3}/h for several halocarbons known to be present in natural waters (Table V). These estimates were derived by using previously measured rate constants for reaction of solvated electrons with the halocarbons (48).

Photoreductions of Sorbed Halocarbons. Comparisons of computed rates of halocarbon photoreduction by e_{aq}^- with observed rates in natural water samples indicate that other reaction pathways are more important. For example, recent results obtained with continuous irradiations indicate that chlorinated acetates produce chloride more efficiently than chloroethanol in solutions of dissolved organic matter that was isolated from the Suwannee River. Observed quantum yields (355 nm) for chloride production at pH 6.2 in aque-

Table V. Computed Values for Reductions of Selected Halocarbons by Solvated Electrons in Sunlight

Compound	$k\ (M^{-1}\ s^{-1})^a$	$k_r\ (h^{-1})^b$
Trichloroacetate	2.1×10^{10}	9.4×10^{-4}
Carbon tetrachloride	3.1×10^{10}	1.4×10^{-3}
Chloroform	3.0×10^{10}	1.3×10^{-3}
Trichloroethylene	1.9×10^{10}	8.5×10^{-4}
Methylchloroform	1.4×10^{10}	6.3×10^{-4}
Methyl iodide	1.7×10^{10}	7.6×10^{-4}

[a] Second-order rate constant for reaction of solvated electron with compound (38).
[b] Computed assuming $[e_{aq}^-]_{ss} = 1.2 \times 10^{-17}$ M, the average value estimated for a Swiss lake (Greifensee) during July, near the water surface.

ous solutions of Suwannee DOM containing 0.05 M halocarbon were 1.5×10^{-4} for trichloroacetate, 0.9×10^{-4} for chloroacetate, and 0.22×10^{-4} for 2-chloroethanol. For trichloroacetate, this difference is partly attributable to subsequent thermal reactions of the chlorinated free radical formed in the primary photoprocess.

In addition to the chloroacetates, as shown by comparisons with the estimates in Table V, the half-lives for both lindane and mirex in the natural water samples would have been considerably longer than those observed, had reaction with solvated electrons in bulk solution been the dominant mechanism for photoreaction. The higher efficiency of these halocarbon reactions may be attributable to sorption of the chloroacetates on the NOM, which permits more facile electron capture. Other possible pathways for reactions of sorbed halocarbons include direct photoreduction by excited states of the NOM, which, like solvated electrons, also are quenched by oxygen. Alternatively, the enhancement may involve other direct electron-transfer mechanisms such as amine–halomethane reactions. These alternative possibilities are examined in the following section.

Photoreductions of Halocarbon Complexes. Another possible mechanism for photoreduction involves complex formation between halocarbons and electron-donating natural substances (eq 7). Halocarbons form weak charge-transfer and excited-state complexes with electron donors such as amines and anilines (*49–55*). The complexes absorb at much longer wavelengths than the halocarbons themselves. On irradiation, these complexes photoreact via electron transfer from the amine to the halocarbon, and the result is dehalogenation of the halocarbon and oxidation of the amine. For example, halomethanes (*49–53, 55*) and haloaromatic compounds, such as DDT (*52*) and hexachlorobenzene (*54*), are reductively dechlorinated when irradiated in solutions of alkyl and aromatic amines. Nitrogen bases are widely observed substances in aquatic environments and are known components of NOM and humic substances (*56*). However, almost all of the studies of photoinduced halocarbon–amine reactions have been conducted in organic solvents, and little is known about the applicability of these studies to environmental conditions.

Ionic halocarbons, including halogenated carboxylic acids, may form photoreactive complexes with transition metals in the aquatic environment. Indeed, complexes of carboxylates with dissolved Fe(III) and iron oxides are very photoreactive under solar radiation (*28, 29*). Photoreactions of such complexes may help to explain the enhanced photoreactivity of chlorinated acetates in natural water samples.

Photooxidations of Halocarbons. In addition to the photoreduction pathways, other indirect photochemical reactions are significant in the oxidation of halogenated organic compounds. These reactions were reviewed elsewhere (57–59) and will not be discussed in detail here. Nitrate and nitrite ions photolyze in sunlight to form hydroxyl radicals that react rapidly with most halocarbons (60, 61). Reactions of photochemically produced reduced iron or copper species with hydrogen peroxide also produce hydroxyl radicals (62, 63). Studies by Mopper and Zhou (64) indicate that photoreaction of marine DOM produces much greater concentrations of OH radicals in the sea than was originally believed. These workers suggest that the NOM chromophores are the main source of the OH radicals (Table VI). Reactions of halocarbons with other transients such as singlet molecular oxygen, carbonate radicals, and organoperoxyl radicals also provide aquatic sinks in a more selective fashion.

Aquatic Sinks for Methylchloroform. Methylchloroform distributions in the troposphere have been used to estimate the concentrations of tropospheric hydroxyl radicals (65). The estimates assume that reaction with tropospheric OH radicals is the dominant sink. Aquatic sinks have been ignored. Wine and Chameides (66), however, presented model computations that indicated that hydrolysis and reaction with solvated electrons may be a significant sink for methylchloroform and other halocarbons in the open ocean.

Subsequent field studies by Butler et al. (11) showed "extensive negative saturation anomalies" for methylchloroform in surface waters of the Pacific Ocean. CFC-11, an unreactive gas that also dissolves in the upper layers of the sea, did not exhibit such anomalies. The methylchloroform data were used to compute fluxes of methylchloroform for various locations. As shown in Figure 4, the methylchloroform fluxes were variable but almost uniformly negative with a mean flux of -3.1 nmol/m^2 per day, close to the computed value for hydrolysis (67). Figure 4, however, shows that the negative fluxes were much greater than the mean at a number of locations.

Table VI. Measured and Estimated OH Steady-State Concentrations and Fluxes in Sunlight-Irradiated Seawater and Fresh Water

Sample	$10^{18} \times [OH]_{ss}$ (M)	Total Flux from DOC (%)
Open ocean surface water (Sargasso Sea)	1.1	>95%
Subtropical coastal water (Biscayne Bay, FL)	9.7	96%
Temperate coastal water (Vineyard Sound, MA)	10.6	>95%
Equatorial upwelled water (estimated)	7.4	65%
DOM-rich fresh water (Everglade)	840	>95%

SOURCE: Reproduced with permission from reference 64. Copyright 1990.

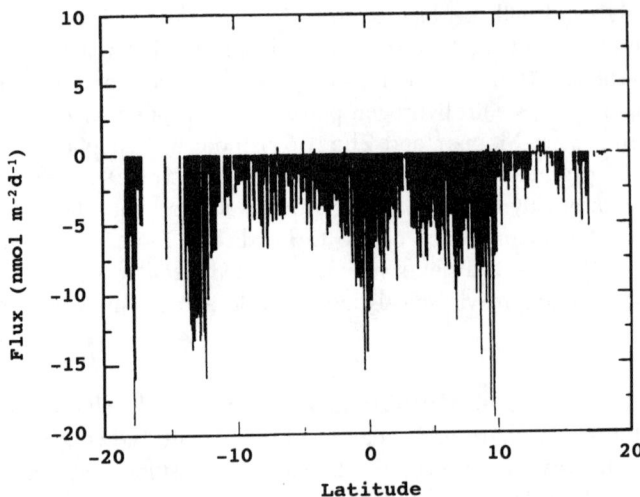

Figure 4. Computed fluxes of CH_3CCl_3 in the Pacific Ocean as a function of latitude. Butler et al. reported that the most negative fluxes corresponded to upwelling regions. (Reproduced with permission from reference 11. Copyright 1991 American Geophysical Union.)

Because the high values coincided with upwelling regions, Butler et al. (11) suggested that biological consumption may be involved. As illustrated by the data of Mopper and Zhou (Table VI) (64), upwelling and coastal regions of the sea also exhibit much higher photochemical activity, suggesting that photoreactions may be involved.

To further examine the possible role of various processes in the consumption of methylchloroform, pseudo-first-order rate constants were approximately computed for the mixed layer of different oceanic regions (Table VII).

The following assumptions were made:

Table VII. Estimated Pseudo-First-Order Loss Rates for Hydrolytic and Selected Light-Induced Processes Affecting Methylchloroform in the Surface Layer

Oceanic Type	k_h	k_e	k_{OH}	Σk
Open ocean[a]	2×10^{-9}	2×10^{-9}	4×10^{-11}	5×10^{-9}
Equatorial upwelling[b]	2×10^{-8}	1×10^{-8}	3×10^{-10}	3×10^{-8}
Near coastal[a]	2×10^{-9}	2×10^{-8}	4×10^{-10}	2×10^{-8}

NOTE: k_h represents hydrolysis rate constants, k_e denotes rate constants for reaction with solvated electrons, and k_{OH} symbolizes rate constants for reaction with OH radicals. All values are given in reciprocal seconds.
[a] Temperature assumed to be 285 K.
[b] Temperature assumed to be 298 K.

- Variations in solvated electron photoproduction approximately parallel the variations in OH radical production reported by Mopper and Zhou (Table VI) (dissolved organic matter appears to be the source of both transients).
- The surface mixed-layer depth approximately equals the photic zone.
- Light attenuation effects were computed by using equation 6, assuming typical diffuse attenuation coefficients at 350 nm for the different oceanic types (20, 24).
- Sea-surface temperatures were selected to reflect measurements of Butler et al. (11) in the tropics and, for comparison, the assumed temperature used by Wine and Chameides (66) in their earlier calculations.

The hydrolysis rates were computed by using data of Jeffers and co-workers (67). Though imprecise, these calculations reinforce the original conclusion of Wine and Chameides (66) that aquatic photochemical processes may be an important sink for volatile halocarbons, especially in upwelling and coastal regions.

Aquatic Sources of Halocarbons. One of the main pathways for the natural formation of halocarbons involves peroxidase enzymes (8, 68–70). Chloroperoxidases can catalyze the formation of chlorinated, brominated, and iodinated organic compounds; bromoperoxidases can produce only the latter two types of halocarbons (8, 69). The Fe–heme peroxidase enzymes, which are activated by hydrogen peroxide and organic peroxides, produce methyl halides as well as polyhalogenated compounds (Scheme II).

The net reaction is shown in equation 16.

$$AH + X^- + H^+ + H_2O_2 \longrightarrow AX + 2H_2O \qquad (16)$$

Other vanadium-containing enzymes appear to be particularly important in the production of polybrominated compounds in the sea (8, 70). The vanadium peroxidases utilize hydrogen peroxide to oxidize bromide or iodide, probably to corresponding hypohalite ions, which then react with NOM to produce polyhalogenated compounds (eq 17) (8, 70).

$$Br^- + H_2O_2 \xrightarrow{\text{V–BrPO}} HOBr \text{ or } Br_2 + DOM \longrightarrow RBr_x \qquad (17)$$

where RBr_x represents brominated 1,3-dicarbonyls, $CHBr_3$, and other polybrominated compounds. These enzymes are known to be present in macroalgae (8, 71) and are probably present in other biota as well (6, 8).

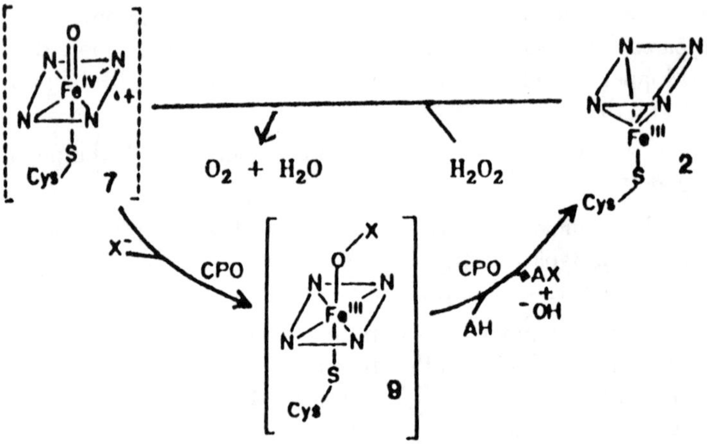

Scheme II. Mechanism for Fe–heme haloperoxidase-catalyzed production of halocarbons (59).

The peroxides that activate these enzymes are produced internally by the biota or, alternatively, are produced externally, mainly by photochemical processes in the sea (72). Gschwend et al. (71), having found very little halocarbon in the interior tissues of macroalgae that produce these compounds, concluded that much of the activity must be located in the surface tissues. This finding is consistent with the idea that extracellular peroxides are involved in the activation of these haloperoxidases. Moffett and Zafiriou (73) and Cooper and Zepp (74) provided evidence that peroxidases associated with aquatic particles in fresh water and coastal waters account for a large fraction of the decay of hydrogen peroxide, although it was not shown that haloperoxidases were responsible for the observed activity.

Bromide and iodide also can be oxidized by reactive oxidizing species that are produced photochemically (ref. 57 and references therein). The resulting halide radicals may react with NOM to produce halocarbons, although few field or laboratory studies have examined these reactions.

Conclusions

These discussions indicate that aquatic photochemical processes play an important role as sinks for halogenated pollutants and as a source of certain natural halocarbons, including volatile halocarbons that escape from the sea to the atmosphere. Those photoreactions that provide sinks often result in dehalogenation. Direct photoreactions such as photohydrolysis are likely to be the dominant photoreactions of aromatic halocarbons that strongly absorb solar radiation.

Natural substances, especially natural organic matter, have important effects on halocarbon photoreactions in the environment. These effects include the initiation of indirect photoreductions of dissolved or sorbed halocarbons via the intermediacy of solvated electrons or excited states. Evidence is presented here that sorption enhances the quantum efficiencies of these indirect photoreactions, although more studies are required to better define these processes.

The enhanced photoreactivity of sorbed nonionic halocarbons may involve photoreactive complexes with amines and other electron-donating substances. The enhanced photoreactivity of ionic halocarbons (e.g., chloroacetates) may involve complexes with DOM and transition metals. Additional studies are needed to examine the role of complexation in the aquatic photochemistry of halocarbons.

Natural chromophores that participate in indirect photoreactions or complexation tend to be highest in concentration in aquatic environments that are most biologically productive. Thus, the highest rates of indirect photoreactions of halocarbons are likely to be in fresh waters, coastal waters, and upwelling regions of the sea. Field studies looking for halocarbon sinks are most likely to be rewarded by focusing on these aquatic environments.

Photoreactions in the upper layers of the sea are continuously producing large amounts of hydrogen peroxide. A steady-state amount of approximately a teramole (10^{12} moles) of hydrogen peroxide are present in the upper mixed layer. This peroxide is decomposed mainly through various interactions with the biota. These interactions, in part, involve activation of enzymes called haloperoxidases that produce halocarbons, some of which volatilize to the atmosphere. Current evidence suggests that these reactions may be an important source of polyhalogenated methanes and ethanes, such as bromoform. Moreover, reactive oxygen species produced photochemically in the sea may react to produce halide radicals that can halogenate the organic matter.

Acknowledgments

We thank P. Freeman for helpful discussions of photoinduced electron transfer from amines to halocarbons. We thank R. Wever and A. Butler for useful information and exchanges related to haloperoxidase-catalyzed reactions. We thank J. Hoigné and J. Leenheer for providing samples of NOM isolated from the Greifensee in Switzerland and from the Suwannee River in southern Georgia.

References

1. Gribble, G. W. *J. Nat. Prod.* **1992**, *55*, 1353–1395.
2. Asplund, G.; Grimvall, A. *Environ. Sci. Technol.* **1991**, *25*, 1346–1350.
3. Singh, H. B.; Salas, L. J.; Stiles, R. E. *Environ. Sci. Technol.* **1982**, *16*, 872–880.

4. Harper, D. B. *Nature (London)* **1985**, *315*, 55–57.
5. Cicerone, R. J.; Elliot, S.; Turco, R. P. *Science (Washington, DC)* **1991**, *254*, 1191–1194.
6. Singh, H. B.; Salas, L. J.; Stiles, R. E. *J. Geophys. Res.* **1983**, *88*, 3684–3690.
7. Penkett, S. A.; Jones, B. M. R.; Rycrofft, M. J.; Simons, D. A. *Nature (London)* **1985**, *318*, 550–553.
8. Wever, R. In *Microbial Production and Consumption of Greenhouse Gases: Methane, Nitrogen Oxides, and Halomethanes;* Rogers J. E.; Whitman, W. B., Eds.; American Society for Microbiology: Washington, DC, 1991; pp 277–287.
9. Zepp, R. G. In *Pesticide Chemistry: Advances in International Research, Development, and Legislation;* Frehse, H., Ed.; VCH: New York, 1991; pp 329–345.
10. Crossland, N. O.; Wolff, C. J. M. *Environ. Toxicol. Chem.* **1985**, *4*, 73–86.
11. Butler, J. H.; Elkins, J. W.; Thompson, T. M.; Hall, B. D.; Swanson, T. H.; Koropalov, V. *J. Geophys. Res.* **1991**, *96*, 22347–22356.
12. Zepp, R. G.; Braun, A. M.; Hoigné, J.; Leenheer, J. A. *Environ. Sci. Technol.* **1987**, *21*, 485–490.
13. Gamble, D. S. *Can. J. Chem.* **1970**, *48*, 2662–2669.
14. Leenheer, J. A.; Noyes, T. I. *U.S. Geol. Surv. Water-Supply Pap.* **1984**, No. 2230. (Copies are available from U.S. Geological Survey, Books and Open-File Reports Section, Box 25425, Federal Center, Denver, CO 80225–0425.)
15. Fuchs, F.; Raue, B. *Vom Wasser* **1981**, *57*, 95–106.
16. Hatchard, C. G.; Parker, C. A. *Proc. Roy. Soc. Lon. Ser. A* **1956**, *235*, 518–536.
17. Stookey, L. L. *Anal. Chem.* **1970**, *42*, 779–781.
18. Cohen, Y. *Environ. Sci. Technol.* **1986**, *20*, 538–544.
19. Burns, L. A.; Cline, D. M.; Lassiter, R. R. *Exposure Analysis Modeling System Exams: User Manual and System Documentation;* EPA-600/3-82-023; U.S. Environmental Protection Agency: Athens, GA, 1982; 443 pp.
20. Baker, K. S.; Smith, R. C. *Limnol. Oceanogr.* **1982**, *27*, 500–509.
21. Jerlov, N. *Marine Optics;* Elsevier: Amsterdam, The Netherlands, 1976.
22. Miller, G. C.; Zepp, R. G. *Water Res.* **1979**, *13*, 453–459.
23. Plane, J. M. C.; Zika, R. G.; Zepp, R. G.; Burns, L. A. In *Photochemistry of Environmental Aquatic Systems;* Zika, R. G.; Cooper, W. J., Eds.; ACS Symposium Series 327; American Chemical Society: Washington, DC, 1987; pp 250–267.
24. Smith, R. C. *Photochem. Photobiol.* **1989**, *50*, 459.
25. Zepp, R. G.; Cline, D. M. *Environ. Sci. Technol.* **1977**, *11*, 359–366.
26. Sikorski, R. J.; Zika, R. G. *J. Geophys. Res. Oceans* **1993**, *98*, 2315–2328.
27. Hoigné, J. In *Aquatic Chemical Kinetics;* Stumm, W., Ed.; Wiley: New York, 1990; pp 43–70.
28. Sulzberger, B. In *Aquatic Chemical Kinetics;* Stumm, W., Ed.; Wiley: New York, 1990; pp 401–430.
29. Faust, B. C. In *Aquatic and Surface Photochemistry;* Helz, G. R.; Zepp, R. G.; Crosby, D. G., Eds.; Lewis Publishers: Chelsea, MI, 1994; pp 3–37.
30. Gauthier, T. D.; Shane, E. C.; Guerin, W. F.; Seitz, W. R.; Grant, C. L. *Environ. Sci. Technol.* **1986**, *20*, 1162–1166.
31. Mudambi, A. R.; Hassett, J. P. *Chemosphere* **1988**, *17*, 1133–1146.
32. Ackman, D. R.; Hornbuckle, K. C.; Eisenreich, S. J. *Environ. Sci. Technol.* **1993**, *27*, 75–87.
33. Backus, D. A.; Gschwend, P. M. *Environ. Sci. Technol.* **1990**, *24*, 1214.
34. Schlautman, M. A.; Morgan, J. J. *Environ. Sci. Technol.* **1993**, *27*, 961–969.
35. Zepp, R. G.; Wolfe, N. L. In *Aquatic Surface Chemistry: Chemical Processes at the Particle–Water Interface;* Stumm, W., Ed.; Wiley: New York, 1987; pp 423–455.
36. Zika, R. G.; Gidel, L. T.; Davis, D. D. *Geophys. Res. Lett.* **1984**, *11*, 353–356.

37. Kropp, P. J. *Acc. Chem. Res.* **1984**, *17*, 131–137.
38. Zepp, R. G.; Wolfe, N. L.; Azarraga, L. V.; Cox, R. H.; Pape, C. W. *Arch. Environ. Contam. Toxicol.* **1977**, *6*, 305–314.
39. Peijnenburg, W. J. G. M.; de Beer, K. G. M.; de Haan, M. W. A.; den Hollander, H. A.; Stegeman, M. H. L.; Verboom, H. *Environ. Sci. Technol.* **1992**, *26*, 2116–2121.
40. Wagner, P. J. *Acc. Chem. Res.* **1989**, *22*, 300.
41. Green, S. A.; Morel, F. M. M.; Blough, N. V. *Environ. Sci. Technol.* **1992**, *26*, 294–302.
42. Wagner, P. J. *J. Am. Chem. Soc.* **1966**, *88*, 5672.
43. Zepp, R. G.; Schlotzhauer, P. F. *Chemosphere* **1981**, *10*, 453–460.
44. Zepp, R. G.; Wolfe, N. L.; Gordon, J. A.; Fincher, R. A. *J. Agric. Food Chem.* **1976**, *24*, 727–733.
45. Fischer, A. M.; Winterle, J. S.; Mill, T. In *Photochemistry of Environmental Aquatic Systems;* Zika, R. G.; Cooper, W. J., Eds.; ACS Symposium Series 327; American Chemical Society: Washington, DC, 1987; pp 141–156.
46. Power, J. F.; Sharma, D. K.; Langford, C. H.; Bonneau, R.; Joussot-Dubien, J. In *Photochemistry of Environmental Aquatic Systems;* Zika, R. G.; Cooper, W. J., Eds.; ACS Symposium Series 327; American Chemical Society: Washington, DC, 1987; pp 157–173.
47. Claridge, R. F. C.; Willard, J. E. *J. Am. Chem. Soc.* **1965**, *87*, 4992–4997.
48. Buxton, G. V.; Greenstock, C. L.; Helman, W. P.; Ross, A. P. *J. Phys. Chem. Ref. Data* **1988**, *17*, 513–886.
49. Stevenson, D. P.; Coppinger, G. M. *J. Am. Chem. Soc.* **1962**, *84*, 149–152.
50. Davis, K. M. C.; Farmer, M. F. *J. Chem. Soc. B* 1967, 28–32.
51. Lautenberger, W. J.; Jones, E. N.; Miller, J. G. *J. Am. Chem. Soc.* **1968**, *90*, 1110–1115.
52. Miller, L. I.; Narang, R. S. *Science (Washington, DC)* **1970**, *169*, 368–370.
53. Wryzkowski, K.; Grodowski, M.; Weiss, M.; Latowski, T. *Photochem. Photobiol.* **1978**, *28*, 311–318.
54. Freeman, P. K.; Srinivasa, R.; Campbell, J.-A.; Deinzer, M. L. *J. Am. Chem. Soc.* **1986**, *108*, 5531–5536.
55. Boszczyk, W.; Latowski, T. *Z. Naturforsch* **1989**, *44b*, 1589–1592.
56. Thurman, E. M. *Organic Geochemistry of Natural Waters;* Martinus Nijhoff–Dr. W. Junk Publishers: Dordrecht, The Netherlands, 1985.
57. Waite, T. D.; Sawyer, D. T.; Zafiriou, O. C. *Appl. Geochem.* **1987**, *3*, 9–17.
58. Hoigné, J.; Faust, B. C.; Haag, W. R; Zepp, R. G. In *Aquatic Humic Substances: Influence on Fate and Treatment of Pollutants;* MacCarthy, P.; Suffet, I. H., Eds.; Advances in Chemistry 219; American Chemical Society: Washington, DC, 1988; pp 363–384.
59. Blough, N. V.; Zepp, R. G. In *Reactive Oxygen Species in Chemistry and Biochemistry;* Foote, C. S.; Valentine, J. S., Eds.; Chapman & Hall: New York, 1995.
60. Zafiriou, O. C.; Bonneau, R. *Photochem. Photobiol.* **1987**, *45*, 723–727.
61. Zepp, R. G.; Hoigné, J.; Bader, H. *Environ. Sci. Technol.* **1987**, *21*, 443–450.
62. Zepp, R. G.; Faust, B. C.; Hoigné, J. *Environ. Sci. Technol.* **1992**, *26*, 313–319.
63. Hayase, K.; Zepp, R. G. *Environ. Sci. Technol.* **1991**, *25*, 1273–1279.
64. Mopper, K.; Zhou, X. *Science (Washington, DC)* **1990**, *250*, 661–664.
65. Prinn, R.; Cunnold, D.; Simmonds, P.; Alyea, F.; Boldi, R.; Crawford, A; Fraser, P.; Gutzler, D.; Hartley, D.; Rosen, R.; Rasmussen, R. *J. Geophys. Res.* **1992**, *97*, 2445–2462.
66. Wine, P. H.; Chameides, W. L. In *Scientific Assessment of Stratospheric Ozone;* Global Ozone Research and Monitoring Project; World Meteorology Organization:

Geneva, Switzerland, 1989; Appendix of the AFEAS Report, Vol. 2, Rep. 20, pp 273–295.
67. Jeffers, P. M.; Ward, L. M.; Woytowitch, L. M.; Wolfe, N. L. *Environ. Sci. Technol.* **1989,** *23,* 965–969.
68. Neidleman, S. L.; Geigart, J. *Biohalogenation: Principles, Basic Roles, and Applications;* Halstead: New York, 1986.
69. Dawson, J. H. *Science (Washington, DC)* **1988,** *240,* 433–439.
70. Everett, R. R.; Butler, A. *Inorg. Chem.* **1989,** *28,* 393–395.
71. Gschwend, P. M.; MacFarlane, J. K.; Newman, K. A. *Science (Washington, DC)* **1985,** *227,* 1033–1035.
72. Zika, R. G. In *Effects of Solar Ultraviolet Radiation on Biogeochemical Dynamics in Aquatic Environments;* Blough, N. V.; Zepp, R. G., Eds.; Woods Hole Oceanographic Institute Technical Report WHOI-90-09; Woods Hole Oceanographic Institute: Woods Hole, MA, 1990.
73. Moffett, J. W.; Zafiriou, O. C. *Limnol. Oceanogr.* **1990,** *35,* 1221–1229.
74. Cooper, W. J.; Zepp, R. G. *Can. J. Fish. Aquat. Sci.* **1990,** *47,* 883–893.
75. Hwang, H-M.; Hodson, R. E.; Lee, R. F. In *Photochemistry of Environmental Aquatic Systems;* Zika, R. G.; Cooper, W. J., Eds.; ACS Symposium Series 327; American Chemical Society: Washington, DC, 1987; pp 27–43.
76. Dulin, D.; Drossman, H.; Mill, T. *Environ. Sci. Technol.* **1986,** *20,* 72–76.
77. Miller, G. C.; Miile, M. J.; Crosby, D. G.; Sontum, S.; Zepp, R. G. *Tetrahedron* **1979,** *35,* 1797–1800.
78. Skurlatov, Y. I.; Zepp, R. G.; Baughman, G. L. *J. Agric. Food Chem.* **1983,** *31,* 1065–1071.
79. Wolfe, N. L.; Zepp, R. G.; Schlotzhauer, P. F.; Sink, R. M. *Chemosphere* **1982,** *11,* 91–99.

RECEIVED for review October 23, 1992. ACCEPTED revised manuscript August 11, 1993.

14

Photochemical Reductive Dissolution of Lepidocrocite

Effect of pH

Barbara Sulzberger and Hansulrich Laubscher

Swiss Federal Institute for Environmental Science and Technology (EAWAG), CH–8600 Dübendorf, Switzerland

> *The kinetics of the photochemical reductive dissolution of lepidocrocite (γ-FeOOH) with oxalate as the reductant depends strongly on pH; both the rate and the overall rate constant, k_o, decrease with increasing pH. This behavior means that the pH dependence of the rate does not simply reflect the pH dependence of oxalate adsorption at the lepidocrocite surface. Between pH 3 and 5, the log k_o values can be fitted with a straight line. The dependence of k_o on the concentration of surface protons, $\{>FeOH_2^+\}$, can be estimated from the slope of this line and from the protonation curve of lepidocrocite: $k_o \propto \{>FeOH_2^+\}^{1.6}$. The value of 1.6, which can be considered only a rough estimate, is not too different from the theoretically expected value of 2 for the proton-catalyzed detachment of reduced surface iron centers (i.e., of surface metal centers with the formal oxidation state of II).*

THE RATE OF REDUCTIVE DISSOLUTION of oxide minerals such as iron(III) and manganese(III,IV) (hydr)oxides depends strongly on pH; it generally increases with decreasing pH (1–3). Several phenomena may contribute to this pH dependence.

1. The surface concentration of a reductant depends on pH. The pH dependence of adsorption of weak acids and anions can, with the help of the ligand-exchange model, be predicted from the acid–base equilibria of both the anion and the (hydr)oxide;

0065–2393/95/0244–0279$08.00/0
© 1995 American Chemical Society

the extent of adsorption is usually maximal around the pK_a value of the corresponding weak acid (4, 5).

2. Protonation of the (hydr)oxide surface accelerates the nonreductive dissolution of (hydr)oxide minerals (6, 7). This effect is explained in terms of polarization of the bonds between surface metal centers and the neighboring oxygen ions that are additionally protonated. In turn, the polarization causes detachment of the surface metal centers (8). Similarly, surface protonation may accelerate detachment of reduced surface metal centers. The efficiency of detachment is a key parameter in the overall photochemical reductive dissolution kinetics, because in oxic environments detachment of reduced surface metal centers may be in competition with their oxidation (9, 10).

3. Readsorption of Fe_{aq}^{2+} and Mn_{aq}^{2+}, which are formed from the reductive dissolution of Fe(III) and Mn(III,IV) (hydr)oxides, becomes more important with increasing pH (11). At higher pH, these ions may block the (hydr)oxide surface and thus limit adsorption of a reductant.

4. Because of the pH-dependent solubility of (hydr)oxides, the thermodynamic driving force of reductive dissolution increases with decreasing pH. Thus, the rate constant of reductive dissolution may follow the same trend.

To evaluate the predominant pH effect on the kinetics of reductive dissolution of oxide minerals, it is indispensable to examine the pH dependence of both the rate, R, and the overall rate constant, k_o, of reductive dissolution. The rate constant differs from the rate by the surface concentration of the reductant, $R = k_o \{Red_{ads}\}$. If the rate constant k_o is independent of pH, then the pH dependence of the rate is solely due to the pH dependence of adsorption of the reductant at the (hydr)oxide surface. If, on the contrary, the rate constant is dependent on pH, then other pH effects have to be considered.

This chapter reports on laboratory experiments on the pH dependence of the kinetics of the photochemical reductive dissolution of lepidocrocite (γ-FeOOH) as the solid phase and oxalate as the reductant–ligand. Oxalate, an analog for natural organic acids, occurs at considerably high concentrations in atmospheric waters (12). Furthermore, oxalate is a convenient model compound because it does not undergo redox reactions with surface Fe(III) of lepidocrocite in the absence of light. Furthermore, the ligand-promoted, nonreductive dissolution of γ-FeOOH with oxalate as the ligand is much slower than the reductive dissolution. Thus, the surface concentration of oxalate as a function of pH can easily be determined experimentally.

The evaluation of the effect of pH on the light-induced reductive dissolution of iron(III) (hydr)oxides is a key to the identification of the important parameters that control the kinetics of dissolved iron(II) formation in atmospheric water. Because of the comparatively low pH values and the presence of reductants in atmospheric water, a high fraction of total iron is present as dissolved iron(II). Maximal concentrations of 0.2 mM Fe(II) were reported in fog water; the concentration of Fe(II) increased with decreasing pH and with increasing light intensity (*13*). Iron(II) is an important intermediate for the formation of monomeric Fe(III) species, which may control iron uptake by phytoplankton (*14, 15*). Most likely, the atmospheric transport of continental weathering products is responsible for much of the mineral material and Fe entering the open ocean, and it is probably the dominant source of the nutrient Fe in the photic zone (*16*).

Background

Rate Expression of the Photochemical Reductive Dissolution of γ-FeOOH by Oxalate.

The kinetics of the light-induced reductive dissolution of oxide minerals obey the general rate expression of surface-controlled reactions. The rate is proportional to the concentration of the adsorbed reductant, as in the case of adsorbed oxalate:

$$\frac{d[Fe(II)_{aq}]}{dt} = k_o \{ > Fe^{III}C_2O_4^- \} \quad (1)$$

where $\{>Fe^{III}C_2O_4^-\}$ represents the concentration of the oxalato surface complex formed from the specific adsorption of oxalate at the lepidocrocite surface. k_o is the overall rate constant of the photochemical reductive dissolution; it depends on the specific rate of light absorption by the chromophore, on the quantum yield of electron transfer, and on the efficiency of detachment of reduced surface iron from the crystal lattice, which in turn may depend on pH (*10*).

Rate Expression of the Proton-Catalyzed Dissolution of Oxide Minerals.

Surface protonation may accelerate detachment of reduced surface metal centers. The process is similar to the acceleration of detachment of nonreduced surface metal centers by additional protonation of their nearest-neighbor hydroxo and oxo groups. Therefore we briefly discuss the theory of the proton-catalyzed dissolution of oxide minerals (*8*).

Combining concepts of surface coordination chemistry with established models of lattice statistics and activated complex theory, Wieland et al. (*8*) proposed a general rate expression for the proton-catalyzed dissolution of oxide minerals:

$$R_H = k_H \cdot x_a \cdot P_j \cdot S_T \quad (2)$$

where R_H is the proton-catalyzed dissolution rate (mole per square meter per second), k_H is the rate constant per second, x_a is the mole fraction of dissolution active sites (−), P_j is the probability of finding a specific site in the coordinative arrangement of the precursor complex (−), and S_T is the concentration of total surface sites (mole per square meter). The probability of finding a surface metal center surrounded by j protonated hydroxo and oxo groups, assuming small mole fractions of protonated surface groups $x_H \ll 1$ (8), is

$$P_j = \binom{4}{j} x_H^j \tag{3}$$

Combination of equations 2 and 3 yields

$$R_H = k_H \cdot x_a \cdot \binom{4}{j} \cdot x_H^j \cdot S_T \tag{4}$$

The mole fraction x_H is defined as the concentration of protonated surface sites, $\{>MOH_2^+\}$, divided by the concentration of total surface sites, S_T:

$$x_H = \frac{\{>MOH_2^+\}}{S_T} \tag{5}$$

Then equation 4 becomes

$$R_H = k'_H \{>MOH_2^+\}^j \tag{6}$$

where

$$k'_H = k_H \cdot x_a \cdot \binom{4}{j} \cdot S_T^{-(j-1)} \tag{7}$$

Experimentally, the rate of the proton-catalyzed dissolution is determined as a function of pH and as fractional order from proton activity:

$$R_H = k_H \cdot [H^+]^n \quad n < 1 \tag{8}$$

or

$$\log R_H = \log k_H - n \cdot pH \quad n < 1 \tag{9}$$

As an example, $n = 0.17$ for BeO (17). In order to assess experimentally the dependence of the rate, R_H, from the surface concentration of protons, the surface protonation as a function of pH has to be known.

As described by Wieland et al. (8), the same protonation curve applies for various oxide minerals if the surface concentration of protons, $\{>\text{MOH}_2^+\}$, is plotted as a function of $\text{pH}_{zpc} - \text{pH} = \Delta\text{pH}$, where the index zpc denotes the zero point of charge caused by binding or dissociation of protons. In the range $\Delta\text{pH} > 1$, a straight line fits all the experimental data within a factor of 2. Therefore, a suitable approximation can be given by a Freundlich master isotherm:

$$\log \{>\text{MOH}_2^+\} = \log K_F + m(\text{pH}_{zpc} - \text{pH}) \tag{10}$$

$$\text{pH} = \frac{1}{m} \log K_F + \text{pH}_{zpc} - \frac{1}{m} \log \{>\text{MOH}_2^+\} \tag{11}$$

The pH_{zpc} of lepidocrocite was determined to be 7.5 (18). We fitted the protonation curve of lepidocrocite (18) with a Freundlich isotherm (eq 10) with a slope of $m \approx 0.17$.

The pH in equation 9 can be replaced by equation 11, yielding

$$\log R_H = \log k_H - \frac{n}{m} \log K_F - n\text{pH}_{zpc} + \frac{n}{m} \log \{>\text{MOH}_2^+\} \tag{12}$$

and with

$$\log k'_H = \log k_H - \frac{n}{m} \log K_F - n\text{pH}_{zpc} \tag{13}$$

$$\log R_H = \log k'_H + \frac{n}{m} \log \{>\text{MOH}_2^+\} \tag{14}$$

The fraction n/m is equal to the exponent j in equation 6. It corresponds approximately to the oxidation number of the surface metal center; $j = 2.1$ for BeO (17), $j = 3.25$ for α-FeOOH (6), and $j = 3.95$ for SiO_2 (quartz) (19, 20). The following question arises: What is the value of j in reductive, proton-catalyzed dissolution of oxide minerals?

Experimental Section

All chemicals were reagent grade, and the solutions were prepared with high-purity water from a Millipore system. The pH measurements were carried out with a combined glass electrode (Metrohm) standardized with pH-buffer solutions (Merck). The lepidocrocite suspensions were prepared according to the procedure developed by Brauer (21) by oxidation of a FeCl_2 solution with NaNO_2 as the oxidant, in presence of hexamethylenetetramine at 60 °C for 3 h. To remove excess chloride, the lepidocrocite suspensions were washed several times by centrifugation and resuspension in 0.01 M NaOH. X-ray powder diffraction patterns examined with a high-resolution Guinier–NONIUS camera (Mark IV) and Feα1 radiation were identical with pure lepidocrocite standards. The specific surface area as determined by Brunauer–Emmett–Teller measurements was 89 m^2/g.

The concentrations of ferrous iron were determined colorimetrically with ferrozene by measuring the absorbance of the Fe(II)–ferrozene complex at 562 nm according to a modified method described by Stookey (22). The concentrations of dissolved oxalate were determined by measuring the β counts of ^{14}C-labeled oxalate. The ionic medium used for the adsorption and dissolution experiments was 5 mM NaClO$_4$. The pH values were established with HClO$_4$ and NaOH solutions. The lepidocrocite concentration used in the adsorption and dissolution experiments was 0.5 g/L.

All the experiments were carried out at constant temperature (25 °C) by using Pyrex glass vessels with a water jacket or by keeping bottles in a water bath whose temperature was controlled with a thermostat. For the photochemical experiments, an experimental setup was used as described by Siffert and Sulzberger (9). The experiments were carried out with white light from a 1000-W high-pressure xenon lamp (OSRAM) that was filtered by the bottom window of the Pyrex glass vessel, which acts as a cutoff filter ($\lambda_{1/2}$ = 350 nm). The incident light intensity was $I_o \approx 0.5$ kW/m^2. The reaction volume was 300 mL.

The suspensions were purged with nitrogen that had previously passed a Jones reductor and were vigorously stirred throughout the dissolution experiments to prevent settling of the particles. The ligand was added from a solution of Na$_2$C$_2$O$_4$ with trace concentrations (~1%) of radiolabeled (^{14}C) oxalic acid. For the Fe(II) determination 1-mL aliquots were mixed with 0.06 mL of H$_2$SO$_4$ (3.6 M) immediately after filtration through 0.1-μm Millipore filters. The extent of adsorption of oxalate at the lepidocrocite surface was established by equilibrating 10-mL aliquots of lepidocrocite suspension containing 1 mM oxalate at various pH values for 15 min. The suspensions were then filtered through 0.1-μm Millipore filters, and the concentration of oxalate was determined in the filtrates.

Results

In deaerated lepidocrocite suspensions, the photochemical reductive dissolution with oxalate as the reductant occurs according to the following overall stoichiometry (Figure 1):

$$2\gamma\text{-FeOOH} + \text{C}_2\text{O}_4^{2-} + 6\text{H}^+ \xrightarrow{h\nu} 2\text{Fe}^{2+} + 2\text{CO}_2 + 4\text{H}_2\text{O} \quad (15)$$

Thus, two dissolved iron(II) are formed per oxidized oxalate. This ratio exists

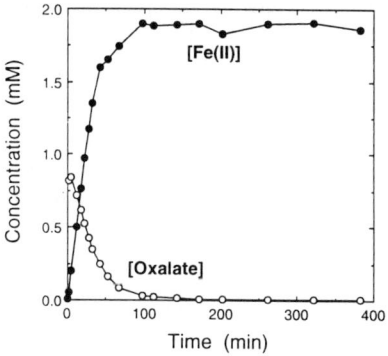

Figure 1. Concentration of dissolved Fe(II) and of oxalate as a function of time upon photochemical reductive dissolution of γ-FeOOH in a deaerated suspension at pH 3. Initial oxalate concentration was 1 mM. (Reproduced with permission from reference 10. Copyright 1994 Lewis Publishers.)

because the oxalate radical undergoes a fast decarboxylation reaction, yielding CO_2 and the CO_2^- radical (23), which is a strong reductant and can, in a thermal redox reaction, reduce a second surface iron(III).

Figure 2 shows the concentration of dissolved Fe(II) as a function of time at various pH values upon the photochemical reductive dissolution of lepidocrocite (deaerated suspensions) with oxalate as the reductant. The rate of dissolved Fe(II) formation (i.e., the slope of the straight lines through the experimental points) decreases strongly with increasing pH.

To calculate the overall rate constants of the photochemical reductive dissolution of lepidocrocite, we experimentally determined the surface con-

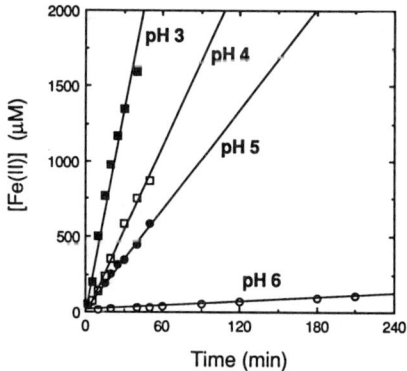

Figure 2. pH dependence of the rate of the photochemical reductive dissolution of γ-FeOOH with oxalate as the reductant in deaerated suspensions. Initial oxalate concentration was 1 mM. (Reproduced with permission from reference 10. Copyright 1994 Lewis Publishers.)

centration of the specifically adsorbed oxalate at several pH values (Figure 3). As expected, the extent of oxalate adsorption at the lepidocrocite surface reaches a maximum near 3.5, the second pK_a value of oxalate.

According to equation 1, the overall rate constant is the rate of the photochemical reductive dissolution of lepidocrocite divided by the surface concentration of oxalate. The overall rate constant as a function of pH is shown in Figure 4. The three experimental points between pH 3 and 5 can be fitted reasonably well with a straight line:

$$\log k_o = \log k_{RED} - n \cdot pH \tag{16}$$

with a slope of $n = 0.27$. k_{RED} is the pH-independent part of the overall rate constant. The experimental value of log k_o at pH 6 lies below the straight line drawn through the experimental points between pH 3 and 5. Possible reasons for this deviation are discussed in the next section.

Discussion

Under our experimental conditions, the overall rate constant of the photochemical reductive dissolution of lepidocrocite in the presence of oxalate is pH-dependent. Thus, the pH dependence of the rate reflects more than the pH dependence of oxalate adsorption at the lepidocrocite surface. Various pH effects may account for this observed pH dependence of k_o. One possibility is catalysis of detachment of the reduced surface iron centers by protonation of their neighboring hydroxo and oxo groups. The following question then arises: How does the observed rate constant, k_o, depend on surface protonation? The general rate expression of the proton-catalyzed dissolution of oxide

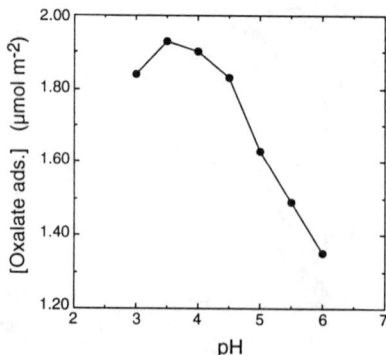

Figure 3. Surface concentration of oxalate as a function of pH upon the specific adsorption of oxalate at the lepidocrocite surface. (Reproduced with permission from reference 10. Copyright 1994 Lewis Publishers.)

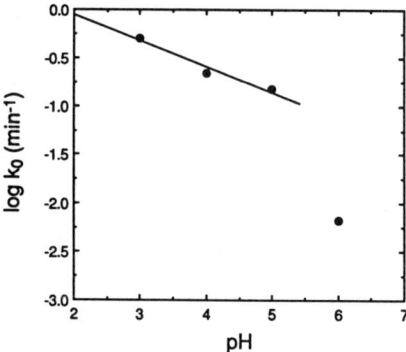

Figure 4. *Overall rate constant of the photochemical reductive dissolution of lepidocrocite with oxalate as the reductant, as a function of pH.*

minerals, equation 14, may be applied to the overall rate constant of the photochemical reductive dissolution:

$$\log k_o = \log k'_{RED} + \frac{n}{m} \log \{>FeOH_2^+\} \quad (17)$$

where

$$\log k'_{RED} = \log k_{RED} - \frac{n}{m} \log K_F - n \cdot pH_{zpc} \quad (18)$$

With the experimentally determined values $n = 0.27$ and $m \approx 0.17$, j can be estimated: $j = n/m \approx 1.6$. This result indicates that, between pH 3 and 5, catalysis of detachment of the reduced surface iron centers by protons may be the predominant pH effect in this heterogeneous photoredox process, because the estimated j value of 1.6 is not too different from the theoretically expected value of 2 for reduced surface iron centers. However, readsorption of the photochemically formed Fe(II), blocking surface sites for the adsorption of oxalate, also has to be taken into account. Because the extent of adsorption of Fe(II) at the surface of lepidocrocite is expected to increase with increasing pH (*11*), this effect becomes increasingly important. Thus, at pH 6 a large fraction of the photochemically formed Fe(II) may become readsorbed at the reconstituted lepidocrocite surface before oxalate gets adsorbed. This process may explain the relatively low value of $\log k_o$ at pH 6.

Last but not least, the pH dependence of the thermodynamic driving force may account for the pH dependence of the observed rate constant. Reductive dissolution of lepidocrocite with oxalate as the reductant is an exergonic process in the investigated pH range. The Gibbs free energies of the overall dissolution reaction, ΔG_R, are −216.1, −181.4, −146.7, and −113.9 kJ/mol for pH 3, 4, 5, and 6, respectively, with a 1 mM iron(II) concentration. In using the Gibbs free reaction energy to evaluate the dependence of the free activation energy, the individual steps of the overall reaction must be considered (Figure 5).

The activation barrier of the one-electron-transfer step, ΔG_{ET}^F, is expected to decrease with negatively increasing ΔG_{ET}. The free activation energy of electron transfer, however, is largely overcome by light absorption by the chromophore, which is either the surface complex formed from the specific adsorption of oxalate at the lepidocrocite surface or the lepidocrocite bulk phase (9). Thus, most likely not the activation barrier of electron transfer but the activation barrier of detachment of the reduced surface iron centers from the crystal lattice determines the rate of the overall photochemical dissolution process. The latter rate depends on the coordinative environment of the surface iron(II) (i.e., protonation of the neighboring hydroxo and oxo groups may lower the free activation energy of the detachment step).

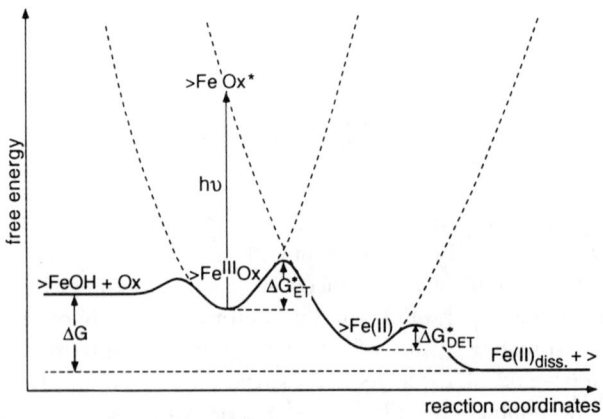

Figure 5. Qualitative representation of the energetics of the photochemical reductive dissolution of lepidocrocite with oxalate as the electron donor. >Fe^{III}Ox is the iron(III) oxalato surface complex (i.e., the precursor complex) in its electronically ground state and >FeOx* is the precursor complex in its electronically excited state. ΔG is the free energy of the overall reductive dissolution process; $\Delta G_{ET}^\ddagger$ is the free energy of activation of formation of a reduced surface iron, >Fe(II), and the oxidized oxalate, $C_2O_4^{.-}$; and $\Delta G_{DET}^\ddagger$ is the free energy of activation of detachment of the reduced surface iron from the crystal lattice. For the sake of simplicity, the oxidized product is omitted in this figure. (Adapted from reference 9. Copyright 1991 American Chemical Society.)

Conclusions

Both the rate and the rate constant of the photochemical reductive dissolution of lepidocrocite with oxalate as the electron donor are pH-dependent. This observation suggests that other pH effects, in addition to the pH dependence of oxalate adsorption at the lepidocrocite surface, have to be considered. Such effects are catalysis of detachment of the reduced surface iron centers by protonation of their neighboring hydroxo and oxo groups, and the pH-dependent readsorption of the photochemically formed Fe(II), thus blocking surface sites for the adsorption of oxalate.

Our results do not allow us to decide which of these pH effects is predominant. The experimental observation, however, that in the pH range between 3 and 5 the overall rate constant depends on the concentration of surface protons to the power 1.6 may be an indication that proton catalysis of the detachment of surface Fe(II) is an important factor. In this case, the detachment of surface Fe(II) would be the rate-determining step of the overall process. The experiments presented here serve as an illustrative example, pointing out that reductive dissolution of oxide minerals may be catalyzed by protons, and hence that the rates of proton-catalyzed and of reductive dissolution may not be merely additive. However, more experimental evidence is needed to evaluate the validity of applying the rate expression of the proton-catalyzed dissolution to the overall rate constant of reductive dissolution.

Acknowledgments

I thank Werner Stumm, EAWAG, for teaching me surface chemistry.

References

1. LaKind, J. S.; Stone, A. T. *Geochim. Cosmochim. Acta* **1989**, *53*, 961–971.
2. Suter, D.; Banwart, S.; Stumm, W. *Langmuir* **1991**, *7*, 809–813.
3. Xyla, A. G.; Sulzberger, B.; Luther, G. W.; Hering, J. G.; van Cappellen, P.; Stumm, W. *Langmuir* **1992**, *8*, 95–103.
4. Stumm, W.; Kummert, R.; Sigg, L. *Croat. Chem. Acta* **1980**, *53*, 291–312.
5. Stumm, W. *Chemistry of the Solid–Water Interface;* Wiley Interscience: New York, 1992.
6. Zinder, B.; Furrer, G.; Stumm, W. *Geochim. Cosmochim. Acta* **1986**, *50*, 1861–1870.
7. Stumm, W.; Furrer, G. In *Aquatic Surface Chemistry;*, Stumm, W., Ed.; Wiley-Interscience: New York, 1987; pp 197–219.
8. Wieland, E.; Wehrli, B.; Stumm, W. *Geochim. Cosmochim. Acta* **1988**, *52*, 1969–1981.
9. Siffert, C.; Sulzberger, B. *Langmuir* **1991**, *7*, 1627–1634.
10. Sulzberger, B.; Laubscher, H. U.; Karametaxas, G. In *Aquatic and Surface Photochemistry;* Helz, G.; Zepp, R.; Crosby, D., Eds.; Lewis Publishers: Chelsea, MI, 1994.

11. Stone, A. T.; Ulrich, H.-J. *J. Colloid Interface Sci.* **1989**, *132*, 509–522.
12. Zuo, Y.; Hoigné, J. *Environ. Sci. Technol.* **1992**, *26*, 1014–1022.
13. Behra, P.; Sigg, L. *Nature (London)* **1990**, *344*, 419–421.
14. Rich, H. W.; Morel, F. M. M. *Limnol. Oceanogr.* **1990**, *35*, 652–662.
15. Morel, F. M. M.; Hudson, R. J. M.; Price, N. M. *Limnol. Oceanogr.* **1991**, *36*, 1742–1755.
16. Duce, R. A.; Tindale, N. W. *Limnol. Oceanogr.* **1991**, *36*, 1715–1726.
17. Furrer, G.; Stumm, W. *Geochim. Cosmochim. Acta* **1986**, *50*, 1847–1860.
18. Bondietti, G. C. Ph.D. Thesis, ETH Zurich, 1992, No. 9723.
19. Guy, C.; Schott, J. *Chem. Geol.* **1989**, *78*, 181–204.
20. Knauss, K. G.; Wolery, T. J. *Geochim. Cosmochim. Acta* **1989**, *53/7*, 1493–1501.
21. Brauer, G. *Handbuch der Präparativen Anorganischen Chemie;* Zweite Auflage, Enke; Verlag: Stuttgart, Germany, 1962.
22. Stookey, L. L. *Anal. Chem.* **1970**, *42*, 779–781.
23. Prasad, D. R.; Hoffman, M. Z. *J. Chem. Soc. Faraday Trans. 2* **1986**, *82*, 2275–2289.

RECEIVED for review October 23, 1992. ACCEPTED revised manuscript July 22, 1993.

15

Photocatalytic Degradation of 4-Chlorophenol in TiO$_2$ Aqueous Suspensions

Chengdi Dong and Chin-Pao Huang*

Department of Civil Engineering, University of Delaware, Newark, DE 19716

> *Photocatalytic degradation of 4-chlorophenol in TiO$_2$ aqueous suspensions produces 4-chlorocatechol, an ortho hydroxylated product, as the main intermediate. This result disagrees with data reported by other researchers, who proposed the formation of a para-hydroxylated product, hydroquinone, as the major intermediate. Results also indicated that further oxidation of 4-chlorocatechol yields hydroxyhydroquinone, which can readily be oxidized and mineralized to carbon dioxide. Complete dechlorination and mineralization of 4-chlorophenol can be achieved. In contrast, direct photolysis of 4-chlorophenol produces hydroquinone and p-benzoquinone as the main reaction products. The photocatalytic oxidation reaction, initially mediated by TiO$_2$, is generated by an electrophilic reaction of the hydroxyl radical attacking the benzene ring.*

THE INITIAL STEPS OF PHOTOCATALYTIC OXIDATION of aromatic compounds in aqueous suspensions of TiO$_2$ or other semiconductors have been described in terms of several mechanisms (1–13). The most important and widely accepted of these mechanisms is the generation from water decomposition of hydroxyl radicals that attack the aromatic ring.

*Corresponding author

$$\text{TiO}_2 + h\nu \longrightarrow h^+ + e^- \quad (1)$$

$$\text{H}_2\text{O} + h^+ \longrightarrow {}^\bullet\text{OH} + \text{H}^+ \quad (2)$$

$$\text{C}_6\text{H}_6 + {}^\bullet\text{OH} \longrightarrow \text{C}_6\text{H}_5\text{OH} \quad (3)$$

A similar degradation mechanism was proposed for homogeneous oxidation reactions involving hydroxyl radicals (14–20). Therefore, one would intuitively expect to find similar reaction intermediates in heterogeneous oxidation systems. However, the degradation pathways of chloro-substituted phenols are far from conclusive. For example, Barbeni et al. (16) proposed that the initial steps of the photocatalytic degradation of 4-chlorophenol in the presence of TiO_2 would form hydroquinone through an attack by hydroxyl radicals.

$$\text{4-chlorophenol} + {}^\bullet\text{OH} \longrightarrow \text{hydroquinone} + \text{Cl}^\bullet \quad (4)$$

In contrast, Lipczynska-Kochany and Bolton (19) reported that in photodegradation of 4-chlorophenol mediated by hydrogen peroxide the primary product was 4-chlorocatechol and hydroquinone was only a minor product.

$$\text{4-chlorophenol} + 2\,{}^\bullet\text{OH} \longrightarrow \text{4-chlorocatechol} + \text{H}_2\text{O} \quad (5)$$

The formation of hydroquinone implies that the initial hydroxylation step takes place with dechlorination at the para position. In contrast, the formation of 4-chlorocatechol implies that the initial hydroxylation reaction occurs without dechlorination at the ortho position. Because of the different physical and chemical properties of these two intermediates, the overall reaction kinetics and mechanisms can be much different.

Information about the degradation pathway and oxidation products is important from the viewpoint of water quality control. The objective of this study was to determine the intermediate products during the photocatalytic oxidation of 4-chlorophenol in TiO_2 aqueous suspensions. To verify the extent of 4-chlorophenol detoxification, the formation of carbon dioxide and chloride ions was also measured. On the basis of the results obtained, we proposed a reaction pathway for the photocatalytic oxidation of 4-chlorophenol.

Materials and Methods

Materials. 4-Chlorophenol, 4-chlororesorcinol, hydroquinone, and hydroxyhydroquinone were purchased from Aldrich Chemical Co., Milwaukee, WI; 4-chlorocatechol was purchased from TCI American, Portland, OR. Acetonitrile, hexane, methylene chloride, methanol, and pyridine were obtained from the Fisher Scientific Co., Fulton, CA. The photocatalyst was Degussa P–25 titanium dioxide [mainly anatase form; surface area 50 m^2/g; pH_{zpc} (pH at zero point of charge) 6.3; contains some impurities such as Al_2O_3 (<0.3%), SiO_2 (<0.2%), Fe_2O_3 (<0.01%), and HCl (<0.3%)].

Experimental Procedure. All experiments were conducted in Pyrex test tubes, each containing 15 mL of an oxygenated solution. Typically, the solutions contained 1 g/L of TiO_2, 10^{-3} M 4-chlorophenol, and 0.05 M $NaNO_3$ unless otherwise indicated. The initial solution pH was adjusted to the desired value with nitric acid (1 M) and sodium hydroxide (1 M). The solution was mixed vigorously over a reciprocal shaker throughout the reaction. Irradiation was done by a 1600-W medium-pressure, mercury-vapor discharge lamp (American Ultraviolet Co.). A light intensity of 17.0 W/m^2, detected by a Spectronics model DM–365H UV radiometer, was used. The temperature was controlled over the entire irradiation period by a thermostat pump in conjunction with a water-cooling system. At preselected reaction times, samples were taken, and the residual concentrations of 4-chlorophenol and its intermediates were analyzed.

Chemical Analysis. Gas chromatography–mass spectrometry (GC–MS) and high-performance liquid chromatography (HPLC) techniques were used to analyze 4-chlorophenol and its oxidation intermediates. For GC–MS analysis, the samples were acetylated in pyridine. The samples were first evaporated to dryness. Then 200 μL of pyridine and 200 μL of acetic anhydride were added to the dry residue. The samples were heated at 65 °C for 2–3 h to ensure the complete acetylation reaction, and then gently evaporated to dryness in a nitrogen stream. Finally, the residue was redissolved in 0.1 mL of hexane for GC analysis. A GC (HP model 5890) equipped with mass selective detector (HP model 5971) and SPB–5 capillary column (Supelco Co., PA., 25- × 0.2-mm i.d. × 0.33-μm film thickness) was used. To separate different intermediate products, various oven-temperature programs were performed. The GC–MS interface line was maintained at 300 °C. The mass-selective detector was scanned at a rate of 1.2 s per decade, and mass spectra were produced by using the standard electron ionization (70 eV) at the electron-impact mode.

An HPLC (Waters Associates model 6000 gradient system), used to analyze the irradiated samples, was fitted with two model 501 pumps, a model U6K injector, a Digital Professional 350 computer with a system interface module (Waters Associates), and an auto scan diode array UV light detector (Perkin–Elmer model LC–480). A Vydac 201TP54 polymeric C-18 reversed-phase column (5 μm, 4.6 mm × 25 cm) was employed to separate the compounds in the samples, and the mobile phase used was HPLC-grade acetonitrile and water. The gradient condition was 5 min of equilibration at 30% acetonitrile in water, linear gradient to 100% acetonitrile in water for 20 min, and holding for 5 min. Identification of the eluting compounds was accomplished by comparing the UV light spectra to their standard compounds with a diode array detector. Portions of the samples were analyzed by another HPLC system (Dionex model Bio LC) equipped with a Dionex model VDM–1 UV–visible detector and an Alltech C-8 reversed-phase column (5 μm, 4.6 mm × 15 cm) under the gradient conditions described.

Chloride ion concentration was measured by ion chromatography with a Dionex Bio LC chromatograph equipped with a Dionex pulsed electrochemical detector and a Dionex PAX–100 metal-free anion column (25 cm long, 4.6 mm i.d.). The eluent was a mixture of 80% H_2O, 10% acetonitrile, and 10% 191-mM NaOH; the flow rate was 1 mL/min, and the injection loop volume was 50 μL.

The evolution of carbon dioxide was monitored by gas chromatograph (Hewlett–Packard model 5890) equipped with an HP thermal-conductivity detector and a Hayesep 100–120-mesh packed column (10 m × 3.2 mm). Helium was the carrier gas, and its flow rate was 28 mL/min. In each run, 0.5 mL of the gas mixture in the headspace was withdrawn from the test tube by using a pressure-lock gas syringe and injected into the GC column. The calibration curve was made by acidifying a known amount of potassium bicarbonate in the reaction tube under conditions similar to those of the corresponding experimental conditions. Standard carbon dioxide gas (Supelco) was used to confirm the formation of carbon dioxide.

Results and Discussion

Effect of Process Parameters. Figures 1A and 1B show the effect of UV light and TiO_2 on the degradation of 4-chlorophenol (10^{-3} M) and the generation of chloride ions. UV light alone caused ~15% 4-chlorophenol degradation and production of an equal amount of chloride ions (15%) after 2 h of reaction time. The presence of both TiO_2 and UV light resulted in greater than 95% 4-chlorophenol decomposition and 60% chloride ion formation after

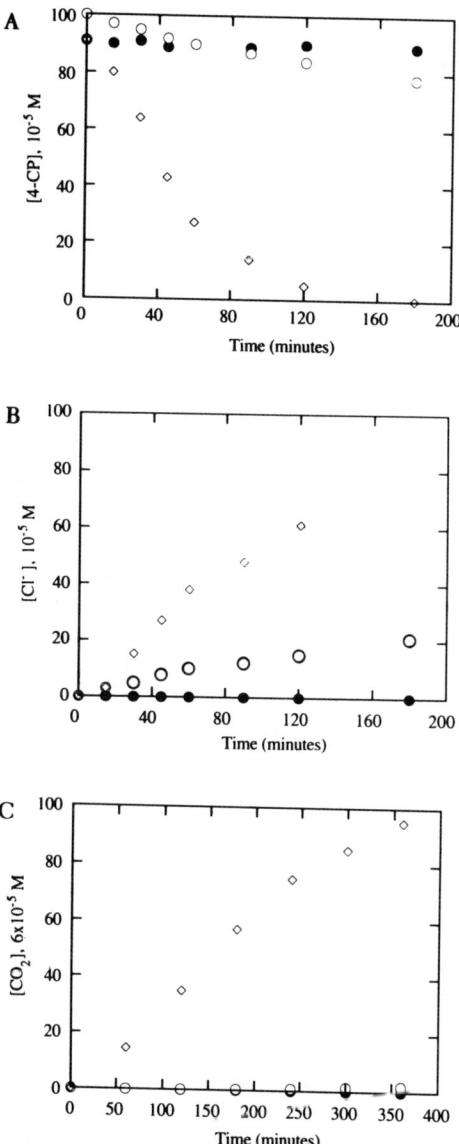

Figure 1. Degradation of (A) 4-chlorophenol, (B) formation of chloride ions, and (C) carbon dioxide versus time at UV light alone (○), TiO^2 alone (●), and UV light plus TiO_2 (◊). Experimental conditions: 4-chlorophenol = 10^{-3} M, TiO^2 = 1 g/L, pH = 4.0, I = 5×10^{-2} M $NaNO_3$, oxygen atmosphere, and temperature = 25 °C.

2 h. Apparently, 4-chlorophenol is sensitive to UV light photolysis. Tseng and Huang (12, 13) reported a similar finding.

The production of carbon dioxide under these three experimental conditions is shown in Figure 1C. UV light alone yields only 2% carbon dioxide, whereas the presence of both TiO_2 and UV light produces greater than 95% carbon dioxide under similar experimental conditions. This product clearly indicates that the photocatalytic oxidation process is effective in completely mineralizing 4-chlorophenol to carbon dioxide and chloride ions.

Table I gives the pseudo-first-order rate constants (in reciprocal minutes) for 4-chlorophenol degradation (k_{4CP}), chloride ion production (k_{Cl}), and carbon dioxide generation (k_{CO_2}) in the presence of UV light alone and TiO_2 plus UV light, respectively. In the presence of UV light alone, the 4-chlorophenol degradation rate is almost the same as the rates of formation of chloride ions. This similarity implies that the initial step of the direct photolysis reaction may be mainly the dechlorination reaction. However, when TiO_2 is added to the same system, the dechlorination rate constant (2.63×10^{-3} min^{-1}) is only half of that of the 4-chlorophenol degradation (1.21×10^{-3} min^{-1}). This change indicates that reactions other than dechlorination occur. Apparently the non-dechlorination reactions occurred during the early stage of the photocatalytic oxidation reaction of 4-chlorophenol, yielding chlorinated intermediates. Results from GC–MS analysis clearly confirm this reaction step.

Figure 2A shows the TiO_2-mediated photocatalytic oxidation of 4-chlorophenol under various initial organic concentrations, $[4CP]_0$; the lower the concentration, the higher the percentage of 4-chlorophenol decomposition. For example, a 3-h irradiation of 10^{-3} M 4-chlorophenol leads to complete 4-chlorophenol decomposition. The results cannot be fit with simple first-order kinetics. As shown in Figure 2B, the first-order rate constants obtained vary with the initial 4-chlorophenol concentration. Figure 2B also gives initial rate constants, r_0, which are calculated from the products of the calculated first-order rate constants and the corresponding initial 4-chlorophenol concentrations, that is, $r_0 = k_{4CP} \cdot [4CP]_0$ in millimolar per minute (21).

Figure 2B also shows that the first-order rate constants of 4-chlorophenol degradation are inversely proportional to the initial 4-chlorophenol concentration, whereas the initial rate constants are proportional to the initial 4-chlorophenol concentration. When the reciprocals of the initial rate constants are

Table I. Kinetic Results of the Photodegradation of 4-Chlorophenol

Compounds	UV Light Alone		UV Light Plus TiO_2	
	k (min^{-1})	$t_{1/2}$ (min)	k (min^{-1})	$t_{1/2}$ (min)
4-Chlorophenol	1.35×10^{-3}	500.0	2.63×10^{-2}	40.0
Chloride ion	1.26×10^{-3}	520.0	1.21×10^{-2}	92.0
Carbon dioxide	$<10^{-6}$	—	4.1×10^{-3}	163.0

15. DONG & HUANG *Photocatalytic Degradation* 297

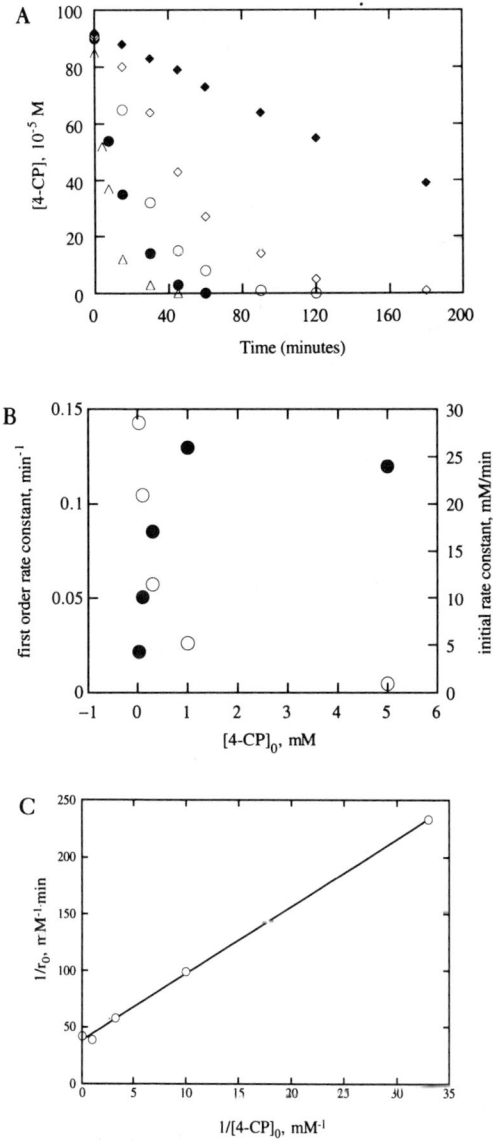

Figure 2. Part A: Degradation of 4-chlorophenol versus time at 3.0 (♦), 1.0 (◇), 0.3 (○), 0.10 (●), and 0.03 (△) mM 4-chlorophenol. Experimental conditions: TiO_2 = 1 g/L, pH = 4.0, I = 5 × 10^{-2} M $NaNO_3$, oxygen atmosphere, and temperature = 25 °C. Part B: The first-order rate constants (○) and the initial rate constants (●) versus the initial 4-chlorophenol concentration. Part C: The reciprocals of the initial rate constant versus reciprocals of the initial 4-chlorophenol concentration.

plotted against the reciprocals of the initial 4-chlorophenol concentrations by using the Langmuir–Hinshelwood equation (eq 6) (2, 9–13, 22–24), a linear relationship is obtained with $r^2 = 0.99$ (Figure 2C).

$$\frac{1}{r_0} = \frac{1}{k} + \frac{1}{kK[4\text{CP}]_0} \tag{6}$$

Results from regressional analysis show that the rate constant (k) is 2.62×10^{-2} mM/min and the adsorption coefficient (K) is 6.43 mM^{-1}.

The effect of TiO$_2$ concentration on the 4-chlorophenol degradation is presented in Figure 3. Increasing the amount of TiO$_2$ increases the rate of 4-chlorophenol degradation until a maximum rate is reached at 1 g/L of TiO$_2$. Further increase in TiO$_2$ decreases the 4-chlorophenol degradation rate, resulting in part from the blockage of UV light.

The effect of pH on 4-chlorophenol oxidation is shown in Figure 4. The results show that oxidation is favored under both acidic and basic conditions. This fact implies that different reaction mechanisms may be operative and that photocatalytic oxidation is affected by both H$^+$ and OH$^-$ ions. At high pH, the number of hydroxyl ions on the TiO$_2$ surface increases because of the abundance of OH$^-$ ions, thereby increasing the population of ·OH radicals. Hickling and Hill (25) suggested that, at high pH, the adsorbed OH$^-$ group can be readily converted to ·OH radical upon irradiation.

$$\text{Ti}^{\text{IV}} - \text{OH}^- + \text{H}^+ \longrightarrow \text{Ti}^{\text{IV}} - \text{·OH}$$

$$E° = 2.0 \text{ V(NHE)} \tag{7}$$

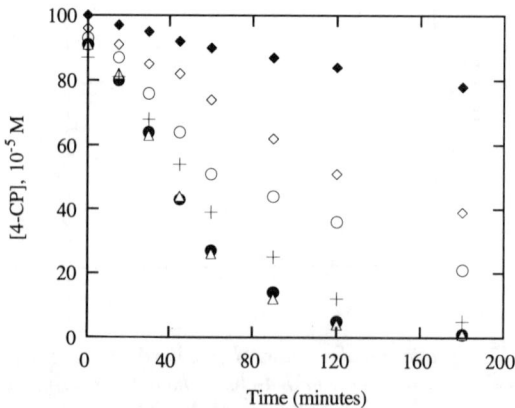

Figure 3. Degradation of 4-chlorophenol versus time at 0 (♦), 0.1 (◊), 0.3 (○), 1 (●), 2 (△), and 5 (+) g/L of TiO^2. Experimental conditions: 4-chlorophenol = 10^{-3} M, pH = 4.0, I = 5×10^{-2} M NaNO$_3$, oxygen atmosphere, and temperature = 25 °C.

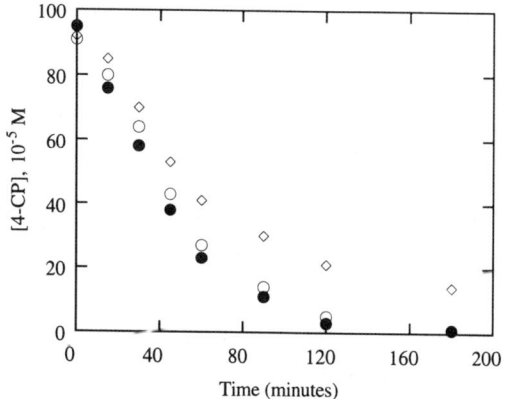

Figure 4. Degradation of 4-chlorophenol versus time at pH 4.0 (○), 8.6 (◇), and 11.0 (●). Experimental conditions: 4-chlorophenol = 10^{-3} M, TiO_2 = 1 g/L, I = 5 × 10^{-2} M $NaNO_3$, oxygen atmosphere, and temperature = 25 °C.

where $E°$ is the standard potential versus the normal hydrogen electrode (NHE). At low pH, surface-water molecules may be oxidized to hydroxyl radicals.

$$Ti^{IV} - H_2O + H^+ \longrightarrow Ti^{IV} - {}^{\bullet}OH + H^+$$

$$E^{size-2o} = 2.8 \text{ V (NHE)} \qquad (8)$$

Both $E°$ values shown in equations 7 and 8 are lower than the anodic valence band energy of TiO_2 (3.2 V, NHE) upon irradiation. Therefore, the generation of •OH radicals may be thermodynamically possible under both high and low pH conditions. Moreover, the relatively low $E°$ value at high pH for the formation of •OH radicals suggests that the oxidation reaction will be more favorable under the high pH condition.

Results shown in Figure 4 indicate that the photocatalytic reaction of 4-chlorophenol is favored under both alkaline and acidic pH. This phenomenon implies that a reaction other than equation 8 must be taking place. D'Oliveira et al. (6) suggested that a photocatalytic oxidation reaction involving oxygen species would be favorable at low pH. Tseng and Huang (12) also reported that photocatalytic oxidation of phenol is strongly affected by dissolved oxygen concentration. Apparently oxygen plays an important role in trapping electrons on the TiO_2 surface, thereby minimizing the extent of electron-hole recombination. As a result, the number of active holes, which are important for the generation of surface •OH radicals, is maximized. Moreover, the adsorbed O_2^- can also react with H^+ to form $HO_2^{\bullet}$, which is subsequently converted to H_2O_2, as another pathway to produce •OH radicals (2, 6, 11–13).

However, the effect of oxygen on the photocatalytic oxidation of 4-chlorophenol was not within the scope of this research undertaking.

Intermediates and Reaction Pathway. *Identification of Intermediates.*

The derivatization technique and GC–MS analysis were successfully used to detect organic pollutants in the aquatic environment (26, 27). This same technique was chosen to identify the partial oxidation products of 4-chlorophenol in TiO_2 aqueous suspensions. Figure 5 and Table II show the mass spectra of the derivatized reaction products of 4-chlorophenol. Four derivatized intermediate peaks [**A**, **B**, **C**, and **D**, with retention times (RTs) at 10.32, 12.50, 13.72, and 14.11 min, respectively] were observed. Peak **A** is the acetylated parent compound, 4-chlorophenol. Peaks **B**, **C**, and **D** are acetylated intermediate products. The identification of peaks **B**, **C**, and **D** was based on the following observations:

- The mass spectra of peak **B** show a molecular ion peak (M^+) at m/e 194 and a base peak ($M-CH_2CO^+$ or $[M-AC_2]^+$) at m/e 110. The mass spectra of peaks **C** and **D** show a M^+ peak at m/e 228 and a $[M-AC]^+$ peak at m/e 144. The difference in m/e value between the M^+ peak and the $[M-AC_2]^+$ peak in peaks **B**, **C**, and **D** results from the loss of two acetyl radicals (CH_2CO). This difference indicates the presence of two hydroxyl ions in these two intermediates because the acetyl groups are derivatized from the hydroxyl group of corresponding organic substrates after acetylation.

- The mass spectra of peaks **C** and **D** have a near-3:1 ratio based on the isotope ion peaks at m/e 144 and 146. This ratio indicates the possible presence of a chlorine atom, which also has a near-3:1 ratio based on the isotope ion peaks at m/e 35 and 37. These characteristics are not observed at peak **B**, which indicates the absence of chlorine atoms in peak **B**.

- The m/e value of the molecular ion peak and the base peak of peak **B** differs from that of peaks **C** and **D** by 34. This difference indicates that compounds **C** and **D** have one more Cl atom than compound **B** has.

- A slight difference between peaks **C** and **D** is observed in other fragment ion peaks such as m/e 51, 61, 63, 79, 80, 115, and 128. This comparison implies that these two intermediates have similar chemical structures.

- Considering the chemical structure of 4-chlorophenol, there are three possible modes of attack by ·OH radicals; all three paths lead to the production of dihydroxybenzenes. An attack at the

Table II. Mass Spectral Data for Acetylated 4-Chlorophenol and Its Intermediates

Peak[a]	RT (min)[b]	m/e Fragment				ID[d]
		[HAc]$^+$	[M–Ac$_2$]$^+$	[M–Ac]$^+$	M^{+c}	
A	10.32	0.49[e]	—[f]	1.00	0.09	4-chlorophenol
B	12.50	0.67	1.00	0.16	0.15	hydroquinone
C	13.72	0.81	1.00	0.09	0.02	4-chlorocatechol
D	14.11	0.88	1.00	0.09	0.03	4-chlororesorcinol

[a] Order of elution of peaks.
[b] Retention time (min).
[c] Molecular ion peak.
[d] Compound identification based on retention time and mass spectral data in comparison with authentic compound.
[e] Relative intensity of mass spectra peak.
[f] indicates not present.

C-4 position will result in a dechlorination reaction to generate hydroquinone, an attack at the C-2 or the C-6 position will generate 4-chlorocatechol, and an attack at the C-3 or the C-5 position will produce 4-chlororesorcinol. These three dihydroxybenzenes can be identified by using authentic standards.

- The GC–MS retention time and mass spectra of peak **B** are closely matched with the retention time and mass spectra of hydroquinone standard. The mass spectra of hydroquinone also show the same M$^+$ peak at m/e 194 and [M–AC$_2$]$^+$ peak at m/e 110 as peak **B**. In addition, the mass spectra of peak **B** also closely match those of the acetylated hydroquinone in the computer database of the National Institute of Science and Technology (NIST) mass spectral library. That is, peak **B** corresponds to acetylated hydroquinone.

- The GC–MS retention times and mass spectra of peaks **C** and **D** are closely matched with the retention times and mass spectra of 4-chlorocatechol and 4-chlororesorcinol standards, respectively. The mass spectra of 4-chlorocatechol and 4-chlororesorcinol show the same M$^+$ peak at m/e 228 and [M–AC$_2$]$^+$ at m/e 144 as peaks **C** and **D**. That is, peak **C** corresponds to an acetylated 4-chlorocatechol and peak **D** is equivalent to an acetylated 4-chlororesorcinol.

To gain insight into the quantitative nature of the reaction, concentration changes of intermediate products were determined. Figure 6 shows the concentration changes of 4-chlorophenol (4CP), 4-chlorocatechol (4CCA), 4-chlororesorcinol (4CRE), hydroquinone (HQ), and hydroxyhydroquinone (HHQ)

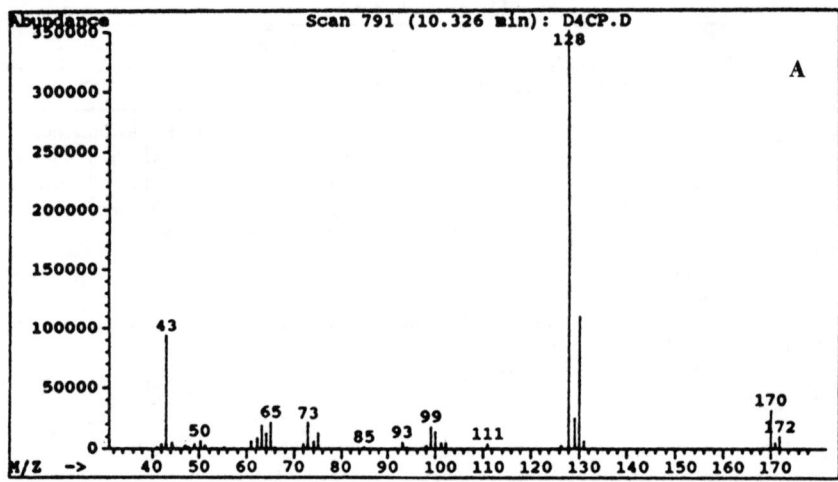

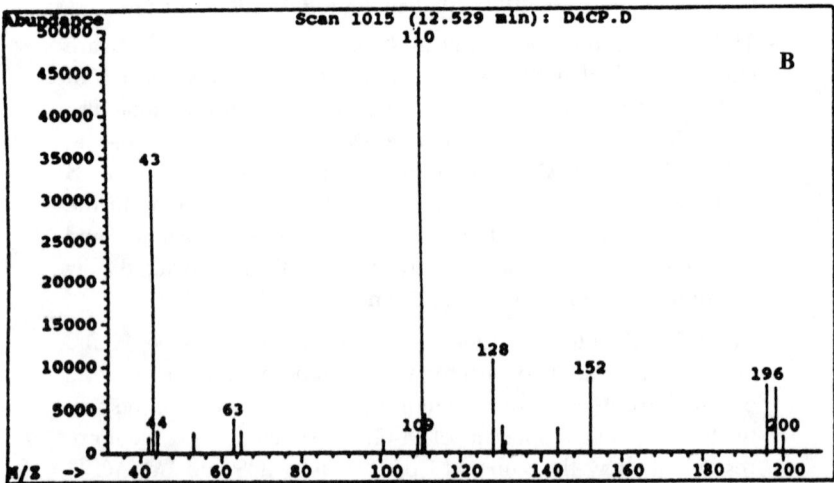

*Figure 5. Mass spectra of derivatized intermediates from the photocatalytic degradation of 4-chlorophenol. Experimental conditions: 4-chlorophenol = 10^{-3} M, TiO^2 = 1 g/L, pH = 4.0, I = 5 × 10^{-2} M $NaNO_3$, oxygen atmosphere, temperature = 25 °C. Peak **A** is 4-Chlorophenol, peak **B** is hydroquinone, peak **C** is 4-chlorocatechol, and peak **D** is 4-chlororesorcinol.*

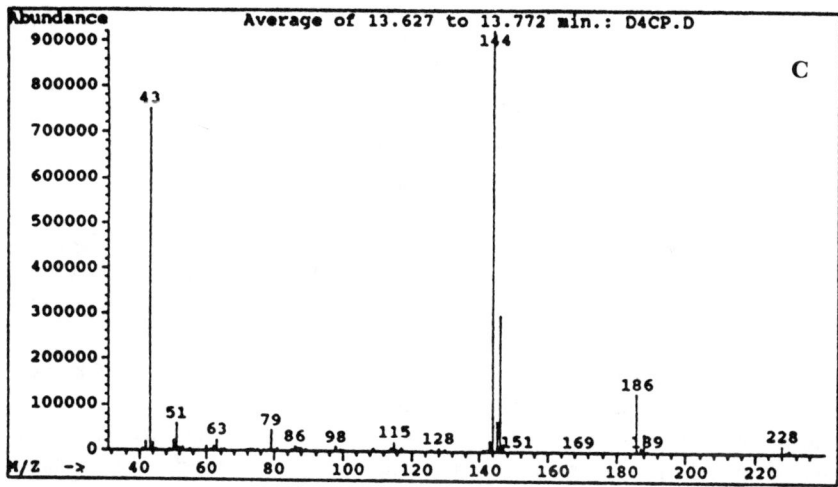

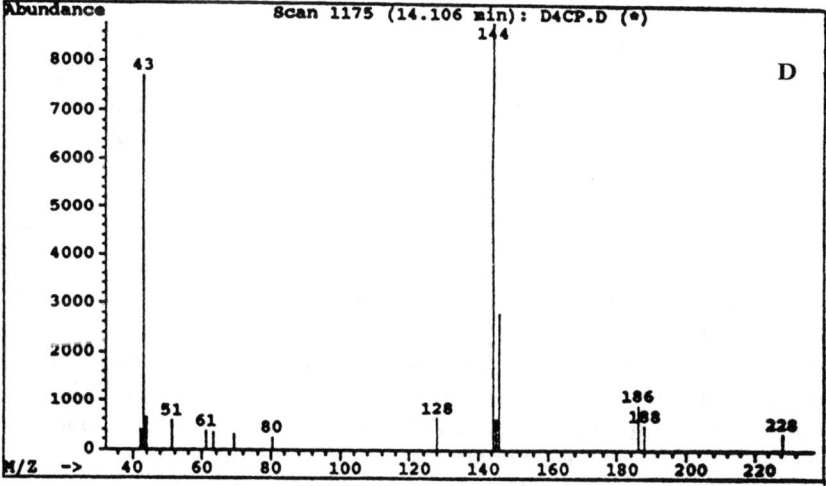

Figure 5.—Continued.

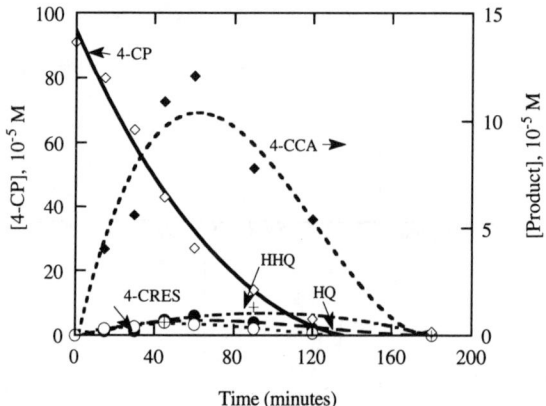

Figure 6. Concentration changes of 4-chlorophenol and its intermediates. Experimental conditions: 4-chlorophenol = 10^{-3} M, TiO^2 = 1 g/L, pH = 4.0, I = 5 × 10^{-2} M $NaNO_3$, oxygen atmosphere, and temperature = 25 °C.

as a function of reaction time. Results show that only 4-chlorocatechol is present in a significant amount, whereas hydroquinone, 4-chlororesorcinol, and hydroxyhydroquinone concentrations are at least 10 times less than that of 4-chlorocatechol. The maximum concentration of 4-chlorocatechol obtained during the reaction was about 12% of the initial 4-chlorophenol concentration at about 1 h of reaction time. After that, the concentrations of 4-chlorocatechol, 4-chlororesorcinol, hydroquinone, and hydroxyhydroquinone gradually decrease. These organic compounds completely disappear at a reaction time of 3 h.

Further degradation of these hydroxylated aromatic intermediates will lead to the formation of low-molecular-weight species that cannot be detected by current analysis methods. However, as shown in Figure 1C, the evidence of high carbon dioxide yield indicates that the reaction will eventually proceed to a total mineralization of 4-chlorophenol.

Reaction Mechanisms. Our analysis of intermediates and reactions reported by other researchers leads to proposed reaction pathways describing the photocatalytic oxidation of 4-chlorophenol in TiO_2 aqueous suspensions. The photocatalytic oxidation reaction is brought about by ˙OH radicals, which are formed mainly from water decomposition on the TiO_2 surface upon UV light irradiation (9–13). The ˙OH radicals can either directly react with the adsorbed organic species on the TiO_2 surface or diffuse to the solution and then react with the dissolved organic species in the solution phase. Both reactions lead to formation of hydroxylated products such as 4-chlorocatechol, hydroquinone, 4-chlororesorcinol, and hydroxyhydroquinone as the initial products (Figure 6). Eventually, the reaction will mineralize these intermediates to end products such as carbon dioxide and chloride ions (Figures 1B and 1C).

Scheme I shows one of the possible reaction pathways involving ˙OH radicals as the main reaction species in the TiO_2-mediated photocatalytic oxidation of 4-chlorophenol. Only nonradical species were detected in this reaction scheme.

The first step of the reaction is an attack by the ˙OH radical to yield the chlorodihydroxycyclohexadienyl (ClDHCD) radical (step 1) (2, 20). This unstable ClDHCD radical can decay to the chlorophenoxy (ClPO) radical by the elimination of one water molecule before an attack by another ˙OH radical (step 2). The reduced ClPO radical may undergo an electron rearrangement to obtain a relatively stable intermediate radical. This electron rearrangement reaction may stabilize the ClPO radical at the ortho and para positions by resonance (28, 29). However, the presence of one chlorine atom at the para position of 4-chlorophenol inhibits occurrence of the reaction at this position. Therefore, only one relatively stable ClPO radical structure can be obtained (step 3). Further reaction of the ClPO radical with ˙OH radicals will generate 4-chlorocatechol as the main intermediate product (step 4). The ClPO radical may also result from hydrogen abstraction by the ˙OH radical directly from the hydroxyl group of 4-chlorophenol.

4-Chlorocatechol is the main intermediate and hydroquinone is only a minor product from the degradation of 4-chlorophenol. This fact indicates that the dechlorination reaction is not an important step at the beginning of

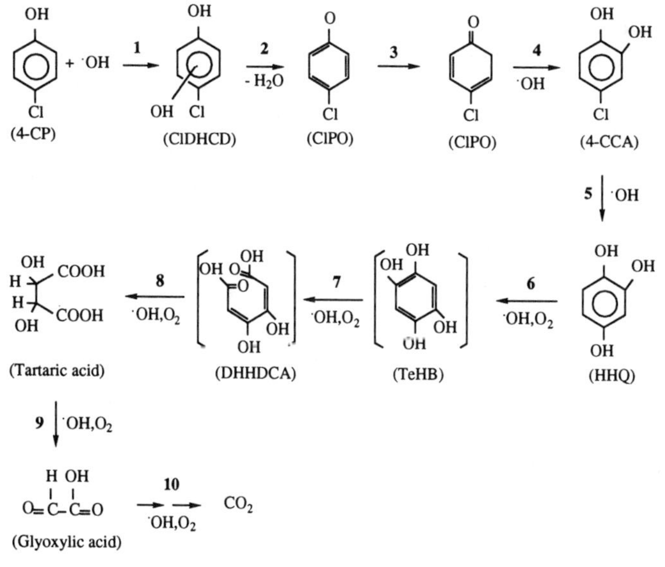

Scheme I. Possible reaction pathways involving ˙OH radicals.

the photocatalytic oxidation reaction. One possible explanation is that the electron-withdrawing property of the chlorine atom of 4-chlorophenol significantly reduces the opportunities of attack of the electrophilic ·OH radical at the chlorine site of 4-chlorophenol. As a result, the amount of hydroquinone formed is small compared with 4-chlorocatechol, which is generated from ·OH radical attack at the hydrogen atom of 4-chlorophenol. These results are different from the mechanism proposed by Barbeni et al. (16) for the TiO_2-mediated photocatalytic oxidation of 4-chlorophenol, as shown in reaction 4. According to Barbeni et al., hydroquinone is the major intermediate. However, our results agree with the mechanism proposed by Lipczynska-Kochany and Bolton (19) for the H_2O_2-mediated photodegradation of 4-chlorophenol (eq 5).

Further degradation of the 4-chlorocatechol intermediate involving ·OH radical attack is proposed. Intuitively, one can also suggest that a repeated attack of 4-chlorocatechol by the ·OH radical may take place according to steps a–d in the formation of chlorotrihydroxybenzene (CTHB) (Scheme II).

However, GC–MS and HPLC analysis indicate that the only detectable trihydroxybenzene is hydroxyhydroquinone, and no CTHB is detected. This result may indicate either that CTHB is not stable (it quickly degenerates to other species after it is formed) or that this pathway is not favorable. 4-Chlorocatechol has fewer hydrogen atom sites than 4-chlorophenol. Therefore, the chance of ·OH radical attack on the hydrogen atom site of 4-chlorocatechol will be smaller than the chance of a similar attack on the hydrogen atom sites of 4-chlorophenol. Conversely, there are more chances for ·OH radical attack on the chlorine atom site. The experimental results suggest that the dechlorination reaction is the main pathway to formation of hydroxyhydroquinone, although the formation of chlorotrihydroxybenzene still cannot be excluded.

A separate experiment using 4-chlorocatechol as the starting compound was performed to verify this reaction step. Figure 7 shows the oxidation of 4-chlorocatechol compared with that of 4-chlorophenol. The concentrations of hydroxyhydroquinone from the oxidation of 4-chlorocatechol and 4-chlorophenol as a function of time are also shown in Figure 7, which indicates that the photocatalytic oxidation of 4-chlorocatechol yields hydroxyhydroquinone as the only significant intermediate. Step 5 of Scheme I shows that a contin-

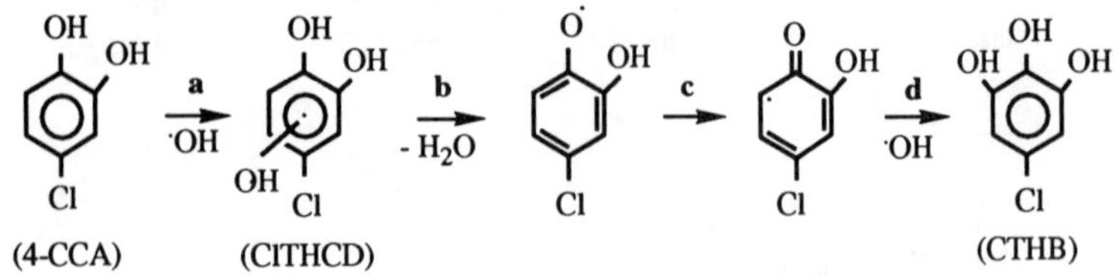

Scheme II. Formation of chlorotrihydroxybenzene.

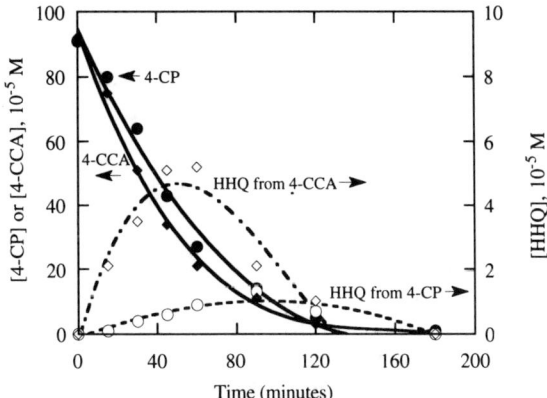

Figure 7. Concentration changes of 4-chlorophenol, 4-chlorocatechol, and their intermediates. Experimental conditions: 4-chlorophenol = 10^{-3} M, TiO^2 = 1 g/L, pH = 4.0, I = 5 × 10^{-2} M $NaNO_3$, oxygen atmosphere, and temperature = 25 °C.

uous degradation of 4-chlorocatechol yields nonchlorinated hydroxyhydroquinone with a release of chlorine atom (i.e., dechlorination).

Further degradation of hydroxyhydroquinone to the end product, carbon dioxide, has not yet been investigated. However, several investigators have reported the formation of carbonyl compounds and acid through a ring-cleavage reaction (14, 30). In a separate study we observed two organic acids, tartaric acid and glyoxylic acid, as the ring-cleavage products of oxidation of 2-chlorophenol by Fenton's reagent (31). These products are generated through hydroxyhydroquinone and then 1,2,4,5-tetrahydroxybenzene (TeHB). Because ·OH radicals are also involved in the Fenton's-reagent oxidation, it can be assumed intuitively that this same degradation step will be applicable to the TiO_2-mediated photocatalytic oxidation. That is, hydroxyhydroquinone will degrade to TeHB (step 6, Scheme I). Further oxidation of TeHB will produce 3,4-dihydroxy-2,4-hexadiene-1,6-dicarboxylic acid (DHHDCA) (step 7), which can be further oxidized to tartaric acid (step 8). Oxidation of tartaric acid will yield glyoxylic acid (step 9). The final step of the oxidation reaction is the conversion of glyoxylic acid to carbon dioxide (step 10).

Comparison with Direct Photolysis Process. The TiO_2-mediated photocatalytic oxidation reaction involves a complex free-radical reaction mechanism in which ·OH radicals are responsible for the oxidation of 4-chlorophenol. The initial reaction step produces 4-chlorocatechol as the main product. In contrast, the direct photolysis of 4-chlorophenol produces a different set of reaction products. Figure 8 shows that the direct photolysis of

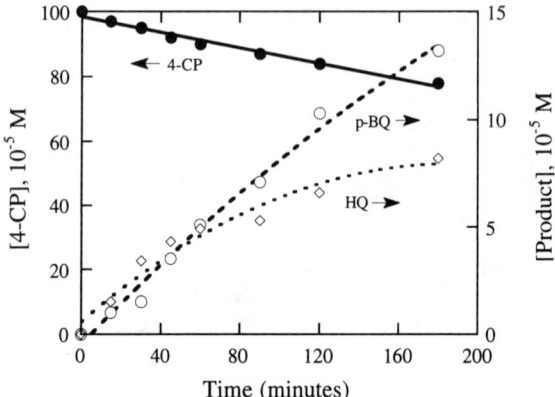

Figure 8. Direct photolysis of 4-chlorophenol. Hydroquinone and 4-chlorocatechol were analyzed by GC–MS; p-benzoquinone was analyzed by HPLC with UV–visible detector at a wavelength of 254 nm. Experimental conditions: 4-chlorophenol = 10^{-3} M, pH = 4.0, I = 5×10^{-2} M $NaNO_3$, oxygen atmosphere, and temperature = 25 °C.

4-chlorophenol produces significant amounts of hydroquinone (HQ) and p-benzoquinone (p-BQ), whereas no 4-chlorocatechol is observed. The total amount of hydroquinone and p-benzoquinone resembles the reduction in the amount of 4-chlorophenol during the 3 h of direct photolysis of pure solution of 10^{-3} M 4-chlorophenol. Scheme III is proposed to describe the possible degradation pathway. According to this reaction scheme, the photolysis of 4-chlorophenol leads to cleavage of the carbon–chlorine bond, which leads to the formation of hydroquinone and p-benzoquinone.

In addition to the difference in forms and the distribution of reaction products, the photocatalysis and photolysis processes exhibit different extents of mineralization. As shown in Figure 1C, the generation of carbon dioxide in direct photolysis of 4-chlorophenol is almost negligible. In contrast, photocatalytic oxidation leads to total mineralization of 4-chlorophenol to carbon dioxide. There is, however, no evidence that further ring-cleavage reaction takes place in direct photolysis of 4-chlorophenol.

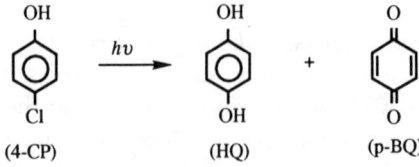

Scheme III. Photolysis of 4-chlorophenol.

Hydroxylated compounds can be more toxic than their starting monohydroxylated compounds, such as 4-chlorophenol. For example, hydroquinone (LD_{50} = 320 mg/kg for rat) and p-BQ (LD_{50} = 130 mg/kg for rat) are more toxic than their parent compound 4-chlorophenol (LD_{50} = 670 mg/kg for rat) (32). (LD_{50} is the dose that is lethal to 50% of test subjects.) Great care must be exercised in the application of a direct photolysis process for the treatment of organic pollutants such as phenolic compounds. Thus, photocatalytic oxidation can be more feasible for the treatment of toxic organic waste than the direct photolysis process.

Dynamics of the Reaction Network. Information on the concentration changes of 4-chlorophenol and its intermediates (4-chlorocatechol, 4-chlororesorcinol, hydroquinone, and hydroxyhydroquinone) allows us to perform some dynamic analysis of the photocatalytic oxidation of 4-chlorophenol. Scheme IV shows a reaction network that can be established to describe the overall mineralization of 4-chlorophenol. According to this reaction network, 4-chlorophenol (4-CP) first decomposes to 4-chlororesorcinol (4-CRE), 4-chlorocatechol (4-CCA), or hydroquinone (HQ); 4-chlorocatechol is the major primary product. Further oxidation of the primary intermediates yields hydroxyhydroquinone (HHQ), as the secondary intermediate, which is readily mineralized to carbon dioxide. For simplicity, a pseudo-first-order expression was used to model the dynamics of the reaction network (eqs 9-14).

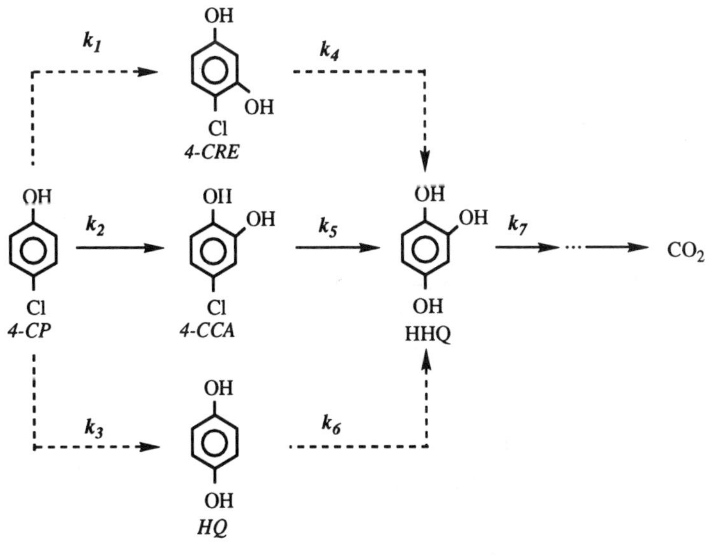

Scheme IV. Mineralization of 4-chlorophenol.

$$\frac{d[4CP]}{dt} = k_A[4CP] \qquad (9)$$

where

$$k_A = k_1 + k_2 + k_3 \qquad (10)$$

$$\frac{d[4CRE]}{dt} = k_1[4CP] - k_4[HHQ] \qquad (11)$$

$$\frac{d[4CCA]}{dt} = k_2[4CP] - k_5[HHQ] \qquad (12)$$

$$\frac{d[HQ]}{dt} = k_3[4CP] - k_6[HHQ] \qquad (13)$$

$$\frac{d[HHQ]}{dt} = k_4[4CRE] + k_5[4CCA] + k_6[HQ] - k_7[HHQ] \qquad (14)$$

Seven reaction constants (k_1–k_7) are used to describe the formation and degradation of 4-chlorophenol and its intermediates. Equations 15–19 list the analytical solutions of equations 9–14, respectively.

$$[4CP] = [4CP]_0 e^{-k_A t} \qquad (15)$$

$$[4CRE] = \frac{k_1[4CP]_0}{k_4 - k_A}(e^{-k_A t} - e^{-k_4 t}) \qquad (16)$$

$$[4CCA] = \frac{k_2[4CP]_0}{k_5 - k_A}(e^{-k_A t} - e^{-k_5 t}) \qquad (17)$$

$$[HQ] = \frac{k_3[4CP]_0}{k_6 - k_A}(e^{-k_A t} - e^{-k_6 t}) \qquad (18)$$

$$[HHQ] = [4CP]_0 \left(\frac{k_1 k_4 e^{-k_A t}}{(k_7 - k_A)(k_4 - k_A)} - \frac{k_1 k_4 e^{-k_A t}}{(k_7 - k_4)(k_4 - k_A)} \right.$$
$$\left. - \frac{k_1 k_4 e^{-k_7 t}}{(k_7 - k_4)(k_7 - k_A)} \right) + [4CP]_0 \left(\frac{k_2 k_5 e^{-k_A t}}{(k_7 - k_A)(k_5 - k_A)} \right.$$
$$\left. - \frac{k_2 k_5 e^{-k_5 t}}{(k_7 - k_5)(k_5 - k_A)} - \frac{k_2 k_5 e^{-k_7 t}}{(k_7 - k_5)(k_7 - k_A)} \right)$$

$$+ [4CP]_0 \left(\frac{k_3 k_6 e^{-k_A t}}{(k_7 - k_A)(k_6 - k_A)} - \frac{k_3 k_6 e^{-k_6 t}}{(k_7 - k_6)(k_6 - k_A)} \right.$$

$$\left. - \frac{k_3 k_6 e^{-k_7 t}}{(k_7 - k_6)(k_7 - k_A)} \right) \tag{19}$$

where $[4CP]_0$ is the initial concentration of 4-chlorophenol. With an iterative computation scheme and the data on the concentrations of intermediates, it is possible to calculate the rate constants. For example, k_A in equation 15 was obtained (Table I) when 10^{-3} M initial 4-chlorophenol concentration was used. The constant k_A can then be substituted into equations 16–18, in which there are only two rate constants. With a best fit with least square root technique, k_1–k_6 can be calculated. Finally, k_7 can be solved from equation 19 with a substitution of k_A and k_1–k_6 into equation 19. The calculated first-order rate constants (in reciprocal minutes) are $k_1 = 1.30 \times 10^{-3}$, $k_2 = 2.32 \times 10^{-2}$, $k_3 = 2.10 \times 10^{-3}$, $k_4 = 6.50 \times 10^{-2}$, $k_5 = 2.80 \times 10^{-2}$, $k_6 = 8.70 \times 10^{-2}$, and $k_7 = 9.90 \times 10^{-1}$. As shown in Figure 7, the fit of first-order rates gives a good approximation of the dynamics of the photocatalytic oxidation of 4-chlorophenol in TiO_2 suspensions.

On the basis of the rate constants determined, it is possible to estimate the percent yields of the primary intermediates. That is, the ratio of $k_1/(k_1 + k_2 + k_3)$ gives the percent production of 4-chlororesorcinol; the ratio of $k_2/(k_1 + k_2 + k_3)$ gives the percent production of 4-chlorocatechol; and the ratio of $k_3/(k_1 + k_2 + k_3)$ gives the percent production of hydroquinone. About 87% of 4-chlorophenol is converted to 4-chlorocatechol, which gives evidence that the hydroxylation probably occurred at the ortho position. The percent yield of 4-chlororesorcinol, the para oxidation product, is only 8%, and the percent yield of hydroquinone, the meta oxidation product, is about 5%. Apparently the presence of chlorine group at the para position greatly inhibits the oxidation reaction at this position; the presence of hydroxy group, an ortho and para director, lessens the oxidation reaction at the meta position. As a result, hydroxyl radicals exhibit a strong preference for the ortho position, with the formation of 4-chlorocatechol as the major intermediate product.

Conclusions

The TiO_2-mediated photocatalytic oxidation process can readily degrade 4-chlorophenol in aqueous solutions, with a complete mineralization to carbon dioxide and chloride ions, whereas the direct photolysis of 4-chlorophenol generates only a small amount of carbon dioxide. The distribution of intermediates during the course of the reaction shows that the reaction mechanism of the photocatalytic oxidation process is clearly different from that of the direct photolysis reaction.

The TiO_2-mediated photocatalytic oxidation reaction can be described by the radical mechanism involving $^{\bullet}OH$ as the major reaction species. The reaction mechanism follows the ortho pathway, so that the main intermediate found is 4-chlorocatechol, whereas the formation of 4-chlororesorcinol and hydroquinone is only a minor pathway. Further degradation of 4-chlorocatechol leads to production of hydroquinone, which can be further oxidized and mineralized to carbon dioxide. In contrast, the direct photolysis of 4-chlorophenol follows the para pathway, which leads to the formation of hydroquinone and p-benzoquinone as the major products.

A complete pathway for the mineralization of 4-chlorophenol can be described by the hydroxylation reaction through dechlorination, ring cleavage, and mineralization. Reaction kinetics of the 4-chlorophenol and its intermediates can be reasonably well approximated by using a complex parallel and consecutive first-order reaction mechanism.

Acknowledgments

This work was supported in part by Grant R815081 from the Exploratory Research Program of U.S. Environmental Protection Agency. Contents of this publication do not necessarily reflect the views and policies of U.S. Environmental Protection Agency. Any conclusions, mentions of chemicals, and processes are made solely by the authors and should not be implied as their endorsement by the funding agency. We thank Robert S. Ehrlich for his editorial help and comments.

References

1. Barbeni, M.; Morello, M.; Pramauro, E.; Pelizelli, E.; Borgarello, E.; Graetzel, M.; Serpone, N. *Nouv. J. Chim.* **1984**, *8*, 547–550.
2. Okamoto, K.; Yamamoto, Y.; Tanaka, H.; Tanaka, M.; Itaya, A. *Bull. Chem. Soc. Jpn.* **1985**, *58*, 2015–2022.
3. Okamoto, K.; Yamamoto, Y.; Tanaka, H.; Tanaka, M.; Haya, A. *Bull. Chem. Soc. Jpn.* **1985**, *58b*, 2023–2028.
4. Peral, J.; Casado, J.; Domenech, J. *J. Photochem. Photobio.* **1988**, *44*, 209–217.
5. Sakata, T. In *Photocatalysis: Fundamentals and Applications;* Serpone, N.; Pelizzetti, E., Eds.; John Wiley & Sons: New York, 1989; Chapter 10, pp 311-338.
6. D'Oliveira, J.; Al-Sayyed, G.; Pichat, P. *Environ. Sci. Technol.* **1990**, *24(7)*, 990–996.
7. Sclafani, A.; Palmisano, L.; Schiavello, M. *J. Phys. Chem.* **1990**, *94*, 829–832.
8. Sclafani, A.; Palmisano, L.; Davi E. *New J. Chem.* **1990**, *14*, 265–268.
9. Ollis, D. F.; Pelizzetti, E.; Serpone, N. In *Photocatalysis: Fundamentals and Applications;* Serpone, N.; Pelizzetti, E., Eds.; John Wiley & Sons, New York, 1989; pp 6603–6637.
10. Ollis, D. F.; Pelizzetti, E.; Serpone, N. *Environ. Sci. Technol.* **1991**, *25(9)*, 1523-1529.
11. Minero, C.; Aliberti, C.; Pelizzetti, E.; Terzian, R.; Serpone, N. *Langmuir* **1991**, *7*, 928–936.

12. Tseng, J. M.; Huang, C. P. *Water Sci. Tech.* **1991**, *23*, 377–387.
13. Tseng, J. M.; Huang, C. P. In *Emerging Technologies in Hazardous Waste Management;* Tedder, D. W.; Pohland, F. G., Eds.; ACS Symposium Series 422; American Chemical Society: Washington, DC; Chapter 2.
14. Ho, P. C. *Environ. Sci. Technol.* **1986**, *20*, 260.
15. Glaze, W. H.; Kang, J. W.; Chapin, D. H. *Ozone Sci. Eng.* **1987**, *9*, 335–352.
16. Barbeni, M. B.; Minero, C.; Pelizzetti, E.; Borgarello, E.; Serpone, N. *Chemosphere* **1987**, *10-12*, 2225–2237.
17. Sundstorm, D. W.; Weir, B. A.; Klei, H. E. *Environ. Prog.* **1989**, *8*, 6–11.
18. Sundstorm, D. W.; Klei, H. E.; Nalette, T. A.; Reidy, D. J.; Weir, B. A. *Hazard. Waste Hazard. Mater.* **1986**, *3*, 101.
19. Lipczynska-Kochany, E.; Bolton, J. R. *Environ. Sci. Technol.* **1992**, *26(2)*, 259–262.
20. Sedlek, D. L.; Andren, A. W. *Environ. Sci. Technol.* **1991**, *25(4)*, 777–782.
21. Glaze W. H.; Kenneke, J. F.; Ferry, J. L. *Environ. Sci. Technol.* **1993**, *27*, 177–184.
22. Hussain, A.-E.; Serpone, N. *J. Phys. Chem.* **1988**, *92*, 5726–5731.
23. Pelizzetti. E.; Minero, C.; Pramauro, E.; Serpone, N.; Borgarello, E. In *Photocatalysis and Environment;* Schiavello, M., Ed.; Kluwer Academic: Hingham, MA, 1988; pp 469–497.
24. Turchi, C. S.; Ollis, D. F. *J. Catal.* **1990**, *122*, 178–192.
25. Hickling, A.; Hill S. *Disc. Faraday Soc.* **1950**, *46*, 557–559.
26. Knuutinen, J.; Korhonen, I. O. O. *Organic Mass Spectrometry* **1983**, *18(10)*, 438–441.
27. Knuutinen, J.; Tarhanen, J.; Lahtiperä M. *Chromatographia* **1982**, *15(1)*, 9–12.
28. Huyser, E. S. *Free-Radical Chain Reactions;* John-Wiley & Sons: New York, 1970; pp 194–195.
29. Liberles, A. *Introduction to Theoretical Organic Chemistry;* Macmillan: New York, 1973; pp 423–473.
30. Hashimoto, K.; Kawai, T.; Sakata, T. *J. Phys. Chem.* **1984**, *88*, 4083–4088.
31. Dong, C. D.; Huang, C. P., submitted for publication in *Environ. Sci. Technol.*
32. *The Merck Index*, 7th ed.; Merck & Co. Inc.: Rahway, NJ, 1989; pp 332, 1286.

RECEIVED for review November 25, 1992. ACCEPTED revised manuscript June 16, 1993.

16

From Algae to Aquifers

Solid–Liquid Separation in Aquatic Systems

Charles R. O'Melia

Department of Geography and Environmental Engineering, 313 Ames Hall, The Johns Hopkins University, Baltimore, MD 21218-2686

> *Interfacial chemistry and system hydrodynamics control the aggregation, deposition, and separation of particles and particle-reactive substances in natural aquatic environments and in many technological systems. Hydrodynamics (particle transport) are particularly sensitive to particle size and size distribution; colloidal stability is usually determined by the presence of macromolecular natural organic substances. Recent theoretical and experimental studies of the effects of these two classes of variables on solid–liquid separation in aquatic systems are presented and discussed.*

SOLID–LIQUID SEPARATION occurs in virtually all natural and technological systems and affects environmental quality in these systems. Particle aggregation and deposition phenomena contribute significantly to and can dominate solid–liquid separation in most aquatic environments and in many water- and wastewater-treatment technologies. Physical, chemical, and biological treatment processes are all affected, including coagulation, filtration, various membrane separations, activated sludge, thickening, biofilm systems, and others. All natural aquatic systems are influenced by solid–liquid separation: groundwaters, mountain springs, rivers, lakes, estuaries, and the oceans, a sequence that is described later in this chapter in the words of James Joyce, "a long the riverrun" (1).

This chapter was written with three objectives: (1) to review the chemical aspects of particle–particle interactions in aquatic systems; (2) similarly, to review physical or hydrodynamic characteristics of these interactions; and (3) on the basis of this information, to speculate about solid–liquid separation in aquatic systems. Chemical information is taken from the work of Stumm and co-workers in Switzerland, Lyklema and co-workers in The Netherlands, and Tiller at Johns Hopkins University. Hydrodynamic assessment is taken largely from the work of Han and Lawler at University of Texas at Austin. Simulations are taken from the efforts of Jackson and co-workers with algal blooms in the ocean and from studies of filtration in treatment technologies and aquifers at Johns Hopkins University. Speculations of colloidal stability along the riverrun are introduced here.

A schematic diagram of forces between two particles in a fluid is presented in Figure 1. Gravity (F_{grav}) is always involved, although it is not always important. Some form of fluid motion is also always involved, although at times the motion of the fluid is induced by the motion of the particles; this fluid motion results in drag forces (F_{drag}) on the particles. These two forces are

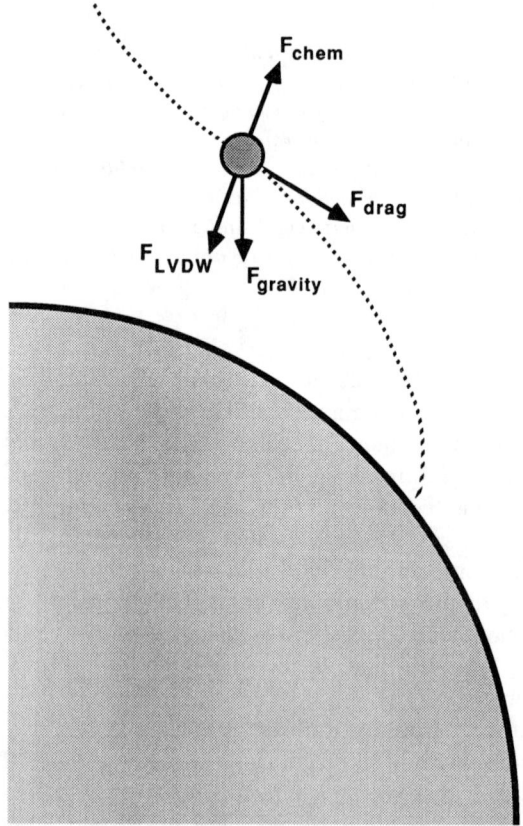

Figure 1. Schematic diagram of forces acting on a particle in a flowing fluid as it approaches a stationary sphere. The moving particle is sufficiently large (viz., diameter larger than 1 μm) that Brownian motion is negligible.

termed physical forces; they are well characterized for simple geometries in the engineering and fluid mechanics literature.

The other two forces depicted in Figure 1 are called colloid–chemical forces. London–van der Waals forces (F_{LVDW}) are present between all solids; they are usually attractive and are fairly well characterized theoretically and experimentally for several systems. The remaining force arrow in Figure 1, F_{chem}, represents a collection of chemical forces that may be repulsive or attractive. These chemical forces may arise from electrostatic, steric, and other interactions between particles in water. They are described primarily in the colloid chemical literature. Our ability to express these forces quantitatively and accurately may be written as $F_{grav} \approx F_{drag} > F_{LVDW} > F_{chem}$.

Chemical Aspects

The focus here is on the effects of dissolved natural organic matter (NOM) on the colloidal stability of particles in aquatic systems and, in particular, on the importance of the macromolecular nature of NOM in these effects. The approach used here has three components: (1) modeling studies with mathematical polyelectrolytes, surfaces, and solvents; (2) laboratory studies with well-characterized polyelectrolytes and particles; and (3) laboratory studies with aquatic NOM, also using well-characterized particles.

Idealized polyelectrolytes and surfaces are termed "mathematical chemicals" here. They exist only in the model and in the mind of the modeler. They are used to study and to simulate the effects of the charge and reactivity of solid surfaces and polyelectrolytes, of the pH and ionic strength of the aqueous solution, and of the size (molecular weight, number of monomer units) of the polyelectrolytes on the configurations of polyelectrolytes at solid-solution interfaces and also to suggest the effects of polyelectrolytes on colloidal stability.

Well-characterized particles and polyelectrolytes used in laboratory experiments are termed "model chemicals". They exist in the laboratory, but they may or may not be present in significant concentrations in the field. They are selected as models for aquatic environments and are used in laboratory experiments to determine the strengths and the limitations of the model simulations using mathematical chemicals.

Experiments using "natural chemicals" are designed to represent actual aquatic environments and to test the applicability to actual natural and technological systems of studies using mathematical and model chemicals. As is often or even usually the case, the speciation of chemicals in real aquatic systems is not known, so the identity of the natural chemicals is described operationally.

This three-component approach to studying the influence of NOM on colloidal stability in aquatic environments was used by Tiller (2). Some of the results presented here are adapted from this work; additional information about the models and the experiments can be found in this thesis.

Scheutjens–Fleer (SF) Theory. A conceptual model for the effects of NOM on colloidal stability can be developed by using existing theoretical and experimental investigations of polymer and polyelectrolyte adsorption on solid surfaces and of the effects of macromolecules on colloidal stability. The modeling approach begins with the work of Scheutjens and Fleer for uncharged macromolecules, termed here the SF theory (3–5). This approach has been extended to the adsorption of linear flexible strong polyelectrolytes by van der Schee and Lyklema (6), adapted to weak polyelectrolytes (7–9), and applied to particle–particle interactions (8, 10).

A schematic picture of a mathematical chemical system containing a solid surface, a solvent, and a linear, flexible, uncharged polymer is shown in Figure 2, which is taken from Lyklema (11). The equilibrium adsorbed state is determined by a balance of conformational entropy changes (such molecules lose conformational entropy on adsorption), intermolecular interactions, and surface–solute interactions. The surface–solute–macromolecule interactions may be specific, hydrophobic, or electrostatic in origin (12). Specific interactions include van der Waals forces, hydrogen bond formation, and (especially in aquatic systems) chemical reactions such as complexation and ligand exchange (13).

SF theory is a statistical thermodynamic model in which chain conformations are formulated as step-weighted random walks in an interfacial lattice (Figure 2). A simple case involves the adsorption of a flexible, linear, homodisperse, uncharged molecule at a uniform planar surface. Interactions among

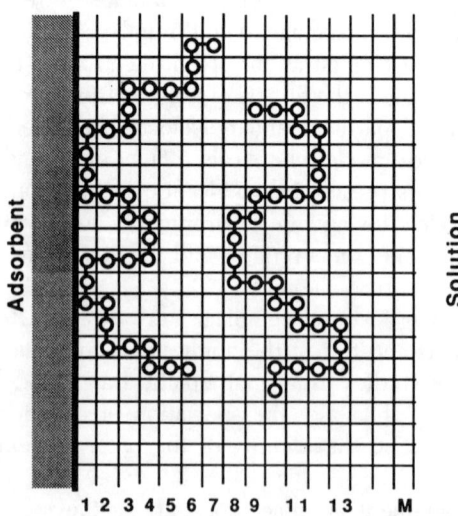

Figure 2. Schematic lattice representation of a macromolecular adsorbed layer showing adsorbed and nonadsorbed uncharged macromolecules. (Reproduced with permission from reference 11. Copyright 1985.)

polymer segments and solvent molecules are characterized by the Flory–Huggins parameter, χ; interactions among surface sites, polymer segments, and solvent molecules are described by a more empirical term denoted as χ_s. The equilibrium configuration of the macromolecular adsorbed layer (MAL) is determined by minimizing the free energy of the system. That is, the canonical partition function is maximized with respect to the number of chains in each possible configuration, and this step allows specification of the distribution of chains, loops, and tails in the interfacial region.

For uncharged polymers, SF theory predicts an exponential decay in the segment density distribution close to a surface. At great distances, tails dominate the segment density and the decay with distance is much less. For moderately sized polymers, a typical root-mean-square adsorbed layer thickness is 5–10 nm, considerably larger than the diffuse layer thickness or Debye length in ocean waters. Tails may extend as far as 20 nm and can have a significant influence in particle–particle interactions. The extent of adsorption increases with the molecular weight and concentration of the polymer, with surface affinity (χ_s), and with decreasing solvent quality (increasing χ).

For charged macromolecules, electrostatic interactions among charged segments, ions, and surfaces are added to the free energy calculations. In many cases, as a result of electrostatic repulsions between charged segments, polyelectrolytes adsorb in flat configurations on uncharged or oppositely charged surfaces. With increasing ionic strength, electrical interactions between segments are screened by the electrolytes, so that polyelectrolytes behave more like uncharged polymers, adsorption is enhanced, and adsorbed layer thicknesses increase. Because polymer and surface charge are usually dependent on pH, adsorption is often a strong function of pH. Specific interactions of polyelectrolytes and particle surfaces with divalent metal ions such as Ca^{2+} and Mg^{2+} may also affect polyelectrolyte adsorption and configuration, with consequent effects on colloidal stability. These latter effects have not yet been incorporated into SF-based models for macromolecules at interfaces and for colloidal stability.

Experimental Results. *Effects of pH and Ionic Strength.* Experiments showing the effects of pH and ionic strength on the configuration of NOM in solution are presented in Figure 3, taken from the work of Cornel et al. (*14*). Experiments were conducted with the 50–100 K apparent molecular weight fraction of a humic acid (HA). Results are expressed in terms of the equivalent Stokes–Einstein or hydrodynamic radius (r_h) calculated from measurements of the diffusion coefficients of the HA fraction.

At low ionic strength, intramolecular repulsion among negatively charged groups on the HA molecules causes the molecules to have an extended shape and a large hydrodynamic radius. At neutral pH and an ionic strength of 3×10^{-4} M, r_h in solution is about 7.5 nm. At high ionic strength, the electrolyte screens these intramolecular repulsions and thereby allows the molecules to

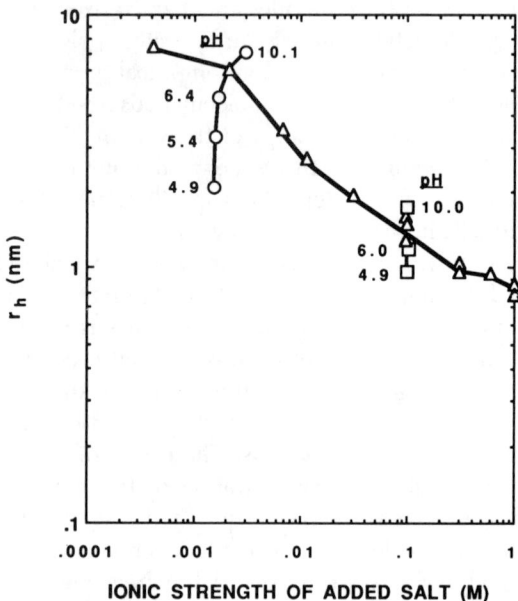

Figure 3. Effect of ionic strength and pH on the Stokes-Einstein (hydrodynamic) radius of the 50–100 K apparent molecular weight fraction of a humic acid. (Reproduced with permission from reference 14. Copyright 1986.)

assume a more coiled shape. At neutral pH and an ionic strength of 1 M, r_h of these molecules is only about 0.85 nm. The charge on the HA depends on pH, so that pH has a considerable effect on the size and configuration of the HA molecules. At high pH, HA molecules have a substantial negative charge arising from the loss of protons from carboxyl and other functional groups on the molecules. At low pH, these functional groups accept protons and the charge on the HA is greatly reduced. At an ionic strength of 0.1 M, the hydrodynamic radius of the HA is reduced from about 1.8 to 0.95 nm as the pH is lowered from 10.0 to 4.9. The effects of pH on molecular size are greater at lower ionic strength.

The effects of pH on the conformations of a moderately anionic polyelectrolyte, both in solution and adsorbed on positive and negative surfaces, are shown in Figure 4, which is taken from the work of Yokoyama et al. (15). The polyelectrolyte used was polygalacturonic acid (PGUA), a linear polysaccharide; the sample used was monodisperse, with a molecular weight of 540,000. Two types of polystyrene latex particles were used, one negatively charged by sulfate groups and the other positively charged by amidine groups. The size of the PGUA in solution at constant ionic strength, taken as twice the radius of gyration ($2r_g$), varied from 150 nm at pH 8.2 (where the mole-

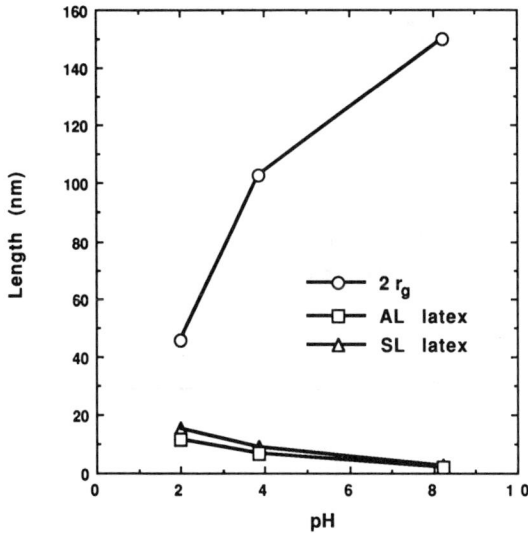

Figure 4. Effects of pH on the radius of gyration (r_g) of polygalacturonic acid (PGUA) in solution and on the adsorbed-layer thicknesses of PGUA on positively charged (AL) and negatively charged (SL) latex particles. (Reproduced from reference 15. Copyright 1989 American Chemical Society.)

cules were fully deprotonated) to under 50 nm at pH 2 (where most of the carboxyl groups on the PGUA, pK_a 2.9, were protonated). This reduction in molecular size as the pH decreased was attributed to reduced intramolecular electrostatic repulsion as the charge on the PGUA was lowered by the pH decrease.

Various effects were observed for the PGUA adsorbed on both types of latex surfaces.

- The thickness of the adsorbed layers of PGUA ranged from 2 to 17 nm; adsorbed PGUA layers were much thinner than the dimensions of the PGUA in bulk solution. These experiments were conducted at an ionic strength of 0.085 M, so the Debye length characterizing the thickness of the diffuse ionic layer was 1 nm, smaller than all of the observed adsorbed layer thicknesses.

- The thickness of the adsorbed layers increased with decreasing pH (i.e., with decreasing molecular charge).

- The negatively charged adsorbed PGUA layers were thicker on the negatively charged latex particles than on the positively charged ones.

Each of these effects has an electrostatic component originating from intra- or intermolecular interactions of the PGUA and between the particle surfaces and the adsorbed PGUA molecules. Specific chemical interactions were also important; negatively charged PGUA adsorbed on negatively charged latex and also stabilized these particles with respect to aggregation.

Effects of NOM. Laboratory experiments by Tiller (2) on the effects of NOM on the colloidal stability of hematite particles (α-Fe_2O_3) are presented in Figure 5. Colloidal stability characterized by the sticking probability (α) is plotted as a function of the concentration of NOM (mg of C/L) added to a

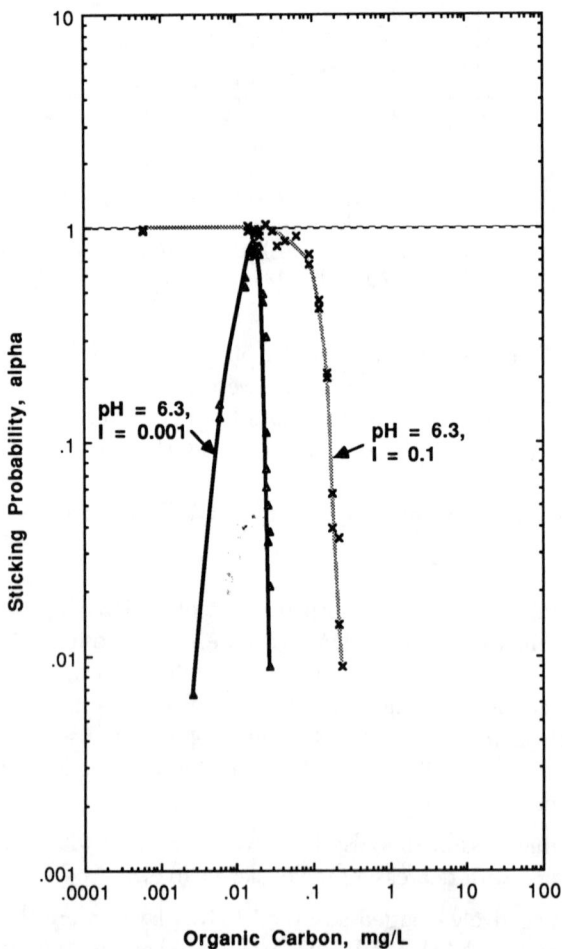

Figure 5. Effect of NOM on colloidal stability (α) of hematite particles. (Reproduced with permission from reference 2. Copyright 1992.)

suspension containing 10 mg/L of hematite particles with an initial diameter of 60 nm.

Hematite was synthesized by following the procedure of Penners (16). NOM was obtained from the Great Dismal Swamp (Virginia); the only treatment was filtration through 0.22-μm Millipore filters to remove debris. Characteristics of this material are presented elsewhere (17–19). The sticking probability, α, is defined as the ratio of the experimentally determined aggregation rate observed by light-scattering techniques to the mass-transport limited rate described by Smoluchowski (20) for fast Brownian aggregation. The solution pH was constant in these experiments at 6.3. Two ionic strengths were studied, 0.001 and 0.1. In natural aquatic systems these values represent a fresh water and a section of an estuary.

At low concentrations of dissolved organic matter (DOM) (<0.01 mg of C/L) and at low ionic strength (10^{-3} M), the hematite particles are positively charged at this pH and are stabilized electrostatically by interacting diffuse layers with characteristic (Debye) lengths of 10 nm. As the ionic strength is increased to 10^{-1} M at these low DOM concentrations, the diffuse layers are compressed to 1 nm, and attractive van der Waals forces promote attachment in classical Derjaguin–Landau–Verwey–Overbeek (DLVO) destabilization by what has been termed double-layer compression.

At intermediate NOM concentrations (from 0.01 to 0.1 mg of C/L, depending on the ionic strength), specific chemical interactions such as ligand exchange allow adsorption of the NOM in sufficient amounts to neutralize the positive charge on the hematite and provide complete destabilization ($\alpha \rightarrow 1$). At higher NOM concentrations, these chemical interactions between the hematite and the NOM produce additional adsorption; the hematite particles become negatively charged, and the suspensions again are kinetically stable ($\alpha \ll 1$).

At low ionic strength (10^{-3} M), this stabilization can be explained by invoking electrostatic diffuse-layer interactions as before, because the thick diffuse layers (Debye length = 10 nm) can prevent the particles from approaching closely enough for attractive van der Waals forces to induce attachment. The hematite is stabilized at high NOM and low ionic strength by specific adsorption of negatively charged NOM and by diffuse layer interaction.

The situation is somewhat different at the higher ionic strength (10^{-1} M). Here, in the presence of adsorbed NOM, compression of the so-called diffuse layer to a thickness of 1 nm is no longer sufficient to induce coagulation. Specific adsorption of NOM produces negatively charged hematite particles, with a fixed layer of negative charges that extends from the surface. If this fixed layer of charge has a thickness that is similar to or greater than the Debye length (1 nm in this case), then an electrosteric stabilization may result.

A schematic representation of the effects of pH and ionic strength on the configuration of anionic polyelectrolytes such as humic substances at solid–

water interfaces is presented in Figure 6. In fresh waters at neutral and alkaline pH values, charged macromolecules assume extended shapes (large hydrodynamic radius, r_h) in solution as a result of intramolecular electrostatic repulsive interactions. When adsorbed at interfaces under these solution conditions, they assume flat configurations (small hydrodynamic thickness, δ_h). At high ionic strength or at low pH, the polyelectrolytes have a coiled configuration in solution (small r_h) and extend further from the solid surface when adsorbed (large δ_h).

Physical Aspects

Aggregation is a kinetic phenomenon that can be represented by a second-order equation.

$$\frac{dn}{dt} = -kn^2 \tag{1}$$

Here n is the number concentration of particles in the system (mol/L) at time t, and k is a second-order rate constant. The chemical and physical aspects of the rate constant are sometimes separated, and k is then written as follows.

$$k = \alpha\beta \tag{2}$$

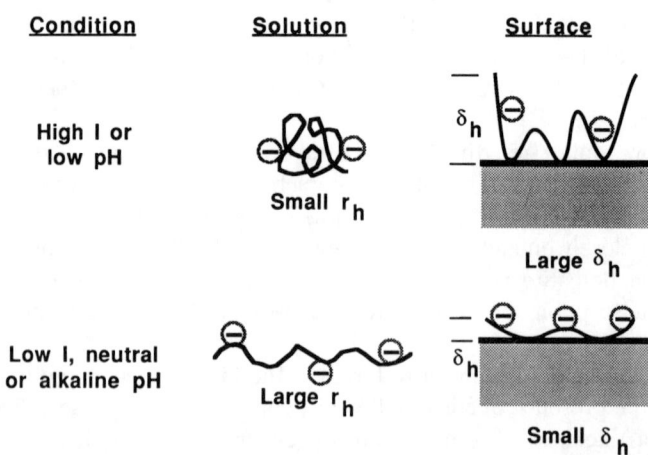

Figure 6. Schematic description of the effects of ionic strength (I) and pH on the conformations of polyelectrolytes such as humic molecules in solution and at a solid-water interface. r_h denotes the hydrodynamic radius of the polyelectrolyte in solution and (δ_h denotes the hydrodynamic thickness of the adsorbed polyelectrolyte. (Adapted from reference 15. Copyright 1989 American Chemical Society.)

Here α is the attachment efficiency or sticking probability discussed previously; it reflects the chemical aspects of the aggregation process and is dimensionless. The term β is a coefficient that describes the physical or mass-transport aspects of the system; typical dimensions are liter seconds per mole. The focus of this section of the chapter is on the physical aspects of aggregation (β).

Three physical mechanisms are involved in particle-aggregation processes: Brownian diffusion, fluid shear, and differential settling. These mechanisms are illustrated schematically in Figure 7. Brownian or molecular diffusion is the result of thermal motion of solvent molecules and surface drag forces on small particles; collisions by this mechanism are often termed perikinetic flocculation (Figure 7a). Particle collisions by fluid shear occur because velocity differences or gradients occur in all real moving fluids, whether the flow is laminar or turbulent. Particles that simply follow the motion of the suspending

a. Brownian diffusion

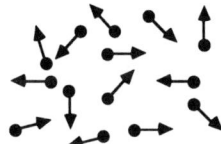

b. Velocity gradients

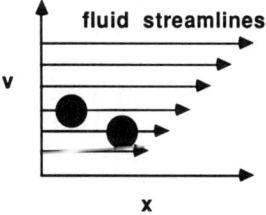

c. Differential sedimentation

Figure 7. Schematic representation of three physical processes involved in aggregation processes in aquatic environments. (Adapted with permission from reference 21. Copyright 1991 Lewis.)

fluid will then travel at different velocities, resulting in collisions between them (Figure 7b). Collisions by this mechanism are often termed orthokinetic flocculation. The third mechanism illustrated here results from the gravity force, which produces vertical motion of particles and depends on the buoyant weight of the particles. Large or dense particles can contact smaller or lighter particles by a transport process termed differential sedimentation (Figure 7c). Analogous physical processes in particle deposition in porous media are: convective diffusion, interception, and sedimentation (22). These processes are not discussed here.

Expressions for the transport coefficients of these physical processes can be written as follows.

$$\beta_{Br}(i,j) = \frac{2k_B T}{3\mu} \left[\frac{1}{d_i} + \frac{1}{d_j} \right] [d_i + d_j] \tag{3a}$$

$$\beta_{Sh}(i,j) = \frac{1}{6} [d_i + d_j]^3 G \tag{3b}$$

$$\beta_{Ds}(i,j) = \frac{\pi g}{72\mu} [\rho_p - \rho_l][d_i + d_j]^3 |d_i - d_j| \tag{3c}$$

Here $\beta_{Br}(i,j)$, $\beta_{Sh}(i,j)$, and $\beta_{Ds}(i,j)$ are the transport coefficients for interparticle contacts between particles of diameters d_i and d_j by Brownian diffusion, fluid shear, and differential sedimentation, respectively; k_B is Boltzmann's constant; T is the absolute temperature; μ is the viscosity of the liquid; G is the mean velocity gradient of the liquid; g is the gravity acceleration; and ρ_p and ρ_l are the densities of the particles and the liquid, respectively.

Smoluchowski Approach to Aggregation. This approach, taken from the works of Smoluchowski (20) and Findheisen (23), has been called the Smoluchowski approach to aggregation. It has two important assumptions. First, the paths or trajectories of the particles are straight lines (Figure 7); it is a rectilinear approach. Second, it is assumed that aggregates are coalesced spheres; particle volume is conserved as aggregation proceeds. A third factor, although not an assumption, is also important in using these results. Chemical effects are collected in α, the sticking probability included in equation 2 and described here.

An important improvement in Smoluchowski's approach was to consider hydrodynamic interactions between two particles as they approach each other. These interactions are of two types and result in curvilinear models. First are deviations from rectilinear flow paths that occur as two particles approach each other. Second is the increasing hydrodynamic drag that occurs as two particles come into close proximity.

These effects are illustrated in a comparison of the rectilinear and curvilinear approaches to differential sedimentation presented in Figure 8, adapted from Han and Lawler (24). In both approaches, the upper, larger, faster particle is settling by gravity toward the lower, smaller, slower particle. In the rectilinear or Smoluchowski approach, all small particles with size d_i that reside below the larger particle (size d_j) within the area A_r with diameter $(d_i + d_j)$ come into contact with the larger particle; $\beta_{DS}(i,j)$ is given by equation 3c.

In the curvilinear case where all hydrodynamic interactions between the two particles are considered, only those small particles in the shaded area denoted as A_c can come into contact with the larger particle. The area A_c can be determined numerically. The result is that the actual collision rate is less than the rectilinear rate, and the actual mass-transport coefficient is equal to $[(A_c/A_r)\beta_{DS}(i,j)]$. Han and Lawler (24) calculated reductions in the rectilinear transport rate by differential sedimentation ranging from about 0.3 to 0.001, so the effects of hydrodynamic interactions on this transport process can be substantial in many cases.

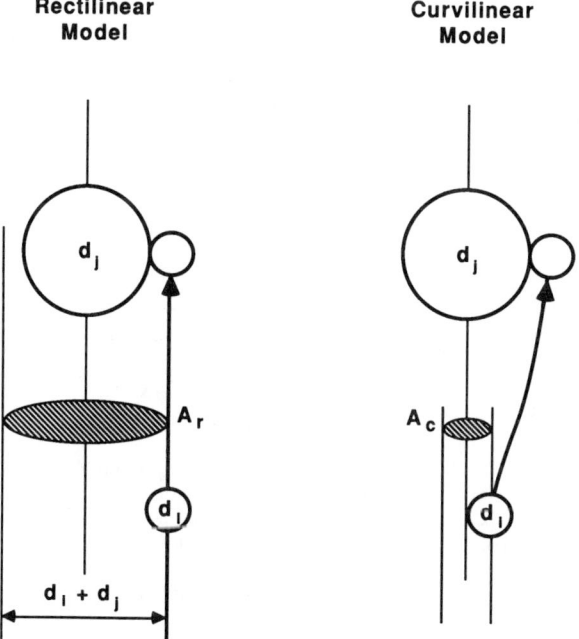

Figure 8. Schematic presentation of rectilinear and curvilinear trajectories in particle collisions by differential sedimentation. (Adapted with permission from reference 24. Copyright 1991.)

Numerical results obtained by Han and Lawler (25) for the effects of hydrodynamic interactions on particle transport by fluid shear are summarized graphically in Figure 9. These results are based on the work of Adler (26). The effects of particle size, velocity gradient, and van der Waals interaction are characterized by a dimensionless group, H_A, defined as follows:

$$H_A = \frac{A}{18\pi\mu d_j^3 G} \qquad (4)$$

in which A is the Hamaker constant (joules) and d_j is the diameter of the larger of two interacting particles. Retarded van der Waals interactions are used; A is selected as $10kT$. The ratio of the curvilinear transport coefficient to the rectilinear case is plotted as a function of the size ratio (d_i/d_j).

Hydrodynamic interactions in aggregation by fluid shear are relatively small for monodisperse suspensions and become increasingly important as the

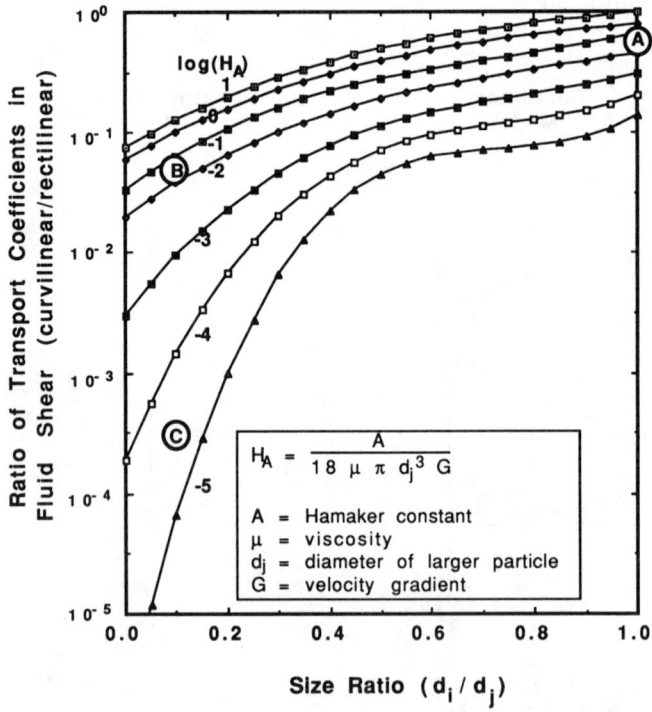

Figure 9. Effect of particle size on physical aspects of particle collisions by fluid shear. The ratio of rectilinear to curvilinear transport coefficients is plotted as a function of the ratio of the size of the smaller particle (d_i) to a larger one (d_j). The circled A, B, and C refer to cases discussed in the text. (Reproduced with permission from reference 25. Copyright 1992.)

systems become more heterodisperse (i.e., as the size ratio decreases). The hydrodynamic effects become more important as H_A becomes smaller. Increasing the velocity gradient or the size of the larger particle increases the effects of the interactions and the deviations from the rectilinear model.

Illustrative Cases. Three cases are illustrated in Figure 9, marked by the circles labeled A, B, and C. Case A refers to classical experiments by Swift and Friedlander (27) on the coagulation of monodisperse latex particles (diameter = 0.871 μm) in shear flow and in the absence of repulsive chemical interactions. Considering a velocity gradient of 20 s^{-1}, H_A is 0.0535, log H_A is −1.27, and d_i/d_j is 1.0 for these experimental conditions. The circle labeled A in Figure 9 marks these conditions and indicates that the hydrodynamic corrections to Smoluchowski's model predict a reduction of about 40% in the aggregation rate by fluid shear. The experimental measurements by Swift and Friedlander showed a reduction of 64%. This observed reduction from Smoluchowski's rectilinear model was therefore primarily physical or hydrodynamic and consistent with the curvilinear model.

Case B in Figure 9 is a representation of shear coagulation in a surface water, such as a lake or the ocean. Considering the interactions between a bacterium (d_i = 1 μm) and an algal cell (d_j = 10 μm) in a shear field with G = 0.1 s^{-1} at 20 °C, H_A is 0.007, log H_A is −2.15, and the size ratio is 0.1. Under these circumstances the curvilinear approach predicts contact rates by fluid shear that are on the order of 3% of the Smoluchowski model. Case C is a representation of aggregation in a water- or wastewater-treatment plant, again considering contacts between a small particle (1 μm) and a larger one (10 μm). In this case, fluid mixing is provided to achieve a velocity gradient of 20 s^{-1}. For this situation H_A is 3.6 × 10^{-5}, log H_A is −4.45, and the size ratio is again 0.1. Hydrodynamic interactions are predicted to reduce the contact rate to only about 0.03% of the rectilinear case, a result indicating that the effectiveness of fluid shear in producing interparticle contacts in treatment systems has been overrated (25).

These three examples show that Smoluchowski's rectilinear model can provide a useful representation of the initial contact rates in laboratory studies using monodisperse suspensions (case A in Figure 9), that this model can overpredict contacts in lakes and oceans by 1–2 orders of magnitude (case B in Figure 9), and that it overestimates initial contact rates in water- and wastewater-treatment systems by 3 orders of magnitude or more (case C, Figure 9).

Assumptions. The first assumption in the Smoluchowski approach, that of rectilinear particle motion, can lead to significantly overestimated collision rates in some aggregation processes. The second assumption, that of volume conservation or the formation of coalesced spheres, can lead to an underestimation of collision opportunities and aggregation rates. As the coagulation of a suspension of solid particles proceeds, fluid is incorporated into pores in the aggregates that are formed. Aggregate density decreases, and total aggregate volume increases as the process continues. The result is that the target cross sections or collision diameters of the aggregates increase, and thereby increase the rates of interparticle contacts brought about by Brownian diffusion, fluid shear, and differential sedimentation. Models for this process using empirical size-density relationships and fractal geometry are under development.

In the absence of better information, it is sometimes assumed that errors from the two assumptions in the Smoluchowski approach compensate for each other. Reductions in collisions resulting from hydrodynamic effects are assumed to be offset by increases in collision rates as aggregate volume increases while coagulation proceeds. The Smoluchowski approach modified to include hydrodynamic interactions is useful at the onset of aggregation processes, when the inclusion of fluid within aggregate pores is small.

Simulations and Speculations

Two simulations of the effects of solid–liquid separation are considered here. From algae to aquifers, these are, first, a consideration of the effects of coagulation on algal populations and sedimenting algal fluxes in marine waters taken from Jackson and Lochmann (28) and, second, evaluation of the effects of deposition or filtration on the transport of particles in groundwater systems taken from Tobiason (29).

The results (28) of simulations of an algal bloom in a marine environment such as the coastal ocean are presented in Figure 10. The vertical flux of particulate nitrogen leaving the mixed surface layer by sedimentation is plotted as a function of time after a pulse input of nitrate from the bottom waters stimulates algal growth in surface waters.

In the model that yielded these results, algal growth is influenced by nutrients and light; algal cells coagulate by Brownian motion, fluid shear, and differential sedimentation; and algal biomass is removed from the surface layer by sedimentation. A rectilinear interparticle transport model is used. These results are for an algal diameter of 20 μm, $\alpha = 0.25$, $G = 0.1$ s^{-1}, and an initial concentration of inorganic nitrogen of 20 μM. Zooplankton grazing is not included.

These simulations indicate that coagulation can reduce peak algal populations in the mixed surface layer and substantially increase the peak flux of biomass leaving the mixed layer by the formation of large, fast-settling aggregates during rapid coagulation. These results suggest that vertical particle transport associated with aggregation has the potential to be an important mechanism for removing biological material from eutrophic marine waters (28).

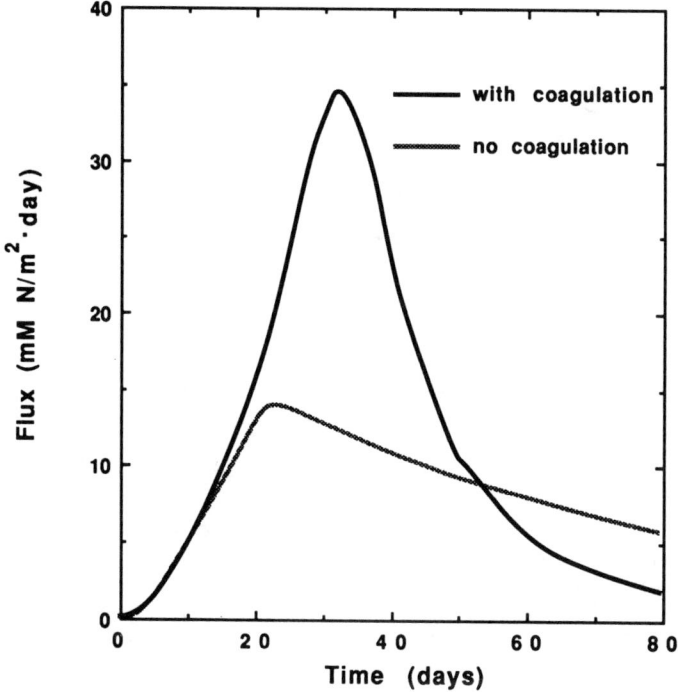

Figure 10. Vertical fluxes of biomass (expressed as millimoles of particulate N leaving the surface layer of a marine system per square meter of ocean surface per day) as a function of time after the introduction of nutrients (N) in days. Two cases are considered, one without coagulation ($\alpha = 0$) and the other with rapid coagulation ($\alpha = 0.25$). (Reproduced with permission from reference 28. Copyright 1992.)

By using theory and experimental results concerning particle deposition in porous media, predictions can be made of the removal of viruses, bacteria, and other microorganisms in aquifers. For a "clean" aquifer (i.e., one that has not received significant inputs of colloidal particles), the effects of particle size and surface chemistry on colloid transport are illustrated in Figure 11. The travel distance required to deposit 99% of the particles in a suspension (L_{99}) is plotted as a function of the size (radius) of those particles for two chemical conditions.

In these results the effects of surface and solution chemistry on particle deposition are represented by the sticking coefficient or attachment probability (α), defined in this case as the rate at which particles adhere to a grain of aquifer media divided by the rate at which they contact it. In Figure 11, α is assumed as 1 (perfect sticking) and also as 10^{-3} (one collision in every 1000

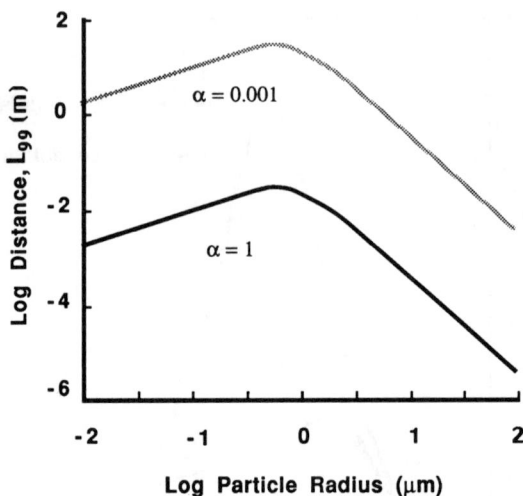

Figure 11. Distance required to remove 99% of particles from suspension (L_{99}) as a function of suspended particle radius for two chemical conditions (attachment probabilities). Flow rate = 0.1 m/day, media radius = 0.025 cm, temperature = 25 °C, particle density = 1.05 g/cm³, and aquifer porosity = 0.4. (Reproduced with permission from reference 29. Copyright 1987.)

that occur results in deposition). Conditions of flow rate, media size, and porosity are assumed to be representative of a sandy aquifer. Theoretical background for the results in Figure 11 was presented elsewhere (22, 30–32). Experimental evidence for the successful application of filtration theory to particle deposition in porous media was provided by Martin et al. (33).

Three points are made here from the results shown in Figure 11. First, viruses (with radii from 0.01 to 0.05 μm) are easier to deposit in porous media or filter than bacteria (radii on the order of 1 μm) for comparable attachment probabilities. This comparison suggests, for example, that coliform bacteria could serve as a conservative indicator for the presence of viruses because they are easier to transport through porous aquifers (questions of attachment probabilities and die-off rates would also need examination). Second, solution and surface chemistry, which often dominate attachment probability (α), can control the passage of pathogens and other microorganisms in groundwater environments. Bacteria can be effectively removed within a few centimeters if they will stick ($\alpha = 1$), whereas they can travel several tens of meters if $\alpha = 10^{-3}$. Other studies dealing with the effects of natural organic matter (NOM) and Ca^{2+} on colloidal stability in natural waters (34–36) indicate that it is plausible that soft waters high in NOM will yield low sticking probabilities and permit extensive passage of microorganisms. However, hard waters containing little NOM will produce relatively sticky particles and aquifer materials and will enhance particle deposition. Third, these results have several limitations that require attention, including the assumption of "clean" aquifer media and the neglect of particle detachment or entrainment.

A Long the Riverrun

Let us consider the colloidal stability of particles as one's view moves from a mountain spring to a river, through an estuary, and into the ocean. Solid particles and natural organic matter are everywhere in aquatic environments. Here we consider that NOM adsorbs on particles and affects their colloidal stability. We discuss the origins of this stability "a long the riverrun" (1). Our conclusions are based on results with both mathematical and laboratory chemicals (2). Speculation is extensive.

Of the many changes in solution composition that occur a long the riverrun, the focus here is on ionic strength, which affects the range of electrostatic interactions in classical DLVO views of colloidal stability, and which also affects the sizes and conformations of macromolecules in solution and adsorbed at interfaces. Here the range of colloidal electrostatic interactions is described by the Debye length, denoted as κ^{-1}, and the thickness of an adsorbed layer of polyelectrolyes on a particle surface is described by a hydrodynamic thickness, termed δ_h. The Debye length varies inversely with the square root of the ionic strength (I); at 20 °C it can be written as follows:

$$\kappa^{-1} = \frac{0.28}{I^{1/2}} \quad (5)$$

in which I is expressed in moles per liter and the Debye length has dimensions of nanometers. The thickness of an organic adsorbed layer depends on many factors such as molecular weight, solvent quality (χ), specific particle–polymer segment interactions (χ_s), surface charge, polyelectrolyte charge, and ionic strength; it is difficult to predict or to measure.

Speculations about colloidal stability a long the riverrun are presented in Figure 12. Debye length (κ^{-1}), adsorbed organic layer thickness (δ_h), and colloidal sticking probability (α), all on logarithmic scales, are plotted as functions of ionic strength, also on a logarithmic scale. The Debye length is calculated with equation 5. The hydrodynamic thickness of the adsorbed organic layer is an interpretation of laboratory experiments such as those of Yokoyama et al. (15, presented in Figure 4) and calculations using mathematical chemicals such as those by Tiller and O'Melia (2, 37). The colloidal sticking probabilities are an interpretative summation of experimental measurements in laboratory and field systems, including results such as those shown in Figure 5, laboratory measurements of the colloidal stability of marine algae (38), and others (31–36, 39).

In all of the aquatic systems sketched a long the riverrun in Figure 12, it is considered that the adsorption of natural organic matter, including humic

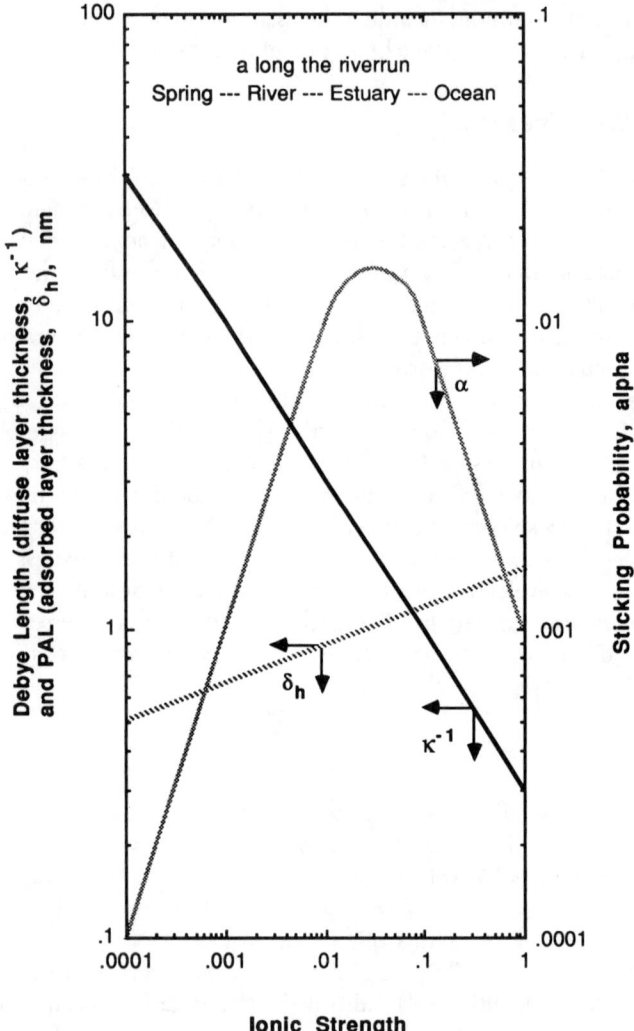

Figure 12. Illustration of Debye length, adsorbed layer thickness, and colloidal sticking coefficient in natural aquatic environments, a long the riverrun.

substances, establishes the surface charge on the particles in suspension. (In an aquifer, adsorbed NOM could also establish the surface charge on the aquifer materials.) The principal factors in this adsorption are the macromolecular nature of the adsorbing organic matter and specific chemical interactions between functional groups on the NOM and sites on the particle surfaces.

In fresh waters, the thickness of this adsorbed layer is small, primarily because of intra- and intermolecular electrostatic repulsions among negatively charged groups on the NOM (Figure 6). The result is a thin adsorbed layer, with a thickness of 1 nm or less at ionic strengths of 10^{-2} or lower. The electrostatic diffuse layers that surround the colloidal particles and their adsorbed layers are much thicker, ranging from some 30 nm when I is 10^{-4} to 3 nm when I is 10^{-2}. The ratio of the Debye length to the adsorbed layer thickness, or κ^{-1}/δ_h, ranges from an estimate of 60 in the mountain stream with $I = 10^{-4}$ to the order of 3 at the entrance to an estuary where I may be 10^{-2}. This ratio suggests that particles in these fresh waters are stabilized by interacting diffuse layers of similar charge. In summary, the surface charge on the particles in fresh waters is set by specific adsorption of NOM, and the colloidal stability that results derives from the electrostatic interactions of the thick diffuse layers that are present.

In the ocean, adsorption of NOM is expected to increase as a result of reduced intra- and intermolecular electrostatic repulsions among the negatively charged groups on the NOM. Surface charge may increase, as will the adsorbed layer thickness (δ_h, Figure 6). Adsorbed layer thicknesses of 1 to a few nanometers seem likely, although measurements in such systems are not yet available. In contrast, the Debye length in marine waters is on the order of 0.5 nm, significantly smaller than δ_h. The ratio of these thicknesses, κ^{-1}/δ_h, can be 1 or smaller.

As in fresh waters, the charge on particles in marine environments can result from the adsorption of negatively charged NOM. However, in ocean waters the diffuse layers surrounding particles are too thin to prevent attractive van der Waals forces from causing attachment when interparticle contacts occur. Where particles are observed to be stable in marine environments, as in the stable marine diatom suspensions observed by Kiørboe et al. (38), it is proposed that this can be due to the presence of thick, charged, relatively fixed adsorbed layers of organic matter that induce a stabilization that may be termed electrosteric (40).

The speculation continues here into estuarine waters, bounded in this analysis at both inlet and outlet by waters that can, in the presence of suitable NOM, yield stable colloids. Diffuse layer thicknesses in these waters are small, on the order of 1 nm at $I = 0.1$. There can be sufficient salt in these waters to prevent classical DLVO electrostatic stabilization; this is the conventional view. There may also be insufficient salt to form a thick layer of adsorbed NOM by screening of intra- and intermolecular repulsive interactions of the molecules of NOM. The result would then be a region of ionic strength or salinity in an estuary within which colloidal particles have a minimum stability and a maximum sticking probability. This possibility is shown by the proposed relationship between α and ionic strength shown in Figure 12.

The accuracy of the estimates of the three variables shown in Figure 12 is $\kappa^{-1} > \delta_h \gg \alpha$. There is much to be learned a long the riverrun.

Acknowledgments

The writer gratefully acknowledge the contributions of Werner Stumm to his study and understanding of coagulation and deposition phenomena over some three decades. The help of Christine Tiller, Menachem Elimelech, John Tobiason, and Desmond Lawler was particularly valuable in this present work. This work was supported in part by the U.S. National Science Foundation under Grant BCS–9112766 and by the U.S. Office of Naval Research under Grant N00014–92–J–1811.

References

1. Joyce, J. A. *Finnegans Wake;* Mandarin Paperbacks: London, 1992; pp 628–3.
2. Tiller, C. L. Ph.D. Thesis, The Johns Hopkins University, Baltimore, MD, 1992.
3. Scheutjens, J. M. H. M.; Fleer, G. J. *J. Phys. Chem.* **1979,** *83,* 1619–1635.
4. Scheutjens, J. M. H. M.; Fleer, G. J. *J. Phys. Chem.* **1980,** *84,* 178–190.
5. Scheutjens, J. M. H. M.; Fleer, G. J. *Macromolecules* **1985,** *18,* 1882–1900.
6. van der Schee, H. A.; Lyklema, J. *J. Phys. Chem.* **1984,** *88,* 6661–6667.
7. Evers, O. A.; Fleer, G. J.; Scheutjens, J. M. H. M.; Lyklema, J. *J. Colloid Interface Sci.* **1986,** *111,* 446–454.
8. Böhmer, M. R.; Evers, O. A.; Scheutjens, J. M. H. M. *Macromolecules* **1990,** *23,* 2288–2301.
9. Böhkmeer, J.; Bo°:hmer, M. A.; Cohen Stuart, M. A.; Fleer, G. J. *Macromolecules* **1990,** *23,* 2301–2309.
10. Lyklema, J.; Fleer, G. J. *Colloids Surf.* **1987,** *25,* 357–368.
11. Lyklema, J. In *Flocculation Sedimentation and Consolidation;* Moudgil, B. M.; Somasunsaran, P., Eds.; Engineering Foundation: New York, 1985; pp 3–21.
12. Westall, J. C. In *Aquatic Surface Chemistry: Chemical Processes at the Particle–Water Interface;* Stumm, W., Ed.; Wiley Interscience: New York, 1987; pp 3–32.
13. Stumm, W.; Wollast, R. *Rev. Geophys.* **1990,** *128,* 53–69.
14. Cornel, P. K.; Summers, R. S.; Roberts, P. V. *J. Colloid Interface Sci.* **1986,** *110,* 149–164.
15. Yokoyama, A.; Srinivasan, K. R.; Fogler; H. S. *Langmuir* **1989,** *5,* 534–538.
16. Penners, N. H. G. Ph.D. Thesis, Wageningen Agricultural University, Wageningen, The Netherlands, 1985.
17. Liao, W. Ph.D. Thesis, University of North Carolina, Chapel Hill, NC, 1981.
18. Dempsey, B. A. Ph.D. Thesis, University of North Carolina, Chapel Hill, NC, 1981.
19. Hundt, T. A. Ph.D. Thesis, The Johns Hopkins University, Baltimore, MD, 1985.
20. Smoluchowski, M. *Kolloid Z. Z. Polym.* **1917,** *92,* 129–168.
21. Pankow, J. K. *Aquatic Chemistry Concepts;* Lewis: Chelsea, MI, 1991; p 646.
22. Yao, K.-M.; Habibian, M. T.; O'Melia, C. R. *Environ. Sci. Technol.* **1971,** *5,* 1105–1112.
23. Findheisen, W. *Meteorol. Z.* **1939,** *56,* 365–368.
24. Han, M.; Lawler, D. L. *J. Hydraul. Eng.* **1991,** *117,* 1269–1289.
25. Han, M.; Lawler, D. L. *J. Am. Water Works Assoc.* **1992,** *84(10),* 79–91.
26. Adler, P. M. *J. Colloid Interface Sci.* **1981,** *84,* 461–474.

27. Swift, D. L.; Friedlander, S. K. *J. Colloid Sci.* **1964**, *19*, 621–647.
28. Jackson, G. A.; Lochmann, S. E. *Limnol. Oceanogra.* **1992**, *37*, 77–89.
29. Tobiason, J. E. Ph.D. Thesis, The Johns Hopkins University, Baltimore, MD, 1987.
30. Rajagopalan, R.; Tien, C. *Am. Inst. Chem. Eng. J.* **1976**, *22*, 523–533.
31. Tobiason, J. E.; O'Melia, C. R. *J. Am. Water Works Assoc.* **1988**, *80(12)*, 54–64.
32. Elimelech, M.; O'Melia, C. R. *Environ. Sci. Technol.* **1990**, *24*, 1528–1536.
33. Martin, R. E.; Bouwer, E. J.; Hanna, L. M. *Environ. Sci. Technol.* **1992**, *26*, 1053–1058.
34. Tipping, E.; Higgins, D. C. *Colloids Surf.* **1982**, *5*, 85–92.
35. Ali, W.; O'Melia, C. R.; Edzwald, J. K. *Water Sci. Technol.* **1985**, *17*, 701–712.
36. Weilenmann, U.; O'Melia, C. R.; Stumm, W. *Limnol. Oceanogra.* **1989**, *34*, 1–18.
37. Tiller, C. L.; O'Melia, C. R. *Colloids Surf.* **1993**, *73*, 89–102.
38. Kiørboe, T.; Andersen, K. P.; Dam, H. G. *Mar. Biol.* **1990**, *107*, 235–245.
39. Liang, L.; Morgan, J. J. *Aquat. Sci.* **1990**, *52*, 32–55.
40. Hunter, R. J. *Foundations of Colloid Science;* Clarendon: Oxford, England, 1987; Vol. 1.

RECEIVED for review March 10, 1993. ACCEPTED revised manuscript November 29, 1993.

17

Surfactant Solubilization of Phenanthrene in Soil–Aqueous Systems and Its Effects on Biomineralization

Shonali Laha, Zhongbao Liu, David A. Edwards, and Richard G. Luthy*

Department of Civil and Environmental Engineering, Carnegie Mellon University, Pittsburgh, PA 15213

> *A series of related experiments investigated nonionic surfactant sorption onto soil, mechanisms of nonionic surfactant solubilization of polycyclic aromatic hydrocarbon (PAH) compounds from soil, and microbial mineralization of phenanthrene in soil–aqueous systems with nonionic surfactants. Surfactant solubilization of PAH from soil at equilibrium can be characterized with a physicochemical model by using parameters obtained from independent tests in aqueous and soil–aqueous systems. The microbial degradation of phenanthrene in soil–aqueous systems is inhibited by addition of alkyl ethoxylate, alkylphenyl ethoxylate, or sorbitan- (Tween-) type nonionic surfactants at doses that result in micellar solubilization of phenanthrene from soil. Available data suggest that the inhibitory effect on phenanthrene biodegradation is reversible and not a specific toxic effect.*

EXPERIMENTAL INVESTIGATIONS AND MODELING of the interactions in soil–aqueous systems among nonionic surfactants, polycyclic aromatic hydrocarbon (PAH) compounds, and microorganisms are presented here. These interactions affect the potential for surfactant-facilitated hydrophobic organic compound (HOC) transport in soil and groundwater systems, and the feasibility of engineered surfactant-aided cleanup of contaminated sites (1). In low

*Corresponding author

0065–2393/95/0244–0339$08.54/0
© 1995 American Chemical Society

ionic strength aqueous fluid at 25 °C, most nonionic surfactants at sufficiently high concentrations form regular micelles in single-phase solutions. In contrast, certain surfactants, such as a dodecyl ethoxylate with four ethoxylate groups, $C_{12}E_4$, form bilayer lamellae or other types of aggregates in more complex two-phase solutions (2, 3). The critical concentrations for the onset of micelle and aggregate formation are termed the critical micelle concentration (CMC) and the critical aggregation concentration (CAC), respectively. Important changes occur in surfactant sorption, surfactant solubilization of HOCs, and microbial mineralization of HOCs in the presence of nonionic surfactants at or near these critical surfactant concentrations.

Surfactant Solubilization of HOCs from Soil

When nonionic surfactant is applied to a soil–aqueous system, the surfactant can exist as dissolved monomers, sorbed molecules on the soil, or aggregated groups of molecules called micelles. Molecules of HOCs in such a system can be solubilized in surfactant micelles, dissolved in the surrounding solution, sorbed directly on the soil, or sorbed in association with sorbed surfactant. The presence of nonionic surfactant micelles in the bulk solution of the system results in the partitioning of the HOC between two bulk solution compartments, commonly referred to as pseudophases. The micellar pseudophase consists of the hydrophobic interiors of surfactant micelles, whereas the aqueous pseudophase consists mainly of dissolved surfactant monomers and water. Micelles form when the bulk solution concentration exceeds the surfactant CMC.

In soil–aqueous systems, the hydrophobic interiors of nonionic surfactant micelles can compete strongly with soil organic matter as a compartment for the partitioning of HOCs. Surfactant micelles in such systems can markedly increase the bulk solution fraction of the total HOC mass, and micellar surfactant flushing has consequently been considered as a potential means for remediating soils contaminated with sorbed HOCs (4–6). Relatively little is known, however, about the physicochemical interactions of surfactants with HOCs and soil.

Several of the most important equilibrium relationships affecting HOC distribution in soil–aqueous systems with surfactants were discussed by Edwards et al. (7) in terms of a mathematical model. Figure 1 illustrates these interactions in terms of characteristic parameters for the equilibrium distribution of HOC in a closed system of soil and micellar surfactant solution, assuming that pure-phase HOC is not present. The distribution of HOC between the sorbed phase and the aqueous solution surrounding the micelles is represented by $K_{d,cmc}$. This parameter gives the ratio of the number of moles of HOC sorbed per gram of soil to the number of moles of HOC dissolved per liter of surfactant solution at the CMC. The distribution of HOC between the micelles and the aqueous solution surrounding the micelles is given by K_m, which represents the ratio of X_m, the mole fraction of HOC in the micellar

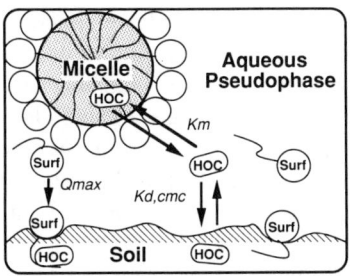

Figure 1. Distribution of HOC and nonionic surfactant in soil–aqueous systems.

pseudophase, to X_a, the mole fraction of HOC in the nonmicellar aqueous solution (8, 9).

The distribution of surfactant between soil and the aqueous solution surrounding the micelles appears to depend on the type of surfactant and the type of soil on which it sorbs. The sorption of certain nonionic surfactants onto soil may attain a maximum value at or near a bulk solution concentration equal to the CMC. Micelle-forming, nonionic alkylphenyl ethoxylate surfactants exhibited a maximum value for sorption, Q_{max}, at the CMC with a soil of 1% initial organic carbon content (10). The nonionic alkyl ethoxylate surfactant $C_{12}E_4$ is a less common type of surfactant; at 25 °C it forms a two-phase surfactant bilayer system. Sorption greater than that attained at the CAC occurred for $C_{12}E_4$ at supra-CAC bulk solution surfactant concentrations (10). Sorption of micelle-forming, nonionic surfactants at supra-CMC bulk solution concentrations may be greater than that attained at the CMC for low-organic-carbon aquifer sediments (11).

The effect of surfactant sorption on HOC sorption can be represented by a modified sorption partition coefficient $K_{d,cmc}$ (7).

$$K_{d,cmc} = K_d \left(\frac{S}{S_{cmc}}\right)\left(\frac{f_{oc}^*}{f_{oc}}\right) \qquad (1)$$

where K_d is the soil–water partition coefficient (L/g) in the absence of surfactant, S is the HOC solubility limit in pure water (mol/L), S_{cmc} is the enhanced, apparent solubility limit for HOC in surfactant solution at the CMC (mol/L), f_{oc} is the weight fraction of organic carbon in the soil before surfactant sorption, and f_{oc}^* is the equivalent weight fraction of organic carbon in the soil after surfactant sorption. The latter quantity is given by equation 2.

$$f_{oc}^* = f_{oc} + \epsilon Q_{surf} MW_{surf} f_{carbon} \qquad (2)$$

where f_{carbon} is the weight fraction of carbon in surfactant, MW_{surf} is the surfactant molecular weight, Q_{surf} is the number of moles of nonionic surfactant sorbed per gram of soil, and ϵ is an index characterizing the relative effectiveness of organic carbon in the surfactant to organic carbon in the humic matter as a sorbent for HOC. It is assumed for soils with moderate-to-high initial organic carbon content that ϵ has a value close to unity. Recent published research (11) suggests that greater values of ϵ may be appropriate for low-organic-carbon aquifer sediments, possibly due to the nature of surfactant sorption on such media and the interaction of HOC with surface micelles (admicelles) or bilayers.

The value for K_m can be obtained from the MSR (9), the molar solubilization ratio, which is the number of moles of HOC solubilized per mole of surfactant in micellar form.

$$K_m = \left(\frac{1}{S_{cmc}V_w}\right)\left(\frac{MSR}{1 + MSR}\right) \qquad (3)$$

where V_w is the molar volume of water. The MSR can be determined from the slope of an apparent HOC solubility curve plotted against micellar surfactant concentration in the absence of soil and in the presence of solid-phase HOC (9).

The solubilization of HOC from soil in the presence of surfactant can be plotted as the fraction of total HOC mass in the bulk solution versus the surfactant dose added to the system. In this representation, F is the ratio of the number of moles of HOC dissolved and solubilized to the number of moles dissolved, solubilized, and sorbed. A simplified equilibrium model (7) for predicting F is

$$F = \frac{A}{A + B} \qquad (4)$$

where $A = 1 + K_m V_w C_{mic}$, $B = K_{d,cmc} w_{soil}/v_{aq}$, w_{soil} is the weight of the soil (g), v_{aq} is the volume of the bulk solution (L), and C_{mic} is the concentration of nonionic surfactant in micelle form (mol/L). Equation 4 is an approximation appropriate for HOCs having a value of the octanol–water partition coefficient greater than about 10^4. The concentration of nonionic surfactant in micelle form is equal to the total dose of surfactant added to the system less the sum of the CMC and the concentration of sorbed surfactant, each expressed in moles per liter.

The model outlined predicts the equilibrium distribution of HOC in a closed system of soil and micellar surfactant solution as a function of surfactant dose. The model requires values for the parameters K_m, Q_{max}, f_{oc}, CMC, K_d, S, and S_{cmc}. If values for these parameters can be obtained independently from separate tests, no fitting coefficients are required in theory to match model results to data. The validity of this conclusion has been shown (12) and will be demonstrated for phenanthrene and two nonionic surfactants, Triton X-100 (a $C_8PE_{9.5}$ alkylphenyl ethoxylate), and Brij 30 (a $C_{12}E_4$ alkyl ethoxylate).

HOC Availability and Biodegradation in Soil

Soil bioremediation involves the use of microorganisms and their capacity to degrade organic pollutants. Although this technology may be cost-effective for the cleanup and restoration of contaminated sites, not much is known about the physicochemical mechanisms that may control rates of bioremediation. This is especially true for HOCs that may sorb onto soil or that may exist as non-aqueous-phase liquids (NAPLs). The low solubilities of HOCs and their slow desorption rates from soil organic matter, or dissolution rates from NAPLs, may limit bioavailability and biodegradation rates. Understanding of rate-controlling processes in bioremediation is complicated by physicochemical phenomena governing interfacial processes affecting both the microorganisms and the organic compounds.

In natural environments, surfaces have often been found to be major sites of microbial activity (13). However, reduced substrate utilization rates have been observed for substrates sorbing onto a solid phase (14). This reduction is probably a consequence of decreased substrate concentration in solution. Much has been learned in recent years individually about microbial transformation of HOCs in soil and in the subsurface and about the processes of NAPL contaminant behavior in soil and aquifer materials. However, comparatively little is known about the combined effects of desorption–solubilization and microbial degradation on HOC fate and transformation (14, 15).

Some examples for PAH compounds suggest that decreased solubility and non-aqueous-phase partitioning and sorption processes are restrictive toward microbial degradation. Wodzinski and Bertolini (16) and Wodzinski and Coyle (17) concluded that bacteria utilize naphthalene, biphenyl, and phenanthrene as dissolved solutes, with the rate of biodegradation independent of the total amount of solid-phase hydrocarbon. Stucki and Alexander (18) found that the dissolution rate of phenanthrene may limit the biodegradation rate.

The rate of degradation of naphthalene in soil–water suspensions was modeled by assuming that soil-associated substrate is inaccessible to microorganisms and that microbial degradation of the naphthalene in solution creates a concentration gradient, inducing sorbed substrate to desorb and diffuse through pore water to the bulk liquid (15). Investigations involving other organic compounds with soils, clays, or silicate also suggest that sorption of organic substrates either slows biodegradation, renders the compound poorly accessible to microorganisms, or results in the compound being degraded at a rate proportional to the solution concentration only and independent of the mass sorbed (14, 19–21). The treatment of contaminated soil in laboratory systems under ideal conditions indicates that a range of PAHs may biodegrade when present in aqueous solution; however, removal from a solid matrix is less predictable. On the basis of such observations, it is hypothesized that soil–waste matrix effects prevent the release of PAHs into the aqueous phase, where they may undergo biodegradation. Thus, the mass-transfer limitations associated with the release of contaminants from the soil–waste matrix, rather than the explicit aqueous-phase biodegradation kinetics, limit the rate of removal of PAHs from the soil.

Surfactant Amendments

The application of surfactants or emulsifying agents may decrease interfacial tension and assist solubilization of sorbed HOCs from soils, thereby making the hydrophobic compounds more available for microbial degradation (22, 23). However, the experimental observations of the effects of surfactant addition on microbial degradation of HOCs are not always consistent, nor has a general explanation been advanced for their influence. Some of the varied observations of surfactant effects on biodegradation are summarized in Table I (24–37).

Figure 2 presents a schematic of one scenario in which nonionic surfactant may assist biomineralization. In this situation micellar nonionic surfactant has solubilized HOC from soil. As microorganisms deplete aqueous-phase HOC via mineralization, the micelle releases HOC to solution. HOC exit rates from micelles may be significantly faster than HOC desorption rates from soil, and this condition thereby potentially enhances the availability of HOC to the microorganism. Other researchers (25, 28) suggested that surfactants may make HOCs more available for microbial attack in soil by decreasing the interfacial tension between the compound and water.

Many organisms growing on hydrocarbons are able to produce substances that lower the interfacial tension of the growth medium, and may serve to emulsify oil in water (30, 38–41). Such biosurfactant production is believed to facilitate microbial uptake of hydrocarbon by increasing the substrate surface area via emulsification. Thus, it permits greater contact between hydrocarbon and bacteria and enhances the substrate dissolution rate. Alternatively, biosurfactant production may increase the solubility of the hydrocarbons, which are utilized only in solution.

Dispersants were used with some success in remediation programs to control marine oil spills (42). The increased rate of hydrocarbon biodegradation by emulsification with a suitable chemical agent is believed to be a consequence of the increased interfacial area (32). However, not all dispersants enhance degradation of hydrocarbons, and some may be toxic to microorganisms (36, 43, 44). Several fermentation studies (45–49) employed hydrocarbons emulsified with commercial surfactant, and lubricating-oil-degrading bacteria were grown in liquid culture as long as an oil-in-water emulsion was maintained through the application of a dispersant (28).

Table I. Reported Effects of Surface-Active Agents on Microbial Degradation

Overall Effect	Observation	Explanation	Reference
+	Enhanced bacterial growth rate and increased rate of n-alkane consumption	Surfactant solubilization increases aqueous solubility of hydrocarbon	24
+	Extent of biodegradation of phenanthrene in soil increased by low doses of surfactants in absence of surfactant-induced desorption	No explanation	25
+	Triton X-100 increased both the rate and extent of mineralization of naphthalene	Although bacterial adherence prevented, the aqueous naphthalene concentration was sufficient	26
+	Increased hydrocarbon degradation rate and extent with biosurfactant addition	Reduction in interfacial tension	27
+	Increased degradation rate of oil in soil–water slurries	Reducing interfacial tension promotes formation of more interfacial area	28
+	Different Tween-type nonionic surfactants enhanced phenanthrene biodegradation	Surfactant solubilization of HCs increases bioavailability	29
+	Nonionic detergents stimulated growth on hexadecane	Emulsifying action permits effective contact between cells and substrate	30
+	Enhanced rate of biodegradation of PCBs in ligninsulfonate emulsion	Emulsifying action overcomes interfacial area limitation	31
+	Increased rate of hydrocarbon degradation	Emulsifying action increases interfacial area	32
0	No substantial effect on phenanthrene mineralization in soil–water systems at low surfactant dose	Surfactant sorbed onto soil	33
0	Aromatic biodegradation by pure cultures either unaffected or slightly stimulated by emulsification of oil	Microbial uptake only of solubilized substrate, implying prior adherence not required	34
0	Bioemulsifier addition has no significant effect on PAH degradation	No explanation	35

Continued on next page.

Table I—Continued

Overall Effect	Observation	Explanation	Reference
−	Mineralization of phenanthrene inhibited at higher surfactant doses	Possible bacteria–surfactant interactions	33
−	Reduced effectiveness or inhibition observed at higher surfactant concentrations	Toxicity of surfactants to microorganisms	25
−	Triton X–100 completely prevented mineralization of hexadecane dissolved in heptamethylnonane	Surfactant prevents bacterial adherence to solvent–water interface or affects cell membranes	26
−	Decreased biodegradation of HCs in emulsan-treated oil	Surfactant prevents adherence to hydrophobic surfaces	34
−	Crude oil biodegradation retarded by chemical dispersants	Nontoxic surfactants used as preferential substrate	36
−	Decrease in growth rate of yeast cells on alkanes in the presence of surfactant Tween 20	Surfactant prevented large fraction of cells from adhering to hydrocarbon	37

Guerin and Jones (29) reported that the use of various Tween-type nonionic surfactants in aqueous media solubilized phenanthrene to different degrees and enhanced phenanthrene utilization. The order of enhancement did not correlate directly with increased solubility, a result suggesting physiological as well as physicochemical effects of surfactants. However, Aiba et al. (37) and Mimura et al. (50) observed that the growth rates of certain yeast strains actually decreased in the presence of artificial surfactants, presumably by in-

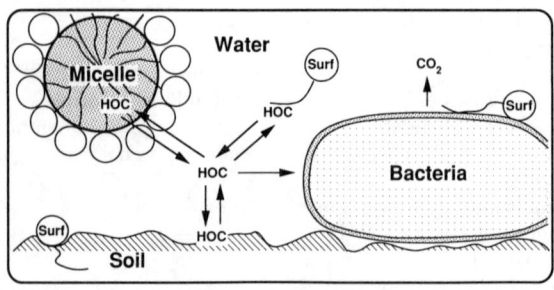

Figure 2. Schematic representation of physicochemical phenomena affecting microbial mineralization of nonionic-surfactant-solubilized HOC in soil–aqueous systems (not drawn to scale).

terfering with the direct interaction between cells and hydrocarbons. Similar reasons were given by Foght et al. (34) to explain inhibition of microbial degradation of hydrocarbon when oil was pretreated with the biosurfactant emulsan, and by Efroymson and Alexander (26) to explain the inhibition of microbial mineralization of hexadecane in the presence of the nonionic surfactant Triton X–100 ($C_8PE_{9.5}$), whereas Breuil and Kushner (30) observed that some nonionic surfactants produced a stimulatory effect on hexadecane utilization.

Although it was suggested that solubilization of a sorbed organic compound may make it readily available for microbial utilization (51), few studies addressed the effect of surfactant addition on microbial degradation in the presence of soil. One investigation (27) reported that the degradation rates for a number of hydrocarbons in a model system with 10% soil could be doubled by addition of a biosurfactant. A slight stimulation of microbial degradation of lubricating oil was noted upon adding a dispersant to a slurry of contaminated soil (28). In another laboratory test, surfactants at low concentrations promoted the mineralization of aromatic compounds with soil, even in the absence of significant surfactant-induced solubilization, whereas higher concentrations were inhibitory (25).

Laha and Luthy (33) concluded from soil-slurry studies with three nonionic surfactants at sub-CMC (or sub-CAC) surfactant levels in the aqueous phase that the mineralization of phenanthrene in the absence of surfactant proceeded at least as fast as mineralization in the presence of surfactant. Low (sub-CMC) levels of surfactant were not significantly inhibitory, but neither did they enhance mineralization. At nonionic surfactant doses in excess of the aqueous-phase CMC, these surfactants exhibited an inhibitory effect on phenanthrene mineralization. Subsequent work by Laha and Luthy (52), summarized here in part, was performed to

- explore a larger group of nonionic surfactants of varying structures and properties to determine whether the inhibitory effect on microbial degradation of phenanthrene was specific to the nonionic surfactants used previously
- assess the relationships between biomineralization of phenanthrene, the surfactant CMC, and the dose to attain the CMC in soil–water systems
- assess the bioavailability of micellar phenanthrene
- verify some results from earlier experiments

Experimental Methods

Batch tests using radiolabeled techniques were employed for both the solubilization and biodegradation tests (33, 52). The soil–water suspensions used soil:water ratios of 1:8 (g/mL). Three PAH compounds (phenanthrene, anthracene, and pyrene) were considered in the initial solubilization tests (53); phenanthrene was the only compound tested for degradation. The amount of PAH added to each system was such that the initial aqueous-phase concentration of the PAH in the absence of surfactant would be near the solubility limit, assuming equilibrium sorption of PAH onto soil (54).

^{14}C-Phenanthrene with a specific activity of 13.1 mCi/mmol and >99% chemical purity was obtained from Sigma Chemical Company. Phenanthrene spiking solutions, consisting of mixtures of labeled and unlabeled compound, were prepared in methanol. Samples were counted for ^{14}C on a liquid scintillation counter (Beckman LS 5000 TD), using quench monitoring with automatic quench compensation. The soil used was an A horizon Hagerstown silt loam, air-dried and screened to pass a U.S. standard no. 10 mesh (2-mm) sieve prior to use, with fractional organic carbon determined by the Walkley–Black method to be 1.5% (55).

A wide range of nonionic surfactants was examined for both the solubilization and biodegradation work. These include the alkyl ethoxylate surfactants Brij 30 ($C_{12}E_4$) and Brij 35 ($C_{12}E_{23}$) used individually and also as a 1:1 surfactant mixture; the alkyl ethoxylate surfactant mixture, Neodol 25–3 ($C_{12-15}E_3$) and Neodol 25–9 ($C_{12-15}E_9$) at a 1:3 ratio; the alkylphenyl ethoxylate surfactant, Triton X–100 ($C_8PE_{9.5}$); the Tween surfactants, Tween 20 and Tween 80, that are polyoxyethylene sorbitan esters commonly used as food and pharmaceutical emulsifiers; and two surfactants with relatively high values of CMC and low micellar size or aggregation number, CHAPS (3-[(3-cholamidopropyl)dimethylamino]-1-propane sulfonate) and octylglucoside (52). Aqueous surfactant solutions were prepared in biochemical oxygen demand (BOD) dilution water (56) with addition of excess ammonia–nitrogen. Surfactant concentrations are reported as either moles per liter of aqueous solution or percent volume of surfactant per volume of solution (v/v). Biodegradation tests measured the mineralization of phenanthrene by monitoring the evolution of $^{14}CO_2$ with biometer flasks (33). A PAH-degrading inoculum was used in the mineralization tests because indigenous soil bacteria were unable to mineralize phenanthrene over a period of 10–12 weeks. The biometers were purged with oxygen each week to preclude the possibility of oxygen limitation to the systems. Enumeration of phenanthrene-degrading organisms was performed by plating dilutions of inoculum on phenanthrene spread plates. Abiotic controls were set up by adding 100 mg of $HgCl_2$ to the biometer contents. These biometers were periodically sampled to assess any abiotic phenanthrene mineralization.

Following mineralization of ^{14}C-phenanthrene, the contents of the biometer flasks were analyzed for residual ^{14}C before disposal to ascertain that mass-balance constraints were satisfied. The soil–water mixture was combusted in a biological material oxidizer (R. J. Harvey Instrument Corp., Hillsdale, NJ). $^{14}CO_2$ from combustion was captured directly in a scintillation cocktail containing carbamate (Carbon 14 Cocktail OX161, R. J. Harvey Instrument Corp.) and analyzed for ^{14}C. ^{14}C-material balances by combustion analysis for a variety of biometer systems showed satisfactory recovery efficiencies ranging from 82 to 97%, with an average recovery efficiency of ~90%. Similar mass-balance analyses, performed for the soil–water slurries used in solubilization tests, showed an average ^{14}C recovery efficiency of ~100%.

Surface-tension experiments to evaluate the CMC or CAC values of the surfactants were performed at 24–25 °C with a ring tensiometer (Fisher Tensiomat model 21 DuNouy). Surfactant solutions of varying concentration were made in deionized water amended with 200 mg/L of $HgCl_2$ as microbial inhibitor and used with the Hagerstown silt loam at soil:water ratios of 1 g of soil:8 mL of solution. Soil–water suspensions were equilibrated for 24 h prior to analysis. Multiple testing of each surfactant solution was performed to ensure that consistent readings were obtained, and standard corrections were made for ring geometry to report surface-tension values (9, 52).

In PAH solubilization tests with soil, batch-test soil–aqueous samples with nonionic surfactant and ^{14}C-PAH were rotated on a tube rotator periodically to maintain the soil in suspension during equilibration. The samples were centrifuged, and aliquots were expressed through preconditioned 0.22-μm Teflon filters to reduce soil-derived colloidal substances. The extent of PAH solubilization in nonionic surfactant solution without soil was assessed in batch tests as a function of surfactant dose to confirm the value of PAH aqueous solubility (S) and to determine the values of S_{cmc} and MSR. Nonionic surfactant sorption onto soil was evaluated for sub-CMC (or sub-CAC) aqueous-phase concentrations by surface-tension measurements. Supra-CMC sorption of nonionic surfactant to determine Q_{max} or supra-CAC isotherm was assessed either with azo dye solubilization and spectrophotometric analysis, or by measurement of chemical oxygen demand, from which the amount of surfactant in bulk solution could be inferred (10).

Results

Phenanthrene Solubilization. A model characterizing the distribution of HOC in systems of soil and micellar nonionic surfactant solution was described previously (7). In this model HOC is assumed to partition among three distinct compartments: the soil, the micellar pseudophase, and the aqueous pseudophase. The solubilization model accounts for the partitioning of HOC between the micellar pseudophase and the aqueous pseudophase, the increase in apparent HOC solubility associated with nonionic surfactant monomers in the aqueous pseudophase, the sorption of surfactant onto soil, and the increase in fractional organic carbon content of a soil as a result of surfactant sorption. Evaluation of the model with experimental data was described by Edwards et al. (12).

Figure 3 shows experimental data and model results for the nonionic micellar solubilization of phenanthrene in soil–aqueous systems with increasing dose of Triton X–100, $C_8PE_{9.5}$. Figure 4 shows experimental data and model results for solubilization of phenanthrene from soil with the alkyl ethoxylate surfactant, Brij 30, $C_{12}E_4$.

The experimental data in Figures 3 and 4 show the relationship of F, the fraction of total HOC that is either solubilized or dissolved, to the dose of applied surfactant. The model results are calculated with the approach already outlined (7), by using parameter values obtained from independent experiments (viz.: S/S_{cmc} = 0.62 for phenanthrene and $C_8PE_{9.5}$, and 0.63 for $C_{12}E_4$ (9); f_{oc} = 0.015; Q_{max} = 5.7 × 10^{-6} mol/g for $C_8PE_{9.5}$ (10); Q_{surf} for $C_{12}E_4$ from supra-CAC isotherm (10); log K_m = 5.7 for $C_8PE_{9.5}$ and 5.57 for $C_{12}E_4$ (9); v_{aq} = 0.048 L; w_{soil} = 6.0 g; initial PAH concentration in water, C_{int} = S (in water) = 7.2 × 10^{-6} mol/L (57); log K_{ow} = 4.57 (54); CMC = 1.7 × 10^{-4} mol/L for $C_8PE_{9.5}$ (9), CAC = 2.3 × 10^{-5} mol/L for $C_{12}E_4$ (9), and MW_{surf} = 628 g/mol for $C_8PE_{9.5}$ and 363 g/mol for $C_{12}E_4$.

Figures 3 and 4 show good agreement between the experimental data and the solubilization values predicted by the model. Apparently the more important physical and chemical processes can be characterized with parameter values obtained from independent tests in aqueous and soil–aqueous systems. In addition, gross solubilization data alone can be used in an inverse procedure to calibrate the ratio between $K_{d,cmc}$ and K_m (12).

Phenanthrene Mineralization. The mineralization of phenanthrene in a soil–water system with 10 g of soil–80 mL water in the absence of surfactant is shown in Figure 5. Over the course of 10 weeks ~50% of the initial ^{14}C-phenanthrene was mineralized by the PAH-degrading inoculum. The abiotic control that received $HgCl_2$ showed no appreciable mineralization

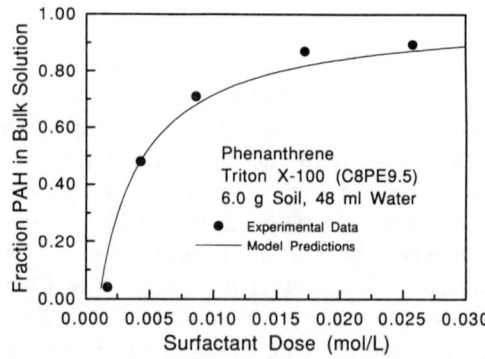

Figure 3. Comparison of model prediction and experimental data for $C_8PE_{9.5}$ solubilization of phenanthrene in the soil–aqueous system.

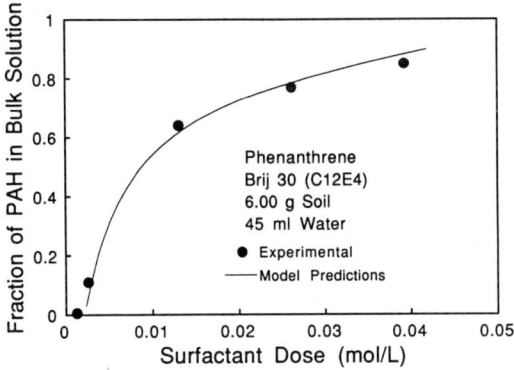

Figure 4. Comparison of model prediction and experimental data for $C_{12}E_4$ solubilization of phenanthrene in the soil–aqueous system.

of phenanthrene, a result indicating that abiotic processes do not contribute to phenanthrene mineralization.

The mineralization of phenanthrene was inhibited substantially in the presence of 0.20% (v/v) of the $C_{12}E_4$ surfactant (Brij 30), as shown in Figure 6. Surfactant inhibition decreased somewhat with a dose of 0.05% (v/v). These results are in agreement with earlier observations that the inhibitory effect for the nonionic surfactants $C_{12}E_4$, $C_8PE_{9.5}$, and $C_9PE_{10.5}$ in soil–water systems was evident at surfactant doses greater than a range of about 0.05–0.10% (v/v) for soil-to-water ratios of 1:8 g of soil/mL of water (33).

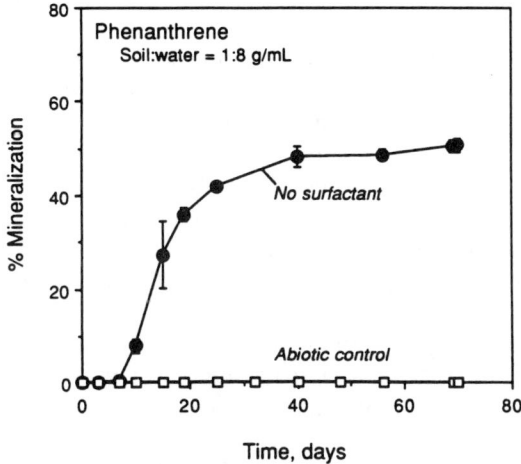

Figure 5. Microbial mineralization of phenanthrene in soil–aqueous system with no surfactant.

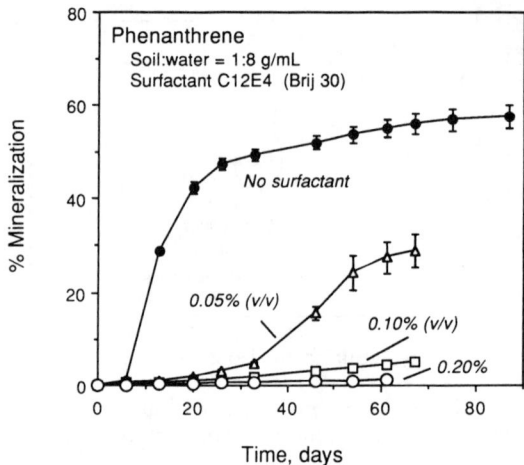

Figure 6. Inhibition of phenanthrene mineralization in soil–aqueous systems receiving various doses of $C_{12}E_4$ nonionic surfactant.

Figure 4 showed the solubilization (i.e., mass fraction in liquid phase) of phenanthrene with the $C_{12}E_4$ surfactant, Brij 30, in 1:8 (g of soil:mL of water) soil–water systems. Phenanthrene solubilization in the soil–water system was observed only at a surfactant dose greater than about 0.0026 mol/L, about 0.1% (v/v), which is many times greater than the pure aqueous CAC value for $C_{12}E_4$ of $\sim 8.3 \times 10^{-4}$% (v/v). The surfactant doses reported in Figures 4 and 6 refer to the bulk addition of surfactant to water, and the liquid-phase surfactant concentrations are smaller than bulk surfactant doses because of surfactant sorption onto soil.

Surface-tension experiments were performed to determine the surfactant dose necessary to form aqueous-phase surfactant aggregates in the presence of soil. Surface-tension measurements for the $C_{12}E_4$ surfactant (Brij 30) are shown in Figure 7, where the surfactant dose necessary to attain the CAC is indicated by the intersection of the two segments on a plot of surface tension versus logarithm of surfactant dose. These data indicate that the surfactant dose at which aggregates form in the soil–water systems is about 0.085% (v/v) (0.0022 mol/L) for $C_{12}E_4$. This dose is consistent with the data in Figure 4, in which solubilization commences with the onset of supra-CAC surfactant concentration in the aqueous phase. A number of other nonionic surfactants were tested in similar fashion, and the results of these investigations were discussed by Laha and Luthy (33, 52). A typical result for one of the other surfactants follows.

Phenanthrene biomineralization, solubilization, and surface-tension data for the polyoxyethylene sorbitan monooleate surfactant, Tween 80, are shown in Figures 8, 9, and 10, respectively. Tween 80 completely inhibits phenan-

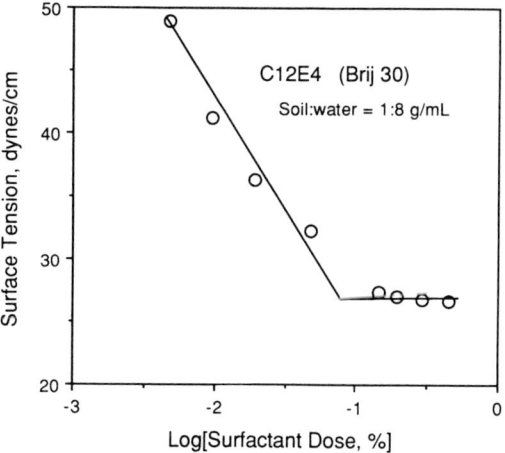

Figure 7. Surface-tension data indicate that micelle formation in the soil–aqueous system occurs at a surfactant dose of ~0.09% (v/v) for the $C_{12}E_4$ nonionic surfactant.

threne mineralization at a dose of 0.20% (v/v) over a period of about 45 days, whereas a dose of 0.02% results in a lower rate of mineralization. The solubilization data presented in Figure 9 indicate that micelle formation for Tween 80 in these soil–water systems occurs at surfactant doses in excess of 0.10% (v/v). The surface-tension data in Figure 10 for Tween 80 suggest that the surfactant dose required to attain aqueous-phase CMC in the presence of soil

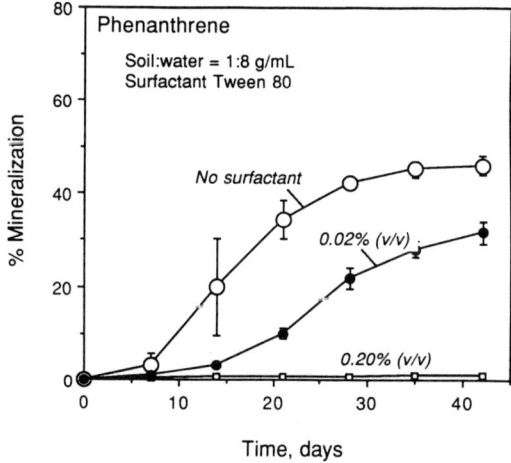

Figure 8. Inhibition of phenanthrene mineralization in soil–aqueous systems with Tween-type sorbitan polyethoxylate surfactant.

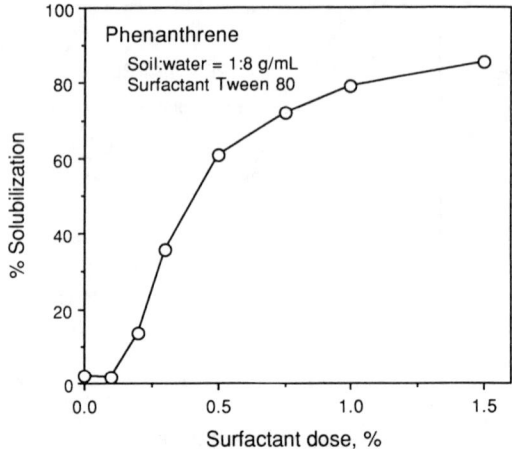

Figure 9. Solubilization of phenanthrene in the soil–aqueous system with Tween-type surfactant.

is about 0.21%. The mineralization, solubilization, and surface-tension data for the Tween 80 sorbitan polyoxyethylene surfactant indicate, as for Brij 30, that microbial mineralization of phenanthrene is somewhat inhibited at surfactant doses less than that necessary to induce solubilization in the presence of soil, with essentially no mineralization occurring for surfactant doses that result in micelle formation in the soil–water systems.

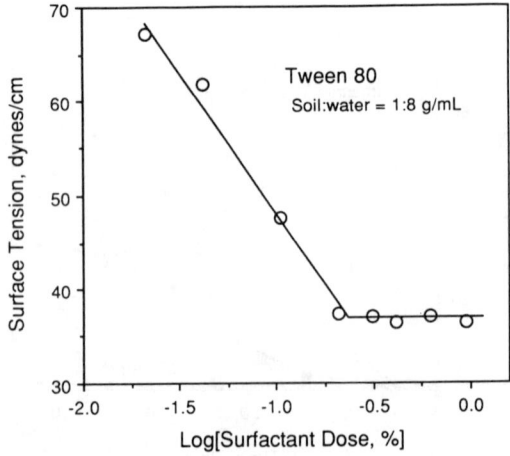

Figure 10. Surface-tension data for the polyoxyethylene sorbitan monooleate surfactant indicate that the surfactant dose required to attain aqueous-phase CMC in the presence of soil is about 0.2% (v/v).

Figure 11 shows the results of mineralization studies involving dilution of surfactant from supra-CMC to sub-CMC aqueous concentrations. These experiments were initially set up with 1 g of soil spiked with 4.5 mg of phenanthrene, and 8 mL of 0.40% (v/v) of the alkylphenyl ethoxylate surfactant $C_8PE_{9.5}$ (Triton X–100). Such soil–water systems exhibited no significant phenanthrene mineralization over a period of 3 months. A subset of test samples was diluted with 9 g of fresh soil and 72 mL of BOD water 4 weeks into the experiment. Following this dilution, mineralization of phenanthrene commenced, reaching ~40% in 9 weeks with the diluted solutions of alkylphenyl ethoxylate surfactant $C_8PE_{9.5}$. The additional volume of dilution water and soil resulted in an effective surfactant dose of 0.04% at a constant soil:water ratio of 1:8.

The effective surfactant dose upon dilution was not inhibitory, and is less than the CMC in the soil–water systems. This dose to attain the CMC is nominally about 0.06% (v/v) or about 0.001 mol/L for Triton X–100 for a soil:water ratio of 1:8 g/mL, as shown in Figure 3 for solubilization data or in reference 52 for surface-tension data. Therefore, the data in Figure 10 demonstrate recovery of phenanthrene biomineralization upon dilution of surfactant to sub-CMC aqueous-phase concentrations in soil–water systems.

Surfactant and Inhibition of Microbial Mineralization. A comparison of solubilization, surface-tension, and mineralization data suggests that the inhibitory effect on microbial degradation of phenanthrene is related in part to the CMC (or CAC) of the surfactant in the presence of soil (i.e., surfactant inhibition appears to be substantial at doses that result in HOC

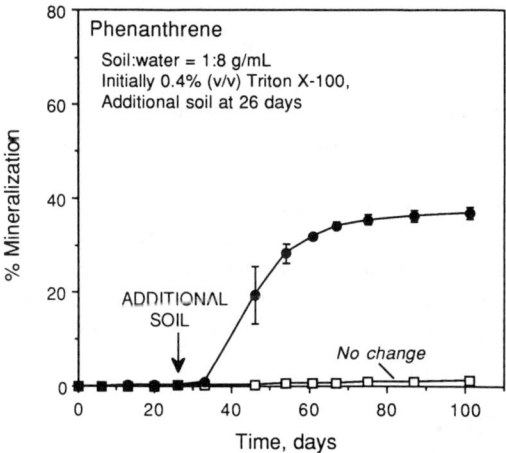

Figure 11. The biomineralization of phenanthrene commenced following dilution of the nonionic surfactant $C_8PE_{9.5}$ to sub-CMC aqueous-phase concentration.

micellization) (52). For all the surfactants tested, no significant phenanthrene mineralization was observed at surfactant doses that exceeded the CMC in the presence of soil. Sub-CMC surfactant doses had less effect on the biodegradation of phenanthrene. For example, partial inhibition was observed for surfactant doses less than that required to attain the CMC (or CAC); inhibitory effects were evident for $C_{12}E_4$ at about one-half the dose required to attain the CAC, for Tween 80 at about one-fifth to one-tenth the necessary CMC dose, for CHAPS at about one-third the dose, and for octylglucoside at about one-fourth the dose (52).

The likely causes for the inhibition of microbial degradation of phenanthrene at higher surfactant doses are

- a toxicity effect
- a lower bioavailability of micellar phenanthrene (e.g., due to low exit rates from micelles)
- surfactant interactions with microbial membranes and proteins
- competitive substrate utilization of surfactant in preference to phenanthrene

Assessment of Toxicity. Dilution tests were performed to examine a possible toxicity phenomenon. In these tests surfactant solutions were diluted to concentrations below those resulting in micelle formation by addition of water or soil and water. Such dilution was observed to result in the recovery of the phenanthrene-degrading ability in the soil–water systems. This recovery suggested that the presence of surfactant micelles did not result in cell lysis or destruction, and that the inhibition may be attributable to some reversible surfactant–bacteria interaction.

This result is suggested also by enumerations of phenanthrene-degrading organisms performed on samples from the soil–water liquor at the conclusion of the mineralization tests. Bacterial enumerations were performed for systems without surfactant and systems held at 0.4% (v/v) $C_{12}E_4$ surfactant. Bacterial counts of phenanthrene degraders in excess of 10^6 colony-forming units (CFU)/mL in soil–water liquor were observed at the conclusion of the tests for both systems with and without surfactant, though the soil–water system with 0.4% (v/v) $C_{12}E_4$ surfactant had displayed no phenanthrene mineralization. These observations suggested that the presence of micellar surfactant interferes with the mineralization of phenanthrene without destroying the viability of the microbial population involved in PAH degradation.

Micellar Exit Rates. In micellar solubilization, the dominant factors governing the exit and reentry rates of solubilizates are largely unknown. The exit rates for naphthalene, biphenyl, and 1-methylnaphthalene from ionic micelles are $>5 \times 10^4$ s^{-1}; exit rates for anthracene and pyrene are reported as

$>10^3$ s^{-1}; and exit rates for perylene are $\sim 3 \times 10^2$ s^{-1} (58). The exit rates for benzyl radicals from sodium dodecyl sulfate micelles are reported as $>10^6$ s^{-1} (59). These high rates suggest that micellar exit rates for the PAHs should not limit microbial mineralization.

Experimental results from screening tests to assess phase transfer with $C_8PE_{9.5}$ and phenanthrene partitioning from aqueous surfactant solution to hexane support this conclusion. In such tests >90% of the micellar ^{14}C-phenanthrene transferred to a hexane phase in less than 1 day from aqueous solution having surfactant concentration of 500 × CMC. In such systems there is also the transfer of a fraction of the surfactant to the hexane phase. However, data of Harusawa et al. (60) suggested that for a test with 500 × CMC only 5% of the surfactant would be transferred to the hexane phase.

The extent of phenanthrene transfer far exceeds the amount that may be attributed to dissociation of micelles due to the movement of surfactant monomers to hexane (52). Thus, both reported PAH exit rates from surfactant micelles and screening studies on phase transfer of micellar phenanthrene from water to hexane suggest very rapid exit rates compared to rates of mineralization.

Surfactant Effects on Microbial Membranes and Proteins. Two major factors in the consideration of surfactant toxicity or inhibition of microbial processes are the disruption of cellular membranes by interaction with lipid structural components and reaction of the surfactant with the enzymes and other proteins essential to the proper functioning of the bacterial cell (61). The basic structural unit of virtually all biological membranes is the phospholipid bilayer (62, 63). Phospholipids are amphiphilic and resemble the simpler nonbiological molecules of commercially available surfactants (i.e., they contain a strongly hydrophilic head group, whereas two hydrocarbon chains constitute their hydrophobic moieties). Phospholipid molecules form micellar double layers. Biological membranes also contain membrane-associated proteins that may be involved in transport mechanisms across cell membranes.

Detergents are used to solubilize and study biological membranes and proteins (63–65). Detergents or surfactants disrupt membranes by intercalating into phospholipid bilayers and solubilizing lipids and proteins. Nonionic surfactants are deemed to be milder detergents and do not usually result in protein denaturation. Nonionic surfactants form complexes with membrane proteins that are more or less fixed in the cell membrane (61).

The monomeric species, not the micellar form of surfactant, are involved in the surfactant binding with proteins and membranes (64). The free monomeric concentration of the surfactant, therefore, determines the amount bound. Nonionic surfactants are bound in smaller number and with lower affinity, partly because of the generally low CMCs for these surfactants as compared with the ionic surfactants resulting in limited monomer concentration, and perhaps because their often-rigid and bulky apolar moieties are unable to penetrate the crevices of protein surfaces as effectively.

Usually the binding of nonionic surfactants does not lead to major conformational changes of the protein and loss of activity (i.e., they do not denature the proteins) (64). Saturation of surfactant binding is reached at some point preceding the gradual disintegration of the membrane. Micelles at higher surfactant concentrations may solubilize the cell membranes. For nonionic surfactants there is a correlation between structure and solubilizing potency (64). Almost all effective nonionic surfactants are in the 12.5–14.5 HLB (hydrophile–lipophile balance number) range. Those surfactants with higher HLB values, such as the Tweens, release material (primarily peripheral protein) from many membranes, but fail to dissociate the lamellar membrane structure at the concentrations normally used for solubilization (<5%).

In summary, the exact effects of the nonionic surfactants on microbial membranes are complex and are not understood in the context of the current study. A form of reversible surfactant–membrane binding, and perhaps partial solubilization of membrane proteins without disrupting membrane structures, may occur for at least some of the surfactants considered. A similar conclusion was made by Jafvert et al. (66), who observed the effects of Brij 30 ($C_{12}E_4$) on biological reductive dechlorination of hexachlorobenzene. The presence of Brij 30 totally inhibited dechlorination activity at concentrations greater than the CAC, but not at concentrations less than the CAC. They noted that surfactant interaction with the microorganisms, possibly through alteration of membrane transport, requires further investigation.

Competitive Substrate Utilization. Various experiments with phenanthrene mineralization demonstrated partial inhibition with nonionic surfactants at doses less than that resulting in micellization. Such data suggest an alternative explanation for inhibition, other than surfactant effects on cell membranes and proteins. Possibly PAH-degrading microorganisms, or their competitors, utilize the surfactant as preferential substrate or carbon source. Jafvert et al. (66) made a similar conclusion about the effect of $C_{12}E_4$ on reductive dechlorination of hexachlorobenzene.

With phenanthrene, some indirect evidence for this supposition was demonstrated by adding 100 and 200 mg of glucose (0.13 and 0.25%, w/v) to several phenanthrene and soil–water systems without surfactant to assess whether the presence of a readily degradable substrate would suppress phenanthrene mineralization. In both cases a significant lag period was evident prior to the onset of phenanthrene mineralization. Although not a definitive experiment, this test and the results with nonionic surfactants and phenanthrene (52) and with hexachlorobenzene (66) indicate the need for further investigation.

Summary

Ongoing research is investigating mechanisms of nonionic surfactant sorption onto soil, nonionic surfactant solubilization of hydrophobic organic compounds (HOCs) from soil, and microbial degradation of HOCs in soil–aqueous systems with nonionic surfactants. The equilibrium solubilization of HOC from soil can be described by a physicochemical model with parameters obtained from independent experiments. The microbial degradation of phenanthrene in soil–aqueous systems is inhibited by addition of alkyl ethoxylate, alkylphenyl ethoxylate, and Tween-type surfactants at doses that result in micellization and solubilization of phenanthrene from soil.

Available data suggest that the supra-CMC inhibitory effect on biodegradation is reversible and not a specific toxic effect. Partial inhibition of microbial degradation of phenanthrene was observed for nonionic surfactants at sub-CMC doses. It is not clear whether these effects result from surfactant interactions with microorganisms or from preferential use of surfactant as substrate or source of carbon. The effects of surfactant monomers and micelles on microbial cell surfaces and constituents, and effects related to preferential substrate utilization and mineralization of degradation products, must be better understood in order to evaluate whether synthetic surfactants may be employed advantageously to enhance bioremediation in soil–water systems.

Acknowledgments

This work was supported by the U.S. Environmental Protection Agency, Office of Exploratory Research, under Grants R816113–01–1 and R816729–01–1.

References

1. McCarthy, J. F.; Wober, F. J. *Summary Report: Concepts in Manipulation of Groundwater Colloids for Environmental Restoration;* U.S. Department of Energy, Oak Ridge Laboratory: Oak Ridge, TN, 1991.
2. Mitchell, D. J.; Tiddy, G. J. T.; Waring L.; Bostock, T.; McDonald, M. *J. Chem. Soc. Faraday Trans. 1* **1983**, *79*, 975–1000.
3. Rosen, M. J. *Surfactants and Interfacial Phenomena*, 2nd ed.; John Wiley & Sons: New York, 1989.
4. Nash, J. H.; Traver, R. *Proceedings of the 12th Annual Research Symposium;* Report No. 600/9–86–022; Environmental Protection Agency: Cincinnati, OH, 1986; pp 208–217.
5. Vigon, B. W.; Rubin, A. J. *J. Water Poll. Control Fed.* **1989**, *61*, 1233–1240.
6. Abdul, A. S.; Gibson, T. L.; Ang, C. C.; Smith, J. C.; Sobczynski, R. E. *Ground Water* **1992**, *30*, 219–231.
7. Edwards, D. A.; Liu, Z.; Luthy, R. G. *J. Environ. Eng.* **1994**, *120*, 5–22.
8. Hayase, K.; Hayano, S. *Bull. Chem. Soc. Jpn.* **1977**, *50*, 83–85.
9. Edwards, D. A.; Luthy, R. G.; Liu, Z. *Environ. Sci. Technol.* **1991**, *25*, 127–133.
10. Liu, Z.; Edwards, D. A.; Luthy, R. G. *Water Res.* **1992**, *26*, 1337–1345.

11. Edwards, D. A.; Adeel, Z.; Luthy, R. G. *Environ. Sci. Technol.* **1994**, *28*, 1550–1560.
12. Edwards, D. A.; Liu, Z.; Luthy, R. G. *J. Environ. Eng.* **1994**, *120*, 23–41.
13. van Loosdrecht, M. C. M.; Lyklema, J.; Norde, W.; Zehnder, A. J. B. *Microbiol. Rev.* **1990**, *54*, 75–87.
14. Alvarez-Cohen, L., Ph.D. Thesis, Stanford University, Stanford, CA, 1991.
15. Mihelcic, J. R.; Luthy, R. G. *Environ. Sci. Technol.* **1991**, *25*, 169–177.
16. Wodzinski, R. S.; Bertolini, D. *Appl. Microbiol.* **1972**, *23*, 1077–1081.
17. Wodzinski, R. S.; Coyle, J. E. *Appl. Microbiol.* **1974**, *27*, 1081–1084.
18. Stucki, G.; Alexander, M. *Appl. Environ. Microbiol.* **1987**, *53*, 292–297.
19. Ogram, A. V.; Jessup, R. E.; Ou, L. T.; Rao, P. S. C. *Appl. Environ. Microbiol.* **1985**, *49*, 582–587.
20. Rijnaarts, H. H. M.; Bachmann, A.; Jumlet, J. C.; Zehnder, A. J. B. *Environ. Sci. Technol.* **1990**, *24*, 1349–1354.
21. Miller, M. E.; Alexander, M. *Environ. Sci. Technol.* **1991**, *25*, 240–245.
22. Lee, M. D.; Thomas, J. M.; Borden, R. C.; Bedient, P. B.; Ward, C. H.; Wilson, J. T. *CRC Crit. Rev. Environ. Control* **1988**, *18*, 29–89.
23. Mueller, J. G.; Chapman, P. J.; Pritchard, P. H. *Environ. Sci. Technol.* **1989**, *23*, 1197–1201.
24. Bury, S. J.; Miller, C. A. *Environ. Sci. Technol.* **1993**, *27*, 104–110.
25. Aronstein, B. N.; Calvillo, Y. M.; Alexander, M. *Environ. Sci. Technol.* **1991**, *25*, 1728–1731.
26. Efroymson, R. A.; Alexander, M. *Appl. Environ. Microbiol.* **1991**, *57*, 1441–1447.
27. Oberbremer, A.; Muller-Hurtig, R.; Wagner, H. *Appl. Microbiol. Biotechnol.* **1990**, *32*, 485–489.
28. Rittmann, B. E.; Johnson, N. M. *Water Sci. Technol.* **1989**, *21*, 209–219.
29. Guerin, W. F.; Jones, G. E. *Appl. Environ. Microbiol.* **1988**, *54*, 937–944.
30. Breuil, C.; Kushner, D. J. *Can. J. Microbiol.* **1980**, *26*, 223–231.
31. Liu, D. *Water Res.* **1980**, *14*, 1467–1475.
32. Robichaux, T. J.; Myrick, H. N. *J. Petr. Technol.* **1972**, *24*, 16–20.
33. Laha, S.; Luthy, R. G. *Environ. Sci. Technol.* **1991**, *25*, 1920–1930.
34. Foght, J. M.; Gutnick, D. L.; Westlake, D. W. S. *Appl. Environ. Microbiol.* **1989**, *55*, 36–42.
35. Gauger, W. K.; Kibane, J. J.; Kelley, R. L.; Srivastava, V. J. *Proceedings of the Second International IGT/GRI Symposium on Gas, Oil, Coal and Environmental Biotechnology*; Institute of Gas Technology: Chicago, IL, 1989.
36. Mulkins-Phillips, G. J.; Stewart, J. E. *Appl. Microbiol.* **1974**, *28*, 547–552.
37. Aiba, S.; Moritz, V.; Someya, J. I.; Huang, K. L. *J. Ferm. Technol.* **1969**, *47*, 203–210.
38. Goswami, P.; Singh, H. D. *Biotechnol. Bioeng.* **1991**, *37*, 1–11.
39. Harvey, S.; Elashvili, I.; Valdes, J. J.; Kamely, D.; Chakrabarty, A. M. *Biotechnology* **1990**, *8*, 228–230.
40. Neufeld, R. J.; Zajic, J. E. *Microbiol Enhanced Oil Recovery*; Zajic, J. E., Ed.; PennWell: Tulsa, OK, 1983.
41. Gutnick, D. L.; Rosenberg, E. *Annu. Rev. Microbiol.* **1977**, *31*, 379–396.
42. Colwell, R. R.; Walker, J. D. *Crit. Rev. Microbiol.* **1977**, *5*, 423–445.
43. Anderson, J. W.; McQuerry, D. L.; Klesser, S. L. *Environ. Sci. Technol.* **1985**, *19*, 454–457.
44. Linden, O.; Rosemarin, A.; Lindskog, A.; Hoglund, C.; Johansson, S. *Environ. Sci. Technol.* **1987**, *21*, 374–382.
45. Chakravarty, M.; Amin, P. M.; Singh, H. D.; Baruah, J. N.; Iyengar, M. S. *Biotechnol. Bioeng.* **1972**, *14*, 61–73.

46. Reddy, P. G.; Singh, H. D.; Roy, P. K.; Baruah, J. N. *Biotechnol. Bioeng.* **1982**, *24*, 1241–1269.
47. Jaeger, A.; Croan, S.; Kirk, T. K. *Appl. Environ. Microbiol.* **1985**, *50*, 1274–1278.
48. Asther, M.; Lesage, L.; Drapon, R.; Corrieu, G.; Odier, E. *Appl. Microbiol. Biotechnol.* **1988**, *27*, 393–398.
49. Lewandowski, G. A.; Armenante, P. M.; Pak, D. *Water Res.* **1990**, *24*, 75–82.
50. Mimura, A.; Watanabe, S.; Takeda, I. *J. Ferm. Technol.* **1971**, *49*, 255–271.
51. Rosenberg, E. *CRC Crit. Rev. Biotechnol.* **1986**, *3*, 109–132.
52. Laha, S.; Luthy, R. G. *Biotechnol. Bioeng.* **1992**, *40*, 1367–1380.
53. Liu, Z.; Laha, S.; Luthy, R. G. *Water Sci. Technol.* **1991**, *23*, 475–485.
54. Karickhoff, S. W.; Brown, D. S.; Scott, T. A. *Water Res.* **1979**, *13*, 241–248.
55. *Methods of Soil Analysis, Part 2: Chemical and Microbiological Processes*; American Society of Agronomy: Madison, WI, 1965.
56. *Standard Methods for the Examination of Water and Wastewater*, 16th ed.; American Public Health Association: Washington, DC, 1985.
57. Mackay, D.; Shiu, W. Y. *J. Chem. Eng. Data* **1977**, *22*, 399–402.
58. Almgren, M.; Grieser, F.; Thomas, J. K. *J. Am. Chem. Soc.* **1979**, *101*, 279–301.
59. Turro, N. J.; Zimmit, M. B.; Lei, X. G.; Gould, I. R.; Nitsche, K. S.; Cha, Y. *J. Phys. Chem.* **1987**, *91*, 4544–4548.
60. Harusawa, F.; Saito, T.; Nakajima, H.; Fukushima, S. *J. Colloid. Interface Sci.* **1980**, *74*, 435–440.
61. Swisher, R. D. *Surfactant Biodegradation*, 2nd ed.; Marcel Dekker: New York, 1987.
62. Tanford, C. *The Hydrophobic Effect: Formation of Micelles and Biological Membranes*, 2nd ed.; Wiley-Interscience: New York, 1980.
63. Zubay, G. *Biochemistry*, 2nd ed.; Macmillan: New York, 1988.
64. Helenius, A.; Simons, K. *Biochim. Biophys. Acta* **1975**, *415*, 29–79.
65. Lichtenberg, D.; Robson, R. J.; Dennis, E. A. *Biochim. Biophys. Acta* **1983**, *737*, 285–304.
66. Jafvert, C. T.; Van Hoof, P. L.; Heath, J. K. *Abstracts of Papers*, 203rd National Meeting of the American Chemical Society, San Francisco, CA; American Chemical Society: Washington, DC, 1992; pp 914–916.

RECEIVED for review October 23, 1992. ACCEPTED revised manuscript June 15, 1993.

18

Distributed Reactivity in the Sorption of Hydrophobic Organic Contaminants in Natural Aquatic Systems

Walter J. Weber, Jr., Paul M. McGinley[1], and Lynn E. Katz[2]

Environmental and Water Resources Engineering, Department of Civil and Environmental Engineering, University of Michigan, Ann Arbor, MI 48109–2125

> *Particle-scale soil and sediment heterogeneities are addressed in the context of their effects on the sorption of nonpolar organic contaminants. Sorption by heterogeneous solids results from a variety of local processes involving different reaction mechanisms. Processes examined here to illustrate such effects include sorption by evolutionally immature soft-carbon organic matter, resulting in quasilinear sorption isotherms; sorption by common mineral phases in concentration regions where linear behavior is apparent; and sorption by evolutionally mature diagenetically altered hard-carbon organic matter, for which clearly nonlinear behavior is exhibited. The results demonstrate that the importance of different contributions to overall sorption reactivity can vary as the mass fractions of differently sorbing components change. A model predicated on discretely different interaction energies for different components of heterogeneous sorbents is used to characterize contributions to the sorption of nonpolar contaminants by several different types of soils and sediments.*

[1]Current address: Department of Civil Engineering, University of Kentucky, Lexington, KY 40506–0046
[2]Current address: Department of Civil Engineering, University of Maine, Orono, ME 04469

0065–2393/95/0244–0363$08.00/0
© 1995 American Chemical Society

Organic contaminants in surface and subsurface systems are typically distributed by sorption between the aqueous phase and natural solid phases. The extent to which such contaminants are sorbed significantly affects their transport and distribution, their impacts on the ecosystem, and the selection of strategies for their removal. In cases of hydrophobic contaminants, sorption is governed by a complex combination of interactions associated with solute repulsion from the aqueous phase and solute attraction to particular solid phases and interfaces. The variety of thermodynamically driven and kinetic or mass-transport-controlled solute–sorbent interactions that may occur in natural systems were summarized by Weber et al. (1).

In any particular system, some limited subset of these interactions is likely to dominate overall sorption behavior. In saturated subsurface systems, for example, most natural solid phases have only a weak affinity for hydrophobic organic compounds. It is frequently presumed for such systems that the characterization of one major sorption reaction, such as that between organic contaminants and natural organic matter associated with soils, sufficiently characterizes the overall process. Such an approach may be problematic for circumstances in which limited characterizations developed from one set of observations are extended to different system conditions, including different contaminant concentration ranges.

Sorption equilibria are typically quantified by isotherm models of the general functional form

$$q_e = f(C_e) \tag{1}$$

relating the sorbed-phase concentration of solute, q_e, to its solution-phase concentration, C_e. These models can take on a variety of more specific forms, depending upon the particular set of reactions and conditions controlling a given sorption phenomenon.

For example, if the reaction controlling the sorption of each molecule of a contaminant is identical and the capacity of a sorbent for these molecules is operationally limitless, a linear isotherm relationship is prescribed in which the sorbed-phase concentration is a constant proportion of the solution-phase concentration. When the sorption reactions are identical but sorption capacity is limited, an asymptotic approach to a maximum sorbed-phase concentration might be expected. These two limiting-condition models have been described and compared with others for description of the sorption of hydrophobic contaminants on a variety of natural soils, sediments, and suspended solids (1–3).

Such comparisons typically yield varying results for different types of sorbents. This result is not unexpected in that observed behaviors for natural systems may in fact result from the superposition of different types of individual sorption phenomena and relationships. In such cases, simple limiting-condition isotherm models are rigorously applicable only on the local level.

The term "local" is used in the context of a phenomenologically distinct reaction, one in which the distribution of a contaminant between the aqueous phase and a specific solid phase may be controlled by a singular type of reaction, or can be described by a first-principle model.

If a series of local isotherms can be identified, the total sorption at any solution concentration will be the sum of those isotherms produced by the individual local sorption mechanisms.

$$q_{e_r} = \sum_{i=1}^{m} x_i q_{e_i} \qquad (2)$$

where q_{e_r} is the overall solid-phase concentration of solute for the system and q_{e_i} is that part of q_{e_r} attributable to the ith of m individual local isotherms. In this case, q_{e_i} is expressed per unit mass of the solid-phase component responsible for the ith local reaction and x_i is the mass fraction of that component.

The development of accurate sorption relationships for contaminants in natural environments requires characterization of dominant local sorption phenomena operating within a particular system and determination of the extent to which they are expressed. This chapter focuses on the nature of local sorption behavior and associated isotherm relationships commonly observed in natural aquatic systems. These local isotherm relationships are then incorporated into descriptions of overall sorption behavior to demonstrate how distributed local concentration dependencies translate into deviations from sorption predictions predicated on single-mechanism and limiting-condition models.

Local Sorption Isotherms

Local isotherms are defined here as sorption relationships resulting from particular classes or types of sorption reactions. For soil and sediment systems involving heterogeneous mixtures of sorbent surfaces and phases, such local isotherms may be expressed at some level of physical division, such as the grain scale. Alternatively, they may be expressed per unit of some quantifiable component of the solid phase that can be characterized by either mineralogical or elemental analysis, such as organic carbon.

These divisions may each involve several sorption mechanisms, but if they yield a local reaction that can be accurately described with an appropriate isotherm model, they can be incorporated in equation 2. The behavior and resulting local isotherm forms for three major classes of reactions will be examined here: sorption by evolutionally immature soil organic matter; sorption by natural mineral phases; and sorption by diagenetically altered and evolutionally mature organic matter.

Soil Organic Matter. A variety of soils and sediments sorb organic compounds in patterns that can be approximated by linear isotherm models, at least over relatively narrow (e.g., 1 decade) concentration ranges. Such isotherms are generally, although often imprecisely, expressed in terms of a partitioning coefficient, K_p, such that

$$q_e = K_p C_e \qquad (3)$$

Measured values for K_p have been found to correlate reasonably well with the organic contents of soils and sediments. This fact often has been interpreted as indicative of solute partitioning to organic phases associated with soils and sediments (4).

However, repeated cautions have been voiced regarding the indiscriminate interpretation of observed linear sorption behavior and correlations of K_p with the organic content of soils as confirmations of true partitioning processes (1, 3, 5). For cases in which such interpretations are justified by supporting evidence, however, arguments can be made for rewriting equation 3 in terms of a partitioning coefficient, K_{OC}, for the organic carbon content of the sorbent, and the fraction, f_{OC}, of the reactive component of the solid phase comprised by organic carbon.

$$q_e = f_{OC} K_{OC} C_e \qquad (4)$$

Values of K_{OC} reported for the association of 1,2,4-trichlorobenzene (TCB) with a variety of soils and organic macromolecules are shown in Table I (4, 6–8). Such collections of experimental data for sorption of a particular solute by a variety of different soils and sediments have been used to develop correlations with which K_{OC} values are then predicted from certain chemical properties of a solute, such as the octanol–water partition coefficient.

Table I. Reported Values of K_{OC} for Interaction of 1,2,4-Trichlorobenzene with Different Soils and Organic Macromolecules

Sorbent	f_{OC}	$K_{OC}{}^a$ (cm^3/g)	Ref.
Subsoil	0.0073	1986	6
Subsoil	0.0008	3125	6
Silt loam soil[b]	0.0095	1114	4
Soil	0.0011	2100	7
Soil	0.0006	1300	7
Soil	0.012	885	7
Isolated humic acid macromolecules	solution	1288	8

[a] Average K_{OC} = 1752 cm^3/g for six soils listed.
[b] For an assumption that organic carbon represented 50% of the total organic content reported.

Although they are of the same general order of magnitude, the values for K_{OC} tabulated in Table I exhibit a fairly wide range of variability. Partition coefficients normalized to the organic carbon mass fraction of one particular set of soils or sediments may therefore not be appropriate for other soils and sediments. Restrictions to the application of such coefficients have been demonstrated, for example, for soils having natural organic matter mass fractions below $f_{OC} = 0.001$ (6, 7).

Deviations from organic carbon correlations at these low f_{OC} values have been attributed to sorption at mineral surfaces (6, 9). Furthermore, the reliability of K_{OC} predictions at higher f_{OC} values may be affected by differences in sorption behavior attributed to differences in the organic matter associated with different types of natural solid phases and, potentially, with different fractions of any particular natural sorbent (3, 10, 11).

In fact, as evidenced in earlier work by Weber et al. (12), the extent of adsorption may vary not only with the character of the organic matter associated with soils, but also with the nature of the sorption process. Organic matter of more recent geologic origin is likely to have a more amorphous and less rigid structure than organic matter that is more aged and diagenetically altered. The former "soft" forms of organic matter are more likely to function as partitioning media and to exhibit more linear sorption processes.

The extent to which variations in the reactivity of natural organic matter and the importance of other sorption mechanisms can alter apparent K_{OC} values for a given system are largely unpredictable. For systems in which the local isotherms for sorption by different organic fractions are each linear, variations in K_{OC} between different components can still be accommodated by overall linear isotherms for which K_{OC} values reflect mass-weighted averages of the different contributions (9, 13).

Natural Mineral Systems. Organic matter makes up only a minor mass fraction of most natural soils and of many sediments; the bulk of the mass of these systems is mineral. On the other hand, all of the organic fraction may be available for participation in sorption reactions, whereas generally only the exposed surfaces of mineral fractions will participate. Nonetheless, if the reactivity of mineral surfaces is sufficient and the mass fraction of organic carbon is low, such surfaces may contribute significantly to overall sorption behavior.

Rhue et al. (14) reported linear local isotherms for the sorption of several hydrophobic organic contaminants on mineral surfaces at high relative humidity. Perlinger et al. (15) observed that organic contaminant sorption on such minerals initially followed a Langmuir pattern, but eventually became more linear as sorbed-phase concentrations increased to levels high enough to cause partitioning to sorbed organic phases. In both situations, sorption was reported to be greater than that predicted from K_{OC} values on the basis of the residual organic matter associated with the mineral phases.

Table II presents an analysis of several experimental observations of the sorption of TCB, dichlorobenzene (DCB), and trichloroethylene (TCE) by various minerals (6, 16). The analysis employs a coefficient, K_M, the per-unit-area sorption capacity of mineral surfaces. This general estimator of the reactivity of such surfaces for organic solutes is obtained from the relationship

$$K_M = \frac{q_e}{S_M C_e} \qquad (5)$$

where S_M is the specific surface area of the mineral phase. The relationship given in equation 5 is used here only to facilitate a general comparison of mineral surface and organic carbon sorption levels. Its use for this purpose does not imply that sorption by mineral surfaces is a linear phenomenon.

Comparison of the reported K_{OC} values given in Table I with the K_M values given in Table II reveals several interesting points. First, the overall magnitude of sorption of a hydrophobic contaminant such as TCB by soil organic matter is much greater than that by mineral surfaces. For example, if the largest K_M value in Table II (0.02 cm^3/m^2 for TCB) is compared with the lowest K_{OC} values in Table I (885 cm^3/g^2 for TCB), the equivalent surface area of the organic matter associated with the soil for which the latter value of K_{OC} was observed would be 44,250 m^2/g. This value is more than an order of magnitude larger than the surface area of microporous adsorbents such as activated carbon and far in excess of what might be expected for soil organic matter.

Two conclusions might be derived from this observation: First, that sorption of hydrophobic organic solutes by soil organic matter is not strictly a surface phenomenon; and second, that surfaces associated with soil organic matter are much more reactive for such organic solutes than are mineral

Table II. Sorption of Hydrophobic Organic Contaminants by Mineral Surfaces

Sorbent	Contaminant	$K_M{}^a$ (cm^3/m^2)	Ref.
Silica	DCB	0.012	6
Silica	TCB	0.02	6
Alumina	TCB	0.013	6
Fumed silica (pH 4)	DCB[b]	0.029	16
Fumed silica (pH 4)	TCE[b]	0.034	16
Fumed silica (pH 8)	TCE[c]	0.067	16
Fumed alumina (pH 8)	TCE[c]	<0.05	16
Montmorillonite	TCE[c]	0.002	16

[a] Estimates of linear sorption coefficients made from reported data.
[b] Estimated from one data point (from table, C_e = 7–7.5 µg/L).
[c] Estimated (from graph, C_e = 30–48.5 µg/L).

surfaces. More than likely, both of these conclusions are valid to variable extents in different systems.

An additional observation to be made in comparing Tables I and II is that the range of values of K_{OC} for different sorbents is smaller than the range of K_M values for different mineral surfaces. A factor of approximately 3.5 separates the highest and lowest values of K_{OC}, whereas the highest and lowest values of K_M are separated by a factor of 33.5, nearly an order of magnitude larger. This difference suggests that sorption by mineral surfaces is likely to be more specifically related to the surface; that is, likely to be a more specific sorption process.

Another feature distinguishing the sorption values for mineral surfaces given in Table II from those for soil organic matter is an obviously weaker dependence on the hydrophobicity of the solute. This contrast is evidenced by the relatively minor and, in fact, seemingly inverse differences between K_M values for solutes of significantly different hydrophobicity (TCB:log K_{OW} = 4.05; DCB:log K_{OW} = 3.3; and TCE:log K_{OW} = 2.3) on relatively similar surfaces. This lack of a clear-cut dependence on hydrophobicity, or solute rejection from solution, supports the notion that the reactions of organic contaminants with mineral surfaces are more specific than are those with soil organic matter.

Diagenetic Maturation of Soil and Sediment Organic Matter. Certain types of natural sorbents, such as those deriving from sedimentary rocks, may contain significant amounts of diagenetically altered organic matter. Diagenesis is a process by which amorphous and exinous sedimentary organic matter is gradually transformed by thermal maturation into increasingly aromatic and cross-linked vitrinous and fusinous organic structures. The early stages of diagenesis usually occur at lower temperatures in relatively shallow sediments. Later stages of diagenesis, catagenesis and metagenesis, take place under increasingly greater thermal gradients and at increased depth and pressure.

This process may be thought of as the gradual transformation of soft organic matter, such as humic substances or peat at one end of the spectrum, into hard organic matter, such as bituminous or anthracite coal at the other end. The terms soft and hard connote, respectively, higher and lower ratios of atomic carbon to atomic hydrogen and oxygen; degrees of structural order; and resistance to further chemical or biochemical transformation. Because of these differences, soft and hard organic matter might logically be expected to respond differently to different analytical means for measurement of soil organic matter. For example, soft-carbon organic matter may be measurable by a relatively mild oxidation procedure, whereas a more vigorous procedure might be required to oxidize hard-carbon organic matter.

Unlike soft soil organic matter, diagenetically altered organic fractions are not likely to act as partitioning media, but rather as media that present relatively hydrophobic surfaces upon which organic contaminants can adsorb. Local sorption isotherms for hydrophobic organic contaminants on such components are generally nonlinear. Because they are frequently fit well by Freundlich-type models, they may be indicative of nonhomogeneous surfaces containing limited numbers of sites of particular sorption energies (1, 11, 12).

Nonlinear behavior is commonly noted for activated carbons produced from such diagenetically altered organic materials as coals and petroleum residues. Carter et al. (17) demonstrated that such characteristic nonlinearity can be decreased in degree for trichloroethylene adsorption from that reflected by a Freundlich exponent value of $n \approx 0.55$ to a value of $n \approx 0.75$ by coating such surfaces with organic matter of recent geological origin, such as humic substances and other organic matter not altered by diagenesis.

Murphy et al. (18) similarly found sorption of several more hydrophobic compounds by surfaces coated with humic acids to exhibit Freundlich exponent values approaching $n \approx 0.75$. In the Carter et al. work (17), a heterogeneous surface exhibiting highly nonlinear sorptive behavior was rendered less heterogeneous and more linearly sorbing by humic acid coatings, whereas the relatively more homogeneous mineral surfaces studied by Murphy et al. (18) were seemingly rendered more heterogeneous and more nonlinearly sorbing by the same type of coatings. In both cases the resulting degree of nonlinearity was similar.

Figure 1 illustrates an isotherm for sorption of TCB by a shale isolated from a Michigan subsoil designated Wagner soil. This isotherm is nonlinear over the concentration range examined. In contrast, a prediction of TCB sorption by this material, based on equation 4 and using the average K_{OC} value of

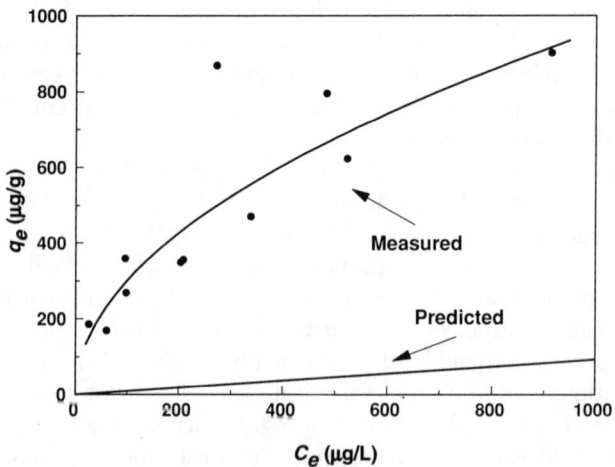

Figure 1. Comparison of a measured isotherm for sorption of 1,2,4-trichlorobenzene by a subsoil shale isolate (f_{OC} = 0.053) to a linear prediction for K_{OC} = 1752 cm³/g and f_{OC} = 0.053.

1752 cm³/g for the six soils listed in Table I and the f_{OC} of the shale (0.053), yields a remarkably different isotherm (also illustrated in Figure 1).

A nonlinear local isotherm model is clearly required for description of sorption reactions between the TCB and the shale isolate. A variety of conceptual and empirical models for representing nonlinear sorption equilibria exists (2). The Langmuir model is one of the ideal limiting-condition-type models cited earlier. It is predicated on a uniform surface affinity for the solute and prescribes a nonlinear asymptotic approach to some maximum sorption capacity.

Natural surfaces generally are too complex to be characterized as having uniform solute interaction energies, however. It is therefore not unexpected that sorption equilibrium data for natural soils and sediments are rarely described adequately by the Langmuir model. Such data are commonly described more satisfactorily by the empirical Freundlich isotherm model, which has the form

$$q_e = K_F C_e^n \qquad (6)$$

Equation 6 can be shown to correspond in mathematical form to a model predicated on a continuous spectrum of sorption interaction energies. If this interpretation is imposed on equation 6, the variable n can be said to reflect both the level and distribution of sorption energies, and K_F the sorption capacity. For most natural solids, n generally ranges in value between 0.5 and 1.0, the upper limit characterizing a linear isotherm. As defined, K_F would logically incorporate the specific reactive surface area, S_H, of the sorbent, which can be abstracted to yield a capacity term, K_{F_H}, expressed per unit surface area ($K_{F_H} = K_F/S_H$). A logarithmic transform of equation 6 can be used to facilitate evaluation of both K_{F_H} and n from observed equilibrium sorption data.

$$\log q_e = \log(S_H K_{F_H}) + n \log C_e \qquad (7)$$

The data shown in Figure 1 for TCB sorption by the shale isolate, after normalizing for surface area and logarithmic transformation, are presented in Figure 2. The K_{F_H} and n values obtained from a linear regression of these data are 2.29 ($\mu g/m^2$)(L/μg)^n and 0.50, respectively. This local isotherm expresses a bulk grain-scale reactivity for the soil sample tested. It probably does not provide more than a gross measure of the concentration dependencies of contributing local sorption isotherms.

Given the deviation of n from unity, this isotherm confirms the mechanistically different behavior of diagenetically altered organic matter from that of softer soil organic matter. In soil systems comprised by admixtures of different types of soil organic matter, it is logical to expect overall sorption to be a composite process involving different combinations of sorption mechanisms.

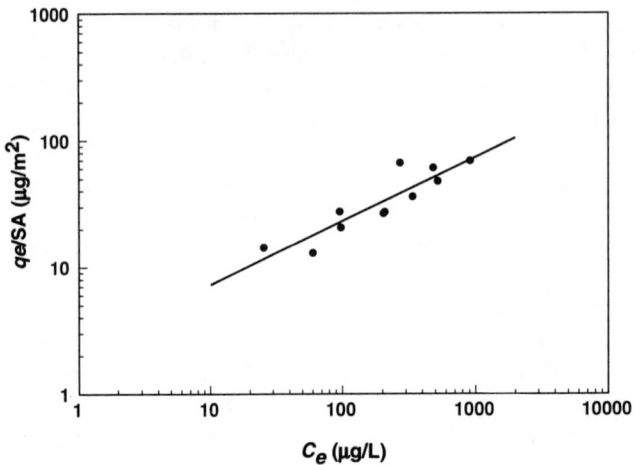

Figure 2. Logarithmic transforms of surface-area-normalized data and the corresponding Freundlich isotherm for sorption of TCB by a shale isolate.

Composite Sorption

Distributed Reactivity. Natural systems commonly are composed of a variety of different solid phases and interfaces, each of which might well yield distinctly individual local sorption behavior with respect to a particular solute and thus an identifiable local isotherm. Figure 3 is a schematic illustration of four different types of local sorption phenomena, ranging from a specific adsorption reaction of a solute molecule with a sorbent surface on the left to the absorption of the molecule into the matrix of a sorbing substance on the right.

Strictly adsorptive reactions are generally expected to manifest nonlinear isotherms, whereas absorption or partitioning processes generally follow the linear relationships prescribed by Henry's and Raoult's laws. Between these two limiting conditions lies a range of mixed phenomenological behavior. The two middle conditions illustrated in Figure 3, for example, depict different degrees of association of contaminant molecules with presorbed molecules of some other organic substance (e.g., soil organic matter). As this association increases from left to right, the isotherm model that best describes the observed condition will become increasingly linear. The exponent in the Freundlich model given in equation 6, for example, will increase gradually from a value somewhere near 0.5 for the left-most condition represented in Figure 3 to a value approaching 1.0 for the right-most condition depicted.

A number of different properties of both the sorbent and the sorbing solute will determine which of the conditions shown in Figure 3 most closely describes their interactions with each other. On the part of the sorbent, these properties include organic carbon content, organic structure (e.g., kerogen

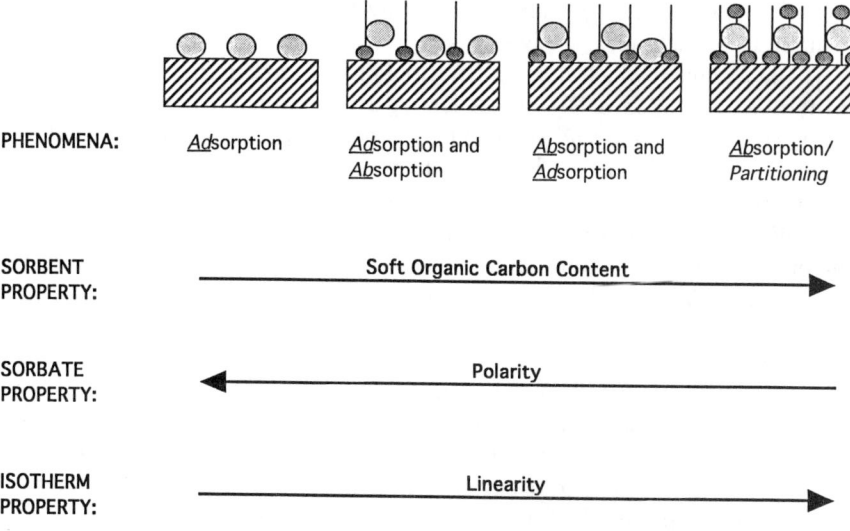

Figure 3. *Representations of various local sorption phenomena and the effects of typical sorbent and sorbate properties.*

versus humic substance), surface polarity, and physical structure (e.g., porous versus nonporous). Molecular size, polarity, polarizability, and chemical functionality are among the determinant features of the sorbing solute.

Distributed Reactivity Model. Isotherm relationships observed for natural systems may well be expected to reflect composite sorption behavior resulting from a series of different local isotherms, including linear and nonlinear adsorption reactions. For example, an observed near-linear isotherm might result from a series of linear and near-linear local sorption isotherms on m different components of soft soil organic matter and p different mineral matter surfaces. The resulting series of sorption reactions, because they are nearly linear, can be approximated in terms of a bulk linear partition coefficient, K_{D_r}; that is

$$q_{e_r} = \sum_{i=1}^{m} x_i f_{OC_i} K_{OC_i} C_e + \sum_{j=1}^{p} x_j S_{M_j} K_{M_j} C_e = K_{D_r} C_e \qquad (8)$$

where x_i and x_j are the mass fractions of the components containing organic and mineral fractions, f_{OC_i} and S_{M_j}, respectively, having per-unit reactivities of K_{OC_i} and K_{M_j}, respectively. Karickhoff (9) and Curtis et al. (19), among others, used composite linear isotherm models of this type to assess the relative roles of organic matter and mineral surfaces in the sorption of hydrophobic contaminants by soils and sediments. The contributions of individual organic and mineral components to overall reactivity are treated as independent of concentration in such linear analyses.

The extent to which mineral contributions to sorption must be included in the determination of K_{D_r} values for typical hydrophobic organic contaminants and natural soils and sediments is suggested by the plots presented in Figure 4 for TCB. These plots indicate how the ratio of K_{D_r} to an organic partitioning coefficient, K_{D_p}, varies as a function of the bulk mass fraction of organic matter of the soil, $\Sigma x_i f_{OC_i}$ for two series of soils containing 1 and 10 m² of silica surface per gram, $\Sigma x_j S_{M_j}$. The K_{D_r} value is predicated on an average K_{OC} value (1752 cm³/g) for TCB taken from Table I and the respective f_{OC} values of the soils in question.

Although sorption of TCB by the silica surfaces can increase values of K_{D_r} over those predicted from f_{OC} and K_{OC} alone, the effect is only weakly apparent until very large quantities of mineral surface and very low f_{OC} values are obtained. Some caution must be exercised in extrapolating these comparisons between the reactions of organic contaminants with organic and mineral surfaces to solutes less hydrophobic than TCB. As noted earlier, specific interactions with mineral surfaces may be more important determinants of sorption for less hydrophobic contaminants.

As indicated clearly for sorption of TCB on diagenetically altered organic matter in natural systems, linear models do not suffice to describe all potentially important local-sorption phenomena. For such nonlinear systems, the relative contributions of different sorbing components and surfaces will vary

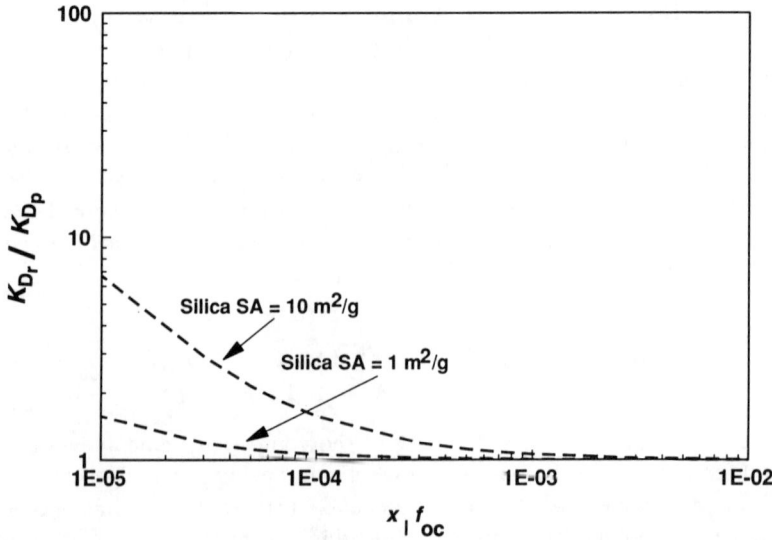

Figure 4. Effects of organic carbon content (f_{OC}) on the ratio of K_{D_r} to K_{D_p} for two soils having different silica surface areas (SA).

with concentration. Composite models for such systems must accommodate both linear and nonlinear sorption phenomena. One such model, the distributed reactivity model (DRM), combines linear local isotherms in a bulk linear term and employs the Freundlich equation to describe discrete nonlinear local isotherms, such that

$$q_e = x_l K_{D_r} C_e + \sum_{j=1}^{n} (x_{nl})_j S_{H_j} K_{F_{H_j}} C_e^{n_j} \qquad (9)$$

where x_l is the summed mass fraction of the solid phase exhibiting linear sorption behavior and $(x_{nl})_j$ is the mass fraction of the jth nonlinearly sorbing component (*12*).

Applications of the Distributed Reactivity Model

Composite Sorption Magnitude and Isotherm Nonlinearity. Accurate assessment of the extent to which the global isotherm for a system is nonlinear is important for accurate portrayal of sorption processes in that system. From a practical point of view, the extrapolation of linear approximations of weakly nonlinear or near-linear sorption isotherms to concentration ranges beyond which they are valid can result in significant errors in projections of contaminant fate and transport (*1*). From a conceptual point of view, observations of isotherm nonlinearity over specific concentration ranges may be employed in conjunction with models such as the DRM to probe and evaluate the extent to which multiple sorption mechanisms are operative in a particular system.

To illustrate one application of the DRM, consider the sorption of TCB by two series of soils having varying levels of organic carbon and two different surface-area components comprised by hard carbon. These organic surfaces are presumed to exhibit the sorption characteristics of the shale isolate for which TCB sorption patterns are shown in Figures 1 and 2. The linearly sorbing component represented by the varying f_{OC} values in this case is considered to have a $q_e - K_{D_r}$ given by the average of the values in Table I.

Figure 5 shows the ratio of sorbed concentration computed by using the DRM, q_e – DRM, to that obtained by using the bulk linear distribution coefficient, K_{D_r}, for varying x_l values for the linearly sorbing carbon components. The values of q_e all correspond to a residual solution concentration of 50 μg/L. Even the very small surface areas of highly reactive diagenetically altered organic material significantly increase the extent of TCB sorption, by more than 2 orders of magnitude for low-f_{OC} soils having as little as 0.01 m²/g of such highly active surface area.

As illustrated in Figure 6, the influence of distributed reactivity on sorption becomes more significant with decreasing residual solution-phase con-

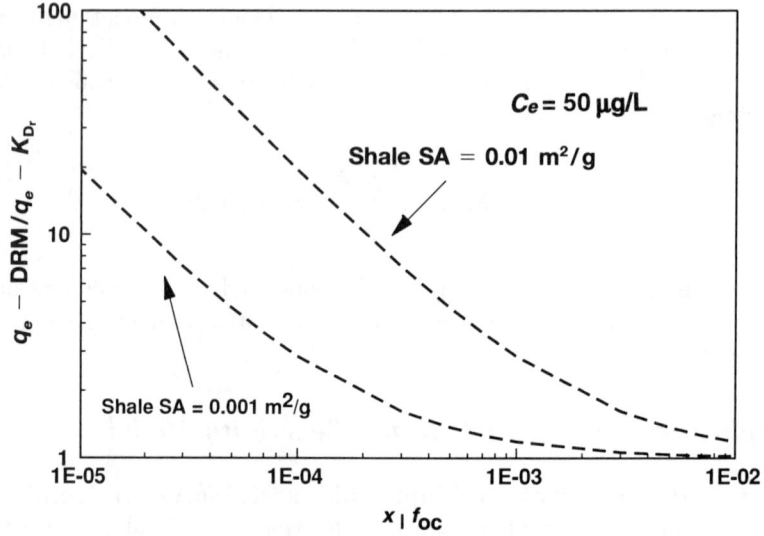

Figure 5. Ratios of q_e values (at C_e = 50 μg/L) based on the DRM to q_e values based on a bulk linear distribution coefficient for varying f_{OC} levels. Ratios are presented for two soils having different shale-isolate surface areas (SA) and K_{OC} = 1752 cm²/g.

centration of solute. As the concentration decreases, the disparity between actual and predicted values for q_e can exceed a factor of 50 for a solution concentration of 5 μg/L. The presence of the nonlinearly sorbing organic component will, of course, affect the f_{OC} value of the bulk soil. Thus, it may act to reduce the disparity between sorption projections using the DRM and K_{D_r} models.

However, for the case presented in Figure 6, which shows a disparity of more than 50-fold in q_e values for a soil having a $\Sigma x_i f_{OC_i}$ value of 10^{-5} for the linearly sorbing soft-carbon components and a nonlinearly sorbing hard-carbon organic surface of 10^{-3} m²/g, translation of the latter number into total carbon content increases the total f_{OC} by only about 50%.

The experimental results given in Figure 7 provide confirmation of the importance of such small but highly reactive nonlinearly sorbing components to the overall sorption behavior of hydrophobic contaminants in subsurface systems. This figure presents a comparison of the carbon-normalized contribution of the nonlinear local TCB isotherm for the shale isolated from the Wagner soil to the composite sorption isotherm for the soil itself.

The nonlinearity (n = 0.77) of the composite isotherm is significant and reflects the dominance by the nonlinearly sorbing components of the overall

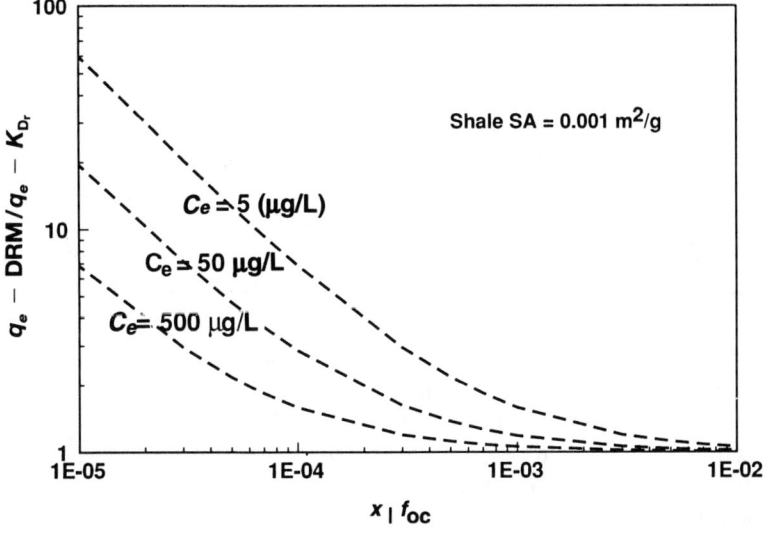

Figure 6. Comparisons of the effects of residual solute concentrations on ratios of q_e values for varying f_{OC} levels.

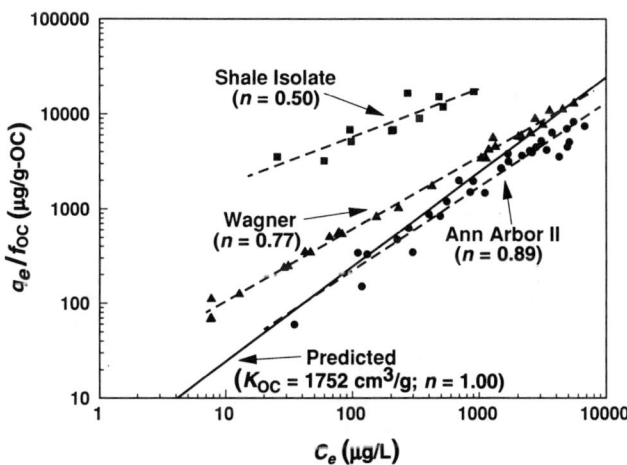

Figure 7. Comparisons of carbon normalized data and Freundlich isotherms for sorption of TCB by different soils with sorption predicted by a linear partitioning model.

reactivity of the system. Comparison of the linear sorption isotherm predicted for the soil with soft carbon, for which sorption can be predicted from a K_{OC}, confirms that such materials can provide only relatively small contributions to the overall sorption characteristics of the subsoil.

The sorption of TCB by another Michigan soil designated Ann Arbor II, which was collected closer to the surface, is also shown in Figure 7. The carbon-normalized sorption here is equal to or lower than that predicted for partitioning into an organic phase, using the average K_{OC} value of 1752 cm^3/g from Table I. The extent of isotherm nonlinearity (n = 0.86) is lower for this near-surface soil than for the isotherms of the subsoil and shale isolate.

Multisolute Effects. The extent to which nonlinearly sorbing components alter the sorption behavior of heterogeneous sorbents is reflected in other characteristics of composite isotherms and in their deviations from the characteristics and behaviors projected by simple partitioning models. In ideal partitioning processes, multiple solutes are expected to behave independently of each other with respect to sorption, at least at low concentrations. Investigations (4) with soils of relatively high organic carbon content showed that hydrophobic organic contaminants sorb independently. For composite systems that are not strictly dominated by partitioning reactions, however, competitive sorption among solutes may result from variations in the affinities of sorption regions available to these compounds (i.e., from sorbent heterogeneities). The extent of nonlinearity of a composite isotherm is a likely indicator of the extent to which competition among sorbing solutes with respect to a particular sorbent may occur.

Figure 8 presents data for the sorption of perchloroethylene (PCE) by the Wagner soil in the presence and absence of TCB. Sorption of the first

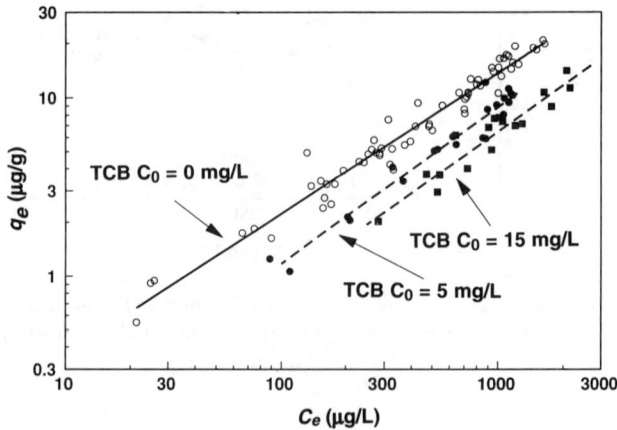

Figure 8. Sorption of PCE by Wagner subsoil in the presence of TCB.

solute is reduced in the presence of the second. Competitive sorption between these two solutes is apparently occurring on the diagenetically altered organic-matter surfaces that dominate the sorption properties of this soil. The nonlinear sorption properties of this system, a behavior attributed earlier to specific surface reactions, appear to correlate with the observed competitive sorption behavior.

In contrast, as illustrated by the isotherms presented in Figure 9 for sorption of PCE in the presence and absence of TCB, sorption by the more linearly sorbing Ann Arbor II soil exhibited little if any competitive interaction. The absence of competitive effects in this case seems to confirm that sorption of PCE by this soil is dominated by partitioning reactions more than similar sorption by the subsurface soil.

Summary

Sorption in natural systems may result from a variety of local processes involving different reaction mechanisms. The extent to which such individual reactions must be considered depends on the accuracy with which the overall sorption process must be characterized, particularly with respect to whether alternative sorption mechanisms will be important determinants of the environmental behavior, transport, and fate of contaminants in a given system.

Several common local sorption processes have been examined here by way of illustrating such effects: sorption by geologically immature soft-carbon organic matter, which results in quasi-linear sorption isotherms, sorption by common mineral phases within concentration regions where linear behavior is exhibited, and sorption by diagenetically altered hard-carbon organic matter

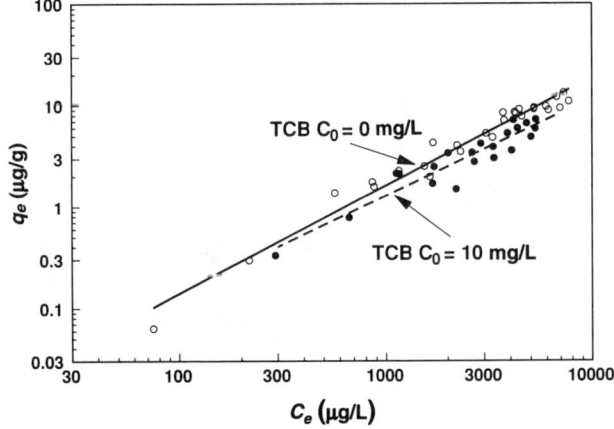

Figure 9. Sorption of PCE by Ann Arbor II near-surface soil in the presence of TCB.

PHYSICAL STRUCTURE	ORGANIC CARBON CONTENT	REACTIVE SURFACE(S)	EXAMPLES	SORPTION EQUILIBRIA* CAPACITY AND ISOTHERM TYPE	SORPTION RATES*
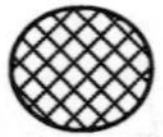	LOW	MINERAL	BEACH SANDS	VERY LOW LANGMUIR	FAST, REACTION CONTROLLED
	LOW TO MODERATE	MINERAL + SOFT CARBON	SANDY SUBSOILS	LOW TO MODERATE FREUNDLICH ($n \approx 0.8 - 0.9$)	MODERATE, REACTION/DIFFUSION CONTROLLED
	MODERATE TO HIGH	SOFT CARBON	SURFACE SOILS	MODERATE TO HIGH NEAR-LINEAR FREUNDLICH ($n \approx 0.9 - 1.0$)	MODERATE, DIFFUSION CONTROLLED
	HIGH	SOFT CARBON	PEATS, SEDIMENTS	MODERATE TO HIGH LINEAR ($n \approx 1.0$)	SLOW, DIFFUSION CONTROLLED
	MODERATE TO HIGH	HARD CARBON	SHALE PARTICLES	MODERATE TO HIGH FREUNDLICH ($n \approx 0.4 - 0.7$)	SLOW, DIFFUSION CONTROLLED
	MODERATE TO HIGH	HARD CARBON + SOFT CARBON	GLACIAL TILLS	MODERATE TO HIGH FREUNDLICH ($n \approx 0.7 - 1.0$)	SLOW, DIFFUSION CONTROLLED

Figure 10. Distributed reactivity and resultant sorption behavior for various types of natural sorbents.

for which nonlinear behavior is exhibited. The results demonstrate how the importance of different contributions to overall sorption reactivity can vary as the mass fractions of differently sorbing components change.

The findings of this work suggest that sorption heterogeneity, even at the particle scale, is not accommodated well by traditional limiting-condition or first-principle isotherm models, such as the linear model and the Langmuir model. Although models of the Freundlich, Langmuir–Freundlich, and generalized Freundlich types may be better able to describe observed sets of sorption data for natural systems, they do not adequately "capture the physics" of composite sorption behavior in such systems. They are predicated on continuously variable site energy assumptions and thus do not provide a satisfactory means for isolating and characterizing discrete component contributions to composite behavior. As such, they are limited in their interpretive value and, potentially, in their utility for extrapolating observations from one system or set of conditions to another. In both these regards the distributed reactivity model (DRM), which provides a means for discretely characterizing varied dominant mechanistic contributions to a composite process, may be more appropriate.

This work focused on the use of the DRM to characterize component contributions to sorption equilibria between hydrophobic organic contaminants and different types of soils and sediments exhibiting particle-scale heterogeneity. These characterizations of equilibrium behavior are based on a foundation of experimental observations. Although no similar level of experimental evidence has been accumulated for nonequilibrium sorption behavior, it is of interest to speculate about how the interpretations structured into the DRM model might allow extrapolation of the model to these conditions.

As detailed in an earlier paper (1), nonequilibrium behavior potentially involves a complex array of reaction-rate and mass-transfer considerations. Any attempt to extrapolate the DRM concept to incorporate such behavior can be no more than pure speculation at this point. That having been said, a suggested matrix of qualitative expectations for both equilibrium and rate behavior of sorption phenomena for nonpolar organic contaminants and natural solids of varying physical structure and surface type is advanced in Figure 10. This figure is a straw man (and a preliminary one at that) presented to elicit analysis and criticism. By this process we hope to nudge our evolving understanding of sorption phenomena in natural systems forward a bit.

Acknowledgments

One acknowledgment stands so much above all others that it will stand alone; namely, that of Werner Stumm for his countless contributions to aquatic chemistry and environmental science. More than 3 decades ago W. J. Weber was fortunate to be an early and direct beneficiary of Professor Stumm's gift for understanding science and for sharing his understanding with others. In the intervening years many others have benefited directly or indirectly from the keen intellectual insights and remarkably clear expositions of aquatic science provided by Professor Stumm. P. M. McGinley and L. E. Katz are honored to be among these beneficiaries.

References

1. Weber, W. J., Jr.; McGinley, P. M.; Katz, L. E. *Water Res.* **1991**, *25*, 499–528.
2. Voice, T. C.; Weber, W. J., Jr. *Water Res.* **1983**, *17(10)*, 1433–1441.
3. Weber, W. J., Jr.; Voice, T. C.; Pirbazari, M.; Hunt, G. E.; Ylanoff, D. M. *Water Res.* **1983**, *17(10)*, 1443–1452.
4. Chiou, C. T.; Porter, P. E.; Schmedding, D. W. *Environ. Sci. Technol.* **1983**, *17*, 227–231.
5. Mingelgrin, V.; Gerstl, Z. *J. Environ. Qual.* **1983**, *12(1)*, 1–11.
6. Schwarzenbach, R. P.; Westall, J. *Environ. Sci. Technol.* **1981**, *15*, 1361–1367.
7. Southworth, G. R.; Keller, J. L. *Water, Air, Soil Pollut.* **1986**, *28*, 239–248.
8. Chin, Y.; Weber, W. J., Jr.; Chiou, C. T. In *Organic Substances and Sediments in Water, Volume 1. Humics and Soils;* Baker, R. A., Ed.; Lewis: Chelsea, MI, 1991; Chapter 14.
9. Karickhoff, S. W. *J. Hydraul. Eng.* **1984**, *110*, 707–735.
10. Garbarini, D. R.; Lion, L. W. *Environ. Sci. Technol.* **1986**, *20*, 1263–1269.
11. Grathwohl, P. *Environ. Sci. Technol.* **1990**, *24*, 1687–1693.
12. Weber, W. J., Jr.; McGinley, P. M.; Katz, L. E. *Environ. Sci. Technol.* **1992**, *26(10)*, 1955–1962.
13. McCarty, P. L.; Reinhard, M.; Rittmann, B. E. *Environ. Sci. Technol.* **1981**, *15*, 40–51.
14. Rhue, R. D.; Pennell, K. D.; Rao, P. S. C.; Reve, W. H. *Chemosphere* **1989**, *18*, 1971–1986.
15. Perlinger, J. A.; Eisenreich, S. J.; Capel, P. D.; Carr, P. W.; Park, J. H. *Water Sci. Technol.* **1990**, *22*, 7–14.
16. Estes, T. J.; Vilker, V. L. *J. Colloid Interface Sci.* **1989**, *133*, 166–175.
17. Carter, M. C.; Weber, W. J., Jr.; Olmstead, K. M. *J. Am. Water Works Assoc.* **1992**, *84(8)*, 81–91.
18. Murphy, E. M.; Zachara, J. M.; Smith, S. C. *Environ. Sci. Technol.* **1990**, *24(10)*, 1507–1516.
19. Curtis, G. P.; Reinhard, M.; Roberts, P. V. In *Geochemical Processes at Mineral Surfaces;* Davis, J. A.; Hayes, K. F., Eds.; ACS Symposium Series 323; American Chemical Society: Washington, DC, 1986; Chapter 10, pp 191–216.

RECEIVED for review October 23, 1992. ACCEPTED revised manuscript March 2, 1993.

19

Interaction of Coagulation–Flocculation with Separation Processes

Hermann H. Hahn

University of Karlsruhe, Postfach 6980, 7500 Karlsruhe, Germany

> *Interaction between aggregate formation and the liquid–solid separation process was explored by systematic investigation of the sedimentation or flotation of coagulated suspensions with various scaled-down continuous-flow separation reactors. Suspensions were coagulated under systematically varied chemical and physical boundary conditions. The pattern of flow (determined by the geometry of the reactor and inflow–outflow configuration) had a marked effect on the removal of aggregated suspensions. Flocs formed under less readily aggregating conditions were generally removed to a lower degree in sedimentation units than those formed with metal salts and polymers. Compensation for these disadvantages is possible through improved tank design or use of flotation. However, suspensions flocculated with polymer alone appear less suitable for flotation.*

CHEMICAL DOSING for improvement of liquid–solid separation is traditionally designed and operated more or less independently of the geometry and hydraulic performance of the actual separation reactor. The widespread use of jar tests in the day-to-day operation of filtration and flotation plants illustrates this fact. Yet it is not difficult to visualize situations in which aggregation processes would furnish flocs unsuitable for the subsequent separation unit, such as voluminous flocs that do not settle out readily.

The aggregation process combines various chemical (destabilization) and physical (collision) reaction steps. In most practical cases, chemical parameters appear to control the process. Separation processes, particularly sedimentation

0065-2393/95/0244-0383$08.00/0
© 1995 American Chemical Society

and flotation (if the step of attachment of gas bubbles to the solid particle is excluded from the analysis), are predominantly controlled by physical parameters. The most important parameters determine the motion of the solid and the surrounding liquid through the reactor.

The interaction of aggregation and subsequent separation step(s) are depicted schematically in Figure 1. For a first analysis, all parameters describing particle aggregation are summarized as chemical boundary conditions and all factors controlling the separation process are described as physical boundary conditions.

The interaction between aggregate formation and the resulting aggregate characteristics on one hand and the separation reaction on the other hand needed to be quantified. Systematic investigations of the liquid–solid separation of suspensions coagulated under defined chemical and physical boundary conditions with various separation reactors of defined geometry and inflow–outflow appurtenances are described and analyzed in the following paragraphs. The study included sedimentation and flotation processes in rectangular and circular basins with moderate to significant depth under systematically varied hydraulic boundary conditions.

Investigations

Experimental Procedure. The investigations were performed in scaled-down continuous-flow reactors (with a scale of about 1:40 relative to the size of technical reactors for aggregation and separation. The hydraulic regime was closely controlled in both coagulation and separation units through exact temperature control ($T = 20 \pm 0.2$ °C) and the maintenance of a constant hydraulic throughput. Suspensions were coagulated in tube-type reactors.

Separation was effected by either sedimentation or flotation. The various shapes of the tanks employed are shown in Figure 2, where characteristic geometric parameters are given for each reactor. The tanks were designed so that the depth:width and depth:length ratios were significantly different. This configuration created differences in the flow regime of different reactors, even at comparable hydraulic loadings. Inflow and outflow structures and baffles also differed among the reactors. Flotation cells were designed as similar to the sedimentation tanks as possible to eliminate problems from differing geometry.

Air bubbles were generated through electroflotation. This procedure allowed easy control of bubble generation and thus of the air–solids parameter, which is crucial for flotation processes. Furthermore, because electroflotation bubbles have a very small diameter, this parameter will no longer be rate-controlling.

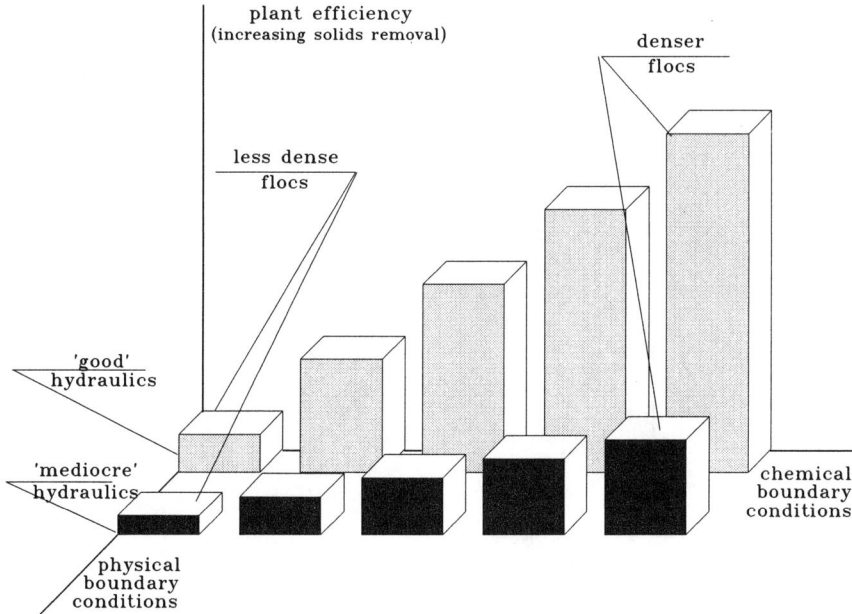

Figure 1. Theoretical effect of aggregation characteristics and reactor geometry on the overall efficiency of liquid–solid separation.

Colloids used for the study (List I) were defined in terms of their size and shape to allow reproducible quantification of aggregate numbers and sizes. As much as possible, the aggregates should resemble the solids encountered in water and wastewater treatment as to size, density, and chemical characteristics like surface charge and its change with variations in the composition of the suspending medium.

Flocs were engineered to show very diverse characteristics, such as strong flocs that could resist destruction in the process of liquid–solid separation and other flocs that could predictably be destroyed under the existing operating conditions of the separation units. This study also aimed to investigate varying floc density, degree of metal hydroxide content, and state of aggregation. The chemical boundary conditions used for floc formation, summarized in List I, correspond to floc formation characteristics as they are reported in the literature (1). Chemicals were added in special mixing cells prior to use in the tube-type coagulation reactor. Hydraulic jump conditions were created in these cells; thus effective mixing was ensured.

Coagulation and flocculation characteristics are described by using a calculated collision efficiency factor. This factor was determined in standardized jar tests (with paddle stirrers) by analyzing particle-counting data according to the von Smoluchowski kinetic model (2). Particle and floc analyses were performed without sampling by repeatedly routing part of the inflowing and out-

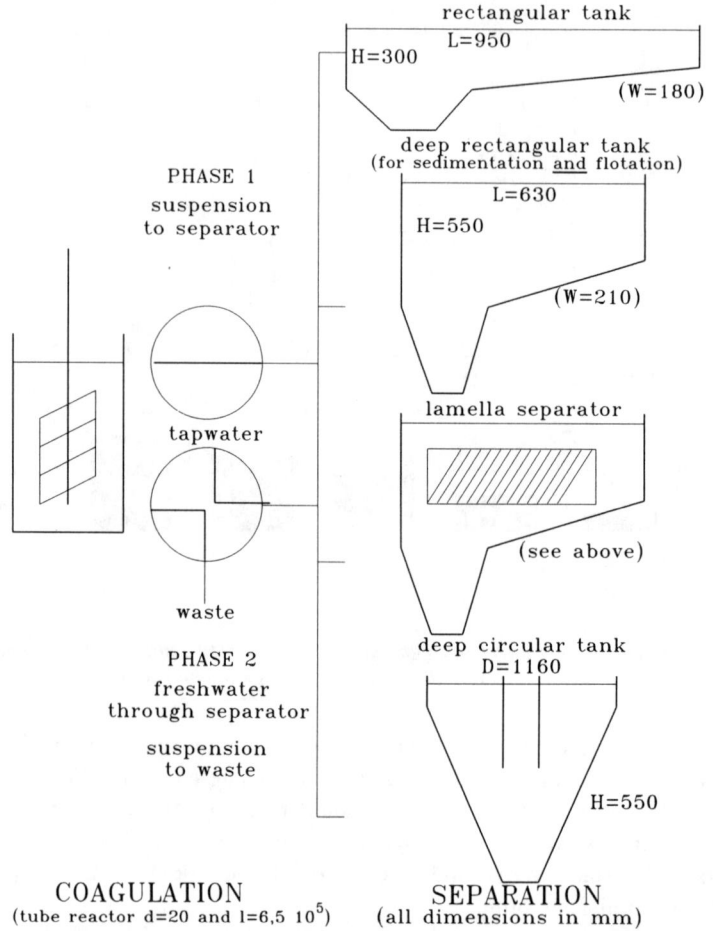

Figure 2. Schematic representation of experimental setup, indicating the delta-pulse loading of the separation reactors and showing the types of separation reactors analyzed.

flowing streams through the cell of a laser particle-analyzing instrument [CIS Galai II from LOT GmbH (3)].

This calculated efficiency factor provides only a relative measure of the properties of the flocs formed. It was used tentatively to quantify the y-axis of the schematic in Figure 1. Other methods of describing floc properties by one parameter, such as floc strength, would be similarly limited in furnishing insight into the coagulating and flocculating properties of the suspensions.

Quantitative description of the hydraulic performance of the separation units (both sedimentation and flotation tanks) was one major aim of this investigation. The flow structure, such as conditions of turbulence (microscopic

List I. Solids, Suspending Medium, and Coagulants Used

Suspension	Suspending Medium	Floc Formation
Silicon-basis micro-glass spheres	Tap water	Ca^{2+}: as $Ca(OH)_2$ 28 mg/L pH 7.3 (data not included)
Density: 2.5 g/cm³	pH 4.5–9.8 (depending on coagulant type)	
Size: 3–5 μm		Al^{3+} as $Al_2(SO_4)_3$ 20 mg/L pH 7.3 (referred to $\alpha = 0.35$)
Concentration: 100 mg/L	Conductivity: 700 μm/cm	
	Total hardness: 350 mg of $CaCO_3$/L	
Composition: 72.5% SiO_2 13.8% CaO 13.7% Na_2O	Temperature: 20 °C (all experiments)	cat. Polymer BC11L (Stockhausen) 5 mg/L pH 6.8 (referred to $\alpha = 0.47$)
$pH_{zpc} = 5.8$		Fe^{3+} as $FeCl_3$ 20 mg/L pH 6.2 (referred to $\alpha = 0.59$)
Specific surface area[a]: 2.6 m²/cm³		Fe + polymer: 12 mg/L + 2 mg/L (referred to $\alpha = 0.84$)

[a] Particle distribution oriented.

parameters), and the flow regime, such as hydraulic throughput and extent of mixing (macroscopic parameters), were determined by tracer experiments.

Rhodamine B was added to the tank inflow, and the resulting effluent concentration was recorded as a function of time in a double-beam (Zeiss) spectral photometer. These measurements were evaluated statistically in order to describe the hydraulic characteristics of the tanks. Effluent concentrations of the tracer, dosed as a delta impulse, observed for different (hydraulic) surface loading (q = 0.28, 0.58, and 1.0 m/h) were evaluated in terms of a characteristic (dimensionless) number, the dispersion number (4), and shown in Figure 3.

The various tanks are characterized by different dispersion numbers. If the overflow rate, which was set to 0.58 m/h for the investigations reported in Figure 3, is increased, then the absolute values of the dispersion number will also change, generally to larger values. Some reactors show a more noticeable effect than others (5).

Reactor theory yields the basis for an evaluation of the meaning of the dispersion number; very small values indicate that the flow regime increasingly

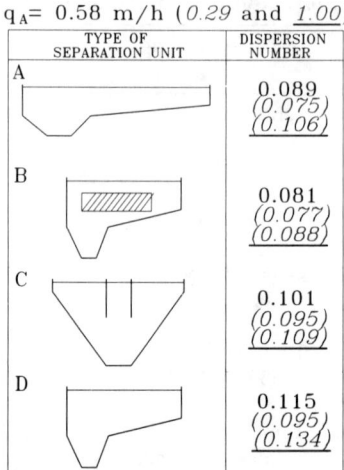

Figure 3. Flow-through characteristics of investigated sedimentation tanks.

resembles that of a plug flow reactor. The other end of the spectrum of possible flow regimes, the completely mixed flow reactor, is characterized by large (theoretically infinite) dispersion numbers. This dispersion number (DN) eventually will have to be used to quantify the x-axis of the schematic in Figure 1. With this concept of quantification correlation, models can be developed for floc properties, reactor characteristics, and removal efficiency.

$$DN = \frac{D}{u \cdot L}$$

where D is turbulent dispersion in the reactor, u is convectional flow velocity, and L is the characteristic length of the reactor.

Data reported in reference 5 show that tanks C and D appear to be less affected by changes in the hydraulic loading than other tank designs. The dispersion number characterizing tanks A and B changes with increasing surface loading from more plug flow to more completely mixed flow conditions.

Sedimentation and flotation tank performance is generally formulated on the basis of a surface-loading principle that stipulates plug flow conditions. Thus, the closer a reactor approaches plug flow conditions, the more likely a high solids-separation performance becomes.

The separation unit's performance was analyzed by introducing a delta impulse. Over a period of less than 1 min the suspension from the aggregation chamber was directed into the sedimentation or flotation unit by means of a three-way switch. The flow pattern in the separation reactor was as little disturbed as possible. The resulting small number of aggregates was then followed on its way through the separation unit. The camera of the counting and image-analyzing device followed the introduced solids on a special tracking system.

In a large-scale operation there is a continuum of particles and flocs in the separation reactor. However, the concentrations of aggregates are not necessarily larger if mixing is neglected, which can occur in a first approximation to the observed dispersion numbers. This technical deviation seemed necessary from an analytical point of view and was considered acceptable.

Additional continuous-load experiments were performed in which the total inflow and effluent were analyzed by routing part of the stream through the cell of the then-stationary counting device. These data furnished the basis for the quantitative evaluation of removal efficiency, information that could be used eventually for the quantification of the z-axis in Figure 1.

Data on particle and aggregate distribution (reported, for instance, in reference 6) are not included here because no currently accepted concept stipulates how these data should be included quantitatively into the analysis described schematically in Figure 1. Solids removal in terms of sludge was not done; sludge characteristics as such were not investigated.

Experimental Results. The effect of wastewater quality and apparatus characteristics on the overall removal efficiency of solids depends on at least two independent variables. Thus, the experiments were performed and reported so that one set of variables was changed systematically while the other was kept constant. Constancy of the parameters reported on the x-axis of Figure 1 (i.e., the hydraulic characteristics of the system) was attained by using one type of separation unit for one experimental series at one given surface loading. Coagulating–flocculating boundary conditions were then changed systematically, exploring the effect of the independent variable reported on the y-axis of Figure 1, the floc characteristic.

The efficiency of the separation process, described as overall solids retention, was derived from particle-counting analyses through summation and integration. An attempt was also made to evaluate and interpret the observed removal behavior by analyzing data on specific microscopic suspension parameters such as floc shape. However, these latter data have not yet been incorporated into mathematical models. They have been used qualitatively to explain incongruencies observed in the performance of sedimentation and flotation tanks.

Aggregate Sedimentation. Figure 4 illustrates the type of results obtained for the sedimentation tanks investigated (cf. Figure 2). The systematically varied coagulation and flocculation conditions are described in detail in List I and denominated in the figure by a coagulation efficiency factor.

The working condition $\alpha = 0.35$ (i.e., Al^{3+} flocs) produced medium-size flocs ($d_{mean} = 160$ μm and $s = 46$ μm, evaluated on the basis of particle

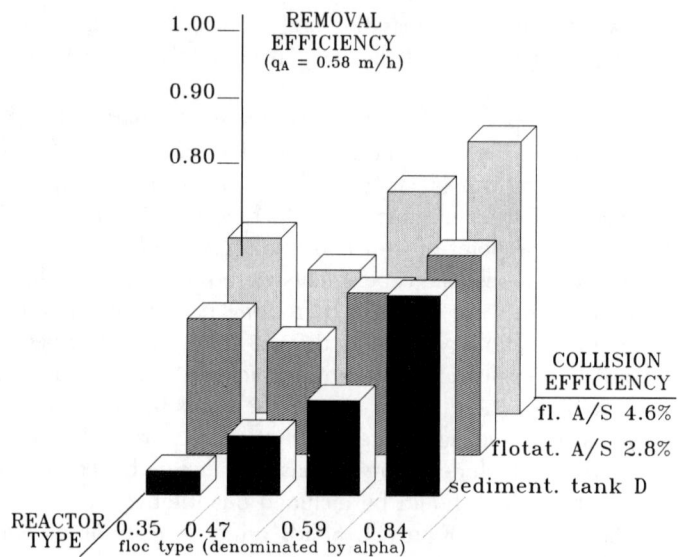

Figure 4. Observed removal efficiencies of specifically aggregated suspensions in varied separation units (sedimentation tanks).

volume distributions) with noticeable heterogeneity in floc-size distribution. Under conditions of α = 0.47 (i.e., polymer flocculation) the suspension is characterized by slightly smaller average floc sizes, as mentioned before, but with a less narrow distribution (d_{mean} = 128 μm and s = 58 μm). These flocs might have overall higher stress resistance. Fe^{3+} flocs (i.e., α = 0.57) produce a faster or more readily coagulating system than those previously mentioned, with a broader distribution (d_{mean} = 120 μm and s = 78 μm). The system α = 0.84 (Fe^{3+} flocs plus polymer) is rapidly coagulating–flocculating, with consequently large average flocs (d_{mean} = 250 μm and s = 84 μm).

These suspensions were separated with varying efficacy in the tanks investigated. Tank C, for instance, a so-called upflow clarifier, has been identified as a reactor with a relatively stable flow structure. It showed little change in dispersion number when the hydraulic surface loading was increased, but very high values of removal efficiency over the whole range of chemical conditions investigated.

Contrary to this finding, tanks A and D showed a relatively lower removal efficiency for lower values of collision efficiency (i.e., for less readily aggregating systems). Tank A is hydraulically characterized by a dispersion number that is closer to plug flow conditions for lower surface loadings, whereas the dispersion number increases noticeably at higher surface loadings. The larger number indicates more mixing. Tank D is included in the analysis because it is geometrically similar to the flotation cell used.

In summary, then, the comparative presentation of Figure 4 shows that

- Some separation units show good efficiency even for less effectively formed aggregates (tank B, lamella supported, and tank C, upflow clarifier). Other types of sedimentation tank designs show unsatisfactory performance with less effective aggregation (tank A, long rectangular basin, and tank D, deep rectangular basin).
- For reactors in which the dispersion number varies significantly with changing hydraulic loading (i.e., less stable hydraulics) the effectiveness of the floc formation process is more significant than for other reactor types.
- Sedimentation tanks that are characterized by more stable hydraulic regimes are more independent of the effectiveness of the preceding floc formation stage. The slight reduction in removal efficiency of tank C with increasing values on the y-axis does not fit too well into this picture. Characteristics other than those described by collision efficiency factors may be responsible for these effects.

Aggregate Flotation. The investigations of the flotation process considered the same colloid–coagulant system as did the sedimentation studies (*see* List I). Likewise, the geometry of the reactors employed for (electro)flotation was kept as similar as possible to the geometry of the sedimentation tanks. And the evaluation of separation efficiency proceeded along the same steps of calculation as in the case of sedimentation.

However, there is one significant difference between sedimentation and flotation: the independent process variable air–solids ratio (A/S ratio). The possibility and necessity of preselecting the amount of air provided for the flotation process introduces an additional degree of freedom that has to be kept constant or controlled in a quantitative way. The observations can be summarized as follows:

- All suspensions, with the exception of the polymer-flocculated one, are removed with noticeably increased efficiency upon increasing the parameter A/S.
- There are differences in the observed removal rate. The Fe^{3+} plus polymer system is separated best, both at lower A/S values and at higher ones.
- The Al^{3+} floc system shows flotation removal efficiencies comparable to those of the Fe^{3+} floc system. Removal rates at low

A/S values are low, and they increase noticeably with increased A/S values.

If A/S is kept constant, thus eliminating the effect of this parameter upon the removal efficiency of coagulated suspensions, one finds relationships between suspension characteristics and removal effect similar to those described for sedimentation units.

Two series of observations are recorded in Figure 5 for A/S ratios of 2.8 and 4.6%, respectively. They are shown with data on sedimentation of the same suspensions in a sedimentation tank of a geometry that closely resembles that of the flotation tank. The y-axis, depicting the independent variable floc formation, is quantified again in terms of the calculated coagulation efficiency factor.

The less rapidly coagulating Al^{3+} floc system (α = 0.35) is separated in flotation nearly as effectively as the better coagulating Fe^{3+} floc system (α = 0.59). In fact, this system appears more suitable for flotation than the quite effective polymer floc formation system (α = 0.47), which has been removed to a more satisfying degree in sedimentation. The Fe^{3+} plus polymer floc system (α = 0.84) shows the best removal efficiency, as it does also in sedimentation units.

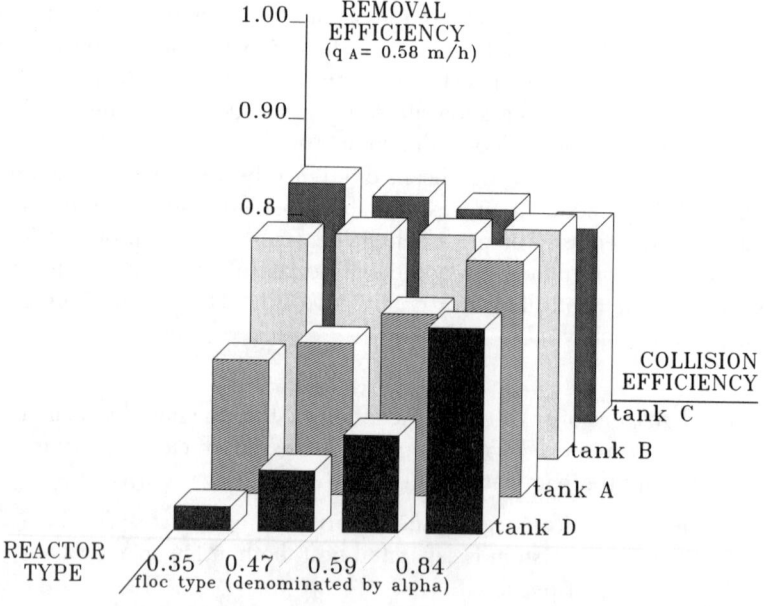

Figure 5. *Removal efficiency of several suspensions in a flotation unit at two different A/S ratios compared to the removal efficiency observed in a geometrically similar sedimentation tank.*

Higher A/S ratios appear to be more effective at higher overall chemical dosages or higher coagulation–flocculation tendencies. This correlation suggests that there might be support for air-bubble attachment to flocs by means of chemicals.

Developing Recommendations for Practice

The investigations reported here were aimed at clarifying the interaction between floc formation tendency or floc properties (y-axis), differently designed and operated separation units (x-axis), and separation efficiency (z-axis). These interactions can be tentatively described in quantitative terms. They can also be illustrated by comparing the efficacy of different separation units for specified suspensions, thus developing rules of thumb for practice.

Comparing the sedimentation or flotation behavior of differently formed flocs, as shown in Figure 5, indicates that the efficiency of a flotation unit of a geometry similar to that of the design D sedimentation tank is noticeably higher for the not so readily coagulating suspension (Al^{3+}). Upgrading inefficient sedimentation seems possible when Al^{3+} salts are used for floc formation.

Suspensions treated under conditions in which good floc formation exist are separated with higher efficiency in flotation units. For intermediary conditions of floc formation, sedimentation in well-designed tanks B and C and flotation tank D show comparable overall efficiency in liquid–solid separation. Flotation appears to be less favored than sedimentation if floc formation is accomplished by lower chemical dosages.

Other authors (7) reported that the Al-type floc is removed better in flotation than in sedimentation. Furthermore, the specific density of aluminum, which is lower than that of iron, is responsible for the better flotation effect. It appears that floc formation in the aluminum system does not provide sufficiently high sedimentation rates, although enough chemicals are available for solid–bubble enmeshment (i.e., good flotation).

Cumulative particle and floc-size distribution functions for these systems are found elsewhere (6). Smaller and larger flocs are removed almost equally well by flotation if the bubble attachment is guaranteed. However, suspensions with heterogeneous size-distribution characteristics are removed less favorably in sedimentation and thus relatively better in flotation units.

Systems with sparingly dosed chemicals might not be separated well in flotation if there are not enough chemicals securing bubble attachment in the case of unfavorable surface-tension relationships. The extent to which the hydraulic regime in tank D is positively affected by bubble movement is still open to speculation.

A logical next step is the formulation of correlation functions between suspension characteristics (as presently described by a calculated collision efficiency value), specific tank parameters (as presently expressed in terms of the dispersion number, derived from tracer analysis), and solids separation efficiency. However, the number of available observations is still not sufficient in terms of necessary statistical evaluation to develop a quasi-continuous domain of aggregation and separation. It is not clear from the available data whether one can interpolate between the discrete data obtained for individual tank designs. For practical purposes the characterization of a tank by the dispersion number alone, as was done in this study, may not be sufficient.

Conclusions

The flow regime in the reactor is determined by the geometry of the reactor and the inflow and outflow configuration. Even if all other parameters are kept constant, the flow regime has a marked effect on the removal of aggregated suspensions. In general, tanks that show a rather constant flow regime perform better, even if the hydraulic loading is increased.

Slowly aggregating suspensions are generally removed to a lower degree in sedimentation units than are rapidly aggregating systems. Removal efficiency becomes more stable (i.e., shows smaller fluctuations) if the aggregating tendency of the system is higher and if the hydraulic characteristics of the separation tank remain stable over wider ranges of hydraulic loading. This correlation was shown elsewhere in detail for sedimentation units (6).

The use of flotation for liquid–solid separation may compensate for some of these limitations of the sedimentation process for suspensions with less optimal floc formation. This effect was observed in particular if suspensions were coagulated with Al^{3+}.

Possibly because of a large average aggregate size or a high aggregate density, systems that sediment readily also show very good flotation characteristics. Flotation does not necessarily require smaller or lighter flocs.

Systems flocculated with polymer only, resulting in a medium collision efficiency factor, appear less suitable for flotation. This effect could result from a lack of chemicals supporting the effective attachment of air bubbles to solids.

Heterogeneity of floc-size distribution, described elsewhere (6), has an effect on the efficiency of the sedimentation process, as can be explained by physical arguments. This effect is particularly true if the aggregation process has not gone very far. The flotation process does not appear to be too much affected by heterogeneity of particle and floc-size distribution.

The better the hydraulic performance of the tank (expressed as dispersion number) and the better the floc formation (expressed as collision efficiency factor), the smaller are the differences in efficiency between the sedimentation and flotation processes.

Finally, the scale of the reactors is significant. The experiments reported here were performed only with one family of models on one scale (about 1:40, relative to real-world reactors). Another investigation (8) of the effect of scale on hydraulic characteristics of coagulation reactors showed that as the scale increased, the flow regime or the energy dissipation became less homogeneous. The same effect could be true for separation reactors. Then the differences in hydraulic characteristics between various reactor types reported here would be even larger. The separation efficiency for suspensions of the type analyzed in this study would be affected by the reactor type and design to an extent even larger than that described here.

References

1. *Proceedings from the 4th Karlsruher Flockungstage 1990;* Schriftenreihe des ISWW-Band 61; Hahn, H. H.; Pfeifer, R., Eds.; Universität Karlsruhe: Karlsruhe, Germany, 1991.
2. von Smoluchowski, M. *Z. Phys. Chem.* **1917,** *92,* 129.
3. *CIS Computerunterstützte Partikelanalyse;* (Manual) LOT GmbH (former ORIEL), Im Tiefen See 58, D-6100, Darmstadt, 1980.
4. Levenspiel, O. *An Introduction to the Design of Chemical Reactors;* John Wiley & Sons: New York, 1962; Chapter 9.
5. Mihopulos, J.; Hahn, H. H. In *Proceedings of the 5th International Gothenburg Symposium on Chemical Treatment;* Klute, R.; Hahn, H. H., Eds.; Springer-Verlag: Berlin, Germany, 1992.
6. Mihopoulos, J.; Hahn, H. H. In *Proceedings of the International Specialized Conference on Pretreatment of Industrial Wastewater;* IAWQ: London, 1993.
7. Rosen, B. In *Proceedings of the 1st International Gothenburg Symposium on Chemical Treatment;* Grohmann, A.; Hahn, H. H.; Klute, R., Eds.; Gustav Fischer-Verlag: Stuttgart, Germany, 1985.
8. Hoffmann, E. *Berichtsheft zum Kolloquium des SFB 210;* Universität Karlsruhe: Karlsruhe, Germany, 1992; p 153.

RECEIVED for review October 23, 1992. ACCEPTED revised manuscript April 21, 1993.

Author Index

Codispoti, Louis A., 157
Cullen, John B., 135
Dong, Chengdi, 291
Dzombak, David A., 59
Edwards, David A., 339
Friederich, Gernot E., 157
Friedl, Gabriela, 111
Häggblom, M. M., 219
Hahn, Hermann H., 383
Hering, Janet G., 95
Hoffmann, Michael R., 233
Huang, Chin-Pao, 291
Hudson, Robert J. M., 59
Jackson, George A., 203
Katz, Lynn E., 363
Kiefer, Elke, 177
Kistler, David, 177
Kotronarou, Anatassia, 233
Kuhn, Annette, 177
Laha, Shonali, 339
Laubscher, Hansulrich, 279
Lidstrom, Mary E., 195
Liu, Zhongbao, 339
Luther, George W., III, 135
Luthy, Richard G., 339
Manceau, Alain, 111
McGinley, Paul M., 363
Morgan, James J., xiv
Murray, James W., 157
O'Melia, Charles R., 315
Ritmiller, Leroy F., 253
Semrau, Jeremy D., 195
Sigg, Laura, 177
Sposito, Garrison, 33
Stumm, Werner, 1
Sulzberger, Barbara, 279
Weber, Walter J., Jr., 363
Wehrli, Bernhard, 111
Wu, Jingfeng, 135
Xue, Hanbin, 177
Young, L. Y., 219
Zepp, Richard G., 253

Affiliation Index

California Institute of Technology, xiv, 195, 233
Carnegie Mellon University, 59, 339
Centre National de la Recherche Scientifique, 111
The Johns Hopkins University, 315
Monterey Bay Aquarium Research Institute, 157
Rutgers, The State University of New Jersey, 219
Swiss Federal Institute for Environmental Science and Technology, 279
Swiss Federal Institute for Water Resources and Water Pollution Control, 111
Swiss Federal Institute of Technology, Zürich, 1, 177
Texas A&M University, 203
U.S. Environmental Protection Agency, 253
University Joseph Fourier, 111
University of California, Berkeley, 33
University of California, Los Angeles, 95
University of California, Santa Cruz, 59
University of Delaware, 135, 291
University of Karlsruhe, 383
University of Michigan, 363
University of Washington, 157

Subject Index

A

Acetylated 4-chlorophenol and its intermediates, mass spectral data, 301t
Acid precipitation, mobilization of aluminum, 96
Acoustic cavitation, chemical effects of ultrasonic irradiation, 234
Activated carbon, nonlinear behavior, 370
Activated complex theory, dissolution rate, 13
Active holes, generation of surface OH radicals, 299
Adsorbate, localized clusters of ions, 53
Adsorbed anions, extended X-ray absorption fine structure spectroscopy, 49–52
Adsorbed metals
  dichotomy, 52
  electron spin resonance spectroscopy, 45–49
Adsorption
  problem in coordination chemistry, 33–57
  surface complex formation equilibria, 5t
  variation, 367
Adsorption of ligands, metal oxides, 6–7
Adsorption processes, surface complexation mechanism, 35–38
Adsorptive reactions, nonlinear isotherms, 372
Aerobic oxidation, methane, 195
Aggregate(s)
  effect of fractal nature, 208
  marine snow, 204–206
  surface-tension experiments, 352
Aggregate flotation, colloid–coagulant system, 391–393
Aggregate formation, interaction with liquid–solid separation process, 383–395
Aggregate sedimentation, experimental results, 389–391
Aggregated suspensions, removal efficiencies, 390f
Aggregation
  second-order equation, 324
  Smoluchowski approach, 326–329
  three physical processes, 325f
Aggregation characteristics and reactor geometry, overall efficiency of liquid–solid separation, 385f
Air–solids ratio, flotation process, 391–392
Algae
  coagulation, 203–217
  concentrations of trace metals, 177–179
  effect of structure on aggregation, 210–212
  importance of coagulation in marine ecosystems, 215
  release of Cu and Zn, 191

Algae—*Continued*
  solid–liquid separation in aquatic systems, 315–337
  Zn:P and Cu:P ratios, 188
Algal biomass, fate of incorporated nutrients, 214
Algal bloom
  aggregates formed from diatoms, 205–206
  biological model, 208
  grazer consumption, 210
  marine environment, simulations, 330
  representative results for coagulation models, 208–214
Algal cells, settling data, 203–204
Alkylbenzenes
  biodegradative pathways, 226–229
  metabolism and degradation of ring structure, 230
Altered aromatic metabolites, biodegradation, 225–226
Aluminum, floc formation, 393
Aluminum oxide
  dissolution, 96
  effect of humic substances on dissolution, 103t
  ligand-promoted dissolution, 99, 101f
Ammonia
  oxic–anoxic interface, 172–173
  suboxic zone, 161
  versus density, 167f
Anaerobes
  characteristics, 220
  diversity and biodegradative capacities, 219–232
Anaerobic processes, use in biological treatment and remediation, 220
Anion, less effective or inert in enhancing dissolution rate, 18
Anion adsorption, hydroxylated surface, 38
Anionic polyelectrolyte
  effects of ionic strength and pH on conformation, 324f
  effects of pH on conformation, 320–322
Anoxic basin, Black Sea, 160
Anoxic environments, characteristics, 158
Aquatic environments
  Debye length, adsorbed layer thickness, and colloidal sticking coefficient, 334f
  photoreactions providing sources of halocarbons, 253–278
Aquatic sinks, methylchloroform, 271–273
Aquatic systems, solid–liquid separation, 315–337
Aqueous reduction potentials, reductants and oxidants, 140t

INDEX

A

Aquifers
  removal of viruses and bacteria, 331
  solid–liquid separation in aquatic systems, 315–337
Aromatic halocarbons
  photoreactions, 261–262
  See also Halocarbons
Aromatic hydrocarbons (BTXs)
  anaerobic biodegradation, 227–228
  biodegradative pathways, 226–229
  depletion under denitrifying conditions, 227f
Atmospheric water, kinetics of dissolved iron(II) formation, 280–281
Attachment probabilities
  pathogens and other microorganisms in groundwater, 332
  removal of particles from suspension, 332f

B

Bacteria, attachment probabilities, 332
Bacterial remineralization, iodide, 145
Bacteriochlorophyll pigments, anaerobic, 171
Benthic boundary layer, particle distribution, 214
Benzene, biodegradative pathways, 226–229
Bidentate ligands, strong surface chelates, 18
Binding mechanism
  copper ion, 191–192
  zinc ion, 192
Binuclear surface complexes, effect on dissolution rate, 19–20
Biodegradation
  altered aromatic metabolites, 225–226
  anaerobic, 219–232
  chloroaromatic compounds, 222–224
  supra-CMC inhibitory effect, 359
  toxic organic compounds, 230
Biodegradative methods, environmental cleanup, advantages and disadvantages, 220
Biological fixation, cycling of iodine in marine environment, 153
Biological materials, specific binding mechanisms, 191
Biomass, vertical fluxes as function of time, 331f
Biomineralization
  effects of surfactant solubilization of phenanthrene in soil–aqueous systems, 339–361
  nonionic surfactant, 344
Bioremediation, rate-controlling processes, 343
Biosurfactant production, microbial uptake of hydrocarbon, 344
Birnessite, local structures, 127f
Bivalent sulfur, ultrasonic irradiation of alkaline oxic aqueous solutions, 235–250
Black Sea, chart, 160f
Breakdown pathways, intermediate of the reaction of I^- and O_3, 148
Brownian diffusion, mechanism in particle-aggregation processes, 325
Brownian motion contact, algal coagulation, 206–207

C

Calcium, adsorption of humic substances, 102
Calcium phosphate, surface nuclei, 53
Carbon
  seasonal variation in composition, 187–188
  sedimentation rate versus time, 186f
  transformation product, 228–229
Carbon dioxide, degradation of 4-chlorophenol and generation of chloride ions, 294–300
Carbonates, surface complex formation, 7
Catalysis, enzymes and coenzymes, 2
Cation(s), surface protonation, 30
Cation-exchange capacity, definition, 61
Cationic surface complexes, interfacial coordination chemistry of hydrous oxides, 35–36
Cavitation bubbles, chemical effects of collapsing, 233–251
Chain conformations, SF theory, 318–319
Charged macromolecules, electrostatic interactions, 319
Chelators, algal cultures, 178
Chemical dosing
  bubble attachment, 393
  liquid–solid separation, 383
Chemical equilibrium models, ion exchange, 67–70
Chemical forces, particles in a fluid, 317
Chemisorption at the surface, ion exchange, 63
Chemocline, Black Sea, 164
Chloride, structure and morphology of oxide film, 28
Chloro-substituted phenols, degradation pathways, 292
Chloroacetates, photoreaction rates, 269–270
Chloroaromatic compounds
  altered aromatic metabolites, 225–226
  anaerobic metabolism, 222
  biodegradability, 222–224
  degradation to inorganic end products, 224
  degradation with sulfate reduction, 224–225
4-Chlorocatechol
  concentration changes, 307f
  photocatalytic degradation of 4-chlorophenol, 292
Chloroethanol, rates and quantum yields for electron production, 267–269

Chloroperoxidases, natural formation of halocarbons, 273–274
4-Chlorophenol
  concentration changes, 307f
  degradation rate, 296
  degradation versus time, 298f, 299f
  direct photolysis, 308f
  dynamic analysis of photocatalytic oxidation, 309–311
  effect of pH on oxidation, 298
  effect of TiO_2 concentration on degradation, 298
  mineralization, 309f, 312
  partial oxidation products, 300–304
  photocatalytic degradation in TiO_2 aqueous suspensions, 291–313
  photocatalytic oxidation and direct photolysis, 311–312
4-Chlorophenol and its intermediates, concentration changes, 301, 304
Chlorophenol degradation, with sulfate reduction, 224–225
Chlorophenol isomers, depletion under methanogenic and sulfate-reducing conditions, 223f
Chlorotrihydroxybenzene, formation, 306f
Chromium, passive oxide films, 29–30
Chromophore, light absorption, 288
Clay subsoil, exchange isotherm data, 77
Co-ion concentration in diffuse layer, versus distance from charged surface, 71f
Coagulation
  dominant effects, 210
  importance of nonalgal particles, 215
  marine algae, 203–217
  marine snow particles, 204
  peak algal populations, 330
  rate and chemical nature of particles, 205
  underdeveloped aspects of theory, 215–216
Coagulation and separation
  experimental setup, 386f
  solids, suspending medium, and coagulants used, 387t
Coagulation efficiency factor, coagulation and flocculation conditions, 389
Coagulation–flocculation, interaction with separation processes, 383–395
Coagulation kernel
  hydrodynamic models, 207
  rectilinear versus curvilinear, 214f
Coagulation rate, particle concentration, 209
Coagulation theory, review of nature of particulate distributions and interactions in ocean, 214
Collapsing cavitation bubbles, chemical effects, 233–251
Collision efficiency factor, coagulation and flocculation characteristics, 385

Colloid aggregation processes, interparticle bridges, 54
Colloidal stability
  adsorption of humic substances on oxides, 102–103
  aquatic environments, 333
  conceptual model for effects of NOM, 318–319
  hematite particles, 322–323
Competition model, binding to ligands, 191
Competitive occupation, metal centers, 14
Complex formation, effect of dissolution rate, 12f
Composite sorption
  distributed reactivity, 372–375
  magnitude, isotherm nonlinearity, 375–378
Conditional dissociation constant, factorization, 36
Constant capacitance model
  encapsulating characteristic, 42
  molecular derivation, 39–41
  molecular interpretation, 41–42
  statistical thermodynamics, 38–43
  thermodynamic equilibrium constants, 37
Coordinate bonds, formation, 2
Coordination chemistry
  adsorption, 33–57
  definition, 35
  solid–water interface, 3t
Coordination environment at solid–water interface, surface-controlled processes, 7
Copper
  concentration in water column, 180–184
  concentration versus depth profiles, 182f, 183f
  cycles in an eutrophic lake, 177–194
  determination and speciation, 179–180
  environmental influence on methane oxidation systems, 199–200
  epilimnion concentrations, 183
  influence on methane oxidation, 195–201
  regulation of sMMO and pMMO, 197–198
  seasonal variation in composition, 187–188, 192
  sedimentation rate versus time, 186f
  toxic effects on marine algae, 178
Copper-scavenging systems, methanotrophs, 200
Corner sharing, adsorbed oxoanions, 20
Corrosion, inhibition of dissolution, 27
Coulombic term, chemical equilibrium models, 75
Counterions
  concentration versus distance from charged surface, 71f
  diffuse and Stern layers of charged surfaces in aqueous suspension, 61

Critical aggregation concentration (CAC), formation, 340
Critical micelle concentration (CMC), formation, 340
Croatica Chemica Acta, series of articles, 35–38
Crystal lattice, activation barrier of detachment of reduced surface iron centers, 288
Cu^{2+}
  adsorbed on δ-Al_2O_3, 48–49
  ESR spectroscopy, 45–49
Cu^{2+} solvation complex, spectral information, 47–48

D

DDE
  photoreaction kinetics, 263t
  photoreaction rate enhancement, 264
Dechlorination, aromatic and aliphatic compounds, 229
Degeneracy, removal, 46f
Degradation of chloroaromatic compounds to inorganic end products, stoichiometry, 224
Dehalogenation, photoreactions, 258
Denitrification zone, oxic–anoxic interface, 172–173
Density
  anomaly, 168t
  seawater relative to freshwater, 163
Density barrier, O_2 and sulfide gradients, 170
Deprotonation reactions, mass law considerations, 35–36
Depth dependence, solar spectral irradiance, 257
Detachment, rate-determining step, 15
Detergents, disruption of biological membranes, 357
Diagenesis
  definition, 369
  local sorption isotherms, 369–371
Diagenetic maturation, soil and sediment organic matter, 369–371
Diatoms, sedimentation of Cu and Zn, 189
Dichlorobenzene, sorption by various minerals, 368–369
p,p'-Dichlorobenzophenone (DCB), enhancement in photoreaction rate, 263
Dichlorophenol
  metabolism under methanogenic conditions, 226f
  sequential dechlorination steps, 225–226
Differential sedimentation
  algal coagulation, 206–207
  mechanism in particle-aggregation processes, 325

Differential sedimentation—*Continued*
  rectilinear and curvilinear approaches, 326–329
Differential sedimentation kernel
  curvilinear model, 213
  rectilinear case, 207
Diffuse-layer sorption
  Gouy–Chapman theory, 70–74
  ion exchange, 59–94
  mixed electrolytes, 74
  surface complexation models, 61, 75–76
Diffusive fluxes
  calculated, 123–124
  measurements, 115–116
Direct photolysis
  4-chlorophenol, 311–312
  comparison with photocatalytic oxidation, 307–309
Direct photoreactions
  definition, 258
  halocarbons in aqueous phase, 260–262
  halocarbons sorbed on NOM, 262–265
  halocarbons sorbed on sediments, 265–266
Dispersants, control of marine oil spills, 344
Dispersion number, reactor theory, 387–388
Dissolution, effect of ligands, 15–23
Dissolution rate
  effect of specifically adsorbable cations, 22
  function of surface speciation, 19f
  overall rate law, 14
Dissolution reaction
  ligand-catalyzed, 17f
  oxide minerals, 12–15
  surface coordination, 13
Dissolved organic matter (DOM)
  correlation with concentrations of dissolved metals, 96
  metal mobilization in soils, 107
  soils, 95–110
  source and composition in soil, 96
Distributed reactivity
  composite sorption, 372–373
  sorption of hydrophobic organic contaminants in natural aquatic systems, 363–382
Distributed reactivity model
  applications, 375–379
  composite sorption, 373–375
  mechanistic contributions to composite process, 381
Divalent sulfur, equilibrium reactions relevant to S(II) system, 236t
Donnal model, ion exchange, 89–90
Drag forces, particles in a fluid, 316–317

E

Edge sharing, adsorbed oxoanions, 20

Electrical potential versus distance from charged surface, 71f
Electrohydraulic cavitation, degradation of chemical contaminants, 234–235
Electron-acceptor properties, molecular orbital diagrams, 143f
Electron-equivalent gradients
 oxic–anoxic interface, 169f
 vertical flux, 169
Electron free energy levels, oxic–anoxic interface, Black Sea, 159f
Electron spin resonance spectroscopy (ESR), adsorbed metals, 45–49
Electron transfer, free activation energy, 288
Electroneutrality constraint, solid–water interface, 77
Electrostatic interactions
 charged macromolecules, 319
 Debye length, 333
Electrostatic sorption
 ion exchange, 70–74
 surface complexation models, 76
Environmental cycling, halides and sulfide, 142
Epilimnion, zinc cycle, 181–184
Equilibria, surface complex formation, 5
Equivalent fraction convention, exchanger phase activities, 68
Estuarine waters, colloidal stability, 335
Ethylxylene, biodegradative pathways, 226–229
Eutrophic lake
 Mn(II) oxidation rates, 111–134
 zinc and copper cycles, 177–194
Exchange equations, insights, 84–87
Exchange isotherm, application of Gouy–Chapman theory, 71
Excited states
 NOM, redox reactions with halocarbons, 258
 quenching and photosensitization, 265
Extended X-ray absorption fine structure spectroscopy (EXAFS)
 adsorbed anions, 49–52
 mechanism, 50
 Mn oxides, 124–128
 spectra of selenite, 51, 52f
Extracellular peroxides, activation of haloperoxidases, 274

F

$FeCO_3$, possible surface groups, 10t
Fictitious species, ion-exchange reaction, 67
Floc(s), liquid–solid separation, 385
Floc formation system, suitable for flotation, 392
Floc-size distribution, efficiency of sedimentation process, 394
Flotation, liquid–solid separation, 394
Flotation unit, efficiency, 393
Flowing fluid, forces acting on a particle, 316–317
Fluid shear
 effect of particle size on physical aspects of collisions, 328f
 mechanism in particle-aggregation processes, 325
Fluorescence, aromatic hydrocarbons, 264–265
Fluoride, structure and morphology of oxide film, 28
Flux
 calculation, 168–169
 diffusive, 122–124
 electrons and concentrations of species, 158
 manganese, 119
 particulate nitrogen, 173
Forest litter, metal mobilization in soils, 105
Fractal nature, aggregates, 208
Free radical(s), photodehalogenations, 259
Free-radical chemistry, sulfur, 235–239
Free-radical mechanism, sonolysis of S(–II), 248–249
Frontier molecular orbital theory, redox chemistry of iodine, 135–155
Functional groups
 interaction to affect surface charge, 11f
 interface of natural particles and minerals with water, 1–31

G

Gaines–Thomas equation
 activities of ions in exchanger phase, 66
 physicochemical ion-exchange model, 4–85
Gapon equation
 isotherms for monovalent–divalent exchange, 6
 physicochemical ion-exchange model, 85–87
Gas bubbles, cavitating, 233–251
Geochemical processes, mineral–water interface, 95–110
Global hydrogeochemical cycle of elements, chemical weathering, 22, 24
Global isotherm, nonlinearity, 375
Goethite
 proton-promoted dissolution, 24f
 reductive dissolution by H_2S, 25f
Gouy–Chapman theory, diffuse-layer sorption, 70–74
Gravity, particles in a fluid, 316–317

H

Half reactions
 equilibrium constant, 68

INDEX

Half reactions—*Continued*
example of use, 90–92
ion-exchange reactions in chemical equilibrium, 67
oxidation–reduction, 158
Halocarbon(s)
aquatic sources, 273–274
aqueous phase, 260–262
complexation with natural substances, 258
direct photoreaction rates in sunlight, 261f
enhanced photoreactivity, 259, 275
indirect photoreactions, 258
mechanism for Fe–heme haloperoxidase-catalyzed production, 274f
natural and industrial sources, 253–254
photoreactions providing sinks and sources in aquatic environments, 253–278
See also Aromatic halocarbons
Halocarbon complexes, photooxidation and photoreduction, 270–271
Halocarbon photoreactions, types, 258–259
Halocarbon sources and sinks, kinetics concepts and data, 256–258
Halogenated pollutants, aquatic photochemical processes, 274
Hazardous waste, bioremediation efforts, 219
Hematite
charge at varying DOM concentrations, 323
effect of NOM on colloidal stability, 322f
sticking probability, 323
surface protonation, 21f
Heterotrophic denitrification, oxic–anoxic interface, 172–173
H_2O_2
inner-sphere process, 144–146
reaction with I^-, 145f
HOI, reaction with OI^-, 151f
HOMO and LUMO energies, reductants and oxidants, 147t
HOMO and LUMO orbitals, energy difference, 141–142
Humic acid, effects of pH and ionic strength on configuration, 319–322
Humic substances
metal mobilization in soils, 106–107
role in complexation, adsorption, and dissolution, 102–104
source and composition, 102
Hydrated passive film, iron oxide, 28
Hydrocarbon, microbial degradation with the biosurfactant emulsan, 347
Hydrodynamic interactions
numerical results on particle transport, 328
two particles, 326–329
Hydrodynamic models, algal coagulation, 207–208
Hydrophobic adsorption, incompatibility of nonpolar substances with water, 2

Hydrophobic cavities, trapping of solutes, 265
Hydrophobic organic compounds (HOC)
availability and biodegradation in soil, 343–344
distributed reactivity, 363–382
distribution in soil–aqueous systems, 340–341
equilibrium solubilization, 359
solute repulsion and attraction, 364
sorption by mineral surfaces, 368t
surfactant solubilization, 340–343
Hydroquinone, photocatalytic degradation of 4-chlorophenol, 292
Hydrous oxide, effect of ligands and metal ions on surface protonation, 21f
Hydroxide minerals, reductive dissolution, 279–290
Hydroxyl (OH) radicals
degradation of aromatic ring, 291–292
homogeneous oxidation reactions, 292
pH conditions, 299
photocatalytic oxidation reaction, 304
possible reaction pathways, 305f
production in seawater, 150
steady-state concentrations and fluxes in water, 271t
tropospheric, 271
ultrasonic irradiation of aqueous solutions, 239
Hydroxylated surface
acid–base behavior, 36–37
anion adsorption, 38
metal cation adsorption, 37–38
Hypolimnion, concentrations of elements, 186–187

I

Indirect photoreactions
mechanisms, 266–271
natural chromophores, 275
photooxidations of halocarbons, 270–271
photoreductions in aqueous solutions, 267–269
photoreductions of sorbed halocarbons, 269–270
Inhibition, dissolution rate, 15
Inner-sphere complexes
adsorbed metal products, 38
ligand exchange, 38
oxide–water interface, 12
surface coordination chemistry, 4–5
surface reactivity, 1–31
Inorganic oxoanions
less effective or inert in enhancing dissolution rate, 19
surface reactivity, 29
Interatomic distances, EXAFS spectra, 51

Interfacial electron transfer, enhancement, 27
Interfacial region, pyrolysis, 234
Intermediates
  concentration changes, 301, 304
  HOI in seawater, 150–151
  identification, 300–304
  photocatalytic degradation of 4-chlorophenol, 302f–303f
Iodide
  conversion to IO^-, 151
  rainwater and aerosols, 152
Iodide oxidation, role of atmosphere, 151–152
Iodide-to-iodate conversion, mechanism in seawater, 139
Iodine in seawater, 135–155
Ion activity ratio, exchange isotherm data, 77, 80
Ion exchange
  chemical equilibrium models, 67–70
  conceptualization, 60
  contributions of diffuse layer sorption and surface complexation, 59–94
  electrostatic sorption, 70–74
  models, 87–88
  physicochemical modeling, 74–84
  surface ion pairing, 80–81
  values and assumptions used in physiocochemical model, 81t
Ion-exchange equations, solid–water interface, 64–66
Ion-exchange half reactions, example of use, 90–92
Ion-exchange isotherms, definition, 62
Ion-exchange mass law expression, Na^+–Ca^{2+} exchange, 73
Ion-exchange process
  data, 61–64
  molecular model, 89–90
Ion sorption on charged mineral surfaces, mechanisms, 64f
Ionic strength
  colloidal stability, 333
  NOM configuration, 319–322
Ionizable halocarbons, complexes with photoreactive transition metals, 258
Iron
  suboxic zone, 161
  versus density, 166f
Iron oxides
  oxic–anoxic interface, 96
  passive film layer, 27–28
  scavenging agent, 173–174
Iron oxyhydroxides, surface-catalyzed oxidation of Mn(II), 129
Isotherm nonlinearity, composite sorption magnitude, 375–378
Isotherm relationships, natural systems, 373

J

Jet fuel spill, anaerobic biorestoration field demonstration project, 230

K

K^+–Ca^{2+} exchange
  Brucedale clay soil, 86f
  model isotherms, 82–83
Kinetic coagulation theory, marine algae, 206–208
Kinetics
  effects of preequilibration, 266f
  halocarbon photoreactions, 256–258
  lindane photoreactions, 267t
  photoreaction of DCB and DDE, 263t

L

Lake Sempach
  average composition of settling particles, 118t
  study of manganese oxidation, 114–134
Lake-water column, ratio of total dissolved Zn to Cu, 183
Langmuir–Hinshelwood equation, 4-chlorophenol concentrations, 298
Leaching studies, metal mobilization in soils, 105
Lepidocrocite
  EDTA-promoted dissolution, 20–23
  photochemical reductive dissolution, 279–290
    energetics, 288f
    overall rate constant, 286, 287f
    rate expression by oxalate, 281
    stoichiometry, 284–286
  surface concentration of oxalate as function of pH, 286f
  surface speciation as a function of pH, 23f
Ligand(s)
  dissolution-promoting and dissolution-inhibiting, 15–23
  effect on strength of the bond that is *trans* to it, 18
  promotion of dissolution of an oxide, 17f
  speciation of copper and zinc, 178
  surface protonation, 20
Ligand-exchange mechanism, inner-sphere surface complex formation, 38
Ligand-promoted dissolution
  oxides, 96, 98–102
  surface protonation, 28–29
Light absorption, chromophore, 256, 288
Lindane, photoreactions, 266, 267t
Linear local isotherms, sorption of HOC on mineral surfaces, 367
Local sorption isotherms
  major classes of reactions, 365–371

INDEX

Local sorption isotherms—*Continued*
  mechanisms, 365
Local sorption phenomena, effects of typical sorbent and sorbate properties, 373f
Local sorption processes, alternative sorption mechanisms, 379
Local structural data, EXAFS spectra, 51
London–van der Waals forces, particles in a fluid, 317
Low-molecular-weight organic ligands
  occurrence in soil DOM, 104
  role in dissolution, 105

M

Macroalgae, iodide oxidation, 153
Macromolecular adsorbed layer
  adsorbed and nonadsorbed uncharged macromolecules, 318f
  equilibrium configuration, 319
Magnesium, adsorption of humic substances, 102
Manganese
  aquatic geochemistry, 112–114
  burial rate, 120–122
  cycling rates, 121t
  diffusive fluxes, 122–124
  oxidation rate near the sediment–water interface, 117–120
  sedimentation rate versus time, 186f
  sedimentation rates in Lake Sempach, 118f
  suboxic zone, 161
  sulfide oxidation, 170–171
  versus density, 166f
Manganese oxidation, reaction rates and products at the sediment–water interface, 111–134
Manganese oxides
  crystal structures, 113–114
  mineralogy, 113
  structural parameters, 126t
Marine algae
  coagulation, 203–217
  copper and zinc, 178
Marine environments, charge on particles, 335
Marine snow
  characterization, 204
  chemical nature of particles, 205
  formation, 204–205
  model system, 205–206
Marine systems, redox end-members, 158
Mathematical chemicals, definition, 317
Mathematical modeling
  chemical effects of collapsing cavitation bubbles, 233–251
  sulfur reactions, 237t–238t

Metal(s)
  catalysis of iodide oxidation, 152
  microbiology, 195–201
  mobility in soils, 107
Metal absorption, chemical modeling, 36–38
Metal catalysis, O_2 as oxidant, 142–144
Metal cations, adsorption by hydroxylated surfaces, 37–38
Metal mobilization in soils
  effects of humic substances, 106–107
  influence of DOM components, 104–106
Metal oxide, adsorption of ligands, 6–7
Metal oxide hypothesis, sulfide oxidation, 170–171
Metal solubilization and translocation in soils, role of DOM, 96
Metal sulfide, electron transfer, 143
Methane, influence of copper on methane oxidation, 195–201
Methane formation, monochlorophenol degradation, 224t
Methane monooxygenase (MMO)
  enzymes, methanotrophic isolates, 196t
  properties, 197t
Methane oxidation systems
  environmental implications of copper availability, 199–200
  role of methanotrophs, 196–197
Methanogenesis, CO_2 production, 221
Methanogenic enrichment, pond sediment, 225–226
Methanotrophs
  copper-scavenging systems, 200
  global cycling of methane, 195–201
  major classes, 195–196
  methane-limited in natural habitats, 199
Methyl bromide, annual global flux, 254
Methylchloroform
  aquatic sinks, 271–273
  computed fluxes, 272f
  estimated pseudo-first-order loss rates, 272–273
Micellar exit rates, solubilization, 356–357
Micelle formation, soil–aqueous system, 353f
Microbe-mediated reactions, 221
Microbial activity
  effect on ligand-promoted dissolution, 105
  formation of metal–humate complex, 106
  substrate concentration, 343
Microbial degradation, hydrocarbons, 347
Microbial membranes, surfactant effects, 357–358
Microbial mineralization
  inhibitory effect of phenanthrene, 355–356
  nonionic-surfactant-solubilized HOC, 346f
Microbial oxidation, species-specific, 221
Microbiological oxidation, Mn(II), 113
Microbiology, metals, 195–102

Mineral dissolution, effects of humic substances, 103–104
Mineral surfaces
  interaction with organic matter, 95–110
  sorption of hydrophobic contaminant, 368–369
Mineral–water interface
  effect on geochemical processes, 95–110
  ion exchange, 59–94
Mineralization
  dilution of surfactant, 355
  phenanthrene, 350–356
Mirex, photoreduction, 266–267
Mn(II)
  half-lives in microbiological oxidation, 128t
  half-lives in surface-catalyzed oxidation, 129t
  oxidation by molecular oxygen, 113
  oxidation kinetics, 128–130
  seasonal concentration variations, 123f
  transport in stratified hypolimnia of lakes, 112–113
Mn oxides, EXAFS spectroscopy, 116
Mn redox cycling, sediment–water interface, 112f
Model(s)
  aggregate formation, 206–214
  chemical effects of collapsing cavitation bubbles, 233–251
  constant capacitance model, 38–43
  equilibrium distribution of HOC in closed system, 342
  physicochemical models of ion exchange, 74–84
  Scheutjens–Fleer theory, 318–319
  soil surface speciation, 43
Model chemicals, definition, 317
Model parameters, sulfur reactions, 238–239
Modeling, metal absorption, 36–38
Molar solubilization ratio, determination, 342
Molecular oxygen, sonolysis of S(–II) at alkaline pH, 250
Monodisperse suspensions
  coagulation in shear flow, 329
  hydrodynamic interactions, 328
Monomeric surfactant, binding with proteins and membranes, 357
Mononuclear ligand surface complexes, effect on dissolution rate, 19–20
Monovalent–divalent ion exchange, physicochemical model, 84–87
Multinuclear surface complexes, mineral dissolution, 104
Multisolute effects, nonlinearly sorbing components, 378–379
Multispecies lattice model, surface complexes, 39–41

N

Na^+–Ca^{2+} exchange
  Brucedale clay soil, 63f, 82f
  chemical model, 78t, 79t
  electrostatic sorption and surface complexation, 76f
  equations, 64–66
  Gouy–Chapman theory, 72–74
  isotherm predicted from Gouy–Chapman theory, 73f
  on a clay platelet, 62f
  soil properties and experimental conditions, 80t
Naphthalene, rate of degradation in soil–water suspensions, 343
Natural mineral systems, local sorption isotherms, 367–369
Natural organic matter (NOM)
  colloidal stability
    hematite particles, 322–323
    particles in aquatic systems, 317
  pH and ionic strength, effect on configuration, 319–322
  variations in reactivity, 367
Natural redox cycling, dissolution of minerals, 22, 24
Natural sorbents, distributed reactivity and resultant sorption behavior, 380f
Nitrate
  algal growth limitation, 208
  anaerobic biorestoration field demonstration project, 230
  suboxic zone, 161
  versus density, 167f
  versus depth, 165f
Nitrite, versus density, 167f
Nitrogen, concentration in different size classes as function of time, 210f
Nitrogen pools, concentrations for algal coagulation, 209f
Nitrogen transformations, oxic–anoxic interface, 172–173
Non-aqueous-phase liquids, bioavailability and biodegradation rates, 343
Noninvasive surface spectroscopies, surface complexes, 44
Nonionic surfactants
  binding with proteins and membranes, 357–358
  sorption and solubilization, 359
Nonlinearly sorbing components, overall sorption behavior of hydrophobic contaminants, 376–378

O

Ocean, source of single-carbon halocarbons, 254

INDEX

Oceanic profiles, iodide and iodate, 137–139
Octanol–water partition coefficient, 1,2,4-trichlorobenzene, 366–367
Organic bidentates, surface ligand orbitals, 29
Organic ligands, adsorption through surface complex formation, 98
Organic matter
  interaction with mineral surfaces, 95–110
  soft and hard, 369
Organic–surface interactions, oxide minerals, 97–98
Organohalogenated compounds, See Halocarbon(s)
Outer-sphere complexes, surface coordination chemistry, 4–5
Oxalate, photochemical reductive issolution of lepidocrocite, 289
Oxalate adsorption, pH dependence at lepidocrocite surface, 286
Oxic–anoxic interface
  Black Sea, 160–161
  distributions of species as function of depth, 162f
  suboxic zone, 157
Oxic environments, characteristics, 158
Oxic–suboxic interface, horizontal transport, 164
Oxidants, reductive dissolution, 29
Oxidation
  iodide to iodate, 135–155
  S(–II), 240–248
Oxidation kinetics, Mn(II), 128–130
Oxidation of S(–II), model, 243–244
Oxidation products, mineralogy, 130–131
Oxidation rates, manganese in bottom water, 120f
Oxidation–reduction environments, suboxic zone in the Black Sea, 157–176
Oxidative biomineralization process, catalytic system, 131
Oxide(s), dissolution, 12–15, 96
Oxide minerals
  ligand-promoted dissolution, 98–102
  organic–surface interactions, 97–98
  oxalate-promoted dissolution, 100f
  rate expression of proton-catalyzed dissolution, 281–283
  reductive dissolution, 279–290
  surface structure and reactivity, 97–102
Oxide surfaces, adsorption of humic substances, 102–103
Oxide–water interface, surface chemistry, 1–31
Oxoanions
  effect of EDTA-promoted dissolution of lepidocrocite, 22
  formation of bi- or multinuclear Fe(III) surface complexes, 20

Oxygen
  distribution in photochemical anoxic interface, 169–172
  sonolysis of S(–II) species, 254
  versus density, 16
Oxygen species, phosaline pH, 250
  reaction, 299
Oxygen transfer, rate-d oxidation
Oxygen-transfer coefficie
  direct-immersion, 242   step, 14–15
  open to atmosphere, 242–
  [S(–II)] profiles, 242f
Ozone
  ocean microlayer, 146–149
  reaction with I$^-$, 147f
  thermal inner-sphere process, 146
  ultraviolet light, 152

P

Particle collisions, rectilinear and curvilinear trajectories, 327f
Particle concentration, coagulation rate, 209
Particle flux spectrum, export of nitrogen by particles, 211f
Particle–particle interactions
  chemical aspects, 317–324
  physical aspects, 324–330
Particulate matter
  concentrations in presence of herbivores, 212f
  vertical flux as algal biomass and fecal pellets, 212f
Particulate methane monooxygenase (pMMO)
  properties, 197t
  regulation by copper, 197–198
  studies, 198–199
Particulate nitrogen, total flux out of euphotic zone, 211f
Partitioning at solid–water interface, physicochemical models, 83–84
Partitioning coefficients, effects of organic carbon content, 374f
Partitioning processes, linear relationships, 372
Passive films
  hydrated passive film on iron, 28f
  inhibition of dissolution, 27
  iron oxide layer, 27–28
Pentachlorophenol, mean concentration–time profile, 262f
Perchloroethylene (PCE) sorption
  Ann Arbor II near-surface soil, 379
  Wagner subsoil, 378–379
Peroxidase enzymes, natural formation of halocarbons, 273–274

408

pH dependence
  kinetics of reductive dissolu...
    minerals, 280
  NOM configuration, 319 ...cite surface,
  oxalate adsorption at l...
    286 ...solution of
  photochemical red... photochemical
    lepidocrocite, ...f lepidocrocite, 289
  rate and rate c...
    reductive d...
Phenanthre...zation, 352f
  biominer...on, 355–356, 359
  inhibiti...zation in soil–aqueous
    micro... surfactant, 351f
    mic...350–356
    ...inhibition, 353f, 358
    ..., 349–350
    ...n model prediction and
    ...ental data, 350f, 351f
    ...tion with Tween-type surfactant,
  ...
  ...tant solubilization, 339–361
  ...ation enhancement, 346
  ...nanthrene transfer, surfactant to hexane
    phase, 357
Phosphate
  dissolution of crystalline iron oxides, 104
  EDTA-promoted dissolution of lepidocrocite,
    20
  oxic–anoxic interface, 173–174
  versus density, 167f
Phospholipids, biological membranes, 357
Phosphorus
  concentration versus depth profiles, 181f
  seasonal variation in composition, 187–188
  sedimentation rate versus time, 186f
Photic zone, solar ultraviolet radiation, 257
Photocatalytic degradation
  4-chlorophenol in TiO_2 aqueous suspensions,
    291–313
  organic pollutants, 27
Photocatalytic oxidation
  4-chlorophenol, reaction mechanisms,
    304–307
  comparison with direct photolysis, 307–309
  mechanisms, 291–293
  TiO_2-mediated, 311–312
Photochemical reactions, seawater, 142
Photochemical reductive dissolution
  lepidocrocite
    concentration of dissolved Fe(II) and
      oxalate, 285f
    effect of pH, 279–290
    pH dependence of rate, 285f
    overall rate constant, 287

Photodegradation of 4-chlorophenol, kinetic
  results, 296t
Photodehalogenation, free radicals, 259
Photoelectron interference process, EXAFS
  spectroscopy, 50f
Photohydrolysis, definition, 258
Photooxidation, halocarbon complexes, 270
Photoproduction rates, relationship to DOC
  content, 269
Photoreaction
  depth-averaged rate, 257
  direct and indirect, 260f
  halocarbon, 258–259
  quantum yield, 256
  sinks and sources of aquatic halocarbons,
    253–278
  solvent effect, 264
  sorption and complexation effects, 259–260
Photoreaction rates, enhancement, 263–265
Photoredox process, efficiency for organic
  compounds, 27
Photoreduction
  aqueous solutions, 267–269
  halocarbon complexes, 270
  sorbed halocarbons, 269–270
Physical factors of solids, ion-exchange
  characteristics, 64
Physicochemical modeling, ion exchange, 60,
  74–84, 88
Phytoplankton
  algal size, 203
  organic iodine compounds, 145
Phytoplankton concentration, modeling of
  algal coagulation, 215
Polycyclic aromatic hydrocarbon (PAH),
  surfactant solubilization, 339–361
Polygalacturonic acid
  effects of pH on conformations, 320–322
  effects of pH on radius of gyration, 321f
  thickness of adsorbed layers, 321
Polyoxyethylene sorbitan monooleate
  surfactant (Tween 80), biomineralization,
    solubilization, and surface-tension data,
    352–354
Pore waters, iodine concentration, 146
Proteins, surfactant effects, 357–358
Proton-promoted dissolution
  oxide minerals, 281–283
  surface protonation, 28–29
Protonation, effect of dissolution rate, 12f
Pyrolysis reactions, gas phase of collapsing
  bubbles, 234

Q

Quenching
  excited states, 265
  fluorescence, 264–265

INDEX

R

Rate-controlled processes, dependence on surface structure, 7–15
Rate constant, sulfur reactions, 236–238
Rate law
 autocatalysis by MnO_2, 129
 Mn(II) oxidation, 128
 surface-controlled dissolution, 13
Reaction network, dynamics, 309–311
Reaction rate, relation to surface structure, 12–13
Reactor
 flow regime, 394
 scale, 394–395
Redox chemistry, iodine in seawater, 135–155
Redox reactions with halocarbons, excited states of NOM, 258
Reductants, reductive dissolution, 29
Reduction, effect of dissolution rate, 12f
Reductive dissolution
 effect of oxoanions and complex formers, 24
 inhibition by oxoanions, 22
 iron oxides in anoxic sediments, 96
 lepidocrocite, 279–290
 manganese oxides, 112
 oxides, 15
Relative dissolution rate, function of pH, 24f
Removal efficiency
 collision efficiency, 390
 suspensions in a flotation unit, 392f
Residual solution-phase concentration of solute, distributed reactivity model, 376, 377f
Resuspension rates, manganese, 119
Rothmund–Kornfeld equation
 isotherms for monovalent–divalent exchange, 66
 physicochemical ion-exchange model, 84–85

S

S-layer, *See* Oxic–anoxic interface, Suboxic zone
S(−II)
 concentration profile and oxidation product distribution, 248f
 constant O_2 at air saturation, 245f
 effect of concentration on initial zero-order oxidation rate, 247f
 free-radical mechanism, 246–247
 oxidation, 240–248
 predicted and observed concentration decrease, 243f, 244f, 249f, 250f
S(−II) + OH system, main pathways, 241f
S(−II) sonolysis, extended chemical mechanism model, 249–250
Salinity, suboxic zone, 163

Scheutjens–Fleer (SF) theory, effects of NOM on colloidal stability, 318–319
Seasonal pattern
 iodine speciation, 141
 Mn oxidation rates, 131
Seawater
 density relative to freshwater, 163
 redox chemistry of iodine, 135–155
Sediment, sorbed hydrophobic halocarbons, 265–266
Sediment cores
 dynamics of Mn within the sediments, 120–122
 total Mn concentration, 122f
Sediment-trap material
 manganese oxides, 124–128
 radial distribution functions, 125f
Sediment–water interface, reaction rates and products of manganese oxidation, 111–134
Sedimentation, copper and zinc in a eutrophic lake, 177–194
Sedimentation rate
 determination, 115
 Lake Sempach, 117f, 118f
 manganese, 117–120
 seasonal variation, 187–188
 settling particles, 184–188
Sedimentation tanks, flow-through characteristics, 388f
Sedimentation units, aggregating suspensions, 394
Selectivity for heterovalent exchange, Gouy–Chapman theory, 73
Selenite
 effect on EDTA-promoted dissolution of lepidocrocite, 23f
 EXAFS spectra, 51, 52f
Semiconductor-mediated photochemical processes, effect of surface complexes, 25–27
Semiempirical ion-exchange reactions, complex systems, 88
Sensitization, excited states, 265
Separation processes, interaction with coagulation–flocculation, 383–395
Separation units, hydraulic performance, 386–387
Settling particles
 composition, 184–188
 possible mechanisms of binding, 191
 ratios of Zn and Cu to P and organic C, 190t
 Zn:P and Cu:P ratios, 188–189
Settling rate, algal cells, 203–204
Shale isolate
 data and isotherm for sorption of TCB, 372f
 TCB sorption reactions, 370–371
Shear coagulation, surface water, 329

Shear contact, algal coagulation, 206–207
Shoaled interface, oxic–anoxic interface, 163–164
Silicates, chemical weathering, 105
Silicon, concentration versus depth profiles, 181f
Singlet O_2, reaction with I^-, 149–150
Sink(s), halocarbons in aquatic environments, 253–278
Sinking flux, particulate N, 173
Smoluchowski approach
  aggregation, 326–329
  assumptions, 329–330
$S_2O_3^{2-}$ formation, prediction, 245
Software
  EPISODE, 239
  MICROQL, 77
Soil
  correlation of dissolved Al and DOC with forest litter, 105f
  dissolved organic matter and solubilization of metals, 95–110
  effects of humic substances on metal mobilization, 106–107
  metal mobilization, 104–107
Soil–aqueous systems
  distribution of HOC and nonionic surfactant, 341f
  surfactant solubilization of phenanthrene, 339–361
Soil bioremediation, technology, 343
Soil DOM, role in dissolution of soil minerals, 99–104
Soil organic matter, local sorption isotherms, 366–367
Soil surface speciation, models, 43
Soil–waste matrix, rate of removal of PAH from soil, 344
Soil–water chemical equilibrium model, nutrient cycling in forest soils, 68–70
Solar spectral irradiance, light absorption and scattering in stratosphere and troposphere, 256–257
Solid–liquid separation in aquatic systems, algae to aquifers, 315–337
Solid–liquid systems, simulations and speculations, 330–333
Solid–solution interface, redox reactions, 7
Solid–water interface
  coordination chemistry, 3t
  ion-exchange equations, 64–66
Solubilization
  micellar exit rates, 356–357
  partitioning of HOC, 349
  phenanthrene, 349–350
Solubilization of metals, soils, 95–110
Soluble methane monooxygenase (sMMO)
  properties, 197t
Soluble methane monooxygenase (sMMO)—*Continued*
  regulation by copper, 197–198
Solvated electrons
  computed values for reduction of selected halocarbons, 269t
  irradiation of natural water samples, 267
  quantum yields for production, 268t
Sonication
  experimental parameters, 239t
  mathematical model, 240–243
Sonochemical reactions, pyrolysis and radical reactions, 234
Sonolysis data, comparison with calculated profiles, 246f, 247f
Sorbed concentration, ratio computed by using DRM, 375, 376f
Sorption
  hydrophobic halocarbons, 259
  hydrophobic organic contaminants, 363–382
Sorption equilibria, isotherm models, 364
Sorption heterogeneity, isotherm models, 381
Sorption interaction energies, mathematical model, 371
Sorption isotherm, log–log plot, 54f
Sorption model, physicochemical, 88
Sorption values, mineral surfaces, 368–369
Speciation, copper and zinc in a eutrophic lake, 177–194
Species concentrations
  monovalent–divalent exchange model, 83f
  surface complexation, 81
Spectroscopic methods, interactions at the solid–water interface, 11–12
Spectroscopy, noninvasive surface methods, 44
Spines, effect on algal concentrations, 213f
Stability constants, metal–iodide complexes, 143–144
Stabilization, electrostatic diffuse-layer interactions, 323
Statistical thermodynamics, constant capacitance model, 38–43
Stickiness coefficient, algal population dynamics and sedimentation rate, 213
Sticking probability
  hematite, 323
  soft waters high in NOM, 332
Stokes–Einstein (hydrodynamic) radius, effect of ionic strength and pH, 320f
Strain T1, mass-balance determinations, 228–229
Suboxic zone
  Black Sea, 161–163
  interstitial water of marine sediments, 159
  oxic–anoxic interface, 157–176
  ventilation processes, 171–172
Substrate utilization, competitive, 358

Sulfate depletion, monochlorophenol
  degradation, 225t
Sulfide
  depth of first appearance, 163
  distribution in oxic–anoxic interface, 169–172
  model-calculated evolution, 240
  oxidation rate, 244–248
  versus density, 166f
  versus depth, 165f
Sulfide oxidation
  anaerobic, 171
  metal oxides, 170–171
Sulfite, reaction with iodide, 152
Sulfur
  free-radical chemistry, 235–239
  ultrasonic irradiation of alkaline oxic aqueous
    solutions, 235–250
Summer stagnation
  correlation matrix for concentrations, 187t
  settling material composition, 184–188
Surface-active agents, effect on microbial
  degradation, 345t–346t
Surface charge
  adsorption of humic substances on oxides,
    102–103
  specific adsorption of NOM, 335
Surface complex
  concept, 33–57
  definition, 34
  nature, 42–43
  noninvasive methodologies, 44t
  solvated and desolvated metal ions, 52
Surface complex formation
  Fe and Al ions, 9f
  fractional surface coverage of $\equiv$Fe(III), 10f
  function of pH, 8f
  hydrous oxide surface, 4f
Surface complexation
  effect on absorption spectra, 26f
  ion exchange, 59–94
  mechanism, adsorption processes, 35–38
  models
    criteria, 6–7
    diffuse-layer sorption, 75–76
    ion exchange, 60–61, 75
Surface-controlled dissolution, rate law, 13
Surface-controlled reactions
  mechanisms, 2
  dissolution rate, 98
Surface coordination chemistry
  adsorption of ligands on metal oxides, 6–7
  inner- and outer-sphere complexes, 4–5
  surface complex formation on carbonates, 7
Surface nuclei, structure, 53
Surface polymeric structures, growth, 54
Surface precipitate, mineral–water interface,
  55

Surface protonation, adsorption of ligands and
  metal ions, 20
Surface reaction domains, reaction time and
  amount sorbed, 53t
Surface reactivity
  dependence on surface structure, 7–15
  inner-sphere surface complex, 1–31
  rates of oxide dissolution, 99
  reaction types, 16t
Surface site density, Na^+–Ca^{2+} exchange data,
  80
Surface speciation
  effect on surface reactivity, 20
  X-ray absorption spectroscopy, 49
Surface species, spectroscopic probes, 43–45
Surface-tension experiments, aggregates, 352
Surfactant
  distribution between soil and aqueous
    solution, 341
  inhibition of microbial mineralization,
    355–356
Surfactant amendments, microbial
  degradation of HOC, 344–347
Surfactant binding, saturation, 358
Surfactant effects, microbial membranes and
  proteins, 357–358
Surfactant micelles, partitioning of HOCs,
  340
Surfactant solubilization, phenanthrene in
  soil–aqueous systems, 339–361
Surfactant sorption, effect on HOC sorption,
  341–342
Synchrotron-based EXAFS, very small or
  dilute samples, 50

T

Thermal decomposition of H_2S, S(–II)
  sonolysis, 247–248
Thermodynamic equilibrium constants,
  constant capacitance model, 37
Thermodynamic models, ion exchange, 60
Time scale criterion, surface speciation, 52–53
TiO_2, degradation of 4-chlorophenol and
  generation of chloride ions, 294–300
TiO_2 aqueous suspensions, photocatalytic
  degradation of 4-chlorophenol, 291–313
TiO_2-mediated photocatalytic oxidation of 4-
  chlorophenol, rate constants, 296–298
Todorokite, local structures, 127f
Toluene
  biodegradative pathways, 226–229
  dependence of loss and growth on nitrate,
    228t
  metabolism coupled to nitrate reduction,
    227–228
  rate of degradation, 229

Toxic effects, speciation of copper and zinc, 178
Toxicity, assessment, 356
Trace elements
  algae, 178–179
  cycles in a eutrophic lake, 177–194
trans effect, labilizing effect of a ligand on surface bonds, 18
Transition metals, photoreactive complexes with ionic halocarbons, 270
Transition state theory, dissolution rate, 13
Transport
  between bulk solution and surface, 99
  manganese in lakes, 112–113
Transport coefficients, physical processes, 326
Trichloroacetate
  absorption of solvated electrons, 268f
  reaction with solvated electrons, 267
1,2,4-Trichlorobenzene (TCB)
  measured isotherm for sorption, 370f
  partitioning coefficient, 366–367
  sorption by silica surfaces, 374
  sorption by various minerals, 368–369
Trichloroethylene (TCE), sorption by various minerals, 368–369
Triplet ketones, H-atom abstractors, 264
Triplet oxygen ($^3O_2$)
  outer-sphere process, 139–142
  reaction with iodide, 139, 140f
Turbulent shear
  coagulation kernel for rectilinear case, 207
  curvilinear version of coagulation kernel, 207–208

U

Ultrasonic waves, cavitating gas bubbles, 233–251
Ultraviolet light, degradation of 4-chlorophenol and generation of chloride ions, 294–300
Ultraviolet radiation, depth dependence, 257
Uncharged polymers, segment density distribution close to a surface, 319

V

Vanselow equation, activities of species in exchanger phase, 65

Ventilation, source of suboxic zone, 71–172
Vernadite, results of EXAFS spectroscopy, 130–131
Vertical fluxes, calculation, 168–169
Vertical pump profiles, nitrate and sulfide as function of depth, 164–167
Vertical reaction zones, oxic–anoxic interface, 168–169
Viruses, attachment probabilities, 332

W

Wagner soil, isotherm for sorption of TCB, 370–371
Wastewater-treatment plant, aggregation, 329
Wastewater-treatment technologies, types, 315
Water column, typical iodide and iodate profiles, 138f
Weathering
  dissolution of minerals, 22, 24
  effect of microbial inoculum on rates, 106t
  inhibition by oxoanions, 22
  rate variations between laboratory and natural systems, 104

X

X^- species, physical interpretation, 89–90
Xylene, biodegradative pathways, 226–229

Y

Yeast, growth rates in presence of artificial surfactants, 346–347

Z

Zinc
  concentration versus depth profiles, 181f, 182f, 184f
  concentrations in water column, 180–184
  cycles in a eutrophic lake, 177–194
  determination and speciation, 179–180
  seasonal variation in composition, 187–188, 192
  sedimentation rate versus time, 186f
  toxic effects on marine algae, 178

Copy editing and indexing: Colleen P. Stamm
Production: Margaret J. Brown
Acquisition: Rhonda Bitterli
Cover design: Michele Telschow

Typeset by PRO-Image Corporation, York, PA
Printed by United Book Press, Inc., Baltimore, MD
Bound by American Trade Bindery, Baltimore, MD

Highlights from ACS Books

Good Laboratory Practice Standards: Applications for Field and Laboratory Studies
Edited by Willa Y. Garner, Maureen S. Barge, and James P. Ussary
ACS Professional Reference Book; 572 pp; clothbound ISBN 0–8412–2192–8

Silent Spring Revisited
Edited by Gino J. Marco, Robert M. Hollingworth, and William Durham
214 pp; clothbound ISBN 0–8412–0980–4; paperback ISBN 0–8412–0981–2

The Microkinetics of Heterogeneous Catalysis
By James A. Dumesic, Dale F. Rudd, Luis M. Aparicio, James E. Rekoske, and Andrés A. Treviño
ACS Professional Reference Book; 316 pp; clothbound ISBN 0–8412–2214–2

Helping Your Child Learn Science
By Nancy Paulu with Margery Martin; Illustrated by Margaret Scott
58 pp; paperback ISBN 0–8412–2626–1

Handbook of Chemical Property Estimation Methods
By Warren J. Lyman, William F. Reehl, and David H. Rosenblatt
960 pp; clothbound ISBN 0–8412–1761–0

Understanding Chemical Patents: A Guide for the Inventor
By John T. Maynard and Howard M. Peters
184 pp; clothbound ISBN 0–8412–1997–4; paperback ISBN 0–8412–1998–2

Spectroscopy of Polymers
By Jack L. Koenig
ACS Professional Reference Book; 328 pp;
clothbound ISBN 0–8412–1904–4; paperback ISBN 0–8412–1924–9

Harnessing Biotechnology for the 21st Century
Edited by Michael R. Ladisch and Arindam Bose
Conference Proceedings Series; 612 pp;
clothbound ISBN 0–8412–2477–3

From Caveman to Chemist: Circumstances and Achievements
By Hugh W. Salzberg
300 pp; clothbound ISBN 0–8412–1786–6; paperback ISBN 0–8412–1787–4

The Green Flame: Surviving Government Secrecy
By Andrew Dequasie
300 pp; clothbound ISBN 0–8412–1857–9

For further information and a free catalog of ACS books, contact:
American Chemical Society
Distribution Office, Department 225
1155 16th Street, NW, Washington, DC 20036
Telephone 800–227–5558

Bestsellers from ACS Books

The ACS Style Guide: A Manual for Authors and Editors
Edited by Janet S. Dodd
264 pp; clothbound ISBN 0–8412–0917–0; paperback ISBN 0–8412–0943–X

The Basics of Technical Communicating
By B. Edward Cain
ACS Professional Reference Book; 198 pp;
clothbound ISBN 0–8412–1451–4; paperback ISBN 0–8412–1452–2

Chemical Activities (student and teacher editions)
By Christie L. Borgford and Lee R. Summerlin
330 pp; spiralbound ISBN 0–8412–1417–4; teacher ed. ISBN 0–8412–1416–6

Chemical Demonstrations: A Sourcebook for Teachers,
Volumes 1 and 2, Second Edition
Volume 1 by Lee R. Summerlin and James L. Ealy, Jr.;
Vol. 1, 198 pp; spiralbound ISBN 0–8412–1481–6;
Volume 2 by Lee R. Summerlin, Christie L. Borgford, and Julie B. Ealy
Vol. 2, 234 pp; spiralbound ISBN 0–8412–1535–9

Chemistry and Crime: From Sherlock Holmes to Today's Courtroom
Edited by Samuel M. Gerber
135 pp; clothbound ISBN 0–8412–0784–4; paperback ISBN 0–8412–0785–2

Writing the Laboratory Notebook
By Howard M. Kanare
145 pp; clothbound ISBN 0–8412–0906–5; paperback ISBN 0–8412–0933–2

Developing a Chemical Hygiene Plan
By Jay A. Young, Warren K. Kingsley, and George H. Wahl, Jr.
paperback ISBN 0–8412–1876–5

Introduction to Microwave Sample Preparation: Theory and Practice
Edited by H. M. Kingston and Lois B. Jassie
263 pp; clothbound ISBN 0–8412–1450–6

Principles of Environmental Sampling
Edited by Lawrence H. Keith
ACS Professional Reference Book; 458 pp;
clothbound ISBN 0–8412–1173–6; paperback ISBN 0–8412–1437–9

Biotechnology and Materials Science: Chemistry for the Future
Edited by Mary L. Good (Jacqueline K. Barton, Associate Editor)
135 pp; clothbound ISBN 0–8412–1472–7; paperback ISBN 0–8412–1473–5

For further information and a free catalog of ACS books, contact:
American Chemical Society
Distribution Office, Department 225
1155 16th Street, NW, Washington, DC 20036
Telephone 800–227–5558

Other ACS Books

Biotechnology and Materials Science: Chemistry for the Future
Edited by Mary L. Good
160 pp; clothbound, ISBN 0–8412–1472–7; paperback, ISBN 0–8412–1473–5

Chemical Demonstrations: A Sourcebook for Teachers
Volume 1, Second Edition by Lee R. Summerlin and James L. Ealy, Jr.
192 pp; spiral bound; ISBN 0–8412–1481–6
Volume 2, Second Edition by Lee R. Summerlin, Christie L. Borgford, and Julie B. Ealy
229 pp; spiral bound; ISBN 0–8412–1535–9

The Language of Biotechnology: A Dictionary of Terms
By John M. Walker and Michael Cox
ACS Professional Reference Book; 256 pp;
clothbound, ISBN 0–8412–1489–1; paperback, ISBN 0–8412–1490–5

Cancer: The Outlaw Cell, Second Edition
Edited by Richard E. LaFond
274 pp; clothbound, ISBN 0–8412–1419–0; paperback, ISBN 0–8412–1420–4

Chemical Structure Software for Personal Computers
Edited by Daniel E. Meyer, Wendy A. Warr, and Richard A. Love
ACS Professional Reference Book; 107 pp;
clothbound, ISBN 0–8412–1538–3; paperback, ISBN 0–8412–1539–1

Practical Statistics for the Physical Sciences
By Larry L. Havlicek
ACS Professional Reference Book; 198 pp; clothbound; ISBN 0–8412–1453–0

The Basics of Technical Communicating
By B. Edward Cain
ACS Professional Reference Book; 198 pp;
clothbound, ISBN 0–8412–1451–4; paperback, ISBN 0–8412–1452–2

The ACS Style Guide: A Manual for Authors and Editors
Edited by Janet S. Dodd
264 pp; clothbound, ISBN 0–8412–0917–0; paperback, ISBN 0–8412–0943–X

Personal Computers for Scientists: A Byte at a Time
By Glenn I. Ouchi
276 pp; clothbound, ISBN 0–8412–1000–4; paperback, ISBN 0–8412–1001–2

Chemistry and Crime: From Sherlock Holmes to Today's Courtroom
Edited by Samuel M. Gerber
135 pp; clothbound, ISBN 0–8412–0784–4; paperback, ISBN 0–8412–0785–2

For further information and a free catalog of ACS books, contact:
American Chemical Society
Distribution Office, Department 225
1155 16th Street, NW, Washington, DC 20036
Telephone 800–227–5558